Birkhäuser

Applied and Numerical Harmonic Analysis

Eugenio Hernández • Marco Maria Peloso •
Fulvio Ricci • Fernando Soria • Anita Tabacco
Editors

The Mathematical Heritage of Guido Weiss

Birkhäuser

Editors
Eugenio Hernández
Departamento de Matemáticas
Universidad Autónoma de Madrid
Madrid, Spain

Marco Maria Peloso
Dipartimento di Matematica
Università degli Studi di Milano
Milano, Italy

Fulvio Ricci
Scuola Normale Superiore
Pisa, Italy

Fernando Soria
Departamento de Matemáticas
Universidad Autónoma de Madrid
Madrid, Spain

Anita Tabacco
Dipartimento di Scienze Matematiche
Politecnico di Torino
Torino, Italy

ISSN 2296-5009 ISSN 2296-5017 (electronic)
Applied and Numerical Harmonic Analysis
ISBN 978-3-031-76792-0 ISBN 978-3-031-76793-7 (eBook)
https://doi.org/10.1007/978-3-031-76793-7

This book is published under the imprint Birkhäuser, www.birkhauser-science.com by the registered company Springer Nature Switzerland AG
The registered company address is: Gewerbestrasse 11, 6330 Cham, Switzerland

ANHA Series Preface

The *Applied and Numerical Harmonic Analysis* (ANHA) book series aims to provide the engineering, mathematical, and scientific communities with significant developments in harmonic analysis, ranging from abstract harmonic analysis to basic applications. The title of the series reflects the importance of applications and numerical implementation, but richness and relevance of applications and implementation depend fundamentally on the structure and depth of theoretical underpinnings. Thus, from our point of view, the interleaving of theory and applications and their creative symbiotic evolution is axiomatic.

Harmonic analysis is a wellspring of ideas and applicability that has flourished, developed, and deepened over time within many disciplines and by means of creative cross-fertilization with diverse areas. The intricate and fundamental relationship between harmonic analysis and fields such as signal processing, partial differential equations (PDEs), and image processing is reflected in our state-of-the-art ANHA series.

Our vision of modern harmonic analysis includes a broad array of mathematical areas, e.g., wavelet theory, Banach algebras, classical Fourier analysis, time-frequency analysis, deep learning, and fractal geometry, as well as the diverse topics that impinge on them.

For example, wavelet theory can be considered an appropriate tool to deal with some basic problems in digital signal processing, speech and image processing, geophysics, pattern recognition, biomedical engineering, and turbulence. These areas implement the latest technology from sampling methods on surfaces to fast algorithms and computer vision methods. The underlying mathematics of wavelet theory depends not only on classical Fourier analysis, but also on ideas from abstract harmonic analysis, including von Neumann algebras and the affine group. This leads to a study of the Heisenberg group and its relationship to Gabor systems, and of the metaplectic group for a meaningful interaction of signal decomposition methods.

The unifying influence of wavelet theory in the aforementioned topics illustrates the justification for providing a means for centralizing and disseminating information from the broader, but still focused, area of harmonic analysis. This will be a key role of ANHA. We intend to publish with the scope and interaction that such a host of issues demands.

Along with our commitment to publish mathematically significant works at the frontiers of harmonic analysis, we have a comparably strong commitment to publish major advances in the following applicable topics in which harmonic analysis plays a substantial role:

*Analytic Number theory * Antenna Theory * Artificial Intelligence * Biomedical Signal Processing * Classical Fourier Analysis * Coding Theory * Communications Theory * Compressed Sensing * Crystallography and Quasi-Crystals * Data Mining * Data Science * Deep Learning * Digital Signal Processing * Dimension Reduction and Classification * Fast Algorithms * Frame Theory and Applications * Gabor Theory and Applications * Geophysics * Image Processing * Machine Learning * Manifold Learning * Numerical Partial Differential Equations * Neural Networks * Phaseless Reconstruction * Prediction Theory * Quantum Information Theory * Radar Applications * Sampling Theory (Uniform and Non-uniform) and Applications * Spectral Estimation * Speech Processing * Statistical Signal Processing * Super-resolution * Time Series * Time-Frequency and Time-Scale Analysis * Tomography * Turbulence * Uncertainty Principles *Waveform design * Wavelet Theory and Applications

The above point of view for the ANHA book series is inspired by the history of Fourier analysis itself, whose tentacles reach into so many fields.

In the last two centuries Fourier analysis has had a major impact on the development of mathematics, on the understanding of many engineering and scientific phenomena, and on the solution of some of the most important problems in mathematics and the sciences. Historically, Fourier series were developed in the analysis of some of the classical PDEs of mathematical physics; these series were used to solve such equations. In order to understand Fourier series and the kinds of solutions they could represent, some of the most basic notions of analysis were defined, e.g., the concept of "function." Since the coefficients of Fourier series are integrals, it is no surprise that Riemann integrals were conceived to deal with uniqueness properties of trigonometric series. Cantor's set theory was also developed because of such uniqueness questions.

A basic problem in Fourier analysis is to show how complicated phenomena, such as sound waves, can be described in terms of elementary harmonics. There are two aspects of this problem: first, to find, or even define properly, the harmonics or spectrum of a given phenomenon, e.g., the spectroscopy problem in optics; second, to determine which phenomena can be constructed from given classes of harmonics, as done, for example, by the mechanical synthesizers in tidal analysis.

Fourier analysis is also the natural setting for many other problems in engineering, mathematics, and the sciences. For example, Wiener's Tauberian theorem in Fourier analysis not only characterizes the behavior of the prime numbers, but is a fundamental tool for analyzing the ideal structures of Banach algebras. It also provides the proper notion of spectrum for phenomena such as white light. This latter process leads to the Fourier analysis associated with correlation functions in filtering and prediction problems. These problems, in turn, deal naturally with Hardy spaces in complex analysis, as well as inspiring Wiener to consider communications engineering in terms of feedback and stability, his cybernetics. This latter theory develops concepts to understand complex systems such as learning and cognition and neural networks; and it is arguably a precursor of deep learning and its spectacular interactions with data science and AI.

Nowadays, some of the theory of PDEs has given way to the study of Fourier integral operators. Problems in antenna theory are studied in terms of unimodular trigonometric polynomials. Applications of Fourier analysis abound in signal processing, whether with the fast Fourier transform (FFT), or filter design, or the adaptive modeling inherent in time-frequency-scale methods such as wavelet theory.

The coherent states of mathematical physics are translated and modulated Fourier transforms, and these are used, in conjunction with the uncertainty principle, for dealing with signal reconstruction in communications theory. We are back to the raison d'etre of the ANHA series!

College Park, MD, USA — John Benedetto
College Park, MD, USA — Wojciech Czaja
Boston, MA, USA — Kasso Okoudjou

Preface

Guido Weiss passed away on December 25, 2021, at the age of 92.

With his intensive mathematical production over a 65 year period he has enormously influenced the development of harmonic analysis.

With his strong and inclusive personality, he has supported generations of students and young researchers, favoring the diffusion of this mathematical field to various parts of the world.

The Editors of this volume have been part of this large group and represent two countries, Italy and Spain, where Guido Weiss's influence has been particularly wide and significant. The invited contributors are collaborators, students, and colleagues of Guido, who had a particularly intense relationship with him.

Guido's Mathematics

Guido's mathematical activity can be divided into three periods, with large chronological overlaps and interactions.

The first period (1956–1970) is marked by his collaboration with E. Stein and his first wife Mary Weiss. The main topics are pioneering applications of complex interpolation on Hardy spaces and, most remarkably, the extension of complex-variable methods to real analysis, in particular with the introduction of generalized Cauchy–Riemann systems in real Euclidean spaces. The latter led to the extension of the classic holomorphic Hardy spaces to spaces in several real variables, which preluded the later developments of C. Fefferman and Stein. The collaboration with Stein culminated in the book "Introduction to Harmonic Analysis on Euclidean Spaces," Princeton University Press (1971), which is among the most cited books in this area. It is still used as basic reference in Fourier Analysis.

The second period (1969–1994) is the most articulated for varieties of themes and is marked by his collaboration with R. Coifman. Though many other co-authors were involved, M. Taibleson and R. Rochberg in the first place, the leading force was the strong synergy between him and R. Coifman when both were at "Wash U." The main themes of this period have been

- Atomic characterization of Hardy spaces and their extension to spaces of homogeneous type, with applications to multiplier theory on Lie groups, and other function spaces (Besov, Triebel-Lizorkin, bloch spaces, etc.).
- The transference method, a generalization to group representations of methods already present in the works of Calderón, Herz, and others.
- Analytic families of Banach spaces and complex interpolation theorems.

The third is the wavelet period. Guido was interested in wavelets since its discovery, mainly through the works of Yves Meyer, who visited Washington University often in the 1980s. He gave Graduate courses at Wash U at the beginning or the 1990s, that lead to the book "A First Course on Wavelets" (CRC Press, 1996).

Guido found his personal way to focus in the most theoretical aspects of wavelets, without forgetting his applications. Among his many collaborators in this period, we must single out Ed Wilson, who was his main research partner in this area. Some of the themes of this period are:

- Foundations of the mathematical theory of wavelets in dimension one
- Extensions to several dimensions, introducing the generalized translation invariant systems, the composite dilation wavelets, and the shearlets
- Frames and Riesz systems generated by unitary representations of locally compact abelian groups

Guido's Mathematical Community

Guido had an exuberant personality and very high energy. He influenced the life, professional and personal, of many people. He had many graduate students, all of whom he followed well beyond their graduation and remained in strict contact with. Guido mentored many young students, helping them find the best next step in their professional careers. Perhaps most remarkably, Guido had a very large group of collaborators, young and his senior, and was involved in the activities of many of his colleagues at Washington University.

After graduating from the University of Chicago, he moved to Washington University in St. Louis where he remained for all of his life. The Department of Mathematics at Wash U became a gravitational center for harmonic analysis in the world panorama. With his charisma, he attracted mathematicians from all over the world. With his drive and dedication, Guido invited young students and collaborators from Argentina, Spain, Italy, China, Poland, France, and Croatia, just to name the most numerous groups.

Guido was able to establish a personal relationship with essentially everybody. He organized social events, parties at the department and his own home, naturistic excursions, sports events. With his seemingly endless energy, Guido knitted the connection among the various people he had been interacting with.

To summarize, Guido created a large mathematical community. The significance of this community was very clear to many already, and was exemplified at Guido's Fest, held at Wash U in October 2022, where so many people rediscovered that they had been long-time friends *because of Guido*.

Guido's Personality

We have already mentioned Guido's exuberant personality. Such kind of personality was a necessary requirement in order to create the vast community at Wash U. Guido *loved* to be at the center of the community he created, and gave back a lot in return. Guido spent a lot of his time practically helping his visitors, were they incoming graduate students, short-time visitors, and long-standing collaborators. Guido helped everybody, in one way or another. We could write a book just telling stories of Guido caring for someone.

Guido was very generous, not just in teaching, advising, and sharing his mathematical views, but also in helping *his* community to deal with life events, whether were these daily or serious matters.

Guido's Humanity

Guido has been a very good friend to many people. His desire to help everyone around him led him also to offer his friendship. Guido opened up his home to everyone in his community: dinners, parties, Italian lessons, swimming at the pool, and we could go on. He had a witty sense of humor, and we can safely say that his jokes are legendary. Guido's humanity was warm-hearted, and caring.

Conclusions

We make no pretense of sketching his biography, not to illustrate here all of Guido's achievements. The reader who is interested about Guido's biography is referred to the article of S. Kelly and R. Torres, "Guido Weiss: From Immigrant Boy to International Renowned Research Mathematician" (*The Journal of Geometric Analysis*, 2021, 31: 9146–9179).

Let us mention, however, that he was the recipient of the Chauvenet price in 1971 for his expository writings on Harmonic Analysis. Also, he played an essential role in the academic growth of the Department of Mathematics of Washington University, and held positions in the American Mathematical Society, the National

Research Council, and the National Science Foundation, as well as in the Athletic Department of Wash U.

We are dedicating this volume to Guido's vision of a mathematical community, and ultimately its creation and success. We believe this a very remarkable achievement.

Thank you, Guido!

Madrid, Spain	Eugenio Hernández
Milano, Italy	Marco Maria Peloso
Pisa, Italy	Fulvio Ricci
Madrid, Spain	Fernando Soria
Torino, Italy	Anita Tabacco
May, 2024	

Contents

Contributors

Pascal Auscher Laboratoire de Mathématiques d'Orsay, Université Paris-Saclay, CNRS, Orsay, France

Sorina Barza Department of Mathematics and Computer Science, Karlstad University, Karlstad, Sweden

John J. Benedetto Department of Mathematics, Norbert Wiener Center, University of Maryland, College Park, MD, USA

Árpád Bényi Department of Mathematics, Western Washington University, Bellingham, WA, USA

Aline Bonami Département de Mathématiques, Institut Denis Poisson, Université d'Orléans, Orléans Cedex 2, France

Luca Brandolini Dipartimento di Ingegneria Gestionale, dell'Informazione e della Produzione, Università degli Studi di Bergamo, Dalmine, BG, Italy

María J. Carro Department of Analysis and Applied Mathematics, Complutense University of Madrid, Madrid, Spain

Rama Chellappa Departments of Electrical and Computer Engineering and Biomedical Engineering, Johns Hopkins University, Baltimore, MD, USA

Ronald R. Coifman Department of Mathematics, Program in Applied Mathematics, Yale University, New Haven, CT, USA

Leonardo Colzani Dipartimento di Matematica, Università degli Studi di Milano Bicocca, Milano, Italy

Michael G. Cowling School of Mathematics and Statistics, University of New South Wales, Sydney, NSW, Australia

Wojciech Czaja Norbert Wiener Center for Harmonic Analysis and Applications, Department of Mathematics, University of Maryland, College Park, MD, USA

Zhijie Fan Department of Mathematics, Wuhan University, Wuhan, P. R. China

Bianca Gariboldi Dipartimento di Ingegneria Gestionale, dell'Informazione e della Produzione, Università degli Studi di Bergamo, Dalmine, BG, Italy

Gustavo Garrigós Departamento de Matemáticas, University of Murcia, Murcia, Spain

Giacomo Gigante Dipartimento di Ingegneria Gestionale, dell'Informazione e della Produzione, Università degli Studi di Bergamo, Dalmine, BG, Italy

Philip T. Gressman Department of Mathematics, University of Pennsylvania, Philadelphia, PA, USA

Yongsheng Han Department of Mathematics, Auburn University, Auburn, AL, USA

Cong Hoang Department of Mathematics, Florida Agricultural and Mechanical University, Tallahassee, FL, USA

Ilya Kavalerov Google, Menlo Park, CA, USA

Ilya Krishtal Northern Illinois University, Dekalb, IL, USA

Ji Li Department of Mathematics, Macquarie University, Sydney, NSW, Australia

Yves Meyer Académie des Sciences, Paris, France

Brendan Miller Northern Illinois University, Dekalb, IL, USA

Kabe Moen Department of Mathematics, University of Alabama, Tuscaloosa, AL, USA

Morten Nielsen Department of Mathematical Sciences, Aalborg University, Aalborg, Denmark

Carlos Pérez Department of Mathematics, University of the Basque Country, Ikerbasque and BCAM, Bilbao, Spain

Jacques Peyrière Institut de Mathématiques d'Orsay, CNRS, Université Paris-Saclay, Orsay, France

Andreas Seeger Department of Mathematics, University of Wisconsin-Madison, Madison, WI, USA

Hrvoje Šikić Department of Mathematics, Faculty of Science, University of Zagreb, Zagreb, Croatia

Javier Soria Department of Analysis and Applied Mathematics, Complutense University of Madrid, Madrid, Spain
ICMAT, Madrid, Spain

Rodolfo H. Torres Department of Mathematics, University of California, Riverside, CA, USA

Tino Ullrich Fakultät für Mathematik, Technische Universität Chemnitz, Chemnitz, Germany

Guido Weiss 1928–2021 St. Louis, MO, USA

Lixin Yan Department of Mathematics, Macquarie University, Sydney, NSW, Australia

Ruijie Zheng Department of Computer Science, University of Maryland, College Park, MD, USA

Chapter 1
Guido Weiss: A Few Memories of a Friend and an Influential Mathematician

Pascal Auscher and Aline Bonami

Abstract This contribution starts with an exchange between us on the way we met Guido and he influenced our mathematical lives. Then it is mainly a survey paper that illustrates this influence by describing different topics and their subsequent evolution after his seminal papers and courses. Our main thread is the notion of a space of homogeneous type. In the second section we describe how it became central in pluricomplex analysis and consider particularly the existence of weak factorization for spaces of holomorphic functions. In the last section, one revisits the construction of a basis of wavelets in a space of homogeneous type and the way it allows a Littlewood-Paley analysis.

1.1 Introduction

The authors of this chapter belong to two different generations of harmonic analysts. We both met Guido at a very early stage in our careers. His influence and friendship were important for both of us. We found it interesting to cross our memories of Guido. In a first section, we will tell how we met Guido and try to describe what has been his influence on our mathematical choices. This will be done through a dialog between us.

The rest of the paper will be more classical and look like a survey paper on particular points for which the influence of Guido was like a starting point for us. We will not be exhaustive on the choice of topics and will certainly not cite all the appropriate literature in the fields that we have chosen to describe in more detail. We

P. Auscher
Laboratoire de Mathématiques d'Orsay, Université Paris-Saclay, CNRS, Orsay, France
e-mail: pascal.auscher@universite-paris-saclay.fr

A. Bonami (✉)
Département de Mathématiques, Institut Denis Poisson, Université d'Orléans, Orléans Cedex 2, France

E. Hernández et al. (eds.), *The Mathematical Heritage of Guido Weiss*,
Applied and Numerical Harmonic Analysis,
https://doi.org/10.1007/978-3-031-76793-7_1

apologize in advance for these choices, which are linked with our personal views and memories and may diverge from other choices and memories.

The main thread of this paper is the notion of space of homogeneous type that was introduced by Raphy Coifman and Guido in the early seventies. In the second section, mainly written by Aline, we will speak of its influence in pluricomplex analysis. In the third part, mainly written by Pascal, we will see how it may lead to develop all necessary tools for Littlewood-Paley theory in a very general context.

1.2 The Past, the Past...

Pascal. You start, of course!

Aline. For me, it begins in the academic year 1970–1971, when Guido Weiss and Raphy Coifman gave a course in Orsay, now University Paris-Saclay. Their collaboration had started a few years earlier, and at the time they were interested in developing multiplier theorems related to actions of non-commutative groups. I was in the audience, which was not so numerous: even if harmonic analysis was still one of the main topics in the maths department, there was not such a feverish atmosphere as three or four years before. Moreover, Jean-Pierre Kahane and Yves Meyer were mainly abroad. It was perhaps a chance for us. Guido and Raphy asked for volunteers to help them in writing notes, which I did in collaboration with Jean-Louis Clerc and Bernard Mischler. So I met Guido and Raphy every week. I can still see them at the blackboard, in their very different styles, Guido writing calmly with his magnificent writing long formulas for spherical harmonics, while Raphy tried to convince us of the simplicity of notions. The notes that followed, *Analyse Harmonique Non-Commutative sur Certains Espaces Homogènes* [17] went much further than the initial program and, in a sense, were the first steps into a new paradigm, the world of *spaces of homogeneous type*.

Pascal. Which influence had this course on you?

Aline. The scientific life in Orsay was so intense at that time that it is difficult to sort out influences. It was my third course on real analysis given by foreigners. Eli Stein had given one in 1967–1968, just when I started research, the course that led to his book *Singular integrals...*. I had adored this course and had thought at the time that it was really what I wanted to do. But how? It seemed so difficult and the school of Chicago seemed so much in advance! At the same time I followed a course of Yves Meyer, who started to ask me questions on multipliers that led to my thesis on what is now called hypercontractivity. Working with Yves was a huge luck. At the time of the venue in Orsay of Guido and Raphy, I defended my thesis and even worked during a few months in probability theory. In the mean time, I had also attended a course of N. Riviere, who died prematurely. He talked on singular integrals related to parabolic equations. After the course of Guido and

Raphy, one saw precisely how the two courses I had followed before dealt with two examples of their *Espaces de nature homogène*. One had a very clear description of the geometric objects involved. In fact their course opened the door to a lot of problems in the theory of singular integrals. Afterwards, it was obvious for me that this was the kind of mathematics I preferred.

It would be unfair to limit the influence of Guido and Raphy at Orsay that year to mathematics. From the beginning, and particularly at that time when relations between mathematicians were still somewhat formal in France, we were all stuck by their generosity and kindness, which they expressed directly to everybody.

Pascal. And what influence had this course on the harmonic analysis team in Orsay at the time?

Aline. It had a direct influence on Jean-Louis Clerc, Noël Lohoue and me, both for the content of their course itself and for the way to do mathematics, to interact between us. As I said before, relationships at work in France were then not so natural. There was still the idea one should publish alone. We started to work together, the three of us, in line with the course, trying to extend to other contexts known properties in Fourier Analysis. Inspired by their course, we read together the book of Helgason *Differential geometry and symmetric manifolds*. And with Jean-Louis I studied Cesàro means for expansions in spherical harmonics. I went two months to Washington University (WU) during the spring 1972. Of course Guido and Raphy had a large influence on the harmonic analysts in Orsay, but a little later. In fact, Yves Meyer became a frequent visitor at WU and started to work with Raphy a little later, in 1974. Afterwards the interactions between WU and French harmonic analysts increased.

Again, friendship mixes with mathematics and it is impossible not to speak of this also. I visited regularly WU after 1972, even if much more rapidly. We met also frequently in Europe. To come back to the far past, I particularly remember having been at WU in 1981, at the same time as Jean-Lin Journé. We were both invited in Guido's and Barbara's summer house for a kind of country party. Conversations, canoeing on the small lake, Guido removing splinters in my hand with the same meticulousness he used to show in his classes..., there was such a charm, as in a film. Guido's jokes made a lot to create a friendly atmosphere around him. Nevertheless Guido's life had its share of mystery and suffering, starting with his childhood in Italy and the intellectual heritage of his parents. He was also a man of conviction, deeply attached to the values of human solidarity. He was a model for many of us.

Nearly twenty years later, Pascal spends two and a half academic years at WU in the context of his military service that all French men had to accomplish at that time. He is not the first student of Yves Meyer to do this: Jean-Lin Journé was there in 1981–82. Its influence on Jean-Lin may be measured when looking at the notes of the course that he gave there and which were published under the title Calderón-Zygmund Operators, Pseudo-Differential Operators and the Cauchy Integral of Calderón.

Aline. It is time for me to ask you a question. Did you know Guido before going to WU ? How was your encounter?

Pascal. No, I did not know him, nor, I must say, had read his books or articles. I was working on wavelet theory at the time, during the years 1986–87 and 1987–88. The mathematical theory of wavelets was just starting and I focused on that. For my military service, one possibility was to do it abroad in a teaching or research institution. Yves suggested that I go to WU. He contacted Guido and I recall reading his enthusiastic reply (a mailed letter at the time) dated December 31, 1987. I was just stuck by the date. Then started some exchanges with Guido (by the new email tool!) in the spring to set up things for the next fall (and for arrangements with French military office). I also read his book with Eli Stein [72] before leaving. The encounter was made easy in all means by Guido's welcoming help. With common interests in tennis and bird-watching, it became a strong friendship.

Aline. How was the scientific life in WU at the time? Were there seminars, for instance? How did you interact with others?

Pascal. First, the mathematics department was a highly renowned place for harmonic analysis. Raphy was long gone, but Al Baernstein, Björn Dalhberg, Steven Krantz, Richard Rochberg, Mitch Taibleson were very active members. This department also attracted many doctoral students from abroad (Spain, Italy, Argentina, China..) and post-docs, some became good friends (Estela Gavosto and Rodolfo Torres, Carlo Morpurgo, Marco Peloso, María-Jesús Carro. . .) and became influential mathematicians back home after having benefited from the formidable and stimulating scientific atmosphere. I was the only French person (the French system did not promote much foreign post-docs in those days). Guido wanted me to learn new maths so he proposed to Maria-Jesús, who had done her thesis on interpolation theory, and me to work on some problems on multipliers around the Riesz means, transference and other things. This led to two articles. The activity in the department was intense: regular analysis seminar, colloquiums, lots of master courses, and meetings in Guido's office. I still have the notes of the course of Guido on transference, based on his work with Raphy. All details were concisely given on the board: "too many details" was I thinking, but I realize now that this allowed the topic to profoundly print in my mind. He was always encouraging people so as to make them available for more (integrating the f+ according to the words of A. Zygmund). I recall he brought me to visit Raphy at Yale, and this visit had tremendous impact for me, as an example of how curiosity drives mathematics. After my thesis, I was trying to understand (solve, why not?) the Kato conjecture in any dimension using wavelet methods as Ph. Tchamitchian had just reproved nicely the strongly linked T(b) Theorem of David-Journé-Semmes via regular adapted wavelets [76]. When wandering in the Yale math department, I found a table with help yourself free printed (the pdf's did not yet exist and some references could be hard to find) articles that people were giving away. I discovered the article of McIntosh on bounded holomorphic calculus from the Proceedings of the Center for Mathematical Analysis of the Australian National University and also the article of Jacques-Louis Lions pursuing the interpolation theory of Kato for maximal accretive operators, which inspired me for later research.

Aline. You have a paper with Guido on Wilson bases. How was it?

Pascal. During my stay at WU, Guido was curiously NOT interested to work on wavelets, despite my attempts to suggest problems, like the one which I solved later showing that a mild sufficient condition on the wavelet generating an orthonormal basis suffices to conclude it arises from a multiresolution analysis, introducing at the same time a special series whose values are the dimensions of certain vector spaces, that was next baptised by Guido as the "dimension function" and which became a central tool as beautifully explained in Guido's reference book with Eugenio Hernández [43]. That trip to Yale was when a new interest came in Guido's mind. Raphy explained to us the concept of local Fourier bases and Guido wanted to understand the calculations behind. We started to work out the details, trying in Guido's style to make them as conceptually simple and elegant as possible. The observation that these bases and the Meyer orthonormal wavelet basis were built from the same tricks fascinated him. Victor Wickerhauser joined the discussion and the article was ready for publication [4]. This is how he plunged into the mathematical theory of wavelets, trying to exhibit its finest structures and as always in the most simple and accessible terms. This was when I left WU at the end of 1990...

Pascal. You also have papers with Guido in the nineties. Can you say something on them?

Aline. As you said, Guido started to work on the theory of wavelets in the nineties. Just before, he was particularly interested by characterizations coming from Littlewood-Paley theory. I like very much his notes with Frazier and Jawerth in the CBMS Regional Conference Series (1991), [32]. He had all reasons to be fascinated by wavelets and had questions related to constructions. Yes, we wrote two papers, the first one [8] on band-limited wavelets. Our joint work was pursued by Gustavo Garrigós who was then his PhD student. Gustavo came later to Orléans as a post-doc, which was really great for me. But then Gustavo and I worked on Bergman spaces in tube domains, where of course we used a lot his book with Eli Stein and the properties of Hardy spaces in tube domains given there.

Aline. From your point of view, what is the mathematical legacy of Guido?

Pascal. Of course, the discoveries he made all along his career, some that we are going to elaborate on in this article afterwards, which paved the way to create a strong school in harmonic analysis and linked topics: interpolation, singular integrals, Hardy spaces, spaces of homogeneous type... In all of these, there was always the will to leave no black holes in his way of writing, going into every details even the simplest ones, to make the reading accessible. In that respect, he was particularly proud of the Chauvenet prize he obtained in 1967 [78], the highest American award for mathematical expository writing. For example, he told me that for his book with Eli Stein, the first to propose a synthetic exposition of the extension of harmonic analysis from one to several Euclidean dimensions that occurred in the sixties, he thought at length about notation avoiding as much as possible using coordinates, to present the computations in the most conceptual manner (and with almost no typos). This notation is still accurate as of today. The legacy is then clear. He posed the solid bricks on which many new results can be elaborated. Still, despite there are now many references on each of these topics, my first reaction when I look for a reference or a proof is to browse his masterpieces as I am sure to find what I need (and sometimes understand some new points that I left aside on earlier reading). When I write research articles, I always have in mind the words of Guido: try to stay simple and reader friendly.

Pascal. Same question.

Aline. I will only speak of my own heritage. Of course I mention first his books, primarily his book with Eli Stein and the Notes of Orsay, which played different roles in my personal Pantheon. I had the impression to have the Notes in mind, just as they were engraved during the courses of Guido and Raphy, while I came back to the book with Eli Stein to read it in all details each time I gave a course myself. The book with Eugenio reaches the same degree of perfection. May be you find elsewhere a better source for intuition, for example in Yves's books, but if you want to understand all details, the book of Eugenio and Guido makes the job remarkably. These are three books that I keep on hand in my library. Then there are the works of Guido that inspired me for long. I would mention first the paper in Annals with Raphy and Richard Rochberg on commutators and weak factorization. I have the impression to have turned around this paper part of my life. We will come back to this later. I would like also to mention some parts of his work that surprised me when I heard him speak of them and that I regret not to have understood more deeply, not to have come back to them later. You know, this kind of sensation one has that there, there is something I do not understand enough, there is some new music with which I would like to familiarize myself. For example, I recall how Guido was excited with Block spaces, that is, spaces that are defined from an atomic decomposition [75], but with atoms that are not of mean 0, and for which one has almost everywhere convergence results, for instance. Of course, the subject has been deepened afterwards, in particular by Fernando Soria. I have nevertheless the feeling I should have tried to understand better what was around. The same with interpolation. Being exhaustive is impossible. But I will try to show, in the next section, how the work of Guido influenced mine. And of course, as you said, there is a lot more in the legacy of Guido than a list of results. There is a way to do mathematics, to write and to talk, and to do much more than only mathematics, there is a way to be curious of life and others, there is a lot.

1.3 Orsay's Course and Excursion Inside Complex Analysis

1.3.1 The Notes: A New Paradigm, the One of Space of Homogeneous Type

One of the main new notions of the course was the notion of *space of homogeneous type*. For a topological space X, one first calls *pseudo-distance (or quasi-distance depending on the authors)* a function $\rho : X \times X \mapsto [0, \infty)$ such that there exists $K \geq 1$ such that, for all x, y, z in X,

1. $\rho(x, y) = 0 \iff x = y$
2. $\rho(x, y) = \rho(y, x)$
3. $\rho(x, z) \leq K(\rho(x, y) + \rho(y, z))$.

The pseudo-ball (or quasi-ball, or simply ball) $B(x, r)$ with center x and radius r is the set of points $y \in X$ such that $\rho(x, y) < r$.

We say that X is a *space of homogeneous type* if X is endowed with a pseudo-distance ρ and sith a Borel measure μ such that

1. the balls B centered at x constitute a basis of neighborhoods of x
2. the measure μ is doubling, that is, there exists a constant A such that, for all $x \in X$ and $r > 0$,

$$\mu(B(x, 2r) \leq A\mu(B(x, r)).$$

These definitions are sufficient to define the maximal function of $f \in L^1(d\mu)$ by

$$Mf(x) := \sup_{r>0} \frac{1}{\mu(B(x, r))} \int_{B(x,r)} |f(y)| d\mu(y),$$

to prove the maximal theorem in this context as a consequence of a covering lemma of Vitali type, to develop as well Whitney's covering lemmas and to make it possible to write a Calderón-Zygmund decomposition of a function $f \in L^1(d\mu)$. As in the theory of singular integrals in $\mathbb{R}^n$, it is then possible to prove L^p estimates for singular integrals once one assumes that one already knows them for a particular value $p_0 > 1$. To go ahead and be able to have theorems $T(1)$ or $T(b)$ asks for new tools, and this is what we will tempt to describe later.

Calderón-Zygmund theory was developed at the same time by Coifman and de Guzmán in [15]. This last one studied systematically covering lemmas in relation with differentiation [23], which is always a living field of study (see for instance [69]).

It should be emphasized that, in the seventies, one had in mind some generalizations of Calderón-Zygmund theory, but not all. For example, the idea that this could be used for discrete structures appeared much later. Nevertheless, there is already this idea in the Notes that the doubling property of the measure μ can be replaced by the following one, which is weaker and is expressed only as a geometric property (named nowadays as the geometric doubling property):

2′. There exists a constant N such that, for all $x \in X$ and $r > 0$, the ball $B(x, 2r)$ is covered by at most N balls of radius r.

This will be seen as the right point of view for some constructions later on in this text.

There are also emerging notions in the course, which have been central in the further development of the theory. In particular, the definition of an atom of the Hardy space H^1 and the atomic decomposition of H^1 have only been given by Coifman a little later [14]. But the notion of atoms can already be found in the Notes (page 89 for instance). It is considered there for Riemannian symmetric spaces, and only from examples that are built with approximate identities. The property of singular integrals to transport them into molecules is already present. All this program is developed, this time with precise definitions and statements, in the two fundamental texts [18, 74].

In the Notes, one of the first examples of space of homogeneous type which is given is the unit sphere in $\mathbb{C}^n$ endowed with the Euclidean measure $d\sigma$ and the distance $d_b(z, \zeta) := |1 - \langle z , \zeta \rangle|$. We note $\mathbb{B}_n$ the unit ball in $\mathbb{C}^n$, $\partial\mathbb{B}^n$ its boundary, that is, the unit sphere and $\langle \cdot, \cdot \rangle$ the Hermitian inner product on $\mathbb{C}^n$. The distance d_b (the index b stands for the boundary) is clearly invariant under the action of the group $SU(n)$ and the distance from the point $\mathbf{1} = (1, 0, \cdots , 0)$ satisfies the property

$$d(\mathbf{1}, z) \simeq |\operatorname{Im} z_1| + |z'|^2, \qquad z = (z_1, z'),$$

meaning that the quotient of the two quantities lies between two uniform constants in a neighborhood of $\mathbf{1}$. Moreover, $\sigma(B(z, r)) \simeq r^n$ and

$$|z - \zeta|^2 \lesssim d_b(z, \zeta) \lesssim |z - \zeta|.$$

This metric is usually called the Korányi's metric. It had in particular already been used by A. Korányi for generalizing Fatou Theorem of non tangential convergence for harmonic functions in the unit disc to harmonic functions in $\mathbb{B}_n$ with respect to the Laplace-Beltrami operator for the Bergman metric [49]. Indeed, in the same way that the first question for spaces of homogeneous nature was about generalizations of the Lebesgue Differentiation theorem, the first question for holomorphic functions was about generalizations of Fatou Theorem.

At this point, let us recall definitions before going further. For Ω a smooth bounded domain in $\mathbb{C}^n$ and $0 < p < \infty$, let $L^p(\Omega)$ denote the Lebesgue space with respect to the Lebesgue measure $dV(z)$ and $A^p(\Omega)$ be the corresponding Bergman space, that is, the closed subspace of $L^p(\Omega)$ consisting of holomorphic functions. For $0 < p < \infty$, let $L^p(\partial\Omega)$ denote the Lebesgue space on $\partial\Omega$ with respect to the induced surface measure $d\sigma$ and $H^p(\Omega)$ the Hardy space of holomorphic functions on Ω, with norm given by

$$\|f\|_{H^p}^p := \sup_{0<\varepsilon<\varepsilon_0} \int_{\Omega} |f(w - \varepsilon\nu_w)|^p d\sigma(w),$$

where ν_w is the unit exterior normal vector at the point w, with ν_w seen as a vector in $\mathbb{C}^n$. The question of generalizing non tangential convergence to higher dimension was treated in the small book of E. Stein [71], which has been a major source of inspiration for its adaptation of potential theory to this context and for its list of open questions. For such smooth domains, functions in $H^p(\partial\Omega)$ have admissible limits a.e.: the admissible regions, roughly speaking, are defined in the same way as non tangential regions in the unit disc, except that the ordinary distance on the circle is replaced by the Korányi pseudo-distance on the boundary $\partial\Omega$, that is,

$$d_b(z,\zeta) := |\langle \nu_z, z-\zeta\rangle| + |z-\zeta|^2.$$

It is not symmetric, but $d_b(z,\zeta) \simeq d_b(\zeta,z)$ and can be symmetrized. These admissible regions are optimal for strictly pseudo-convex domains (that is, domains that are locally diffeomorphic to strictly convex domains), but not for weakly pseudo-convex domains in which there may be a larger swelling of admissible regions in complex tangential directions and a need for a definition that takes this into account. See for instance [62]. Because of admissible limits one may consider $H^p(\Omega)$ as a subspace of $L^p(\partial\Omega)$.

The next question concerns Bergman and Szegö projections, which we denote by P_B and P_S and which are, respectively, the orthogonal projection from $L^2(\Omega)$ to $A^2(\Omega)$ and the orthogonal projection from $L^2(\partial\Omega)$ to $H^2(\Omega)$ (identified with a subspace of $L^2(\partial\Omega)$). They are respectively given by the Bergman kernel $B_\Omega(z,\zeta)$ and the Szegö kernel $S_\Omega(z,\zeta)$. For the unit ball they are given by

$$S(z,\zeta) = \frac{c_n}{(1-\langle z\,,\zeta\rangle)^n}, \qquad B(z,\zeta) = \frac{c'_n}{(1-\langle z\,,\zeta\rangle)^{n+1}}.$$

Korányi and Vági, independently of Coifman and Weiss, developed around the same time a theory of singular integrals on homogeneous spaces, which led to the L^p inequalities for the Szegö projection of the unit ball [50]. This one is now seen as an immediate example of singular integral on $\partial\mathbb{B}^n$. Indeed, it is elementary to see that

$$|S(\mathbf{1},z) - S(\mathbf{1},\zeta)| \lesssim \frac{d_b(z,\zeta)^{1/2}}{|1-\mathbf{1}\cdot\zeta|^{n+1}},$$

which, by invariance by the action of $SU(n)$, gives the required inequality for a singular integral on the space of homogeneous type given by $\partial\mathbb{B}^n$, the Korányi distance and the Euclidean measure.

In order to obtain L^p estimates for the Bergman projection, one may also be consider it as a singular integral for the distance that is given by the ordinary distance in the radial direction and by d_b on the boundary $\partial\mathbb{B}_n$. But $L^p(\mathbb{B}_n)$ estimates can be obtained in a much simpler way by using Schur's lemma. These kinds of estimates have been first developed by Forelli and Rudin (see [30]).

This was the beginning of a long story, which we evoke now. From 1970, a considerable work has been done on Hardy and Bergman spaces as well as Szegö and Bergman projections in smooth bounded pseudo-convex domains of $\mathbb{C}^n$. The link between the adequate pseudo-distance at the boundary and the two kernels, which is obvious for the unit ball, asks for long studies in general. Estimates on the Bergman and Szegö kernels have been developed in the context of strictly pseudo-convex domains, then pseudo-convex domains of finite type in $\mathbb{C}^2$, then convex domains of finite type by Fefferman [28], Nagel et al. [63], McNeal and Stein [56] and many others.

The work of Henkin [42] on strictly pseudo-convex domains play a fundamental role for having explicit representation formulas. They lead to the construction of support functions $H \in \mathcal{C}^\infty(\Omega \times U)$ with U a neighborhood of the boundary, such that $H(\cdot, \zeta)$ is holomorphic on Ω for all $\zeta \in U$ and

$$C^{-1}d(z,\zeta) \leq |H(z,\zeta)| \leq Cd(z,\zeta).$$

Here d is defined in a neighborhood of the boundary by

$$d(z,\zeta) = \delta(z) + \delta(\zeta) + d_b(\pi(z), \pi(\zeta)) \tag{1.1}$$

where $\delta(z)$ is the distance of z to $\partial\Omega$ and $\pi(z)$ is the closest point of $\partial\Omega$, which is well-defined in U for U small enough and d_b is the adequate distance on the boundary. When Ω is the unit ball, one can take simply $H(z,\zeta) = 1 - \langle z\,,\zeta\rangle$. It is well adapted to Szegö or Bergman spaces, which have only singularities at the boundary. For convex domains of finite type (with the adequate pseudo-distance on the boundary), such a construction has been given by Diederich and Fornæss [26]. We will use these support functions later on. Unlike what happens in the unit ball, they are not found in connection with the Bergman kernel.

1.3.2 Weak Factorization, Commutators and Hankel Operators

In this subsection we will concentrate on one of the other fundamental papers of Guido, namely the one on commutators [19]. It was written in 1975 with Richard Rochberg and Raphy and, again, had a large influence on problems that have been considered later. Some of them have been solved recently, some are still open. It is among the top cited papers of Guido and has been generalized in many directions during the nearly passed fifty years. One can only find here a personal point of view, which leaves many developments aside.

The first question that is treated in this paper is the characterization of $BMO(\mathbb{R}^n)$ as the class of functions b for which the commutator $[M_b, R_j]$ are bounded in $L^2(\mathbb{R}^n)$. Here M_b stands for the multiplication by b and R_j is the j-th Riesz transform, that is, the convolution by the distribution $c_n p.v.\frac{x_j}{|x|^{n+1}}$. This has been fundamental and, almost twenty years later, revealed itself to be one of the ways to understand the div-curl lemma in relation with compensated compactness [20].

We will consider in more detail the second part of the article, which deals with holomorphic functions and factorization. Let us recall that in dimension 1 every function $f \in H^p(\mathbb{B}_1)$ may be written as the product gh, with $g \in H^{p_1}(\mathbb{B}_1)$, $h \in H^{p_2}(\mathbb{B}_1)$ and $\frac{1}{p} = \frac{1}{p_1} + \frac{1}{p_2}$. This is false in higher dimension (see [35]). In this second part one finds for the first time the idea to replace factorization by weak factorization. Namely, they prove that any function $f \in H^1(\mathbb{B}_n)$ can be written as

$$f = \sum_{j=1}^{\infty} g_j h_j, \qquad \sum_j \|g_j\|_{H^p} \|g_j\|_{H^{p'}} \simeq \|f\|_{H^1}.$$

Their proof uses atomic decomposition of Hardy spaces $H^1(\mathbb{B}_n)$. Namely, a holomorphic atom A is the Szegö projection of an atom a, that is, $A = P_S a$ with a a bounded function of mean zero, which is supported in a pseudo-ball $B \subset \mathbb{B}_n$, and such that $\|a\|_\infty \leq |B|^{-1}$. They prove that H^1 has an atomic decomposition, that is, $f \in H^1(\mathbb{B}_n)$ can be written in terms of atoms a_j related to balls $B_j = B(w_j, r_j)$

$$f = c + \sum_{j=1}^{\infty} \lambda_j a_j, \qquad \sum_j |\lambda_j| \simeq \|f\|_{H^1}.$$

It is then sufficient to factorize an atom a_j related to the ball $B(w_j, r_j)$. At this point their proof may be in some way simplified and can be seen as an easy consequence of the existence of a support function that satisfies (1.1), which makes it easier to find generalizations to other domains. One uses an idea that emerged later on: it is possible to ask for the atoms to have more moments that vanish, so that $|P_S a_j|$, which is not supported in B_j, decreases rapidly outside B_j. Then, one just takes $g_j = H(z, \zeta_j)^{-l}$, where the point ζ_j is chosen at distance r_j of $\partial\mathbb{B}_n$ and such that $\pi(\zeta_j) = w_j$. So for $l > n$ the norm of g_j in H^p is easily estimated, as well as the one of $H(z, \zeta_j)^l P a_j$ in $H^{p'}$ when the atoms have sufficiently vanishing moments.

This kind of proof extends easily to other domains, see [9]. It allows one to generalize weak factorization to H^p for $p < 1$ but does not work when there is no atomic decomposition of H^p, that is, for $p > 1$, even in the unit ball $\mathbb{B}^n$. Up to our knowledge, the existence of weak factorization of H^p is still an open problem in dimension $n > 1$:

For $p > 1$ and $\frac{1}{p} = \frac{1}{p_1} + \frac{1}{p_2}$, is there weak factorization for $H^p(\mathbb{B}^n)$ with products of functions in $H^{p_1}(\mathbb{B}^n)$ and $H^{p_2}(\mathbb{B}^n)$ when $n \geq 2$?

Another way to prove weak factorization consists in proving boundedness of the Hankel operator, which is the anti-linear operator on H^p defined by $H_b(f) = P_S(b\overline{f})$. More precisely, the fact that H^1 may be weakly factorized with products of functions in H^p and $H^{p'}$ is equivalent to the fact that H_b is bounded on H^p if and only if b belongs to the dual of H^1. The same considerations prove that the open problem given above is equivalent to the fact that the Hankel operator H_b maps H^{p_1} into $H^{p_2'}$ if and only if b belongs to $H^{p'}$.

In $\mathbb{R}^n$, such characterizations have been proved recently by Hytönen [45] for commutators $[M_b, R_j]$. It is straightforward that these commutators map L^p into L^q for $b \in L^r$, with $\frac{1}{r} = \frac{1}{p} + \frac{1}{q}$. The challenge was to prove the converse, which is quite intricate. Unfortunately, this result, even if proved for the Szegö projection on $\partial\mathbb{B}_n$, is not sufficient to find weak factorization for $H^p(\mathbb{B}_n)$ for $p > 1$.

Weak factorization for the Hardy space H^1 has also been studied in the polydisc, where there is no atomic decomposition. It was announced in a celebrated paper [29] of Ferguson and Lacey but at this moment there is a gap in their proof as proved in [44]. It is quite curious that the weak characterization of H^1, if true, implies, again on the bidisc, a weak factorization of H^p by products of functions in H^{2p} when $p < 3/2$ (unpublished manuscript of Bonami, Pott, Sehba and Wick).

The paper of Coifman, Rochberg and Weiss finishes with the weak factorization of Bergman spaces. This is by now also completely classical and has been generalized in many contexts. The classical method, now, (see for instance [79]) goes through the atomic decomposition that has been developed by Coifman and Rochberg [16] a little later: for $p > 0$ given, every function in $f \in A^p(\mathbb{B}_n)$ may be written as

$$f(z) = \sum_j \lambda_j \frac{d_j(1 - |w_j|)^N}{(1 - \langle z, w_j\rangle)^{n+N+1}}, \qquad \sum |\lambda_j|^p \simeq \|f\|^p_{A^p},$$

where the points w_j do not depend of the function f but constitute what is called an η−net: they are at distance at most η for the Bergman distance, but balls of radius $\eta/2$ centered at those points are disjoint while balls of radius $2r$ are almost disjoint. The quantities that appear here, $\frac{\omega_j(1-|w_j|)^N}{(1-\langle z, w_j\rangle)^{n+N+1}}$, are the values at (z, w_j) of the Bergman kernel when the Lebesgue measure dV is replaced by the weighted measure $(1 - |z|)^N dV(z)$. The constant d_j is a normalization factor so that its A^p−norm is 1. One has not only the atomic decomposition, but the weak factorization of Bergman spaces for $p \leq 1$. Generalizations are valid in the same context as for Hardy spaces, as well as estimates for related Hankel operators, now defined as $H_b(f) = P_B(b\overline{f})$. But one can go much further in many directions.

First, the fact that such decompositions are valid for $p > 1$ and a tricky use of Rademacher coefficients has allowed Pau and Zhao [68] to get estimates with loss for the Hankel operator, and so to obtain weak factorization of Bergman spaces of $A^p(\mathbb{B}^n)$ for all values of p.

Secondly, the validity of atomic decomposition and weak factorization of Bergman spaces can be proved far beyond the contexts for which one can conclude for Hardy spaces, as outlined in [16]: one can find such decompositions in tube domains over symmetric cones, for instance, but with restrictions that are due to the lack of validity of L^p inequalities for the Bergman projection for all $p > 1$. In particular, if $\Omega = \mathbb{R}^n + i\Gamma$, where Γ is a symmetric cone (think of the forward light cone $y_n > \sqrt{y_1^2 + \cdots + y_n^2}$), results related to the Bergman space may be adapted (see [64]), but with new constraints that are linked to the validity of L^p inequalities for the Bergman projection. The constraints for these L^p inequalities given in [6] are the best possible in the case of the forward light cone as a consequence of the decoupling inequalities of Bourgain and Demeter [11].

We are then far from the geometry of spaces of homogeneous type and singular integrals, but all this finds its origin in the seminal papers written from the years 1970 by Guido and those researchers that gathered around him.

1.4 Littlewood-Paley Decompositions and Wavelets

1.4.1 The Context

The tools based on covering arguments that only need balls allow, in principle, to forget about algebraic contents to focus more on geometry or distributions of points. Still, familiar tool boxes such as convolution and the Fourier transform need to be replaced by something else.

First one needs approximations to the identity in relation to scales replacing usual mollifiers. Second to develop a rich function space theory and the action of operators on them (such as Calderón-Zygmund operators), one needs Littlewood-Paley decompositions. It took some time to develop those tools in the optimal generality offered by the definition of a space of homogeneous type, that is, without extra hypotheses. The aim of this section is to explain the evolution of the ideas and a solution. We use material from Auscher-Hytönen [3] and also the introduction of Han-Li-Ward [38]. The latter reference contains a rich bibliography as well.

Recall that the definition of a space of homogeneous type involves a set X equipped with a quasi-distance d and a positive doubling Borel measure μ (Borel measure is defined in several ways in the literature: either it is a measure on the Borel σ-algebra and one works only with Borel measurable functions or it is an outer measure for which Borel sets are measurable (in the sense of Carathéodory): in the latter case, we add the regularity property that every measurable set is contained in a Borel set with same measure). As said, extra hypotheses had to be imposed in the proofs. However, it seemed that this would not limit the range of applicability thanks to an article of Macías-Segovia that describes the structure of spaces of homogeneous type [53]. This explains why removing these extra conditions remained unexplored for some time.

A popular assumption is that the volume of a ball $B(x,r) = \{y \in X : d(x,y) < r\}$ is comparable to a power r^α of its radius, uniformly in its center (this is called the Ahlfors-David regularity property for the measure). Another one is that the volume of a ball $B(x, Cr)$ exceeds $(1+\varepsilon)$ times the volume of $B(x,r)$ for some constants $1 < C < \infty$, $\varepsilon > 0$ (this is called the reverse doubling property). One or the other are satisfied in many cases. Nevertheless, if one wants one or the other for all $0 < r$ less than a fraction of the diameter of X, this excludes atomic measures, and for example $\mathbb{Z}$ equipped with the induced absolute value and counting measure.

Considering that Hardy-Littlewood maximal theorem's original setup was on p-summable sequences, this would be paradoxical not to have a theory covering all situations.

The balls $B(x, r)$ define the topology on X but they might not be open sets nor even Borel sets when d is a not a distance. This can be remedied by replacing them by their interiors in the definition of doubling. Alternatively, one result of Macías-Segovia [53] says one can change the quasi-distance to a metrically equivalent one (actually, a power of a genuine distance) for which the balls are open. So this is not a major issue as this only affects the quasi-distance and the doubling constants.

A more serious issue is again due to the difference between the metric case and the quasi-metric case: a distance is always Lipschitz-continuous (with respect to itself) from the triangular inequality; it is not necessarily the case of a quasi-distance which might not even be Hölder-continuous. Another result of Macías-Segovia [53] says that one can change the balls so that it becomes true and even the measure becomes Ahlfors-David regular. This is why most articles afterwards claimed that their results were proved in a "general" space of homogeneous type. But this is not quite correct. First, this change preserves the topology but it is not metrically equivalent. Next, it changes the classes of Hölder-continuous functions and of Calderón-Zygmund kernels. An example in [38] shows that doing this change might not be desirable if the quasi-distance has a special interpretation in terms of an explicit operator as one may loose the connection. In complex analysis as above, the quasi-distance is defined according to the geometric context and one does not want to change it. This is, however, in this context (with a new quasi-distance from [53]) that the proof of the T(b) Theorem by David-Journé-Semmes was given [22]. They were able to implement an unpublished method of Coifman to produce approximations to the identity and Littlewood-Paley decompositions. More precisely, the approximations to the identities are families of operators $(S_k)_{k\in\mathbb{Z}}$ ($k \geq k_0$ if X is bounded) with kernels $S_k(x, y)$ enjoying the following properties for some $C < \infty$, $\varepsilon > 0$ and all k and points

$$S_k(x, y) = 0 \text{ if } d'(x, y) \geq C2^{-k}, \text{ and } |S_k(x, y)| \leq C2^k$$

$$|S_k(x, y) - S_k(x, y')| \leq C2^{k(1+\varepsilon)} d'(y, y')^\varepsilon$$

$$|S_k(x, y) - S_k(x', y)| \leq C2^{k(1+\varepsilon)} d'(x, x')^\varepsilon$$

$$\int_X S_k(x, y)\, d\mu(y) = 1 = \int_X S_k(x, y)\, d\mu(x)$$

Here the quasi-distance d' is the one provided by the Macías-Segovia result. If $D_k := S_{k+1} - S_k$, then for all $1 < p < \infty$ and $f \in L^p(X, \mu)$ ($k \geq k_0$ and $\int f\, d\mu = 0$ when X is bounded)

$$C^{-1}\|f\|_p \leq \|(\sum_k |D_k f|^2)^{1/2}\|_p \leq C\|f\|_p.$$

Hence the map $f \mapsto (D_k f)$ acts as a Littlewood-Paley decomposition. From sometime on, these methods have been streamlined and generalized (see [25, 36]) in the same context.

The next development has been to work on spaces with the reverse doubling property. This case is of much interest for example on Lie groups where the volumes of balls may obey two different power laws whether we assume small or large radii (and the exponents are often called local and global dimensions respectively). This covers also the case of some complete Riemannian manifolds. The reverse doubling condition is indeed a quantitative expression of connectedness of the space: annuli between two co-centered balls can never be empty and non-emptyness self improves to having mass at the correct scale. There, the technology of Coifman can be developed nicely (see Han-Müller-Yang, [37]) in a sense that estimates on the kernel family $S_k(x, y)$ go through with slight modification in their formulations. A decisive inequality for the treatment of singular integrals in reverse doubling spaces is that

$$\sum_{k\geq 0}(\mu(B(x, C^k r))^{-1} \lesssim (\mu(B(x, r))^{-1} \tag{1.2}$$

as the series is controlled in a geometric manner. Again, such conditions exclude spaces of homogeneous types of "discrete" nature and more. By the early 2010, about 40 years after the seminal definition of a space of homogeneous type, although the covering arguments and maximal theorem worked beautifully, it was still not known how to develop Littlewood-Paley theory in full generality (again, despite what most authors wrote) and thus the function spaces and operator theory of singular integrals that go with. We shall see that the solution came from the need to work in absence of the doubling condition. Let us explain how.

In parallel to the development of "regular" decompositions, dyadic analysis was made possible thanks to the introduction of "dyadic cubes" on spaces of homogeneous type, first for Ahlfors-David regular measures by G. David [21] and then by M. Christ [13] in full generality. The idea is that containment into a unique parent dyadic cube and having a bounded number of children dyadic subcubes can be obtained from a carefully defined partial order on a family of selected points with conditions only involving the quasi-distance. These points become the "centers" of the cubes, which are sets contained in balls and containing balls with those centers and radii comparable to their diameters. This clever construction allowed a local formulation of the T(b) theorem that was best designed toward an analytic capacity result in mind.

In $\mathbb{R}^n$, the dyadic cubes are linked to the Haar wavelets which form an orthonormal basis of $L^2(dx)$ where dx is Lebesgue measure. The construction of Christ naturally leads to an orthonormal basis with respect to the doubling measure at hand, of course with non regular functions. The beautiful theory of orthonormal wavelets on $\mathbb{R}^n$, launched by Meyer [57] after the first unnoticed discoveries of Strömberg [73], allows one to build orthonormal bases of localized and regular functions: not only they have decay away from a dyadic cube (or even have compact support) but also they enjoy Hölder regularity. Some can even be infinitely smooth.

Explicitly or implicitly, Fourier transform is used. In fact, translation invariance and dilation invariance play a crucial role and indeed regular wavelet bases can be built on stratified nilpotent Lie groups [52].

Coming back to Haar wavelets, it can be shown they provide unconditional bases of $L^p(dx)$ when $1 < p < \infty$. The situation is different when $p = 1$ and $p = \infty$. When $p = 1$, they are unconditional families in the real Hardy space $H^1(\mathbb{R}^n)$ but span a strict subspace called the dyadic Hardy space $H^1_{dyad}(\mathbb{R}^n)$. It had become a challenge to prove that $H^1(\mathbb{R}^n)$ also admits an unconditional basis. B. Maurey [55] solved the question by showing the two spaces are isomorphic, but without providing an explicit basis. Strömberg's orthonormal basis of spline functions (that turn to fit the wavelet theory and were rediscovered by Battle and Lemarié [5, 51]) precisely yields an explicit construction. In the dual range, $BMO(\mathbb{R}^n)$, the dual of $H^1(\mathbb{R}^n)$ from the theorem of C. Fefferman [27], was known to be strictly contained in its dyadic version $BMO_{dyad}(\mathbb{R}^n)$. It has to do with the fact that Euclidean distance $|x - y|$ and the distance induced by the dyadic cubes (the diameter of the smallest dyadic cube containing x and y) are not equivalent. In his celebrated book [34, p. 417], J. Garnett implicitly mentioned that classical $BMO(\mathbb{R})$ is the intersection of three of shifted copies of $BMO_{dyad}(\mathbb{R})$. By shifted, it is meant that one considers dyadic grids $2^k(Q_a + \ell)$, $k \in \mathbb{Z}, \ell \in \mathbb{Z}$, where Q_a is the shifted interval $a + [0, 1[$ by $a \in [0, 1[$. More than an amusing observation, this shifting property can be thought as a random process among dyadic grids and this became a key tool to develop the singular integral theory on subsets X of Euclidean space equipped with a Radon measure satisfying the power law upper estimate $\mu(B(x, r)) \leq Cr^\beta$ for all $x \in X$ and $r > 0$ as evidenced in the work of Nazarov-Treil-Volberg, [65]. The construction of random families of dyadic grids on other contexts would then allow to develop singular integral theory there. But randomness cannot come from translation and dilation invariance anymore. A first construction of T. Hytönen and H. Martikainen on quasi-metric sets with the geometric doubling property [47] was designed to prove a T(b) theorem on such sets equipped with a measure satisfying the upper doubling condition, that is $\mu(B(x, r)) \leq \lambda(x, r)$ where $\lambda(x, 2r) \leq C\lambda(x, r)$ (a condition that covers both the doubling condition for the measure itself and the power law upper estimate). Another possible construction of random grids was proposed in [66] and T. Hytönen and A. Kairema [46] streamlined the construction in [47].

1.4.2 A Regular Wavelet Basis After All

As mentioned, while dyadic decompositions occurred in full generality, neither Littlewood-Paley decomposition, nor regular wavelet bases were available in 2011 in a "general" space of homogeneous spaces. The existence of an orthonormal basis of regular wavelets was too much of a dream to have been conjectured anywhere. In [25], which was one of the most advanced text on the topic at this time, it was even written that they are "out of reach" and frames (a substitute notion where redundancy is allowed) were obtained but using again the Macías-Segovia trick. Still, when Tuomas Hytönen came to Orsay in 2011, I (Pascal) asked him the problem. How to overcome the lack of any algebraic structure? How even can we make sure that for a quasi-distance that is not Hölder-continuous, one can obtain profusion (dense classes) of Hölder-continuous functions? Should we restrict to the metric case? It turned out that the idea of random dyadic grids opened the door to a solution in [3] that required five steps.

1. The first step is the construction of appropriate random dyadic grids using **nested** collections of separated and dense subsets of X at each given scale.
2. The second one is the construction of **spline** functions that are Hölder-continuous of (some) order in (0, 1) depending on the space.
3. In the third step, once the doubling measure is given, the spline functions provide a **multiresolution analysis** on $L^2(\mu)$ (without the translation and dilation requirements, of course).
4. In the fourth step, **construct the wavelets** by transposing Hilbertian algorithms of Y. Meyer [58, 59] for the construction of spline wavelets on $\mathbb{R}^n$ to spaces of homogeneous type.
5. In the fifth step, the **localisation** of the wavelets is obtained by extending from the metric case to the quasi-metric case existing lemmas due to Demko [24] to estimate the entries of the inverse of a band matrix.

Steps 1 and 2 do not require measure information but only the geometric doubling property. The spline functions have interesting properties of interpolation and reproduction which permit to construct the wavelets through a multiresolution structure. The Hölder regularity of the wavelets is the one of the spline functions. The construction of random grids was slightly modified later by Hytönen-Tapiola [48] in the metric case to allow arbitrary Hölder exponent in (0, 1).

This in the end yields the following result (gathering both [3, 48])

Theorem 1.1 *Let (X, d, μ) be any space of homogeneous type and call A_0 its quasi-triangle constant. Let $a := (1 + 2\log_2 A_0)^{-1}$ or $a := 1$ if d is Lipschitz-continuous. Then there exists $\eta \in (0, 1)$ depending only on A_0 (any $\eta \in (0, 1)$ works when $A_0 = 1$), constants $\delta > 0$ and $\gamma > 0$ depending only on A_0, and $0 < C < \infty$ depending on A_0 and the doubling constant, such that one has the following.*

There are non-decreasing sets $\mathscr{X}^k$, $k \in \mathbb{Z}$ (and $k \geq k_0$ if X is bounded) of δ^k-separated and $2A_0\delta^k$-dense points in X, such that if $\mathscr{Y}^k := \mathscr{X}^{k+1} \setminus \mathscr{X}^k$, labelled as $\mathscr{Y}^k = \{y^k_\beta\}_\beta$, there exists an orthonormal basis of real-valued functions ψ^k_β, $k \in \mathbb{Z}$ (and $k \geq k_0$ if X is bounded), corresponding to each $y^k_\beta \in \mathscr{Y}^k$, of $L^2(\mu)$ (or the orthogonal space to constants if X is bounded) having exponential decay

$$|\psi^k_\beta(x)| \leq \frac{C}{\sqrt{\mu(B(y^k_\beta, \delta^k))}} \exp\big(-\gamma(\delta^{-k} d(y^k_\beta, x))^a\big),$$

Hölder-regularity

$$|\psi_\beta^k(x) - \psi_\beta^k(y)| \\ \leq \frac{C}{\sqrt{\mu(B(y_\beta^k, \delta^k))}} \Big(\frac{d(x,y)}{\delta^k}\Big)^\eta \exp\big(-\gamma(\delta^{-k}d(y_\beta^k, x))^a\big), \quad d(x,y) \leq \delta^k,$$

and vanishing mean

$$\int_X \psi_\beta^k(x)\, d\mu(x) = 0, \quad k \in \mathbb{Z}, k \geq k_0, y_\beta^k \in \mathcal{Y}^k.$$

A subset S of X is η-separated if two distinct elements of S have distance at least η, it is η-dense if any point in X has distance to S not exceeding η. This theorem and its argument provide us with the desired Littlewood-Paley decomposition.

Corollary 1.1 *Fix $1 < p < \infty$. The orthogonal projections Q_k on the closed linear span W_k of the ψ_β^k form a Littlewood-Paley decomposition. More precisely, there is a constant $0 < C < \infty$ such that for all $f \in L^p(\mu)$, ($k \geq k_0$ and $\int_X f\, d\mu = 0$ if X is bounded)*

$$C^{-1}\|f\|_p \leq \|(\sum_{k\in\mathbb{Z}} |Q_k f|^2)^{1/2}\|_p \leq C\|f\|_p.$$

Moreover, the family ψ_β^k (together with the indicator of X if X is bounded) forms an unconditional basis of $L^p(\mu)$ and $P_k = \sum_{j<k} Q_j$ ($j \geq k_0$ and add the orthogonal projection onto constants if X is bounded) is a family of approximation to the identity with kernels $P_k(x, y)$, symmetric in x, y, having size estimates

$$|P_k(x,y)| \leq \frac{C}{\sqrt{\mu(B(x,\delta^k))\mu(B(y,\delta^k))}} \exp\big(-\gamma(\delta^{-k}d(x,y))^s\big), \tag{1.3}$$

regularity estimates

$$|P_k(x,y) - P_k(x,y')| \leq C\left(\frac{d(y,y')}{\delta^k}\right)^\eta \\ \times \left(\frac{\exp\big(-\gamma(\delta^{-k}d(x,y))^s\big)}{\sqrt{\mu(B(x,\delta^k))\mu(B(y,\delta^k))}} + \frac{\exp\big(-\gamma(\delta^{-k}d(x,y'))^s\big)}{\sqrt{\mu(B(x,\delta^k))\mu(B(y',\delta^k))}}\right)$$

for some C, γ, s and all $x, y, y' \in X$ and $k \in \mathbb{Z}$ (with $k \geq k_0$ if X is bounded). Moreover

$$\int_X P_k(x,y) d\mu(x) = 1.$$

The kernel $Q_k(x, y)$ of Q_k has similar estimates with s changed to a and the additional multiplicative exponential factor in the size estimate

$$\exp\big(-\gamma((\delta^{-k}d(x,\mathscr{Y}^k))^a + (\delta^{-k}d(y,\mathscr{Y}^k))^a)\big), \tag{1.4}$$

and the cancellation condition

$$\int_X Q_k(x, y)d\mu(x) = 0.$$

The exponential factor in (1.4) is the key new object quantifying the geometry of X. We shall come back to this.

The decay of wavelets is only exponential and an open question is whether one can obtain bounded support up to relaxing orthonormality to bi-orthogonality if necessary.

We shall now indicate the idea of the construction of the splines: the guiding principle is that their interpolation properties, their reproducing properties and their regularity are obtained from a probabilistic argument allowed from the axioms in the construction of random dyadic cubes.

1.4.3 The Key Ideas of the Construction of the Spline Functions

The classical piece-wise linear splines on $\mathbb{R}$ are generated by translation and dilation of the function

$$s(x) = x1_{(0,1]}(x) + (2 - x)1_{(1,2)}(x).$$

Another formula is $s(x) = 1_{[0,1)} * 1_{[0,1)}(x)$ and writing out this convolution yields

$$s(x) = \int_0^1 1_{[0,1)}(x - u)\,\mathrm{d}u = \int_0^1 1_{[u,u+1)}(x)\,du.$$

Random dyadic intervals of sidelength 1, in the sense of Nazarov, Treil and Volberg [65, Sec. 9.1], are precisely defined by translating the standard intervals $[k, k + 1)$ by a random number $u \in [0, 1)$. Thus the unit cube with left end at the origin, $[0, 1)$, is translated to $[u, u + 1)$. So one can think of $s(x)$ as an average of the indicators of random dyadic intervals, or in probabilistic notation

$$s(x) = \mathbb{P}_u\big(x \in [u, u + 1)\big),$$

where $\mathbb{P}_u$ is a fancy notation for the uniform probability measure on $[0, 1[$ (i.e. the Lebesgue measure). This is the basic idea which will guide us in constructing splines for a quasi-metric space X with the geometric doubling property. We next provide some details and key points of the constructions, especially of the splines. Proofs can be consulted in [3].

The **reference dyadic points** are chosen with the additional requirement of **nestedness** compared to the earlier references [13, 46, 47]. For every $k \in \mathbb{Z}$, we choose a set of *reference dyadic points* x_α^k as follows: For $k = 0$, let $\mathscr{X}^0 := \{x_\alpha^0\}_\alpha$ be a maximal collection of 1-separated points. Inductively, for $k \in \mathbb{Z}_+$, let $\mathscr{X}^k := \{x_\alpha^k\}_\alpha \supseteq \mathscr{X}^{k-1}$ and $\mathscr{X}^{-k} := \{x_\alpha^{-k}\}_\alpha \subseteq \mathscr{X}^{-(k-1)}$ be maximal δ^k- and δ^{-k}-separated collections in X and in $\mathscr{X}^{-(k-1)}$, respectively. It is easy to show that for all $k \in \mathbb{Z}$ and $x \in X$, the reference dyadic points satisfy

$$d(x_\alpha^k, x_\beta^k) \geq \delta^k \quad (\alpha \neq \beta), \qquad d(x, \mathscr{X}^k) = \min_\alpha d(x, x_\alpha^k) < 2A_0\delta^k.$$

The second property is the $2A_0\delta^k$-density of the set $\mathscr{X}^k$. Note that $\mathscr{X}^k \subseteq \mathscr{X}^{k+1}$, so that every x_α^k is also a point of the form x_β^{k+1}, and thus of all the finer levels. Note also that $\mathscr{X}^k$ reduces to one point for $k \leq k_0$ for some k_0 if X is bounded. Still we can continue with the full chain of sets for $k \in \mathbb{Z}$. We denote $\mathscr{Y}^k := \mathscr{X}^{k+1} \setminus \mathscr{X}^k$, and relabel these points as $\mathscr{Y}^k = \{y_\beta^k\}_\beta$. These points will be the parameter set of our wavelets.

The **reference partial order** among the pairs (k, α) is the same as in M. Christ [13]. For $\delta > 0$ small enough (the smallness is fixed at the end to make every step work, and depends only on A_0), each $(k+1, \beta)$ satisfies $(k+1, \beta) \leq (k, \alpha)$ for exactly one (k, α), in such a way that

$$\begin{aligned} d(x_\beta^{k+1}, x_\alpha^k) < \frac{1}{2A_0}\delta^k \quad &\Longrightarrow \quad (k+1, \beta) \\ \leq (k, \alpha) \quad &\Longrightarrow \quad d(x_\beta^{k+1}, x_\alpha^k) < 2A_0\delta^k. \end{aligned} \tag{1.5}$$

The pairs $(k+1, \beta)$ with $(k+1, \beta) \leq (k, \alpha)$ are called the children of (k, α). Geometric doubling implies that their number is uniformly bounded.

Randomness is made possible by distinguishing the reference points using **labels** as in [46]. Points (k, α) and (k, β) are called *neighbours*, if they have children $(k+1, \gamma) \leq (k, \alpha)$ and $(k+1, \eta) \leq (k, \beta)$ such that $d(x_\gamma^{k+1}, x_\eta^{k+1}) < (2A_0)^{-1}\delta^k$. In this case, by the quasi-A_0-triangle inequality,

$$\begin{aligned} d(x_\alpha^k, x_\beta^k) &\leq A_0 d(x_\alpha^k, x_\gamma^{k+1}) + A_0^2 d(x_\gamma^{k+1}, x_\eta^{k+1}) + A_0^2 d(x_\eta^{k+1}, x_\beta^k) \\ &< 2A_0^2\delta^k + \tfrac{1}{2}A_0\delta^k + 2A_0^3\delta^k < 5A_0^3\delta^k. \end{aligned}$$

The number of neighbours that any point can have is also uniformly bounded.

Each pair (k, α) is equipped with two *labels*. The primary label, denoted by $\text{label}_1(k, \alpha) \in \{0, 1, \ldots, L\}$, where L is the maximal number of neighbours, is chosen in such a way that any two neighbours have a different label. The secondary label, denoted by $\text{label}_2(k, \alpha) \in \{1, \ldots, M\}$, where M is the maximal number of children, is chosen in such a way that no two children of the same parent have the same label.

As described above, we now want to perform a perturbation of the original x_α^k and $\leq$ so as to obtain a **parametrized family** of similar objects, on which probabilistic statements can later be made. The parameter space is

$$\Omega = \Big(\{0, 1, \dots, L\} \times \{1, \dots, M\}\Big)^{\mathbb{Z}},$$

with a typical point denoted by $\omega = (\omega_k)_{k\in\mathbb{Z}}$, where $\omega_k = (\ell_k, m_k) \in \{0, 1, \dots, L\} \times \{1, \dots, M\}$.

The *random new dyadic points* $z_\alpha^k(\omega)$ are defined by

$$z_\alpha^k(\omega) := \begin{cases} x_\beta^{k+1} & \text{if } \mathrm{label}_1(k,\alpha) = \ell_k,\ (k+1,\beta) \leq (k,\alpha) \\ & \text{and } \mathrm{label}_2(k+1,\beta) = m_k, \\ x_\alpha^k & \text{if } \mathrm{label}_1(k,\alpha) \neq \ell_k \text{ or } \nexists (k+1,\beta) \leq (k,\alpha) \\ & \text{such that } \mathrm{label}_2(k+1,\beta) = m_k. \end{cases}$$

Clearly, this point depends only on ω_k so we write $z_\alpha^k(\omega) = z_\alpha^k(\omega_k) = z_\alpha^k$ and use the most convenient notational choice.

Equip Ω with the natural probability measure $\mathbb{P}_\omega$, which makes all coordinates $\omega_k = (\ell_k, m_k)$ independent of each other and uniformly distributed over the finite set $\{0, 1, \dots, L\} \times \{1, \dots, M\}$. For $(k+1,\beta) \leq (k,\alpha)$ fixed, then

$$\mathbb{P}_\omega(z_\alpha^k(\omega) = x_\beta^{k+1}) \geq \frac{1}{(L+1)M}.$$

In other words, every old point on the level $k+1$ has a positive (and bounded from below) probability of being a new point on the level k. The new points behave qualitatively like the reference points, only with slightly weaker constants in the separation and density:

$$d(z_\alpha^k, z_\beta^k) \geq \frac{1}{2A_0}\delta^k, \qquad \min_\alpha d(x, z_\alpha^k) < 4A_0^2\delta^k.$$

The *new partial order* $\leq_\omega$, $\omega = (\omega_k)_{k\in\mathbb{Z}}$, is set up as follows, declaring that

$$(k+1,\beta) \leq_\omega (k,\alpha)$$

$$\overset{\text{def}}{\Longleftrightarrow} \quad \begin{cases} d(x_\beta^{k+1}, z_\alpha^k(\omega)) < \frac{1}{4}A_0^{-2}\delta^k, & \text{or} \\ (k+1,\beta) \leq (k,\alpha) \text{ and } \nexists\gamma : d(x_\beta^{k+1}, z_\gamma^k(\omega)) < \frac{1}{4}A_0^{-2}\delta^k. \end{cases} \tag{1.6}$$

In other words, to find the new parent of $(k+1,\beta)$ for the new partial order $\leq_\omega$, we first check whether the reference point x_β^{k+1} is close (within distance $\frac{1}{4}A_0^{-2}\delta^k$) to some new dyadic point $z_\alpha^k(\omega)$. If yes, then the corresponding (k,α) will be the new parent of $(k+1,\beta)$. If no such close point exists, then we simply use the original partial order $\leq$ to decide the parent of $(k+1,\beta)$. With this definition, one can observe that for any given k,α,β, the truth or falsity of the relation $(k+1,\beta)\leq_\omega (k,\alpha)$ depends only on the component ω_k of ω. Thus, it is independent of any event that occurs at level $k+1$ and higher. This explicit definition of the original partial order $\leq$ and of $\leq_\omega$ only depends of the geometric configuration of the points, and a condition analogous to (1.5) is a *consequence* of the definition:

$$d(z_\beta^{k+1}, z_\alpha^k) < \tfrac{1}{5}A_0^{-3}\delta^k \implies (k+1,\beta)\leq_\omega (k,\alpha) \implies d(z_\beta^{k+1}, z_\alpha^k) < 5A_0^3\delta^k.$$

Iteration yields that for all $\ell \geq k$,

$$d(z_\beta^{\ell}, z_\alpha^k) < \tfrac{1}{6}A_0^{-4}\delta^k \implies (\ell,\beta)\leq_\omega (k,\alpha) \implies d(z_\beta^{\ell}, z_\alpha^k) < 6A_0^4\delta^k.$$

With the auxiliary objects at hand, the **random dyadic cubes** are easy to define. As in [46], introduce three families of these cubes—the preliminary, the closed, and the open:

$$\hat{Q}_\alpha^k(\omega) := \{z_\beta^\ell(\omega) : (\ell,\beta)\leq_\omega (k,\alpha)\},$$

$$\bar{Q}_\alpha^k(\omega) := \overline{\hat{Q}_\alpha^k(\omega)}, \qquad \tilde{Q}_\alpha^k(\omega) := \text{interior}\, \bar{Q}_\alpha^k(\omega).$$

Note that $\hat{Q}_\alpha^k(\omega)$, and hence $\bar{Q}_\alpha^k(\omega)$ and $\tilde{Q}_\alpha^k(\omega)$, only depends on ω_ℓ for $\ell \geq k$. Of course, the name "cube" solely refers to the analogous situation in Euclidean space without any other geometric meaning for such sets (that same construction in Euclidean space would not necessarily end up with a cube).

The following theorem summarizes the above properties of the random dyadic cubes for a fixed parameter ω (that is, they separately form dyadic grids), and supplements the key statement about their probabilistic behaviour under the random choice of $\omega \in \Omega$.

Theorem 1.2 *For any fixed* $\omega \in \Omega := (\{0,1,\ldots,L\}\times\{1,\ldots,M\})^{\mathbb{Z}}$, *the cubes satisfy the following relations of a dyadic grid: the covering properties*

$$X = \bigcup_\alpha \bar{Q}_\alpha^k(\omega), \qquad \bar{Q}_\alpha^k(\omega) = \bigcup_{\beta:(k+1,\beta)\leq_\omega (k,\alpha)} \bar{Q}_\beta^{k+1}(\omega);$$

the mutual disjointness property

$$\tilde{Q}_\alpha^k(\omega)\cap \tilde{Q}_\beta^k(\omega) = \varnothing \qquad (\alpha\neq\beta);$$

and the comparability with balls:

$$B(z_\alpha^k(\omega), \tfrac{1}{6}A_0^{-5}\delta^k) \subseteq \tilde{Q}_\alpha^k(\omega) \subseteq \bar{Q}_\alpha^k(\omega) \subseteq B(z_\alpha^k(\omega), 6A_0^4\delta^k).$$

Moreover, when Ω *is equipped with the natural probability measure* $\mathbb{P}_\omega$*, we have for some* $\eta \in (0, 1]$ *the small boundary layer property:*

$$\mathbb{P}_\omega\Big(x \in \bigcup_\alpha \partial_\varepsilon Q_\alpha^k(\omega)\Big)$$

$$\leq C\varepsilon^\eta \qquad \Big(\partial_\varepsilon Q_\alpha^k(\omega) := \{y \in \bar{Q}_\alpha^k(\omega) : d(y, {}^c\tilde{Q}_\alpha^k(\omega)) < \varepsilon\delta^k\}\Big); \tag{1.7}$$

and in particular the negligible boundary property:

$$\mathbb{P}_\omega\Big(x \in \bigcup_{k,\alpha} \partial Q_\alpha^k(\omega)\Big) = 0 \qquad \Big(\partial Q_\alpha^k(\omega) := \bar{Q}_\alpha^k(\omega) \setminus \tilde{Q}_\alpha^k(\omega)\Big).$$

Eventually, the original dyadic point x_α^k *may also be viewed as a 'centre' of the random dyadic cubes* $\bar{Q}_\alpha^k(\omega)$ *for all* $\omega \in \Omega$*:*

$$B(x_\alpha^k, \tfrac{1}{8}A_0^{-3}\delta^k) \subseteq \bar{Q}_\alpha^k(\omega) \subseteq \bar{B}(x_\alpha^k, 8A_0^5\delta^k).$$

The small boundary property may not be true for balls in general. It was already observed by G. David (without randomness) and used by M. Christ for singular integrals that dyadic cubes should have this property and that this can be built-in from the partial order. It will be the key to obtain the Hölder-regularity exponent of our splines and wavelets. The **construction of splines** on X, and the proof of their basic properties, is indeed an easy consequence of the preparations above. For every (k, α), we define the **spline function** s_α^k by

$$s_\alpha^k(x) := \mathbb{P}_\omega\Big(x \in \bar{Q}_\alpha^k(\omega)\Big). \tag{1.8}$$

Theorem 1.3 *The splines* (1.8) *satisfy the following properties: bounded support*

$$1_{B(x_\alpha^k, \frac{1}{8}A_0^{-3}\delta^k)}(x) \leq s_\alpha^k(x) \leq 1_{B(x_\alpha^k, 8A_0^5\delta^k)}(x); \tag{1.9}$$

the interpolation and reproducing properties

$$s_\alpha^k(x_\beta^k) = \delta_{\alpha\beta}, \qquad \sum_\alpha s_\alpha^k(x) = 1, \qquad s_\alpha^k(x) = \sum_\beta p_{\alpha\beta}^k \cdot s_\beta^{k+1}(x) \tag{1.10}$$

where $p^k_{\alpha\beta} = \mathbb{P}_\omega\big((k+1,\beta) \leq_\omega (k,\alpha)\big)$ *so that* $\sum_\beta p^k_{\alpha\beta} = 1$*; and Hölder-continuity*

$$|s^k_\alpha(x) - s^k_\alpha(y)| \leq C\Big(\frac{d(x,y)}{\delta^k}\Big)^\eta.$$

Let us prove the **Hölder-continuity of the splines** to show how the probabilistic smallness of the boundary regions, as expressed by (1.7), comes into play. Indeed,

$$\begin{aligned}
|s^k_\alpha(x) - s^k_\alpha(y)| &= \Big|\mathbb{P}_\omega\Big(x \in \bar{Q}^k_\alpha(\omega)\Big) - \mathbb{P}_\omega\Big(y \in \bar{Q}^k_\alpha(\omega)\Big)\Big| \\
&= \Big|\int_\Omega \big(1_{\omega:x\in\bar{Q}^k_\alpha(\omega)} - 1_{\omega:y\in\bar{Q}^k_\alpha(\omega)}\big)\,d\mathbb{P}_\omega\Big| \\
&\leq \int_\Omega \big(1_{\omega:x\in\bar{Q}^k_\alpha(\omega),y\notin\bar{Q}^k_\alpha(\omega)} + 1_{\omega:y\in\bar{Q}^k_\alpha(\omega),x\notin\bar{Q}^k_\alpha(\omega)}\big)\,d\mathbb{P}_\omega \\
&= \mathbb{P}_\omega\Big(x \in \bar{Q}^k_\alpha(\omega),\, y \notin \bar{Q}^k_\alpha(\omega)\Big) + \mathbb{P}_\omega\Big(y \in \bar{Q}^k_\alpha(\omega),\, x \notin \bar{Q}^k_\alpha(\omega)\Big) \\
&\leq \mathbb{P}_\omega\Big(x \in \partial_{d(x,y)\delta^{-k}}\bar{Q}^k_\alpha(\omega)\Big) + \mathbb{P}_\omega\Big(y \in \partial_{d(x,y)\delta^{-k}}\bar{Q}^k_\alpha(\omega)\Big) \\
&\leq C\Big(\frac{d(x,y)}{\delta^k}\Big)^\eta.
\end{aligned}$$

□

So far, there is no measure on X attached to this construction.

Proposition 1.1 *Let* μ *be a non-trivial positive Borel measure on a quasi-metric space with the geometric doubling property* X*, finite on bounded Borel sets. Let* $1 \leq p < \infty$*. Then, the linear span of the spline functions and in particular the space of Hölder-*η*-continuous functions with bounded support where* η *is the Hölder exponent of the splines are dense in* $L^p(\mu)$*.*

Recall that there is a precaution to take in the definition of a Borel measure. Of course, density of Lipschitz-continuous functions with bounded support is well-known in the metric case. Density of Hölder-continuous functions to some order with bounded support has been proved independently in [61] at about the same time by a different argument. It is not clear it had been proved before.

1.4.4 The Meyer Algorithm

We continue the development of the spline theory in the presence of a non-trivial Borel measure μ on (X, d) so that (X, d, μ) is a space of homogeneous type in the most general sense with the exception that we assume that balls are Borel sets. If not, we may replace them by their interior and this only affects constants.

The splines provide a multiresolution analysis of $L^2(d\mu)$. This consists of all properties of a classical multiresolution analysis of Meyer [60, Definition 2.1] and Mallat [54], to the extent that this definition is meaningful in a quasi-metric space context: the classical postulates dealing with translations and dilations, specific to the Euclidean space and the Lebesgue measure, are now meaningless. Let V_k be the closed linear span of $\{s_\alpha^k\}_\alpha$ in $L^2(d\mu)$. Then $V_k \subseteq V_{k+1}$, and

$$\overline{\bigcup_{k\in\mathbb{Z}} V_k} = L^2(d\mu), \qquad \bigcap_{k\in\mathbb{Z}} V_k = \begin{cases} \{0\}, & \text{if } X \text{ is unbounded,} \\ V_{k_0} = \{\text{constants}\}, & \text{if } X \text{ is bounded,} \end{cases}$$

where k_0 is some integer. Moreover, the functions $s_\alpha^k/\sqrt{\mu_\alpha^k}$ form a Riesz basis of V_k: for all sequences of numbers λ_α, we have the two-sided estimate

$$\Big\| \sum_\alpha \lambda_\alpha s_\alpha^k \Big\|_{L^2(d\mu)} \approx \Big(\sum_\alpha |\lambda_\alpha|^2 \mu_\alpha^k \Big)^{1/2},$$

with $\mu_\alpha^k := \mu(B(x_\alpha^k, \delta^k))$. With this at hand, the algorithm is as follows.

1. Define the closed linear span Y_k of the splines s_β^{k+1} that vanish at all the points in $\mathscr{X}^k$. Then, $V_{k+1} = V_k \oplus Y_k$, where the sum is topological and let $S_k : V_{k+1} \to Y_k$ be the bounded projection corresponding to this decomposition.
2. If Q_k is the orthogonal projection of V_{k+1} onto W_k, the orthogonal complement of V_k in V_{k+1}, then $S_k : W_k \to Y_k$ is an isomorphism, with inverse $Q_k : Y_k \to W_k$.
3. Define the L^∞-normalized pre-wavelets

$$\tilde{\psi}_\beta^k := Q_k s_\beta^{k+1}$$

   where the s_β^{k+1} are the splines in Y_k of item 2.
4. Orthonormalize the pre-wavelets.

This algorithm comes with explicit constructions:

1. Construct the bi-orthogonal Riesz basis $\{\tilde{s}_\alpha^k\}_\alpha$ to $\{s_\alpha^k\}_\alpha$ in V_k, where bi-orthogonal means

$$\langle s_\alpha^k, \tilde{s}_\beta^k \rangle_{L^2(\mu)} = \delta_{\alpha,\beta}. \tag{1.11}$$

   One has

$$\sqrt{\mu_\alpha^k}\,\tilde{s}_\alpha^k(x) = \sum_{\beta\in\mathscr{X}^k} M_k^{-1}(\alpha,\beta) \frac{s_\beta^k(x)}{\sqrt{\mu_\beta^k}},$$

with M_k being the infinite matrix indexed by $\mathscr{X}^k \times \mathscr{X}^k$ (write by abuse $\alpha \in \mathscr{X}^k$ for $x_\alpha^k \in \mathscr{X}^k$) with entries

$$M_k(\alpha, \beta) = \frac{\langle s_\alpha^k, s_\beta^k \rangle_{L^2(\mu)}}{\sqrt{\mu_\alpha^k \mu_\beta^k}}.$$

2. Write

$$Q_k f = f - \sum_{\alpha \in \mathscr{X}^k} \langle f, \tilde{s}_\alpha^k \rangle_{L^2(\mu)} s_\alpha^k = f - \sum_{\alpha \in \mathscr{X}^k} \langle f, s_\alpha^k \rangle_{L^2(d\mu)} \tilde{s}_\alpha^k, \tag{1.12}$$

because the sum is the orthogonal projection of f onto V_k. Compute the pre-wavelets.

3. Observe that one can use $\mathscr{Y}^k$ (write by abuse $\beta \in \mathscr{Y}^k$ for $y_\beta^k \in \mathscr{Y}^k$) as a label set for the pre-wavelets. Then

$$\psi_\alpha^k(x) := \sum_{\beta \in \mathscr{Y}^k} \widetilde{M}_k^{-1/2}(\alpha, \beta) \frac{\tilde{\psi}_\beta^k(x)}{\sqrt{\mu_\beta^{k+1}}},$$

where $\widetilde{M}_k^{-1/2}(\alpha, \beta)$ are the entries of the inverse of the square root of the positive self-adjoint matrix

$$\widetilde{M}_k(\alpha, \beta) := \frac{\langle \tilde{\psi}_\alpha^k, \tilde{\psi}_\beta^k \rangle_{L^2(\mu)}}{\sqrt{\mu_\alpha^{k+1} \mu_\beta^{k+1}}}$$

indexed by $\mathscr{Y}^k \times \mathscr{Y}^k$. Here, $\mu_\alpha^{k+1} = \mu(B(y_\alpha^k, \delta^{k+1})) \sim \mu(B(y_\alpha^k, \delta^k))$ and recall that the points y_α^k are the x_α^{k+1} that do not belong to $\mathscr{X}^k$.

1.4.5 Matrices with Exponential Decay on a Quasi-Metric Discrete Set

The algorithm requires to invert matrices in certain classes. The following result is well-known in the metric case. The extension to the quasi-metric case comes with an interesting argument having the spirit of the Macías-Segovia work (see also [67] for a very neat argument). We give the proof to illustrate this.

Lemma 1.1 *Let Ξ be a 1-separated set in a quasi-metric space (X, d) with quasi-triangle constant A_0 having the geometric doubling property with constant N.*

Consider a matrix $M = (M(\alpha, \beta))$ indexed by $\Xi \times \Xi$ such that there exists $c > 0$ for which

$$C = \sup_{(\alpha,\beta)} \exp\big(cd(\alpha, \beta)\big)|M(\alpha, \beta)| < \infty.$$

Then M is bounded on $\ell^2(\Xi)$. If M is invertible, then there exists $c' > 0$ such that

$$\sup_{(\alpha,\beta)} \exp\big(c'(d(\alpha, \beta))^s\big)|M^{-1}(\alpha, \beta)| < \infty$$

with $s = (1 + \log_2 A_0)^{-1}$ or $s = 1$ if d is a Lipschitz-continuous quasi-distance. If, in addition, M is positive self-adjoint, then the same conclusion holds for $M^{-1/2}$.

If d is a genuine distance or a Lipschitz-continuous quasi-distance, then this follows from Theorem 5 in [52] and the remark that follows it, which extends an earlier result in [24] for band-limited matrices, and $s = 1$. The exponent s is best possible in the general case.

Proof That M is bounded on $\ell^2(\Xi)$ follows from a simple application the Schur lemma and we skip it.

We continue with the following observation. For $n \geq 1$, let κ_n be the best constant in the inequality

$$d(\alpha_0, \alpha_n) \leq \kappa_n(d(\alpha_0, \alpha_1) + d(\alpha_1, \alpha_2) + \ldots + d(\alpha_{n-1}, \alpha_n))$$

for every chain $(\alpha_0, \alpha_1, \ldots, \alpha_n)$ of $n + 1$ elements (not necessarily distinct) of Ξ. It is clear that (κ_n) is non-decreasing and $\kappa_1 = 1$, $\kappa_2 \leq A_0$. Moreover, using $d(\alpha_0, \alpha_{m+n}) \leq A_0(d(\alpha_0, \alpha_m) + d(\alpha_m, \alpha_{m+n}))$, it follows that $\kappa_{m+n} \leq A_0 \cdot \max\{\kappa_m, \kappa_n\}$. Thus $\kappa_{2n} \leq A_0\kappa_n$ and therefore, $\kappa_{2^j} \leq A_0^j$. We conclude that $\kappa_n \leq A_0^{1+\log_2 n} = A_0 n^{\log_2 A_0}$. Note also that if d is L-Lipschitz, then $d(\alpha_0, \alpha_n) \leq d(\alpha_0, \alpha_{n-1}) + Ld(\alpha_{n-1}, \alpha_n)$, which by iteration gives $\kappa_n \leq L$ for all n.

Now assume that M is positive self-adjoint and invertible. In this case, one can write $M = h(I - A)$ with $h = (\|M\| + \|M^{-1}\|)/2$ a positive real number and A a matrix with norm $r = (\|M\| - \|M^{-1}\|)/(\|M\| + \|M^{-1}\|) < 1$. Moreover, the coefficients of A have the same decay as those of M. Without loss of generality, we normalize $h = 1$. Develop $(I - A)^{-1}$ in the Neumann series $\sum A^n$ and estimate the coefficients $A^n(\alpha, \beta)$, $n \geq 1$, $\alpha \neq \beta$, in two ways. First $|A^n(\alpha, \beta)| \leq r^n$. Second, we have

$$\begin{aligned} |A^n(\alpha, \beta)| \leq & \sum_{(\alpha_1,\ldots,\alpha_{n-1})\in\Xi^{n-1}} C^n \exp\big(-c(d(\alpha, \alpha_1) + d(\alpha_1, \alpha_2) \\ & + \ldots + d(\alpha_{n-1}, \beta))\big) \\ & \leq C^n \exp\big(-\frac{c}{2\kappa_n} d(\alpha, \beta)\big) \end{aligned}$$

$$\sum_{(\alpha_1,\dots,\alpha_{n-1})\in\Xi^{n-1}} \exp\Big(-\frac{c}{2}(d(\alpha_1,\alpha_2)+\ldots+d(\alpha_{n-1},\beta))\Big)$$
$$\leq \tilde{C}^n \exp\Big(-\frac{c}{2\kappa_n}d(\alpha,\beta)\Big),$$

where we applied $n-1$ times the inequality with $\varepsilon = c/2$,

$$\sup_{\alpha\in X} \exp\big(\varepsilon d(\alpha,\Xi)/A_0\big)\sum_{\beta\in\Xi}\exp\big(-\varepsilon d(\alpha,\beta)\big) \leq c(\varepsilon, A_0, N) < \infty.$$

As κ_n is non decreasing, we have for any integer n_0 using the second estimate for $0 \leq n \leq n_0$ and the first for $n > n_0$,

$$|M^{-1}(\alpha,\beta)| \leq (n_0+1)\tilde{C}^{n_0}\exp\Big(-\frac{c}{2\kappa_{n_0}}d(\alpha,\beta)\Big) + r^{n_0+1}(1-r)^{-1}$$

and $(n_0+1)\tilde{C}^{n_0} \leq D^{n_0}$ for some large constant $D > 0$. Choosing n_0 as the first integer such that the first term dominates, we see that $d(\alpha,\beta) \approx n_0 \cdot \kappa_{n_0} \lesssim n_0^{1/s}$, where $s = 1/(1+\log_2 A_0)$ (or $s = 1$ if d is Lipschitz-continuous, recalling that $\kappa_{n_0} \leq L$ in this case). Hence, for some constant $c' > 0$,

$$|M^{-1}(\alpha,\beta)| \lesssim r^{n_0} \lesssim \exp\big(-c'd(\alpha,\beta)^s\big).$$

For $M^{-1/2}$, we use the power series $(I-A)^{-1/2} = \sum c_n A^n$. As $0 \leq c_n \lesssim n^{1/2}$, the argument is the same. Finally, if M is not positive self-adjoint, then we use $M^{-1} = M^*(MM^*)^{-1}$, the remark that on a 1-separated set $d \geq d^s$ and that the product of matrices having the exponential decay with factor c as in the statement has exponential decay with factor $c' \in (0, c/A_0)$. □

The above lemma applies a first time to the matrix $M_k(\alpha,\beta)$ on the 1-separated set $\mathscr{X}^k$ for the renormalized quasi-distance d_k given by $d_k(x,y) := \delta^{-k}d(x,y)$ on $X\times X$ and then a second time to the matrix $\widetilde{M}_k(\alpha,\beta)$ on the 1-separated set $\mathscr{Y}^k$ for the renormalized quasi-distance d^s_{k+1}. The constants are therefore uniform over all k's. This provides us with the parameters s and a that one can see in the Corollary. This insures the exponential decay in the statements and the preservation of the Hölder exponent throughout the construction.

1.4.6 The Completion of Guido and Raphy's Program

We come back to the importance of the distance to the point set $\mathscr{Y}^k$ in the estimates for the kernel $Q_k(x,y)$ of the projection Q_k. The absence of the reverse doubling property is linked to the presence of holes at certain scales in X. It could be very well be that $\mathcal{Y}^k$ is empty (in which case the distance to it is $-\infty$ by convention and there is no corresponding wavelets) or that $\mathcal{Y}^k$ contains no points in some regions of X where there is an absence of nearby points at scale exactly δ^{k+1}. Therefore the distance to $\mathcal{Y}^k$ is larger than δ^{k+1} in those regions. In other words, the set $\mathcal{Y}^k$ may not be dense at the scale δ^{k+1}. This can be quantified in some organisation for the growth of the measure of balls: there exists $\varepsilon > 0$ depending only on A_0 and the doubling constant such that for every $x \in X$ and $r > 0$, there exists a decreasing sequence, finite or infinite, of integers $\{k_j\}_{j=0}^J$ such that $r \leq \delta^{k_0} < \delta^{k_1} < \ldots$ and

$$\mu(B(x,\delta^k)) \gtrsim (1+\varepsilon)^j \mu(B(x,r)) \quad \text{and} \quad d(x,\mathcal{Y}^k)+\delta^k \gtrsim \delta^{k_{j+1}} \quad \text{if } k_j \geq k > k_{j+1},$$

where we interpret $k_{J+1} := -\infty$ if $J < \infty$. As one cannot tell the growth of this sequence k_j, there is no reason that (1.2) still converges. But with this organisation, one can show the substitute inequality to (1.2): For all $x \in X$ and all positive numbers r, ν, a, γ, we have

$$\sum_{k:\delta^k \geq r} \mu(B(x,\delta^k))^{-\nu} \exp\big(-\gamma(\delta^{-k} d(x,\mathcal{Y}^k))^a\big) \lesssim \mu(B(x,r))^{-\nu}$$

with implicit constant independent of x and r. Such estimates apply to check Calderón-Zygmund kernel estimates. For example, one can deduce

$$\sum_{k,\alpha} |\psi_\alpha^k(x)\psi_\alpha^k(y)| \leq \frac{C}{\mu(B(x,d(x,y)))}.$$

The operators defined by $\psi_\alpha^k \mapsto \varepsilon_\alpha^k \psi_\alpha^k$ where $\varepsilon_\alpha^k = \pm 1$, have kernels formally given by $K(x,y) = \sum_{k,\alpha} \varepsilon_\alpha^k \psi_\alpha^k(x)\psi_\alpha^k(y)$. They are bounded on $L^2(d\mu)$ and, from the estimate above and similar ones for the regularity of $K(x,y)$ from the regularity of the wavelets, are Calderón-Zygmund operators. One applies the extrapolation result on spaces of homogeneous type in full generality to obtain their $L^p(d\mu)$, $1 < p < \infty$, boundedness. As a consequence, the family of wavelets is an unconditional basis in $L^p(d\mu)$. This was done in [3]. Also the wavelet decomposition of BMO via a Carleson measure characterisation and a proof of the T(1) theorem were presented there.

It was clear that this would also open the way to develop the functional spaces and operator theory in this context: Hardy, Besov, Triebel-Lizorkin spaces with various definitions and descriptions. Here is a non-exhaustive list of references where these aspects are treated [12, 31, 33, 38–41]. As said, the construction of regular wavelets in full generality offers the first regular Littlewood-Paley decomposition at the same time. The Corollary above also points out the difference in the kernel estimates for the approximation to the identity ($P_k(x,y)$) and the dyadic block ($Q_k(x,y)$). In the first one the scale is given by $\mathcal{X}^k$ and size estimates (1.3) come with somehow classical (or expected) decay; in the second, in addition to the scale and the (expected) decay, there is an extra term (1.4) measuring the decay in the distance to $\mathscr{Y}^k = \mathscr{X}^{k+1} \setminus \mathscr{X}^k$. Nor this extra decay, neither the role of the difference set $\mathscr{Y}^k$ could be detected under reverse doubling. This tells us how should the operators of a Littlewood-Paley decomposition look like and what should the Calderón reproducing formula be in full generality (see in particular [38–41]). Its convergence in an appropriate distributional sense can be established. Next, the functional spaces can be defined with any such Littlewood-Paley decomposition (the orthogonality being not required) and they are independent of this choice as it gives an equivalent norm with any other choice. For example, this shows the intrinsic character of the spaces; they are independent of the choice of the wavelet basis (Each time we select nested sets $\mathscr{X}^k$ one can construct a wavelet basis. Hence, there are many different choices). Using another point of view (as the one described by Aline above), the set of points $(\delta^{k+1}, y_\alpha^k) \in \mathbb{R}_+ \times X$ furnishes a representation by a hyperbolic network structure in which one should work. From there, one can imagine developing tent space theory and more.

In conclusion, the intuition of Guido and Raphy behind their definition of a space of homogeneous type, continuing a long line of development since Fourier series (see for example the preface by Y. Meyer in [25]) has proved to be fully correct, in the sense that one can build **all** the tool boxes that harmonic analysts rely on in Euclidean spaces only using geometry and distribution of points. This also makes a bridge between continuous and discrete situations where *ad hoc* methods were developed independently. Indeed, the same tools can be **universally** applied to familiar (for analysts) situations such as Ahlfors-David or reverse doubling measures but also to basic examples that were ruled out by these hypotheses like discrete groups $\mathbb{Z}$ or $\mathbb{Z}/p\mathbb{Z}$ and the multidimensional analogs (see, e.g., [70]), the typical discrete metric structures arising in theoretical computer science (trees, graphs, or strings from a finite alphabet), $\mathbb{Q}_p$ from arithmetic [2], or discrete approximations (like those constructed in [1]). The missing ingredients came from the need to develop appropriate technology to work in non-doubling spaces. We would not call it the fall of the doubling condition as did J. Verdera in 2002 [77], but rather the completion of the doubling theory.

And last but not least, one can do this using orthonormal wavelets (which are special molecules, another invention of Guido and Raphy).

The completion of the program launched by Guido and Raphy makes the ground ready for more. For example, another challenge is: can one construct regular wavelets or Littlewood-Paley decompositions that would apply universally in those non-doubling spaces?

References

1. H. Aimar, M. Carena, B. Iaffei, Discrete approximation of spaces of homogeneous type. J. Geom. Anal. **19**(1), 1–18 (2009)
2. S. Albeverio, S. Evdokimov, M. Skopina, p-Adic multiresolution analysis and wavelet frames. J. Fourier Anal. Appl. **16**(5), 693–714 (2010)

3. P. Auscher, T. Hytönen, Orthonormal bases of regular wavelets in spaces of homogeneous type. Appl. Comput. Harmon. Anal. **34**(2), 266–296 (2013)
4. P. Auscher, G. Weiss, V. Wickerhauser, Local sine and cosine bases of Coifman and Meyer and the construction of smooth wavelets. Wavelet Anal. Appl. **2**, 237–256 (1992)
5. G. Battle, A block spin construction of ondelettes. I. Lemarié functions. Commun. Math. Phys. **110**(4), 601–615 (1987)
6. D. Békollé, A. Bonami, G. Garrigós, F. Ricci, B. Sehba, Analytic Besov spaces and Hardy-type inequalities in tube domains over symmetric cones. J. Reine Angew. Math. **647**, 25–56 (2010)
7. A. Bonami, C. Nana, Some questions related to the Bergman projection in symmetric domains. Adv. Pure Appl. Math. **6**(4), 191–197 (2015)
8. A. Bonami, Soria, G. Weiss, Band-limited wavelets. J. Geom. Anal. **3**(6), 543–578 (2015; 1993)
9. A. Bonami, M. Peloso, F. Symesak, Factorization of the Hardy spaces and Hankel operators on convex domains in $\mathbb{C}^n$. J. Geom. Anal. **11**(3), 363–397 (2001)
10. A. Bonami, S. Pott, B. Sehba, B. Wick, Factorization of the Hardy space $H^s(\mathbb{T}^2)$ (unpublished manuscript)
11. J. Bourgain, C. Demeter, Decouplings for curves and hypersurfaces with nonzero Gaussian curvature. J. Anal. Math. **133**, 279–311 (2017)
12. C. Chen, J. Li, F. Liao, Some function spaces via orthonormal bases on spaces of homogeneous type. Abstr. Appl. Anal. (2014). Art. ID 265378, 13 pp.
13. M. Christ, A $T(b)$ theorem with remarks on analytic capacity and the Cauchy integral. Colloq. Math. **60/61**(2), 601–628 (1990)
14. R.R. Coifman, A real variable characterization of H^p. Studia Math. **51**, 269–274 (1974)
15. R.R. Coifman, M. de Guzmán, Singular integrals and multipliers on homogeneous spaces. Rev. Un. Mat. Argentina **25**, 137–143 (1970/1971)
16. R.R. Coifman, R. Rochberg, Representation theorems for holomorphic and harmonic functions in L^p. Astérisque **77**, 11–66 (1980)
17. R.R. Coifman, G. Weiss, *Analyse harmonique non-commutative sur certains espaces homogènes*. Étude de certaines intégrales singulières. Lecture Notes in Mathematics, vol. 242 (Springer, Berlin, 1971)
18. R.R. Coifman, G. Weiss, Extensions of Hardy spaces and their use in analysis. Bull. Am. Math. Soc. **83**(4), 569–645 (1977)
19. R.R. Coifman, R. Rochberg, G. Weiss, Factorization theorems for Hardy spaces in several variables. Étude de certaines intégrales singulières. Ann. Math. (2) **103**(3), 611–635 (1976)
20. R.R. Coifman, P.-L. Lions, Y. Meyer, S. Semmes, Compensated compactness and Hardy spaces. J. Math. Pures Appl. (9) **72**(3), 247–286 (1993)
21. G. David, *Wavelets and Singular Integrals on Curves and Surfaces*, vol. 1465. Lecture Notes in Mathematics (Springer, Berlin, 1991)
22. G. David, J.-L. Journé, S. Semmes, Opérateurs de Calderón-Zygmund, fonctions para-accrétives et interpolation. Rev. Mat. Iberoamericana **1**(4), 1–56 (1985)
23. M. de Guzmán, *Differentiation of Integrals in* $\mathbb{R}^n$. Lecture Notes in Mathematics, vol. 481 (Springer, Berlin-New York, 1975)
24. S. Demko, Inverses of band matrices and local convergence of spline projections. SIAM J. Numer. Anal. **14**(4), 616–619 (1977)
25. D. Deng, Y. Han, *Harmonic Analysis on Spaces of Homogeneous Type*, vol. 1966. Lecture Notes in Mathematics (Springer, Berlin, 2009). With a preface by Yves Meyer
26. K. Diederich, J.E. Fornæss, Support functions for convex domains of finite type. Math. Z. **230**, 145–164 (1999)
27. C. Fefferman, Characterizations of bounded mean oscillation. Bull. Am. Math. Soc. **77**, 587–588 (1971)
28. C. Fefferman, The Bergman kernel and biholomorphic mappings of pseudoconvex domains. Invent. Math. **26**, 1–65 (1974)

29. S. Ferguson, M. Lacey, A characterization of product BMO by commutators. Acta Math. **189**, 143–160 (2002)
30. F. Forelli, W. Rudin, Projections on spaces of holomorphic functions in balls. Indiana Univ. Math. J. **24**, 593–602 (1974/1975)
31. X. Fu, D. Yang, Wavelet characterizations of the atomic Hardy space H1 on spaces of homogeneous type. Appl. Comput. Harmon. Anal. **44**(1), 1–37 (2018)
32. M. Frazier, B. Jawerth, G. Weiss, *Littlewood-Paley Theory and the Study of Function Spaces*. CBMS Regional Conference Series in Mathematics, vol. 79 (American Mathematical Society, Providence, 1991)
33. X. Fu, D. Yang, Y. Liang, Products of functions in $BMO(X)$ and $H^1_{at}(X)$ via wavelets over spaces of homogeneous type. J. Fourier Anal. Appl. **23**(4), 919–990 (2017)
34. J. Garnett, *Bounded Analytic Functions*. Pure and Applied Mathematics, vol. 96 (Academic Press, Inc. [Harcourt Brace Jovanovich, Publishers], New York-London, 1981)
35. M.S. Gowda, Nonfactorization theorems in weighted Bergman and Hardy spaces on the unit ball of $\mathbb{C}^n$ $(n > 1)$. Trans. Am. Math. Soc. **277**(1), 203–212 (1983)
36. Y.S. Han, E.T. Sawyer, Littlewood-Paley theory on spaces of homogeneous type and the classical function spaces. Mem. Am. Math. Soc. **110**(530), vi+126 (1994)
37. Y. Han, D. Müller, D. Yang, A theory of Besov and Triebel-Lizorkin spaces on metric measure spaces modeled on Carnot-Carathéodory spaces. Abstr. Appl. Anal. (2008), Art. ID 893409, 250
38. Y.S. Han, J. Li, L. Ward, Hardy space theory on spaces of homogeneous type via orthonormal wavelet bases. Appl. Comput. Harmon. Anal. **45**(1), 120–169 (2018)
39. Z. He, L. Liu, D. Yang, W. Yuan, New Calderón reproducing formulae with exponential decay on spaces of homogeneous type. Sci. China Math. **62**(2), 283–350 (2019)
40. Z. He, Y. Han, L. Liu, D. Yang, W. Yuan, A complete real-variable theory of Hardy spaces on spaces of homogeneous type. J. Fourier Anal. Appl. **25**(5), 2197–2267 (2019)
41. Z. He, F. Wang, D. Yang, W. Yuan, Wavelet characterization of Besov and Triebel-Lizorkin spaces on spaces of homogeneous type and its applications. Appl. Comput. Harmon. Anal. **54**, 176–226 (2021)
42. G. Henkin, J. Leiterer, *Theory of Functions on Complex Manifolds*. Monographs in Mathematics, vol. 79 (Birkhäuser Verlag, Basel, 1984)
43. E. Hernández, G. Weiss, *A First Course on Wavelets*. Studies in Advanced Mathematics (CRC Press, Boca Raton, 1996)
44. I. Holmes, S. Treil, A. Volberg, Dyadic bi-parameter repeated commutator and dyadic product BMO. arXiv: 2101.00763
45. T. Hytönen, The L^p to L^q boundedness of commutators with applications to the Jacobian operator. J. Math. Pures Appl. (9) **156**, 351–391 (2021). J. Math. Pures Appl. (9) **156**, 351–391 (2021)
46. T. Hytönen, A. Kairema, Systems of dyadic cubes in a doubling metric space. Colloq. Math. **126**(1), 1–33 (2012)
47. T. Hytönen, H. Martikainen, Non-homogeneous Tb theorem and random dyadic cubes on metric measure spaces. J. Geom. Anal. **22**(4), 1071–1107 (2012)
48. T. Hytönen, O. Tapiola, Almost Lipschitz-continuous wavelets in metric spaces via a new randomization of dyadic cubes. J. Approx. Theory **185**, 12–30 (2014)
49. A. Korányi, Harmonic functions on Hermitian hyperbolic space. Trans. Am. Math. Soc. **135**, 507–516 (1969)
50. A. Korányi, S. Vági, Singular integrals on homogeneous spaces and some problems of classical analysis. Ann. Scuola Norm. Sup. Pisa Cl. Sci. (3) **25**, 575–648 (1972)
51. P.G. Lemarié, Ondelettes à localisation exponentielle. J. Math. Pures Appl. (9) **67**(3), 227–236 (1988)
52. P.G. Lemarié, Base d'ondelettes sur les groupes de Lie stratifiés. (Basis of ondelettes on stratified Lie groups). Bull. Soc. Math. Fr. **117**(2), 211–232 (1989)
53. R.A. Macías, C. Segovia, Lipschitz functions on spaces of homogeneous type. Adv. Math. **33**(3), 257–270 (1979)

54. S.G. Mallat, Multiresolution approximations and wavelet orthonormal bases of $L^2(\mathbf{R})$. Trans. Am. Math. Soc. **315**(1), 69–87 (1989)
55. B. Maurey, Isomorphismes entre espaces H^1. Acta Math. **145**(1–2), 79–120 (1980)
56. D. McNeal, E.M. Stein, Mapping properties of the Bergman projection on convex domains of finite type. Duke Math. J. **73**, 177–199 (1994)
57. Y. Meyer, Principe d'incertitude, bases hilbertiennes et algèbres d'opérateurs (French) [The uncertainty principle, Hilbert bases and operator algebras] *Séminaire Bourbaki, Vol. 1985/86. Astérisque No. 145–146* (1987), 4, 209–223
58. Y. Meyer, Ondelettes et fonctions splines, in *Séminaire sur les équations aux dérivées partielles 1986–1987* (École Polytech., Palaiseau, 1987), pp. Exp. No. VI, 18
59. Y. Meyer, Wavelets and operators, in *Analysis at Urbana*, Vol. I (Urbana, IL, 1986–1987), vol. 137. London Math. Soc. Lecture Note Ser. (Cambridge Univ. Press, Cambridge, 1989), pp. 256–365
60. Y. Meyer, *Ondelettes et opérateurs. II*. Actualités Mathématiques. [Current Mathematical Topics] (Hermann, Paris, 1990). Opérateurs de Calderón-Zygmund. [Calderón-Zygmund operators]
61. D. Mitrea, I. Mitrea, M. Mitrea, S. Monniaux, *Groupoid Metrization Theory. With Applications to Analysis on Quasi-Metric Spaces and Functional Analysis*. Applied and Numerical Harmonic Analysis (Birkhäuser/Springer, New York, 2013)
62. A. Nagel, E.M. Stein, S. Wainger, Boundary behavior of functions holomorphic in domains of finite type. Proc. Natl. Acad. Sci. U.S.A. **78**(11)(Part 1), 6596–6599 (1981)
63. A. Nagel, J.P. Rosay, E.M. Stein, S. Wainger, Estimates for the Bergman and Szeg kernels in certain weakly pseudoconvex domains. Bull. Am. Math. Soc. (N.S.) **18**, 55–59 (1988)
64. C. Nana, B.F. Sehba, Toeplitz and Hankel operators from Bergman to analytic Besov spaces of tube domains over symmetric cones. St. Petersbg. Math. J. **30**, 723–750 (2019)
65. F. Nazarov, S. Treil, A. Volberg, The Tb-theorem on non-homogeneous spaces. Acta Math. **190**(2), 151–239 (2003)
66. F. Nazarov, A. Reznikov, A. Volberg, The proof of A_2 conjecture in a geometrically doubling metric space. Indiana Univ. Math. J. **62**(5), 1503–1533 (2013)
67. M. Paluszyński, K. Stempak, On quasi-metric and metric spaces. Proc. Am. Math. Soc. **137**(12), 4307–4312 (2009)
68. J. Pau, R. Zhao, Weak factorization and Hankel forms for weighted Bergman spaces on the unit ball. Math. Ann. **363**, 363–383 (2015)
69. S. Rigot, Differentiation of measures in metric spaces, in *New Trends on Analysis and Geometry in Metric Spaces*. Lecture Notes in Math., vol. 2296 (Springer, Cham, 2022), pp. 93–116
70. G. Steidl, Spline wavelets over $\mathbf{R}$, $\mathbf{Z}$, $\mathbf{R}/N\mathbf{Z}$, and $\mathbf{Z}/N\mathbf{Z}$, in *Wavelets: Theory, Algorithms, and Applications (Taormina, 1993)*, vol. 5. Wavelet Anal. Appl. (Academic Press, San Diego, 1994), pp. 155–177
71. E.M. Stein, *Boundary Behavior of Holomorphic Functions of Several Complex Variables*. Mathematical Notes, No. 11 (Princeton University Press, Princeton, 1972)
72. E.M. Stein, G. Weiss, *Introduction to Fourier Analysis on Euclidean Spaces*. Princeton Mathematical Series, No. 32 (Princeton University Press, Princeton, 1971)
73. J.-O. Strömberg, A modified Franklin system and higher-order spline systems on $\mathbf{R}^n$ as unconditional bases for Hardy spaces, in *Conference on Harmonic Analysis in Honor of Antoni Zygmund, Vol. I, II (Chicago, Ill., 1981)*. Wadsworth Math. Ser. (Wadsworth, Belmont, 1983), pp. 475–494
74. M. Taibleson, G. Weiss, The molecular characterization of certain Hardy spaces. Representation theorems for Hardy spaces. Astérisque **77**, 67–149 (1980)
75. M. Taibleson, G. Weiss, Spaces generated by blocks, in *Probability Theory and Harmonic Analysis* (Cleveland, Ohio, 1983). Monogr. Textbooks Pure Appl. Math., vol. 98 (Dekker, New York, 1986), pp. 209–226
76. Ph. Tchamitchian, Ondelettes et intégrale de Cauchy sur les courbes lipschitziennes. (French) [Wavelets and the Cauchy integral on Lipschitz curves]. Ann. Math. (2) **129**(3), 641–649 (1989)

77. J. Verdera, The fall of the doubling condition in Calderón-Zygmund theory, in *Proceedings of the 6th International Conference on Harmonic Analysis and Partial Differential Equations (El Escorial, 2000)* (2002), vol. Extra, pp. 275–292
78. G. Weiss, *Studies in Real and Complex Analysis*. MAA Stud. Math., vol. 3 (Math. Assoc. America, Washington, 1965), pp. 124–178
79. K. Zhu, *Spaces of Holomorphic Functions in the Unit Ball*. (English) Graduate Texts in Mathematics, vol. 226 (Springer, New York, 2005)

Chapter 2
Optimal Non-absolute Domains for the Cesàro Operator Minus the Identity

Sorina Barza and Javier Soria

Dedicated to the memory of Professor Guido Weiss, an excellent advisor, a brilliant mathematician, and an outstanding person.

Abstract We characterize the optimal non-absolute domain for the Cesàro operator minus the identity $(C - \mathrm{I})$, in the sequence space $\ell^p(\mathbb{N})$, $1 \leq p \leq \infty$, and compare the results obtained with the case of C, showing the different behavior in both cases. We also address this question for the Copson operator C^*.

2.1 Introduction

We are going to consider the classical Cesàro averaging operator acting on the sequence $x = \{x_j\}_{j \in \mathbb{N}} \in \mathbb{R}^{\mathbb{N}}$:

$$Cx(n) = (Cx)_n = \frac{x_1 + \cdots + x_n}{n}$$

The first author "Sorina Barza" acknowledges the support of the GARF Analysis group at the Complutense University of Madrid (https://www.ucm.es/garf/), where most of this work has been developed during a research stay.
The second author "Javier Soria" has been partially supported by the Spanish Government grants PID2020-113048GB-I00 and CEX2019-000904-S, funded by MCIN/AEI/10.13039/501100011033, and Grupo UCM-970966.

S. Barza
Department of Mathematics and Computer Science, Karlstad University, Karlstad, Sweden
e-mail: sorina.barza@kau.se

J. Soria (✉)
Department of Analysis and Applied Mathematics, Complutense University of Madrid, Madrid, Spain

ICMAT, Madrid, Spain
e-mail: javier.soria@ucm.es

E. Hernández et al. (eds.), *The Mathematical Heritage of Guido Weiss*,
Applied and Numerical Harmonic Analysis,
https://doi.org/10.1007/978-3-031-76793-7_2

as well as the Copson operator:

$$C^*x(n) = (C^*x)_n = \sum_{k=n}^{\infty} \frac{x_k}{k}$$

(depending on each particular case, and in order to clarify the calculations, for the elements of a sequence $x \in \mathbb{R}^{\mathbb{N}}$, we will use the notations x_n or $x(n)$, preferably the first one).

Our main goal in this work is the characterization of the optimal domains for C minus the identity on the classical sequence spaces $\ell^p(\mathbb{N})$. Motivations for these considerations come twofold: on the one-hand, there is already a great interest in this topic for the Hardy operator

$$Sf(x) = \frac{1}{x} \int_0^x f(t)\,dt \quad x > 0,$$

(the continuous equivalent version of C), but for non-negative functions, as shown in [5, 7, 12–14]. In particular, if X is a rearrangement invariant (r.i.) Banach function space [3, Definition II.4.1], BFS for short, for which $S : X \to X$ is bounded, then the class of functions for which $S(|f|) \in X$ is known to be much larger than X and, in fact, not even a subspace of $(L^1 + L^\infty)(\mathbb{R}^+)$ [7, Theorem 2.6]. It is also an easy exercise to prove that the only non-negative function $f \in L^1(\mathbb{R}^+)$ such that $Sf \in L^1(\mathbb{R}^+)$ is the zero function (similarly for C and $\ell^1(\mathbb{N})$). Some consideration in the discrete setting of sequence spaces can be also found, for C and C^*, in [1, 2, 6]. On the other-hand, no much is known when positivity is dropped from the definition of the domain (see [10] for some preliminary results, dealing more with duality properties of what the author refers to as the non-absolute domain: the space of sequences x for which $Cx \in \ell^p(\mathbb{N})$).

Moreover, it is also well-known that subtracting the identity from an averaging operator provides some additional regularity and smoothness [9], which is the setting in which we are going to work. In particular, we want to consider the following problem: given a discrete BFS $X = X(\mathbb{N})$ (mostly, $X = \ell^p(\mathbb{N})$), study conditions for a general sequence x so that $(Cx - x) \in X$. That is, determine the conditions to describe the optimal domain:

$$\mathrm{Dom}[C - \mathrm{I}, X] = \{x = \{x_j\}_{j \in \mathbb{N}} \in \mathbb{R}^{\mathbb{N}} : (Cx - x) \in X\}.$$

Observe that, this time, we do not assume a priori that the sequence $\{x_j\}_{j \in \mathbb{N}}$ is non-negative. It is worth noticing that $(Cx - x) \equiv 0$ if and only if x is a constant sequence. Hence, the optimal domain will always be invariant under the addition of constants (the kernel of the operator $C - \mathrm{I}$). Thus, if $C : X \to Y$ is bounded and

$$X + \mathbb{R} = \{x + c : x \in X, c \in \mathbb{R}\},$$

then, since X is r.i., we have that $X \subset Y$ and $X + \mathbb{R} \subset \mathrm{Dom}[C - \mathrm{I}, Y]$.

The paper is structured as follows: in Sect. 2.2 we start by showing some general results for a BFS X. We then establish a useful tool, in Proposition 2.2, for getting precise norm estimates. Our main result is Theorem 2.1, where we fully characterize $\mathrm{Dom}[C - \mathrm{I}, \ell^p(\mathbb{N})]$, $1 \leq p \leq \infty$. As a consequence of these results, we can see that while $\mathrm{Dom}[C, \ell^1(\mathbb{N})] \not\subset \ell^\infty(\mathbb{N})$, however $\mathrm{Dom}[C - \mathrm{I}, \ell^1(\mathbb{N})] \subset \ell^1_0(\mathbb{N}) + \mathbb{R}$. In Sect. 2.3 we briefly describe the behavior at the end-point $p = 1$, in terms of the weak-type space $\ell^{1,\infty}(\mathbb{N})$. Finally, in Sect. 2.4 we consider the study of optimal domains for the Copson operator C^* minus the identity, and show the analogous results in Theorem 2.2.

In what follows, we will use the standard notation $A \lesssim B$ to denote the existence of a positive constant $K > 0$ (independent of the main parameters defining A and B) such that $A \leq KB$ (analogously for the notation $A \gtrsim B$). If both $A \lesssim B$ and $A \gtrsim B$ hold true, we will write $A \approx B$.

2.2 Domain for Cesàro Type Operators on Sequence Spaces

We start with the following auxiliary result (see, e.g., [11, end of page 2]):

Lemma 2.1 *Let X, Y be two BFS and assume T is a positive operator (i.e., $Tf \geq 0$, if $f \geq 0$), such that $Tf \in Y$, whenever $f \in X$. Then, $T : X \to Y$, boundedly.*

We now prove a general result on a discrete BFS X, that completes the observation made in the previous section (we refer to [3, Definition I.2.3] for the definition of X', the associate space of X):

Lemma 2.2 *Let X be a discrete BFS and let X' be its associate space. Then, the following statements are equivalent:*

(i) $C : X \to X$ *and* $C^* : X \to X$ *are bounded.*
(ii) $\mathrm{Dom}[C - \mathrm{I}, X] = X + \mathbb{R}$ *and* $\mathrm{Dom}[C - \mathrm{I}, X'] = X' + \mathbb{R}$.
(iii) $\mathrm{Dom}[C^* - \mathrm{I}, X] = X$ *and* $\mathrm{Dom}[C^* - \mathrm{I}, X'] = X'$.

Proof We start by proving that *(i)* is equivalent to *(ii)*: As we have already observed, the fact that $C : X \to X$ is bounded, immediately gives that $X + \mathbb{R} \subset \mathrm{Dom}[C - \mathrm{I}, X]$. Since $C^* : X \to X$ is equivalent to $C : X' \to X'$, we also get $X' + \mathbb{R} \subset \mathrm{Dom}[C - \mathrm{I}, X']$.

For the reverse inclusions, if $x \in \mathrm{Dom}[C - \mathrm{I}, X]$, let $y = Cx - x \in X$. Then

$$x_2 - x_1 + 2y_2 = 0 \quad \text{and} \quad x_n - x_1 + \frac{n}{n-1} y_n = -\sum_{k=2}^{n-1} \frac{y_k}{k-1}, \; n \geq 3. \tag{2.1}$$

In fact, the equality is trivial for $n = 2$. If $n = 3$, then $-y_2 - 3y_3/2 = -(x_1 + x_2)/2 + x_2 - (x_1 + x_2 + x_3)/2 + 3x_3/2 = x_3 - x_1$. If we now assume (2.1) to hold for a given $n \geq 3$, then

$$-\sum_{k=2}^{n} \frac{y_k}{k-1} = x_n - x_1 + \frac{n}{n-1} y_n - \frac{y_n}{n-1} = \frac{x_1 + \cdots + x_n}{n} - x_1$$

$$= x_{n+1} - x_1 + \frac{x_1 + \cdots + x_{n+1}}{n} - \frac{n+1}{n} x_{n+1}$$

$$= x_{n+1} - x_1 + \frac{n+1}{n} y_{n+1}.$$

Now, observe that $C^* : X \to X$ implies that $\{1/k\}_{k \in \mathbb{N}} \in X'$. In fact, using [3, Theorem I.2.7] and [3, Corollary II.6.8]

$$\left\| \left\{ \frac{1}{k} \right\}_{k \in \mathbb{N}} \right\|_{X'} = \sup_{\|x\|_X = 1} \sum_{k \in \mathbb{N}} \frac{|x_k|}{k} = \sup_{\|x\|_X = 1} \|C^*(|x|)\|_{\ell^\infty(\mathbb{N})}$$

$$\leq \sup_{\|x\|_X = 1} \|C^*(|x|)\|_X = \|C^*\|_{X \to X} < \infty.$$

Thus, if we call $\sigma y(k) = y_{k+1}$ (the left shift operator), we have that $\sigma : X \to X$ and, for $n \geq 3$,

$$\sum_{k=2}^{n-1} \frac{y_k}{k-1} = C^*(\sigma y)(1) - C^*(\sigma y)(n-1),$$

since $y \in X$, using (2.1) we obtain

$$x_n = x_1 - C^*(\sigma y)(1) - \frac{n}{n-1} y_n + C^*(\sigma y)(n-1), \tag{2.2}$$

and taking into account that

$$\left\{ \frac{n+1}{n} y_{n+1} + C^*(\sigma y)(n) \right\}_{n \in \mathbb{N}} \in X^{\mathbb{N}},$$

we finally deduce that $x \in X + \mathbb{R}$. Similarly, since $C^* : X' \to X'$ is equivalent to $C : X \to X$, reversing the role of X and X' in the previous argument we also obtain that $\mathrm{Dom}[C - \mathrm{I}, X'] \subset X' + \mathbb{R}$.

Conversely, to prove that *(ii)* implies *(i)*, if $\mathrm{Dom}[C - \mathrm{I}, X] = X + \mathbb{R}$, pick $x \in X$. Then, $Cx - x \in X$ and hence $Cx \in X$. Thus, using Lemma 2.1, we conclude that $C : X \to X$ is bounded. By a similar and dual argument, we can also get that $C^* : X \to X$.

We now show that *(i)* implies *(iii)*: The fact that $C^* : X \to X$ is bounded gives that $X \subset \mathrm{Dom}[C^* - \mathrm{I}, X]$. Since $C^* : X' \to X'$ we also get that $X' \subset \mathrm{Dom}[C^* - \mathrm{I}, X']$.

For the reversed inclusion, if $x \in \mathrm{Dom}[C^* - \mathrm{I}, X]$, let $y = C^*x - x \in X$. As in (2.1), one can prove by induction that

$$x_n = \frac{1}{n-1}\sum_{k=1}^{n-1} y_k - y_n = (Cy)_{n-1} - y_n, \quad n \geq 2, \tag{2.3}$$

with x_1 arbitrary. Hence $x \in X$. Similarly, since $C : X' \to X'$ we get also that $\mathrm{Dom}[C^* - \mathrm{I}, X'] \subset X'$.

To finish, we prove that *(iii)* implies *(i)*: If $\mathrm{Dom}[C^* - \mathrm{I}, X] = X$, pick $x \in X$. Then $C^*x - x \in X$ and hence $C^*x \in X$. Thus, using Lemma 2.1 we conclude that $C^* : X \to X$ is bounded. By a dual argument we also get that $C : X \to X$. □

Remark 2.1 We observe that, in Lemma 2.2, it is not true that $C : X \to X$ if and only if $\mathrm{Dom}[C - \mathrm{I}, X] = X + \mathbb{R}$, since $C : \ell^\infty(\mathbb{N}) \to \ell^\infty(\mathbb{N})$, but we will prove in Theorem 2.1 that $\mathrm{Dom}[C - \mathrm{I}, \ell^\infty(\mathbb{N})] \neq \ell^\infty(\mathbb{N})$.

Let us now consider the case of the Cesàro operator C on $\ell^1(\mathbb{N})$. Recall that $\ell^1_0(\mathbb{N})$ denotes the subspace of $\ell^1(\mathbb{N})$ sequences with vanishing sum. As we have mentioned above, it is well-known that C is not bounded on $\ell^1(\mathbb{N})$ and, moreover, $x \equiv 0$ is the only non-negative sequence satisfying that $Cx \in \ell^1(\mathbb{N})$. With more generality, we can prove the following (see [10] for some related results):

Proposition 2.1

(i) If x is non-negative sequence such that $Cx \in \ell^1(\mathbb{N})$, then $x \equiv 0$. That is,

$$\ell^1_+(\mathbb{N}) \cap \mathrm{Dom}[C, \ell^1(\mathbb{N})] = \{0\}.$$

(ii) There exists $x \in \ell^1(\mathbb{N}) \setminus \{0\}$ such that $Cx \in \ell^1(\mathbb{N})$. Moreover, any such sequence satisfies that $\sum_{n=1}^\infty x_n = 0$. That is,

$$\ell^1(\mathbb{N}) \cap \mathrm{Dom}[C, \ell^1(\mathbb{N})] = \ell^1_0(\mathbb{N}) \cap \mathrm{Dom}[C, \ell^1(\mathbb{N})] \neq \{0\}.$$

(iii) There exists $x \notin \ell^\infty(\mathbb{N})$ such that $Cx \in \ell^1(\mathbb{N})$ and there exists $y \in \ell^1_0(\mathbb{N})$ such that $Cy \notin \ell^1(\mathbb{N})$. That is,

$$\ell^1_0(\mathbb{N}) \not\subset \mathrm{Dom}[C, \ell^1(\mathbb{N})] \quad \textit{and} \quad \mathrm{Dom}[C, \ell^1(\mathbb{N})] \not\subset \ell^\infty(\mathbb{N}). \tag{2.4}$$

Proof The proof of *(i)* follows from the remark that, for every $j, n \in \mathbb{N}$, we have that $x_n \geq x_j \delta_j(n)$, where

$$\delta_j(n) = \begin{cases} 0, & \text{if } j \neq n \\ 1 & \text{if } j = n. \end{cases} \tag{2.5}$$

Now, for $j \in \mathbb{N}$ fixed:

$$Cx(n) \geq x_j C\delta_j(n) = \begin{cases} 0, & \text{if } n \leq j-1, \\ \dfrac{x_j}{n}, & \text{if } n \geq j. \end{cases}$$

Thus, $Cx \in \ell^1(\mathbb{N})$ if and only if $x_j = 0$, for every $j \in \mathbb{N}$.

In order to find $x \in \ell^1(\mathbb{N})$ such that $Cx \in \ell^1(\mathbb{N})$, we observe that if $y = Cx$, then,

$$x_1 = y_1, x_2 = 2y_2 - y_1, \dots, x_k = ky_k - (k-1)y_{k-1},\ k \in \{2, 3, \dots\}. \tag{2.6}$$

Therefore, the series $\sum_{n=1}^{\infty} x_n$ converges, if and only if there exists $\lim_{k\to\infty} ky_k$. Now, since $y \in \ell^1(\mathbb{N})$, then $\lim_{k\to\infty} ky_k = 0$. In fact, if $\lim_{k\to\infty} |ky_k| = l > 0$, then we could find $k_0 \in \mathbb{N}$ such that $|y_k| > \frac{l}{2k}$, for $k \geq k_0$, contradicting the fact that $y \in \ell^1(\mathbb{N})$. Therefore, we have proved that $\ell^1(\mathbb{N}) \cap \mathrm{Dom}[C, \ell^1(\mathbb{N})] = \ell^1_0(\mathbb{N}) \cap \mathrm{Dom}[C, \ell^1(\mathbb{N})]$.

To finish the proof of *(ii)*, take a positive sequence $y \in \ell^1(\mathbb{N})$ so that $\{ky_k\}_{k\in\mathbb{N}}$ decreases and $\lim_{k\to\infty} ky_k = 0$ (e.g., $y_k = 1/k^2$), and define x as in (2.6). In this case,

$$\sum_{n=1}^{\infty} x_n = \lim_{k\to\infty} ky_k = 0.$$

Moreover, $x_k \leq 0$, for $k \in \{2, 3, \dots\}$ and hence

$$\sum_{n=1}^{\infty} |x_n| = |x_1| - \sum_{n=2}^{\infty} x_n = 2y_1 < \infty.$$

To prove *(iii)*, we consider

$$x_j = \begin{cases} k, & \text{if } j = 2^k, k \in \mathbb{N}, \\ -k, & \text{if } j = 2^k + 1, k \in \mathbb{N}, \\ 0, & \text{otherwise.} \end{cases}$$

Then, $x \notin \ell^\infty(\mathbb{N})$ and

$$Cx(j) = \begin{cases} \dfrac{k}{2^k}, & \text{if } j = 2^k, k \in \mathbb{N}, \\ 0, & \text{otherwise,} \end{cases}$$

which is clearly in $\ell^1(\mathbb{N})$. Finally, with

$$y_1 = \frac{-1}{\log 2} \quad \text{and} \quad y_n = \frac{-1}{\log(n+1)} + \frac{1}{\log n} \geq 0, \quad n \geq 2$$

we have that $y \in \ell^1_0(\mathbb{N})$ and

$$\{Cy(n)\}_{n\in\mathbb{N}} = \left\{\frac{-1}{n\log(n+1)}\right\}_{n\in\mathbb{N}} \notin \ell^1(\mathbb{N}).$$

□

We will now fix our attention to the study of the optimal domain in all different $\ell^p(\mathbb{N})$ spaces, $1 \leq p \leq \infty$. We will see that $C - \mathrm{I}$ enjoys a different behavior at the end-points $\ell^1(\mathbb{N})$ and $\ell^\infty(\mathbb{N})$. Recall that, given a weight w (a sequence with $w_n > 0$, for every $n \in \mathbb{N}$), and $1 \leq p < \infty$, we define

$$\ell^p(w, \mathbb{N}) = \left\{\{x_j\}_{j\in\mathbb{N}} \in \mathbb{R}^{\mathbb{N}} : \|x\|_{\ell^p(w,\mathbb{N})} \stackrel{\text{def}}{=} \left(\sum_{j=1}^{\infty} |x_j|^p w_j\right)^{1/p} < \infty\right\},$$

and analogously if $p = \infty$. With this notation, $\ell^p(\mathbb{N}) = \ell^p(\mathbf{1}, \mathbb{N})$, where $\mathbf{1}$ is the constant weight $\mathbf{1}_n = 1$, for every $n \in \mathbb{N}$. Similarly,

$$\ell^p_0(w, \mathbb{N}) = \ell^p(w, \mathbb{N}) \cap \ell^1_0(\mathbb{N}),$$

with $\|x\|_{\ell^p_0(w,\mathbb{N})} = \|x\|_{\ell^p(w,\mathbb{N})} + \|x\|_{\ell^1(\mathbb{N})}$.

We need first a previous result (which follows using the Closed Graph Theorem), proving the continuity of several inclusions. Observe that, with the previous definition, $\ell^p_0(w, \mathbb{N})$ is a Banach space (i.e., the cancellation property holds true in the limit, since $\ell^1_0(\mathbb{N})$ is closed in $\ell^1(\mathbb{N})$).

Proposition 2.2 *Let w be a weight. Then,*

(i) $\ell^1_0(w, \mathbb{N}) \subset \mathrm{Dom}[C-\mathrm{I}, \ell^1(\mathbb{N})] \iff \|Cx-x\|_{\ell^1(\mathbb{N})} \lesssim \|x\|_{\ell^1(w,\mathbb{N})} + \|x\|_{\ell^1_0(\mathbb{N})}$.

(ii) $(\mathrm{Dom}[C, \ell^1(\mathbb{N})] \cap \ell^1_0(\mathbb{N})) \subset \ell^1_0(w, \mathbb{N}) \iff \|x\|_{\ell^1_0(w,\mathbb{N})} \lesssim \|Cx\|_{\ell^1(\mathbb{N})} + \|x\|_{\ell^1_0(\mathbb{N})}$.

(iii) $\ell^\infty(w, \mathbb{N}) \subset \mathrm{Dom}[C - \mathrm{I}, \ell^\infty(\mathbb{N})] \iff \|Cx - x\|_{\ell^\infty(\mathbb{N})} \lesssim \|x\|_{\ell^\infty(w,\mathbb{N})}$.

Proof Observe that, in all *(i)*, *(ii)*, and *(iii)*, the right-hand side immediately implies the embeddings on the left. Thus, it suffices to prove the implications "⇒".

To prove *(i)*, let us see the continuity of the embedding:

$$\|Cx - x\|_{\ell^1(\mathbb{N})} \lesssim \|x\|_{\ell^1(w,\mathbb{N})} + \|x\|_{\ell^1_0(\mathbb{N})}.$$

Using the Closed Graph Theorem, it suffices to show that

$$\left.\begin{array}{r} x^N \xrightarrow{\ell^1_0(w,\mathbb{N})} 0 \\ Cx^N - x^N \xrightarrow{\ell^1(\mathbb{N})} y \end{array}\right\} \Longrightarrow y \equiv 0. \tag{2.7}$$

Now, since

$$\sum_{n=1}^{\infty} |C(x^N)(n) - x^N(n) - y_n| \to 0 \text{ and } \sum_{n=1}^{\infty} |x^N(n)|w_n \to 0, \text{ as } N \to \infty,$$

then $|C(x^N)(n) - x^N(n) - y_n| \to 0$ and $|x^N(n)| \to 0$, for every $n \in \mathbb{N}$. Hence,

$$|C(x^N)(n) - y_n| \to 0, \text{ as } N \to \infty \text{ and for every } n \in \mathbb{N}.$$

Finally, for a fixed $n \in \mathbb{N}$,

$$C(x^N)(n) = \frac{1}{n}\sum_{k=1}^{n} x^N(k) \longrightarrow y_n = 0,$$

which proves (2.7).

To prove *(ii)*, let us see the continuity of the embedding:

$$\|x\|_{\ell^1_0(w,\mathbb{N})} \lesssim \|Cx\|_{\ell^1(\mathbb{N})} + \|x\|_{\ell^1_0(\mathbb{N})}.$$

Using Proposition 2.1 *(ii)* and the Closed Graph Theorem, it suffices to show that

$$\left.\begin{array}{r} x^N \xrightarrow{\ell^1_0(\mathbb{N})} 0 \\ x^N \xrightarrow{\ell^1_0(w,\mathbb{N})} y \end{array}\right\} \Longrightarrow y \equiv 0,$$

but this follows easily from the fact that $|x^N_n - y_n|w_n \to 0$ and $|x^N_n| \to 0$, as $N \to \infty$ and for very $n \in \mathbb{N}$.

Finally, to prove *(iii)*, let us see the continuity of the embedding:

$$\|Cx - x\|_{\ell^\infty(\mathbb{N})} \lesssim \|x\|_{\ell^\infty(w,\mathbb{N})}. \tag{2.8}$$

Using the Closed Graph Theorem, it suffices to show that

$$\left.\begin{array}{r} x^N \xrightarrow{\ell^\infty(w,\mathbb{N})} 0 \\ Cx^N - x^N \xrightarrow{\ell^\infty(\mathbb{N})} y \end{array}\right\} \Longrightarrow y \equiv 0. \tag{2.9}$$

Given the sequence $\{x^N\}_{N\in\mathbb{N}} \subset \ell^\infty(w,\mathbb{N})$, fix $n \in \mathbb{N}$. Then, for every $N \in \mathbb{N}$,

$$|y_n| \le |y_n - (Cx^N)(n) + x^N(n)| + |(Cx^N)(n)| + |x^N(n)|.$$

Observe that, given $\varepsilon > 0$, we can find $N_1 \in \mathbb{N}$ such that, for all $N \ge N_1$,

$$|y_n - (Cx^N)(n) + x^N(n)| \le \|y - (Cx^N - x^N)\|_{\ell^\infty(\mathbb{N})} < \frac{\varepsilon}{3}.$$

Similarly, there exists $N_2 \in \mathbb{N}$ such that, for all $N \ge N_2$,

$$|x^N(n)| = \frac{1}{w_n}|w_n x^N(n)| \le \frac{1}{w_n}\|x^N\|_{\ell^\infty(w,\mathbb{N})} < \frac{\varepsilon}{3}.$$

Finally, there exists $N_3 \in \mathbb{N}$ such that, for all $N \ge N_3$,

$$\begin{aligned}|(Cx^N)(n)| &\le \frac{1}{\min\limits_{1\le j\le n}\{w_j\}} \frac{w_1|x^N(1)| + \cdots + w_n|x^N(n)|}{n}\\ &\le \frac{1}{\min\limits_{1\le j\le n}\{w_j\}}\|x^N\|_{\ell^\infty(w,\mathbb{N})} < \frac{\varepsilon}{3}.\end{aligned}$$

Thus, with $n \in \mathbb{N}$ fixed, given $\varepsilon > 0$, if $N_0 = \max\{N_1, N_2, N_3\}$, we have that for every $N \ge N_0$,

$$|y_n| < \varepsilon, \qquad \text{for every } \varepsilon > 0,$$

and, hence, $y_n = 0$, for every $n \in \mathbb{N}$, which proves (2.9). □

Theorem 2.1 (Optimal Domain for $C - \mathrm{I}$ on $\ell^p(\mathbb{N})$, $1 \le p \le \infty$)

(i) Case $p = 1$*:* $\mathrm{Dom}[C - \mathrm{I}, \ell^1(\mathbb{N})] = (\mathrm{Dom}[C, \ell^1(\mathbb{N})] \cap \ell^1_0(\mathbb{N})) + \mathbb{R}$. *In addition,*

(a) $\ell^1_0(w,\mathbb{N}) \subset (\mathrm{Dom}[C, \ell^1(\mathbb{N})] \cap \ell^1_0(\mathbb{N})) \iff w_n \gtrsim \log(n+1)$. *Hence, the logarithmic space* $\ell^1_0(\log(n+1),\mathbb{N})$ *is the largest* $\ell^1_0(w,\mathbb{N})$ *space contained in* $(\mathrm{Dom}[C, \ell^1(\mathbb{N})] \cap \ell^1_0(\mathbb{N}))$ *and, moreover,*

$$\ell^1_0(\log(n+1),\mathbb{N}) \subsetneq (\mathrm{Dom}[C, \ell^1(\mathbb{N})] \cap \ell^1_0(\mathbb{N})).$$

(b) $(\mathrm{Dom}[C, \ell^1(\mathbb{N})] \cap \ell^1_0(\mathbb{N})) \subset \ell^1_0(w,\mathbb{N}) \iff w_n \lesssim 1$. *Hence,* $\ell^1_0(\mathbb{N})$ *is the smallest* $\ell^1_0(w,\mathbb{N})$ *space containing* $(\mathrm{Dom}[C, \ell^1(\mathbb{N})] \cap \ell^1_0(\mathbb{N}))$ *and, moreover,*

$$(\mathrm{Dom}[C, \ell^1(\mathbb{N})] \cap \ell^1_0(\mathbb{N})) \subsetneq \ell^1_0(\mathbb{N}).$$

(ii) Case $1 < p < \infty$*:* $\mathrm{Dom}[C - \mathrm{I}, \ell^p(\mathbb{N})] = \ell^p(\mathbb{N}) + \mathbb{R}$.

(iii) Case $p = \infty$*:*

(a) $\ell^\infty(w, \mathbb{N}) \subset \mathrm{Dom}[C - \mathrm{I}, \ell^\infty(\mathbb{N})] \iff w_n \gtrsim 1$. *Hence,* $\ell^\infty(\mathbb{N})$ *is the largest* $\ell^\infty(w, \mathbb{N})$ *space contained in* $\mathrm{Dom}[C - \mathrm{I}, \ell^\infty(\mathbb{N})]$ *and, moreover,*

$$\ell^\infty(\mathbb{N}) \subsetneq \mathrm{Dom}[C - \mathrm{I}, \ell^\infty(\mathbb{N})].$$

(b) $\mathrm{Dom}[C - \mathrm{I}, \ell^\infty(\mathbb{N})] \subset \ell^\infty(w, \mathbb{N}) \iff w_n \lesssim 1/\log(n+1)$. *Hence,* $\ell^\infty(\{1/\log(n+1)\}_n, \mathbb{N})$ *is the smallest weighted* $\ell^\infty(w, \mathbb{N})$ *space containing* $\mathrm{Dom}[C - \mathrm{I}, \ell^\infty(\mathbb{N})]$ *and, moreover,*

$$\mathrm{Dom}[C - \mathrm{I}, \ell^\infty(\mathbb{N})] \subsetneq \ell^\infty(\{1/\log(n+1)\}_n, \mathbb{N}).$$

Proof We start by proving *(i)*. Using the equality:

$$\frac{x_1 + \cdots + x_{n+1}}{n+1} - x_{n+1} = \frac{n}{n+1}\Big(\frac{x_1 + \cdots + x_n}{n} - x_{n+1}\Big),$$

the hypothesis $Cx - x \in \ell^1(\mathbb{N})$ is equivalent to $Cx - \sigma x \in \ell^1(\mathbb{N})$, where, as before, $\sigma x(n) = x_{n+1}$ is the left shift operator. For convenience, we will then work with this second condition.

Assume now that $Cx - \sigma x \in \ell^1(\mathbb{N})$. Let us first see that, necessarily, x is a convergent sequence (to some value $\alpha \in \mathbb{R}$): We define the sequence $d = Cx - \sigma x$. Then, inductively as in (2.1), we have that

$$x_2 = x_1 - d_1, x_3 = x_1 - \frac{d_1}{2} - d_2, \dots, x_{n+1} = x_1 - \frac{d_1}{2} - \cdots - \frac{d_{n-1}}{n} - d_n.$$

Therefore,

$$\lim_{n\to\infty} x_n = x_1 - \sum_{j=1}^{\infty} \frac{d_j}{j+1}.$$

Setting $\alpha = x_1 - \sum_{j=1}^{\infty} d_j/(j+1)$, let us now prove that $x - \alpha = \{x_j - \alpha\}_{j\in\mathbb{N}} \in \ell^1(\mathbb{N})$. In fact,

$$\sum_{n=1}^{\infty} |x_n - \alpha|$$

$$= |x_1 - \alpha| + |x_2 - \alpha| + \sum_{n=3}^{\infty} \Big| x_1 - \sum_{j=1}^{n-2} \frac{d_j}{j+1} - d_{n-1} - \Big(x_1 - \sum_{j=1}^{\infty} \frac{d_j}{j+1}\Big)\Big|$$

$$= |x_1 - \alpha| + |x_2 - \alpha| + \sum_{n=3}^{\infty} \Big| \sum_{j=n-1}^{\infty} \frac{d_j}{j+1} - d_{n-1} \Big|$$

$$\leq |x_1 - \alpha| + |x_2 - \alpha| + \sum_{n=3}^{\infty} \sum_{j=n-1}^{\infty} \frac{|d_j|}{j+1} + \sum_{n=3}^{\infty} |d_{n-1}|$$

$$\leq |x_1 - \alpha| + |x_2 - \alpha| + \sum_{j=2}^{\infty} |d_j| + \sum_{n=3}^{\infty} |d_{n-1}| < \infty.$$

Before proving that $x - \alpha \in \ell_0^1(\mathbb{N})$ (i.e., the cancellation property), let us show that $x - \alpha \in \mathrm{Dom}[C, \ell^1(\mathbb{N})]$:

$$Cx - x = C(x - \alpha) - (x - \alpha) \in \ell^1(\mathbb{N}) \text{ and } x - \alpha \in \ell^1(\mathbb{N}) \Rightarrow C(x - \alpha) \in \ell^1(\mathbb{N}),$$

which means $x - \alpha \in \mathrm{Dom}[C, \ell^1(\mathbb{N})]$. Now, using Proposition 2.1-*(ii)*, we also conclude that $x - \alpha \in \ell_0^1(\mathbb{N})$. Thus, $\mathrm{Dom}[C - \mathrm{I}, \ell^1(\mathbb{N})] \subset (\mathrm{Dom}[C, \ell^1(\mathbb{N})] \cap \ell_0^1(\mathbb{N})) + \mathbb{R}$.

Conversely, if $x - \alpha \in (\mathrm{Dom}[C, \ell^1(\mathbb{N})] \cap \ell_0^1(\mathbb{N}))$, then:

$$C(x-\alpha) \in \ell^1(\mathbb{N}) \text{ and } (x-\alpha) \in \ell^1(\mathbb{N}) \Rightarrow Cx - x = C(x-\alpha) - (x-\alpha) \in \ell^1(\mathbb{N}),$$

i.e., $x \in \mathrm{Dom}[C - \mathrm{I}, \ell^1(\mathbb{N})]$. Observe that, by (2.4), the intersection in *(i)* cannot be simplified.

Let us prove *(i)*-*(a)*. Using Proposition 2.2 *(i)*, if we assume that $\ell_0^1(w, \mathbb{N}) \subset (\mathrm{Dom}[C, \ell^1(\mathbb{N})] \cap \ell_0^1(\mathbb{N}))$, then, with $x^N = \delta_1 - \delta_N$, $N \geq 2$ (see (2.5)), we have

$$\log(N+1) \approx \|Cx^N - x^N\|_{\ell^1(\mathbb{N})} \lesssim \|x^N\|_{\ell^1(w,\mathbb{N})} + \|x^N\|_{\ell^1(\mathbb{N})} = w_1 + w_N + 2 \lesssim w_N.$$

Conversely, since $\ell_0^1(\log(n+1), \mathbb{N}) \subset \ell_0^1(\mathbb{N})$, using the first equality of *(i)*, we only need to show that $\ell_0^1(\log(n+1), \mathbb{N}) \subset \mathrm{Dom}[C, \ell^1(\mathbb{N})]$: If $x \in \ell_0^1(\log(n+1), \mathbb{N})$, then

$$\sum_{n=1}^{\infty} |Cx(n)| = \sum_{n=1}^{\infty} \Big| \frac{1}{n} \sum_{k=n+1}^{\infty} x_k \Big| \leq \sum_{k=1}^{\infty} |x_k| \sum_{n=1}^{k} \frac{1}{n} \approx \sum_{k=1}^{\infty} |x_k| \log(k+1) < \infty.$$

To finish this part, it suffices to observe that, if

$$x_j = \begin{cases} \dfrac{1}{j \log^2 j}, & \text{if } j \text{ is even} \\[2ex] \dfrac{-1}{(j+1)\log^2(j+1)}, & \text{if } j \text{ is odd} \end{cases}$$

then, $\{x_j\}_{j\in\mathbb{N}} \in \ell^1_0(\mathbb{N}) \setminus \ell^1_0(\log(n+1), \mathbb{N})$, and

$$Cx(j) = \begin{cases} 0, & \text{if } j \text{ is even} \\ \dfrac{-1}{j(j+1)\log^2(j+1)}, & \text{if } j \text{ is odd} \end{cases}$$

which clearly satisfies that $Cx \in \ell^1(\mathbb{N})$. Hence,

$$\ell^1_0(\log(n+1), \mathbb{N}) \subsetneq (\mathrm{Dom}[C, \ell^1(\mathbb{N})] \cap \ell^1_0(\mathbb{N})).$$

Let us now see *(i)-(b)*. If $(\mathrm{Dom}[C, \ell^1(\mathbb{N})] \cap \ell^1_0(\mathbb{N})) \subset \ell^1_0(w, \mathbb{N})$, using Proposition 2.2 *(ii)*, with the sequence $x^N = \delta_N - \delta_{N+1}$, we conclude that $Cx^N = \frac{1}{N}\delta_N$, and

$$w_N \le \|x^N\|_{\ell^1_0(w,\mathbb{N})} = w_N + w_{N+1} \lesssim \|Cx^N\|_{\ell^1(\mathbb{N})} + \|x^N\|_{\ell^1_0(\mathbb{N})} = (1 + 1/N) \lesssim 1.$$

Conversely, if $w_n \lesssim 1$, then trivially

$$(\mathrm{Dom}[C, \ell^1(\mathbb{N})] \cap \ell^1_0(\mathbb{N})) \subset \ell^1_0(\mathbb{N}) \subset \ell^1_0(w, \mathbb{N}).$$

That the embedding is strict follows from (2.4).

The proof of *(ii)* is a direct consequence of Lemma 2.2 and the fact that, for $1 < p < \infty$, both operators $C, C^* : \ell^p(\mathbb{N}) \to \ell^p(\mathbb{N})$ are bounded (as was proved in [8, Theorems 326 and 331] and [4]).

We now consider *(iii)*, the end-point case $p = \infty$. It is clear that if $w \gtrsim 1$, then $\ell^\infty(w, \mathbb{N}) \subset \ell^\infty(\mathbb{N}) \subset \mathrm{Dom}[C - \mathrm{I}, \ell^\infty(\mathbb{N})]$. If we now assume the embedding $\ell^\infty(w, \mathbb{N}) \subset \mathrm{Dom}[C - \mathrm{I}, \ell^\infty(\mathbb{N})]$, then necessarily $w_n > 0$, for every $n \in \mathbb{N}$, since, otherwise, we could find $x \in \ell^\infty(w, \mathbb{N})$, with $x_n = \infty$, which clearly does not belong to $\mathrm{Dom}[C - \mathrm{I}, \ell^\infty(\mathbb{N})]$ (we need $x \in \ell^1_{\mathrm{loc}}(\mathbb{N})$ to define C). Let us see that, in fact, $\inf_{n\in\mathbb{N}} w_n > 0$, which is equivalent to $w \gtrsim 1$.

Take now $x = e_n = \{0, \overset{n-1}{\ldots}, 0, 1, 0, \dots\}$, $n \ge 2$. Then, using (2.8),

$$\begin{aligned} w(n) &\gtrsim \|Ce_n - e_n\|_{\ell^\infty(\mathbb{N})} \\ &\approx \left\| \left\{0, \overset{n-1}{\ldots}, 0, \frac{1}{n}, \frac{1}{n+1}, \dots \right\} - \{0, \overset{n-1}{\ldots}, 0, 1, 0, \dots\} \right\|_{\ell^\infty(\mathbb{N})} \\ &\gtrsim \left| \frac{1}{n} - 1 \right| \gtrsim \frac{1}{2}. \end{aligned} \tag{2.10}$$

Hence, $w \gtrsim 1$ and $\ell^\infty(\mathbb{N})$ is the largest $\ell^\infty(w, \mathbb{N})$ space contained in $\mathrm{Dom}[C - \mathrm{I}, \ell^\infty(\mathbb{N})]$. To finish the proof of *(iii)-(a)*, we pick $x = \{\log n\}_{n\in\mathbb{N}} \notin \ell^\infty(\mathbb{N})$. On the other hand,

$$(Cx)(n) = \frac{1}{n}\log(n!) \quad \text{and} \quad (Cx)(n) - x_n = \log\left(\frac{(n!)^{1/n}}{n}\right)$$

But,

$$\lim_{n\to\infty} \frac{(n!)^{1/n}}{n} = \lim_{n\to\infty} \left(\frac{n!}{n^n}\right)^{1/n} = \lim_{n\to\infty} \frac{\frac{(n+1)!}{(n+1)^{n+1}}}{\frac{n!}{n^n}} = \lim_{n\to\infty} \frac{n+1}{(n+1)(1+1/n)^n} = \frac{1}{e}.$$

Thus, $(Cx - x) \in \ell^\infty(\mathbb{N})$; i.e.,

$$x = \{\log n\}_{n\in\mathbb{N}} \in \mathrm{Dom}[C - \mathrm{I}, \ell^\infty(\mathbb{N})] \setminus \ell^\infty(\mathbb{N}). \tag{2.11}$$

We now consider the proof of *(iii)-(b)*. Since $w_n \lesssim 1/\log(n+1)$ implies that $\ell^\infty(\{1/\log(n+1)\}_n, \mathbb{N}) \subset \ell^\infty(w, \mathbb{N})$, it suffices to see that $\mathrm{Dom}[C - \mathrm{I}, \ell^\infty(\mathbb{N})] \subset \ell^\infty(\{1/\log(n+1)\}_n, \mathbb{N})$. In fact, if $x \in \mathrm{Dom}[C-\mathrm{I}, \ell^\infty(\mathbb{N})]$ and we set $y = Cx - x \in \ell^\infty(\mathbb{N})$, using (2.1) we obtain, for $n \geq 2$:

$$\begin{aligned}
\frac{|x_n|}{\log(n+1)} &= \left| x_1 - \frac{n}{n-1} y_n + \sum_{k=2}^{n-1} \frac{y_k}{k-1} \right| \frac{1}{\log(n+1)} \\
&\lesssim \left(|x_1| + |y_n| + \sum_{k=2}^{n-1} \frac{|y_k|}{k-1} \right) \frac{1}{\log(n+1)} \\
&\lesssim |x_1| + \|y\|_{\ell^\infty(\mathbb{N})} (1 + \log n) \frac{1}{\log(n+1)} \lesssim |x_1| + \|y\|_{\ell^\infty(\mathbb{N})}.
\end{aligned}$$

Thus, $x \in \ell^\infty(\{1/\log(n+1)\}_n, \mathbb{N})$. Let us see that the embedding is strict. Consider

$$x = \{x_n\}_{n\in\mathbb{N}} = \{(-1)^n \log n\}_{n\in\mathbb{N}} \in \ell^\infty\left(\frac{1}{\log(n+1)}\right).$$

Then,

$$\begin{aligned}
\frac{x_1 + \cdots + x_{2n}}{2n} - x_{2n} &= \frac{-\log\left(\prod_{j=1}^{n}(2j-1)\right) + \log\left(\prod_{j=1}^{n} 2j\right)}{2n} - \log 2n \\
&= \frac{1}{2n} \log\left(\frac{2^n n!}{\prod_{j=1}^{n}(2j-1)}\right) - \log 2n \\
&= \log\left(\frac{2^n n!}{(2n)^{2n} \prod_{j=1}^{n}(2j-1)}\right)^{\frac{1}{2n}}.
\end{aligned}$$

But,

$$\frac{\dfrac{2^{n+1}(n+1)!}{(2n+2)^{2n+2}\prod_{j=1}^{n+1}(2j-1)}}{\dfrac{2^n n!}{(2n)^{2n}\prod_{j=1}^{n}(2j-1)}} = 2\frac{(2n)^{2n}}{(2n+2)^{2n+2}}\frac{n+1}{2n+1}$$

$$= 2\frac{n+1}{2n+1}\frac{1}{(2n+2)^2}\frac{1}{(1+1/n)^{2n}} \underset{n\to\infty}{\longrightarrow} 0.$$

Hence,

$$(Cx - x)(2n) \underset{n\to\infty}{\longrightarrow} -\infty \quad \text{and} \quad x \notin \mathrm{Dom}[C - \mathrm{I}, \ell^\infty(\mathbb{N})]. \tag{2.12}$$

Finally, let us prove that $\ell^\infty(\{1/\log(n+1)\}_n, \mathbb{N})$ is the smallest $\ell^\infty(w, \mathbb{N})$ space containing $\mathrm{Dom}[C - \mathrm{I}, \ell^\infty(\mathbb{N})]$. In fact, as we have already observed in (2.11), $x = \{\log n\}_{n\in\mathbb{N}} \in \mathrm{Dom}[C - \mathrm{I}, \ell^\infty(\mathbb{N})]$. Therefore, if $x \in \ell^\infty(w, \mathbb{N})$, then

$$w_n \lesssim \frac{1}{\log(n+1)} \quad \Longrightarrow \quad \ell^\infty\left(\frac{1}{\log(n+1)}\right) \subset \ell^\infty(w, \mathbb{N}).$$

□

Remark 2.2 As a consequence of Theorem 2.1, we can prove that, for all values $1 \le p \le \infty$, $\mathrm{Dom}[C - \mathrm{I}, \ell^p(\mathbb{N})]$ is not a space satisfying the lattice property (i.e., if $|y| \le |x|$ and x belongs to the space, so does y). In fact, if $p = 1$,

$$x = e_1 - e_2 \in \mathrm{Dom}[C - \mathrm{I}, \ell^1(\mathbb{N})], \text{ but } y = |e_1 - e_2| = e_1 + e_2 \notin \mathrm{Dom}[C - \mathrm{I}, \ell^1(\mathbb{N})].$$

For $1 < p < \infty$, it suffices to consider

$$x = \left\{\frac{1}{n} + 1\right\}_{n\in\mathbb{N}} \in \ell^p(\mathbb{N}) + \mathbb{R}, \text{ but } y = \left\{(-1)^n\left(\frac{1}{n} + 1\right)\right\}_{n\in\mathbb{N}} \notin \ell^p(\mathbb{N}) + \mathbb{R}.$$

Similarly, as proved in (2.11) and (2.12), for $p = \infty$,

$$\{\log n\}_{n\in\mathbb{N}} \in \mathrm{Dom}[C - \mathrm{I}, \ell^\infty(\mathbb{N})], \text{ but } \{(-1)^n \log n\}_{n\in\mathbb{N}} \notin \mathrm{Dom}[C - \mathrm{I}, \ell^\infty(\mathbb{N})].$$

2.3 Weak-Type Estimates

We are now going to consider some extensions, at the end-point $p = 1$, of the classical weak-type estimates for the Cesàro operator. In particular, we will study the domain $\mathrm{Dom}[C - \mathrm{I}, \ell^{p,\infty}]$, where, for $1 \leq p < \infty$ and x^* being the decreasing rearrangement of the sequence x,

$$\ell^{p,\infty}(\mathbb{N}) = \Big\{ x = \{x_j\}_{j\in\mathbb{N}} : \|x\|_{\ell^{p,\infty}(\mathbb{N})} \overset{\text{def}}{=} \sup_{n\in\mathbb{N}} n^{1/p} x_n^* < \infty \Big\}.$$

Let us first observe that, if $1 < p < \infty$, since both C and C^* are bounded in $\ell^{p,\infty}(\mathbb{N})$, and using Lemma 2.2, then

$$\mathrm{Dom}[C - \mathrm{I}, \ell^{p,\infty}(\mathbb{N})] = \ell^{p,\infty}(\mathbb{N}) + \mathbb{R},$$

as in Theorem 2.1 *(ii)*. Now, if $p = 1$, we have the following results:

Proposition 2.3 *Let w be a weight in $\mathbb{N}$. Then,*

$$\ell^1(w, \mathbb{N}) + \mathbb{R} \subset \mathrm{Dom}[C - \mathrm{I}, \ell^{1,\infty}(\mathbb{N})] \iff w_n \gtrsim 1.$$

Hence, $\ell^1(\mathbb{N})$ is the largest $\ell^1(w, \mathbb{N})$ space contained in $\mathrm{Dom}[C - \mathrm{I}, \ell^{1,\infty}(\mathbb{N})]$ *and, moreover,*

$$\ell^1(\mathbb{N}) + \mathbb{R} \subsetneq \mathrm{Dom}[C - \mathrm{I}, \ell^{1,\infty}(\mathbb{N})].$$

Proof Since $w_n \gtrsim 1$ implies that $\ell^1(w, \mathbb{N}) \subset \ell^1(\mathbb{N})$, using that $\ell^1(\mathbb{N}) \subset \ell^{1,\infty}(\mathbb{N})$ and

$$\|Cx\|_{\ell^{1,\infty}(\mathbb{N})} \leq \|Cx^*\|_{\ell^{1,\infty}(\mathbb{N})} = \sup_{n\in\mathbb{N}} n \bigg(\frac{1}{n} \sum_{k=1}^{n} x_k^* \bigg) = \|x^*\|_{\ell^1(\mathbb{N})} = \|x\|_{\ell^1(\mathbb{N})},$$

we finally get that $\ell^1(w, \mathbb{N}) + \mathbb{R} \subset \mathrm{Dom}[C - \mathrm{I}, \ell^{1,\infty}(\mathbb{N})]$.

Conversely, if we fix $N \geq 2$ and pick $x = \delta_N$, then as in (2.10),

$$\|C\delta_N - \delta_N\|_{\ell^{1,\infty}(\mathbb{N})} = \max \bigg\{ 1 - 1/N, \sup_{k\geq 2} \frac{k}{k+N} \bigg\} = 1 \lesssim w_N.$$

To prove that the inclusion is strict, it suffices to consider the alternating sequence $x = \{(-1)^n/n\}_{n\in\mathbb{N}} \notin \ell^1(\mathbb{N}) + \mathbb{R}$. Then

$$|(Cx - x)(n)| = \frac{1}{n} \bigg| \sum_{k=1}^{n} \frac{(-1)^k}{k} - (-1)^n \bigg| \lesssim \frac{1}{n}.$$

Thus, since $\{1/n\}_{n\in\mathbb{N}} \in \ell^{1,\infty}(\mathbb{N})$, then $(Cx - x) \in \ell^{1,\infty}(\mathbb{N})$ and we conclude the result. □

We end this section with some further results dealing with $\mathrm{Dom}[C - \mathrm{I}, \ell^{1,\infty}(\mathbb{N})]$, and give also a complete description when restricted to the cone of decreasing sequences. We begin with some useful definitions and an interesting weak-type estimate for the Copson operator C^*:

Definition 2.1 Given a positive sequence (a weight) w, we define the weighted weak-type space

$$\ell^{1,\infty}(w, \mathbb{N}) = \{x = \{x_n\}_{n\in\mathbb{N}} \in \mathbb{R}^{\mathbb{N}} : \|x\|_{\ell^{1,\infty}(w,\mathbb{N})} = \sup_{n\in\mathbb{N}} w_n x_n^* < \infty\}.$$

Observe that, with this definition, $\ell^{1,\infty}(\mathbb{N}) = \ell^{1,\infty}(\{n\}_{n\in\mathbb{N}}, \mathbb{N})$. Also, for reasons that will be clarified in the next proposition, we have that

$$\ell^{1,\infty}(\mathbb{N}) \subset \ell^{1,\infty}\Big(\Big\{\frac{n}{1+\log n}\Big\}_{n\in\mathbb{N}}, \mathbb{N}\Big) \subset \ell^p(\mathbb{N}), \text{ for } p > 1.$$

Definition 2.2 Given a sequence space X, we denote by

$$X_{\text{dec}} = \{x = (x_n)_n \in X : x_n \geq 0 \text{ and } x_n \geq x_{n+1},\ n \in \mathbb{N}\}.$$

Lemma 2.3 *Let $p > 1$. Then, $\ell^{1,\infty}(\mathbb{N}) \subset \ell^p(w, \mathbb{N})$, with embedding constant $A > 0$, if and only if $w \in \ell^\infty(\mathbb{N})$. Moreover*

$$\max\left\{\|w\|_\infty^{1/p}, \left(\sum_{j=1}^\infty \frac{w_j}{j^p}\right)^{1/p}\right\} \leq A \leq \left(\sum_{j=1}^\infty \frac{w_j^*}{j^p}\right)^{1/p}.$$

In particular, if w is a decreasing weight, then $\ell^{1,\infty}(\mathbb{N}) \subset \ell^p(w, \mathbb{N})$, with embedding constant $A > 0$, if and only if $A = \big(\sum_{j=1}^\infty \frac{w_j}{j^p}\big)^{1/p} < \infty$.

Proof For the necessity part, taking $x = \delta_j$, we clearly get that

$$\|x\|_{\ell^p(w,\mathbb{N})} = w_j^{1/p} \leq A\|x\|_{1,\infty} = A.$$

Thus, $\|w\|_\infty^{1/p} \leq A$. Now, with $x_n = 1/n$, $\|x\|_{1,\infty} = 1$ and

$$\|x\|_{\ell^p(w,\mathbb{N})} = \left(\sum_{j=1}^\infty \frac{w_j}{j^p}\right)^{1/p} \leq A.$$

Conversely, using Hardy and Littlewood's inequality [3, Theorem II.2.2],

$$\|x\|_{\ell^p(w,\mathbb{N})} = \left(\sum_{j=1}^{\infty} |x_j|^p w_j\right)^{1/p} \le \left(\sum_{j=1}^{\infty} (x_j^*)^p w_j^*\right)^{1/p} \le \|x\|_{1,\infty}\left(\sum_{j=1}^{\infty} \frac{w_j^*}{j^p}\right)^{1/p}.$$

□

Proposition 2.4

(i) $C^* : \ell^{1,\infty}(\mathbb{N}) \longrightarrow \ell^{1,\infty}\left(\left\{\frac{n}{1+\log n}\right\}_{n\in\mathbb{N}}, \mathbb{N}\right)$.
(ii) $C^* : \ell^{1,\infty}(\mathbb{N}) \not\longrightarrow \ell^{1,\infty}(\mathbb{N})$.
(iii) $C^* : \ell^{1,\infty}(\mathbb{N})_{\text{dec}} \longrightarrow \ell^{1,\infty}(\mathbb{N})$.

Proof We start with *(i)*. We observe that, if $x \in \ell^{1,\infty}(\mathbb{N})$, then the series $\sum_{k=1}^{\infty} \frac{x_k}{k}$ converges (absolutely) and $C^*(x)$ is well defined. Now,

$$\left|\sum_{k=n}^{\infty} \frac{x_k}{k}\right| \le \sum_{k=n}^{\infty} \frac{|x_k|}{k}$$

and, hence,

$$\left(\sum_{k=n}^{\infty} \frac{x_k}{k}\right)^* \le \sum_{k=n}^{\infty} \frac{|x_k|}{k}.$$

Now, using Lemma 2.3, with $w = (\overbrace{0, \ldots, 0}^{n-1}, 1/n, 1/(n+1), \ldots)$ and $p = 1$, we have that $w^* = \{1/(k+n-1)\}_{k\in\mathbb{N}}$ and

$$\sum_{k=n}^{\infty} \frac{|x_k|}{k} = \|x\|_{\ell^1(w,\mathbb{N})} \le \sum_{k=1}^{\infty} \frac{1}{k(k+n-1)}\|x\|_{1,\infty} \lesssim \frac{1+\log n}{n}\|x\|_{1,\infty}.$$

Finally,

$$\|C^*(x)\|_{\ell^{1,\infty}\left(\left\{\frac{n}{1+\log n}\right\}_{n\in\mathbb{N}}, \mathbb{N}\right)} = \sup_{n\in\mathbb{N}} \frac{n}{1+\log n}\left(\sum_{k=n}^{\infty} \frac{x_k}{k}\right)^* \lesssim \|x\|_{1,\infty}.$$

The proof of *(ii)* goes as follows. For a fixed odd number $m \in \mathbb{N}$, we consider the sequence $x^m = (1/m, 1/(m-1), \ldots, 1, 0, \ldots, 0, \ldots)$ and set $j = (m+1)/2$. Then, $(x^m)^* = (1, 1/2, \ldots, 1/m, 0, \ldots, 0\ldots)$ and $\|x^m\|_{1,\infty} = 1$. On the other hand, with the same value of $j = (m+1)/2$,

$$\|C^*x^m\|_{1,\infty} \geq jC^*x^m(j) = \frac{m+1}{2} \sum_{k=(m+1)/2}^{m} \frac{1}{k(m-k+1)}$$

$$= \frac{m+1}{2}\frac{1}{m+1}\left(\sum_{k=(m+1)/2}^{m} \frac{1}{k} + \sum_{k=1}^{(m+1)/2} \frac{1}{k}\right) \gtrsim \sum_{k=1}^{(m+1)/2} \frac{1}{k} \underset{m\to\infty}{\longrightarrow} \infty.$$

To finish, if we now select $y \in \ell^{1,\infty}(\mathbb{N})_{\text{dec}}$, then $(C^*y)^*(n) \leq C^*(|y|)(n) = C^*(y^*)(n)$ and

$$n(C^*y)^*(n) \leq n \sum_{k=n}^{\infty} \frac{\|y\|_{1,\infty}}{k^2} \lesssim \|y\|_{1,\infty}.$$

□

Remark 2.3 We observe that Proposition 2.4 *(ii)* shows that $\ell^{1,\infty}(\mathbb{N})$ is not an interpolation quasi-Banach space between $\ell^1(\mathbb{N})$ and $\ell^p(\mathbb{N})$, for every $1 < p < \infty$.

Proposition 2.5

(i) $\text{Dom}[C - \text{I}, \ell^{1,\infty}(\mathbb{N})] \subset \ell^{1,\infty}\left(\left\{\frac{n}{1+\log n}\right\}_{n\in\mathbb{N}}, \mathbb{N}\right) + \mathbb{R}$.

(ii) $\text{Dom}[C - \text{I}, \ell^{1,\infty}(\mathbb{N})] \not\subset \ell^{1,\infty}(\mathbb{N}) + \mathbb{R}$.

(iii) $\text{Dom}_{\text{dec}}[C, \ell^{1,\infty}(\mathbb{N})] = \ell^1_{\text{dec}}(\mathbb{N}) = \text{Dom}_{\text{dec}}[C - \text{I}, \ell^{1,\infty}(\mathbb{N})] \cap \ell^{1,\infty}(\mathbb{N})$.

Proof Using (2.2), if $x \in \text{Dom}[C - \text{I}, \ell^{1,\infty}(\mathbb{N})]$ and $y = Cx - x \in \ell^{1,\infty}(\mathbb{N})$, then

$$x_n = \left(x_1 - \sum_{k=2}^{\infty} \frac{y_k}{k-1}\right) - \frac{n}{n-1} y_n + \sum_{k=n}^{\infty} \frac{y_k}{k-1},$$

and the result of *(i)* follows using Proposition 2.4 *(i)*.

The proof of *(ii)* follows similarly using now Proposition 2.4 *(ii)*.

For the first equality in *(iii)*, we already know that $\ell^1(\mathbb{N}) \subset \text{Dom}[C, \ell^{1,\infty}(\mathbb{N})]$ and hence

$$\ell^1_{\text{dec}}(\mathbb{N}) \subset \text{Dom}_{\text{dec}}[C, \ell^{1,\infty}(\mathbb{N})].$$

Conversely, if $x \downarrow$ and $Cx \in \ell^{1,\infty}(\mathbb{N})$, then

$$\sup_n n(Cx)^*_n = \sup_n \sum_{k=1}^{n} x^*_n = \|x\|_{\ell^1(\mathbb{N})} < \infty.$$

That $\ell^1_{\text{dec}}(\mathbb{N}) \subset \text{Dom}_{\text{dec}}[C - \text{I}, \ell^{1,\infty}(\mathbb{N})] \cap \ell^{1,\infty}(\mathbb{N})$ is trivial. Finally, if

$$x \in \text{Dom}_{\text{dec}}[C - \text{I}, \ell^{1,\infty}(\mathbb{N})] \cap \ell^{1,\infty}(\mathbb{N}),$$

then $Cx = (C - \text{I})x + x \in \ell^{1,\infty}(\mathbb{N})$ and $x \in \text{Dom}_{\text{dec}}[C, \ell^{1,\infty}(\mathbb{N})]$. □

2.4 Domain for Copson Type Operators on Sequence Spaces

In this final section, we are going to study the analogous results of Sect. 2.2, but for the Copson operator C^*. We start with some properties for C^*, similar to those proved in Proposition 2.1 for C.

Proposition 2.6

(i) There exists $x \in \ell^\infty(\mathbb{N}) \setminus \{0\}$ such that $C^*x \in \ell^\infty(\mathbb{N})$; that is,

$$\ell^\infty(\mathbb{N}) \cap \text{Dom}[C^*, \ell^\infty(\mathbb{N})] \neq \{0\}.$$

(ii) There exists $x \notin \ell^\infty(\mathbb{N})$ such that $C^*x \in \ell^\infty(\mathbb{N})$; that is,

$$\text{Dom}[C^*, \ell^\infty(\mathbb{N})] \not\subset \ell^\infty(\mathbb{N}).$$

Proof To prove *(i)*, take $x = (-1)^n$, $n \geq 1$. To show *(ii)*, pick $x = (-1)^n\sqrt{n}$, $n \geq 1$. Then, $x \notin \ell^\infty(\mathbb{N})$ but $C^*x \in \ell^\infty(\mathbb{N})$. □

Proposition 2.7 *Let $w > 0$. Then,*

$$C^* : \ell^\infty(w, \mathbb{N}) \longrightarrow \ell^\infty(\mathbb{N})$$

is bounded if and only if $A = \sum_{n=1}^{\infty} \frac{1}{nw_n} < \infty$.

Proof We prove first the sufficiency of the condition; i.e., if $A = \sum_{n=1}^{\infty} \frac{1}{nw_n} < \infty$ then the operator is bounded. By homogeneity it is sufficient to prove that $\|C^*x\|_\infty \leq A$, for all x such that $|x_n|w_n \leq 1$. Now,

$$\|C^*x\|_\infty = \sup_n \left| \sum_{k=n}^{\infty} \frac{x_k}{k} \right| \leq \sup_n \sum_{k=n}^{\infty} \frac{1}{kw_k} = A.$$

Hence C^* is bounded and $\|C^*\|_\infty \leq A$.

To prove the necessity, take $x_N = (\frac{1}{w_1}, \frac{1}{w_2}, \ldots, \frac{1}{w_N}, 0, \ldots, 0, \ldots)$ (a sequence for every $N \in \mathbb{N}$). Since C^* is bounded, we get

$$\|C^*x_N\|_\infty = \sup_n \sum_{k=n}^{N} \frac{1}{kw_k} = \sum_{k=1}^{N} \frac{1}{kw_k} \leq \|C^*\|_\infty,$$

for any N, which implies that $A = \sum_{k=1}^{\infty} \frac{1}{kw_k} \leq \|C^*\|_\infty$. Hence, A is also the best constant in the inequality. □

Theorem 2.2 (Optimal Domain for $C^* - \mathrm{I}$ on $\ell^p(\mathbb{N})$, $1 \leq p \leq \infty$)

(i) *Case* $p = 1$*:*

(a) $\ell^1(w, \mathbb{N}) \subset \mathrm{Dom}[C^* - \mathrm{I}, \ell^1(\mathbb{N})] \iff w_n \gtrsim 1$. *Hence,* $\ell^1(\mathbb{N})$ *is the largest weighted space* $\ell^1(w, \mathbb{N})$ *contained in* $\mathrm{Dom}[C^* - \mathrm{I}, \ell^1(\mathbb{N})]$ *and the inclusion is strict.*

(b) $\mathrm{Dom}[C^* - \mathrm{I}, \ell^1(\mathbb{N})] \subsetneq \ell^{1,\infty}(\mathbb{N})$.

(ii) *Case* $1 < p < \infty$*:* $\mathrm{Dom}[C^* - \mathrm{I}, \ell^p(\mathbb{N})] = \ell^p(\mathbb{N})$.

(iii) *Case* $p = \infty$*:* $\mathrm{Dom}[C^* - \mathrm{I}, \ell^\infty(\mathbb{N})] = \mathrm{Dom}[C^*, \ell^\infty(\mathbb{N})] \cap \ell^\infty(\mathbb{N})$. *Moreover,*

(a) *if* $W = \left\{(w_n)_n : w_n \gtrsim 1 \text{ and } \sum_{n=1}^{\infty} \frac{1}{nw_n} < \infty\right\}$, *then*

$$\ell^\infty(w, \mathbb{N}) \subset \mathrm{Dom}[C^* - \mathrm{I}, \ell^\infty(\mathbb{N})] \iff w \in W.$$

(b) $\mathrm{Dom}[C^* - \mathrm{I}, \ell^\infty(\mathbb{N})] \subset \ell^\infty(w, \mathbb{N}) \iff w_n \lesssim 1$. *Hence,* $\ell^\infty(\mathbb{N})$ *is the smallest weighted space* $\ell^\infty(w, \mathbb{N})$ *that contains* $\mathrm{Dom}[C^* - \mathrm{I}, \ell^\infty(\mathbb{N})]$ *and the inclusion is strict.*

Proof We prove first *(i)-(a)*. If we assume that $\ell^1(w, \mathbb{N}) \subset \mathrm{Dom}[C^* - \mathrm{I}, \ell^1(\mathbb{N})]$; i.e., $\|C^*x - x\|_1 \lesssim \|x\|_{\ell^1(w,\mathbb{N})}$, taking $x^N = \delta_N$, for $N \geq 2$, we have

$$(C^*x^N - x^N)_n = \begin{cases} \frac{1}{N}, & \text{if } 1 \leq n \leq N-1, \\ -\frac{N-1}{N}, & \text{if } n = N, \\ 0, & \text{if } n > N. \end{cases} \tag{2.13}$$

Hence, we have

$$1 \leq \frac{2(N-1)}{N} \lesssim w_N = \|x^N\|_{\ell^1(w,\mathbb{N})}.$$

Conversely, if $w_n \gtrsim 1$, $\ell^1(w, \mathbb{N}) \subset \ell^1(\mathbb{N}) \subset \mathrm{Dom}[C^* - \mathrm{I}, \ell^1(\mathbb{N})]$, since C^* is bounded from $\ell^1(\mathbb{N})$ into $\ell^1(\mathbb{N})$. That the inclusion is strict follows by considering the sequence $x_n = 1/(n+1)$, $n \geq 1$, which belongs to $\mathrm{Dom}[C^* - \mathrm{I}, \ell^1(\mathbb{N})]$ but not to $\ell^1(\mathbb{N})$. In fact,

$$\begin{aligned}\left(C^* - \mathrm{I}\right)(x) &= \sum_{k=n}^{\infty} \frac{1}{k(k+1)} - \frac{1}{n+1} = \sum_{k=n}^{\infty} \left(\frac{1}{k} - \frac{1}{k+1}\right) - \frac{1}{n+1} \\ &= \frac{1}{n} - \frac{1}{n+1} = \frac{1}{n(n+1)} = y_n,\end{aligned}$$

which shows that $x \in \mathrm{Dom}[C^* - \mathrm{I}, \ell^1(\mathbb{N})]$.

Let us now see *(i)-(b)*. Let $x \in \mathrm{Dom}[C^*-\mathrm{I}, \ell^1(\mathbb{N})]$. Hence, $y = C^*x - x \in \ell^1(\mathbb{N})$ and using (2.3) we get that

$$x_n = (Cy)_{n-1} - y_n, \quad n \geq 2.$$

Thus, using Proposition 2.3 we conclude that $x \in \ell^{1,\infty}$. Finally, to prove that the embedding is strict, we define

$$x_n = \frac{1}{n-1}\sum_{k=1}^{n-1}\frac{(-1)^k}{k} - \frac{(-1)^n}{n}.$$

As above, using (2.3), we have that $y_n = (C^*x - x)(n) = \frac{(-1)^n}{n} \notin \ell^1(\mathbb{N})$ and, hence, $x \notin \mathrm{Dom}[C^* - \mathrm{I}, \ell^1(\mathbb{N})]$, but

$$|x_n| \leq \frac{1}{n-1}\left|\sum_{k=1}^{n-1}\frac{(-1)^k}{k}\right| + \frac{1}{n} \lesssim \frac{1}{n} \in \ell^{1,\infty}.$$

The proof of *(ii)* is a direct consequence of Lemma 2.2 and the fact that, for $1 < p < \infty$, both C and C^* are bounded on $\ell^p(\mathbb{N})$.

We consider now the equality $\mathrm{Dom}[C^*-\mathrm{I}, \ell^\infty(\mathbb{N})] = \mathrm{Dom}[C^*, \ell^\infty(\mathbb{N})] \cap \ell^\infty(\mathbb{N})$ in *(iii)*. Assume firstly that $y = C^*x - x \in \ell^\infty(\mathbb{N})$. As above, using (2.3), since $(Cy)_n \in \ell^\infty(\mathbb{N})$, when $y \in \ell^\infty(\mathbb{N})$, we necessarily get that $x \in \ell^\infty(\mathbb{N})$. Also, the equality

$$\sum_{k=n}^{\infty}\frac{x_k}{k} = x_n + y_n, \quad n \geq 2$$

implies

$$\|C^*x\|_\infty \leq \|y\|_\infty + \|x\|_\infty < \infty;$$

i.e., $x \in \mathrm{Dom}[C^*, \ell^\infty(\mathbb{N})]$.

Conversely, if $x \in \mathrm{Dom}[C^*, \ell^\infty(\mathbb{N})] \cap \ell^\infty(\mathbb{N})$, then trivially $y = C^*x - x \in \ell^\infty(\mathbb{N})$.

For the proof of *(iii)-(a)*, let $w \in W$. Then, $w_n \gtrsim 1$ (i.e., $\ell^\infty(w, \mathbb{N}) \subset \ell^\infty(\mathbb{N})$) and hence, by the previous equality, it is enough to show that $\ell^\infty(w, \mathbb{N}) \subset \mathrm{Dom}[C^*, \ell^\infty(\mathbb{N})]$, which follows from Proposition 2.7.

Conversely, if $\ell^\infty(w, \mathbb{N}) \subset \mathrm{Dom}[C^* - \mathrm{I}, \ell^\infty(\mathbb{N})]$, we get that $\ell^\infty(w, \mathbb{N}) \subset \mathrm{Dom}[C^*, \ell^\infty(\mathbb{N})]$ and $\ell^\infty(w, \mathbb{N}) \subset \ell^\infty(\mathbb{N})$, which as before, and from Proposition 2.7, imply that $w \in W$.

To prove the necessity part of *(iii)-(b)*, we choose $x^N = \delta_N$. The embedding and (2.13) imply $w_N \lesssim 1$.

For the sufficiency part, if $w_n \lesssim 1$, using *(iii)* we get the embeddings

$$\mathrm{Dom}[C^* - \mathrm{I}, \ell^\infty(\mathbb{N})] \subset \ell^\infty(\mathbb{N}) \subset \ell^\infty(w, \mathbb{N}).$$

That $\mathrm{Dom}[C^* - \mathrm{I}, \ell^\infty(\mathbb{N})] \subsetneq \ell^\infty(\mathbb{N})$ follows by taking $x \equiv 1$, which is in $\ell^\infty(\mathbb{N})$ but not in $\mathrm{Dom}[C^* - \mathrm{I}, \ell^\infty(\mathbb{N})]$. □

Acknowledgments We would like to thank the anonymous referees for some important comments and clarifications, which have greatly improved the final exposition of this work.

References

1. G. Bennett, Factorizing the classical inequalities. Mem. Am. Math. Soc. **120**(576), viii, 130 (1996)
2. G. Bennett, K.-G. Grosse-Erdmann, Weighted Hardy inequalities for decreasing sequences and functions. Math. Ann. **334**(3), 489–531 (2006)
3. C. Bennett, R. Sharpley, *Interpolation of Operators* (Academic, Boston, 1988)
4. E.T. Copson, Note on series of positive terms. J. Lond. Math. Soc. **2**, 9–12 (1927)
5. G. Curbera, W.J. Ricker, Optimal domains for kernel operators via interpolation. Math. Nachr. **244**, 47–63 (2002)
6. G. Curbera, W.J. Ricker, Solid extensions of the Cesàro operator on ℓ^p *and* c_0. Integr. Equ. Oper. Theory **80**, 61–77 (2014)
7. O. Delgado, J. Soria, Optimal domain for the Hardy operator. J. Funct. Anal. **244**(1), 119–133 (2007)
8. G.H. Hardy, J.E. Littlewood, G. Pólya, *Inequalities* (Cambridge University Press, Cambridge, 1934)
9. V.I. Kolyada, Rearrangements of functions, and embedding theorems, (Russian). Uspekhi Mat. Nauk **44**, 61–95 (1989). Translation in Russian Math. Surv. **44**(5), 73–117 (1989)
10. P.Y. Lee, Cesàro sequence spaces. Math. Chronicle **13**, 29–45 (1984)
11. J. Lindenstrauss, L. Tzafriri, *Classical Banach Spaces II* (Springer, Berlin, 1979)
12. A. Nekvinda, L. Pick, Optimal estimates for the Hardy averaging operator. Math. Nachr. **283**(2), 262–271 (2010)
13. S. Okada, W.J. Ricker, E.A. Sánchez-Pérez, *Optimal Domain and Integral Extension of Operators Acting in Function Spaces*. Operator Theory: Advances and Applications, vol. 180 (Birkhäuser Verlag, Basel, 2008)
14. J. Soria, P. Tradacete, Optimal rearrangement invariant range for Hardy type operators. Proc. R. Soc. Edinb. Sect. A **146**, 865–893 (2016)

Chapter 3
An Applicable Variant of Spectral Synthesis for Wavelets

John J. Benedetto

I dedicate this paper to the memory of Guido Weiss. Guido was a friend, an inspiring human being, and a great mathematician.

Abstract Beurling's and Wiener's theory of spectral synthesis of Banach algebras can be formulated as an approximation theory where the approximants of a function f are required to be supported within the support of f. This is not the case for L^2 wavelet theory. On the other hand, there are applications in which wavelets can play a role where such support constraints are necessary. For Haar wavelets, it is proved in a constructive, computable way that such approximation can be effected for L^p, $0 < p < 1$ spaces. This is relevant because of the role of sparsity and compressed sensing when $0 < p < 1$. Several diverse applications are given.

3.1 Introduction

3.1.1 Theme and Motivation

Our *theme* is to approximate a given signal f by appropriately defined harmonics whose supports are contained in the support of f. This theme fits naturally into the process of external beam radiation for treating certain cancers, as well as in advances in cellular communications including 5G systems. In Sect. 3.4, we shall comment on these applications, but there are a host of others, perhaps especially in the area of image compression. For example, with regard to image compression and even before the JPEG 2000 wavelet technology was introduced, JPEG image compression began by dividing the given image into 8×8 blocks (subimages) of pixels. These were then

J. J. Benedetto (✉)
Department of Mathematics, Norbert Wiener Center, University of Maryland, College Park, MD, USA
e-mail: jjb@umd.edu

E. Hernández et al. (eds.), *The Mathematical Heritage of Guido Weiss*,
Applied and Numerical Harmonic Analysis,
https://doi.org/10.1007/978-3-031-76793-7_3

processed independently. From the point of view of our theme, the Discrete Cosine Transform harmonics for these subimages were supported in the same block as the corresponding subimage, e.g., [2, 8, 43].

Our *motivation* is from the Wiener and Beurling theory of spectral synthesis. In this case, spectral synthesis is equivalent to a weak $*$ convergence criterion formulated in Theorem 3.1.

3.1.2 Background for Wavelet Theory

Our *background* is wavelet theory. In fact, in 1909, Alfred Haar developed the representation theory of elements of $L^2(\mathbb{T})$, the space of square integrable functions on $\mathbb{R}/\mathbb{Z}$, in terms of the series expansion of certain step functions, now called Haar functions [21]. In the 1980s, Yves Meyer and Stéphane Mallat introduced the notion of a multiresolution analysis (MRA) of $L^2(\mathbb{R}^d)$, see [30, 32]. This is a multiscale process used to construct large classes of wavelet orthonormal bases for $L^2(\mathbb{R})$. A particular example is the Haar wavelet orthonormal basis for $L^2(\mathbb{R})$ using the very same Haar functions as in [21]. The Meyer–Mallat ideas to formulate the notion of an MRA were based on interconnecting Fourier analysis on $\mathbb{R}$ and $\mathbb{R}/\mathbb{T}$, as well as quadrature mirror filters, and models from speech [13, 17, 42] and image processing [30, 31].

A property of the L^2 MRA theory, and, in particular, of the construction of the compactly supported Haar dyadic wavelet orthonormal basis $\{\psi_{m,n}\}$ for $L^2(\mathbb{R})$, is the leakage of the approximants outside of the support of the given function. By this we mean that if $f \in L^2(\mathbb{R})$ is compactly supported, then the supports of the $\psi_{m,n}$ are not generally contained in the support of f, e.g., [14], Chapter 5. This behavior is analogous to what occurs with $L^1(\mathbb{R})$ approximation of f in terms of convolution with an approximate identity, e.g., [6], Chapter 1.

Example 3.1 Formally, the *exchange formula* for the Fourier transform of the convolution of tempered distributions $f, g \in \mathcal{S}'(\mathbb{R})$ is $\widehat{f * g} = \widehat{f}\,\widehat{g}$, where $\widehat{f}$ designates the Fourier transform of f. The underlying theory is complicated. In fact, the convolution $f * g$ does not always exist, and it is not clear, but it is true, that $\widehat{f}$ and $\widehat{g}$ are multiplicable in the sense that $\widehat{f}\,\widehat{g}$ is a well-defined element of $\mathcal{S}'(\mathbb{R})$ when the convolution does exist. The theory and some of its history are outlined in [6], Section 2.5.

On the other hand, if we consider the principle value tempered distribution $f(t) := (1/\pi)\mathrm{pv}(1/t)$, then the exchange formula is valid for $f * f$, and it is straightforward to verify that

$$f * f = \delta, \text{ the Dirac } \delta \text{ measure.} \tag{3.1}$$

Note that $\mathrm{supp}(f) = \mathbb{R}$, and that, even though convolution is intuitively a smoothing operation, we have $\mathrm{supp}(f * f) = \{0\}$ in this case. It is instructive to graph $1/t$ and its translates to see how the cancellation in (3.1) occurs. Of course, in the case of $L^1(\mathbb{R})$ approximate identities, convolution *is* a smoothing operation, and the supports of the approximants do spread beyond the support of the function being approximated, e.g., [6], Section 1.6. This comes back to our theme of keeping the supports of the approximants within the support of the function being approximated.

3.1.3 Background for L^p, $0 < p < 1$

With the importance of utilizing *sparsity* in applications and the associated relevance of the L^0 metric in *compressed sensing*, it is natural to develop wavelet theory for $L^p, 0 < p < 1$. In fact, there was notable work done in this area prior to the advent of compressed sensing. Contributors include Oswald [35], Filippov [18–20], and Laugesen [28]. Our fundamental and elementary observation, that we formulated about 1990 and that was independent of Oswald et al., is that if f is compactly supported and $0 < p < 1$, then the Haar wavelet approximants of f have their supports contained in the support of f itself.

3.1.4 Outline

In Sect. 3.2, we briefly describe the theory of spectral synthesis to demonstrate its equivalence to our *theme*. In the process, we show that approximation by harmonics from within the support of a signal is a fundamental part of this theory, which is often perceived as being the analysis of the ideal structure of Banach algebras. Besides spectral synthesis, aspects of modern potential theory can also be thought of in this way, where determining the closure of function spaces vanishing on various neighborhoods is an essential part of the analysis, e.g., see [1, 22].

Section 3.3.1 gives preliminaries on the Haar function and the setting of $0 < p < 1$. Our main results are proved in Sects. 3.3.2 and 3.3.3. Although Theorem 3.4 is a special case of one of Filippov's and Oswald's results, viz., Theorem 1 [20], they were not concerned with quantifying the weights of the approximants nor their supports. In fact, we prove Theorem 3.4, with its emphasis on support and computation, using standard wavelet techniques developed for L^2 by Meyer [33], Daubechies [14], and Mallat [30], as opposed to the technology of Filippov and Oswald. Also, although MRA is not essential to our calculations, many of its features can be transported to the $p \neq 2$ cases. This is the content of Sect. 3.3.4.

In Sect. 3.4 we discuss the applications mentioned in Sect. 3.1.1, albeit briefly and qualitatively. However, the second application dealing with 5G communication systems is concurrently being quantitatively analyzed with Drs. Dellomo and Nava-Tudela, see our Acknowledgement. This involves new filter design, the construction of useful approximants whose supports are contained within the support of the signals being analyzed, and in the context of the theory of frames and Non-Orthogonal Multiple Access for 5G networks.

Our setting will be on the real line $\mathbb{R}$, and its dual group, which is itself $\mathbb{R}$, see, [39]. However, the theory of spectral synthesis in Sect. 3.2 is naturally formulated on locally compact abelian groups. Further, the main theorems in Sect. 3.3 are valid in $\mathbb{R}^d$ and for wavelets going far beyond the Haar case; and there are simulations for the Haar case in $\mathbb{R}^2$, see [9]. Also, the applications of Sect. 3.4 are naturally for $\mathbb{R}^d$, $d \geq 1$. We have chosen $\mathbb{R}$ and the Haar function for the sake of a succinct, unified exposition.

3.2 Spectral Synthesis

Spectral synthesis is the study of the ideal structure of commutative Banach algebras such as $L^1(\mathbb{R})$. The Banach space of absolutely convergent Fourier transforms on $\mathbb{R}$, taken with the norm induced from $L^1(\mathbb{R})$, is denoted by $A(\mathbb{R})$. Since convolution is the multiplication in $L^1(\mathbb{R})$, and because of the exchange formula, $A(\mathbb{R})$ is a commutative Banach algebra under pointwise multiplication. For the basic harmonic analysis in this section we refer to [6, 27, 39].

Theorem 3.1 is the starting point for the theory of spectral synthesis. Its proof depends on Wiener's Tauberian theorem, see [3], Theorem 1.1.3c, Proposition 1.4.3, and Theorem 1.4.1. Before stating Theorem 3.1, we need the following notation and definitions.

Let $E \subseteq \mathbb{R}$ be closed, and define $M_b(E)$ to be the Banach space of complex regular Borel measures on E, taken with the *total variation* norm. $M_b(\mathbb{R})$ is a commutative Banach algebra under convolution and with unit being the Dirac δ measure. $L^1(\mathbb{R}) \subset M_b(\mathbb{R})$ does not have a multiplicative unit. Define $M_{\text{finite}}(E) := \{f \in M_b(E) : \text{card}(\text{supp}(f)) < \infty\}$.

The dual of $L^1(\mathbb{R})$ is $L^\infty(\mathbb{R})$; and the space of distributional Fourier transforms of elements of $L^\infty(\mathbb{R})$ is the space of *pseudo-measures*; it is denoted by $A'(\mathbb{R})$. Because of the Herglotz–Bochner theorem, we have $M_b(\mathbb{R}) \subset A'(\mathbb{R})$. Define $A'(E) := \{f \in A'(\mathbb{R}) : \text{supp}(f) \subseteq E\}$, where supp denotes the usual support of a distribution.

Let $g \in A(\mathbb{R})$. For convenience, let $Z(g) := \{t \in \mathbb{R} : g(t) = 0\}$ denote the *zero set* of g. Thus, the *zero set* of $X \subseteq A(\mathbb{R})$ is

$$Z(X) := \{t \in \mathbb{R} : \forall g \in X,\ g(t) = 0\} = \bigcap \{Z(g) : g \in X\}.$$

As such, we define $k(E) := \{g \in A(\mathbb{R}) : g = 0 \text{ on } E\} = \{g \in A(\mathbb{R}) : E \subseteq Z(g)\}$ and $j(E) := \{g \in A(\mathbb{R}) : E \cap \text{supp}(g) = \emptyset\} \subseteq k(E)$. $\overline{j(E)}$ and $k(E)$ are closed ideals in $A(\mathbb{R})$. In the context of a generalization of Wiener's Tauberian theorem, we can assert that $\overline{j(E)}$ is the smallest closed ideal and $k(E)$ is the largest closed ideal for which

$$E = Z(\overline{j(E)}) = Z(k(E),$$

see [3], Proposition 1.4.3.

Definition 3.1 Let $E \subseteq \mathbb{R}$. E is a set of *spectral synthesis* if $\overline{j(E)} = k(E)$.

Theorem 3.1 *The following are equivalent for a closed set $E \subseteq \mathbb{R}$.*

(a) E is a set of spectral synthesis, i.e., $\overline{j(E)} = k(E)$.
(b) $M_{\text{finite}}(E)$ is weak $*$ dense in $A'(E)$, with the induced weak $*$ topology from $A'(\mathbb{R})$.
(c) For each closed ideal $I \subseteq A(\mathbb{R})$ for which $Z(I) = E$, we have

$$I = \bigcap_{t \in E} M_t,$$

where the maximal ideals M_t are defined as $M_t := \{g \in A(\mathbb{R}) : g(t) = 0\}$.

Remark 3.1 (Distributions and the Riesz Representation Theorem)

(a) Let $C_0(\mathbb{R})$ be the Banach space of complex-valued continuous functions defined on $\mathbb{R}$ and vanishing at infinity, taken with *sup* norm. A formulation of the Riesz representation theorem is that the dual of $C_0(\mathbb{R})$ is isometrically isomorphic to $M_b(\mathbb{R})$, see [7] Theorem 7.2.7. Further, the Riesz representation theorem for the setting of $\mathbb{R}$ asserts that $g : \mathbb{R} \to \mathbb{C}$ is a right continuous function of bounded variation on $\mathbb{R}$ if and only if its first distributional derivative $f := g'$ is an element of $M_b(\mathbb{R})$, see [7], Theorem 7.5.10.
(b) We mention part *a* because it is well-known that *every $f \in M_b(\mathbb{R})$ is a weak $*$ limit point of the set $M_{\text{finite}}(\text{supp}(f))$ of finitely supported complex measures whose supports are contained in* supp(f), see [10], Chap. III, section 2, n° 4, Corollaire 2. This is in the spirit of our *theme* that approximants of a given function f must be supported within the support of that function, all the more so in the case of our Haar wavelet theorems since the distributional derivatives of Haar functions are sums of δ-measures.
(c) Part *b* is clearly analogous to Theorem 3.1 *b* and to our *theme* of approximation of a signal, in this case the signals are elements of $A'(E)$, from approximants also supported in E.
(d) There is a distributional analogue of part *a* for $\mathbb{R}^d, d > 1$, but it can not be considered a Riesz representation theorem, see [25], cf. [7], Theorem 7.5.11 and Chapter 8.

Remark 3.2 (The Nullstellensatz)

(a) Consider the polynomial ring $\mathbb{C}[z, w]$ in two variables $z, w \in \mathbb{C}$. The fact that the maximal ideals in $\mathbb{C}[z, w]$ are all generated by the two elements, $P_a(z -$

a) and $P_b(w - b)$, where $a, b \in \mathbb{C}$, is a first version of the *Nullstellensatz* of algebraic geometry, e.g., see [26, p. 19]. An ideal $I \subseteq \mathbb{C}[z, w]$, or in any commutative ring, is *prime* if

$$P, Q \in I \text{ and } P \notin I \text{ implies } Q \in I.$$

Since $\mathbb{C}[z, w]$ has a multiplicative unit, its maximal ideals are prime ideals; and it is elementary to construct prime ideals in $\mathbb{C}[z, w]$ that are not maximal. We mention this to compare with the facts that $A(\mathbb{R})$ does not have a multiplicative unit and that the classes of closed prime and maximal ideals are equivalent in $A(\mathbb{R})$, see [3, pp. 41–42].

(b) Let $I \subseteq \mathbb{C}[z, w]$ be an ideal, and define $J := \{P \in \mathbb{C}[z, w] : \exists n > 0 \text{ such that } P^n \in I\}$. Part a is the backdrop for the fact that the full *Hilbert Nullstellensatz* depends on the theorem that J is the intersection of the maximal ideals containing it, see [26]. This theorem on the intersection of maximal ideals is the algebraic geometrical analogue of spectral synthesis as stated in Theorem 3.1 c.

3.3 Haar Approximation for $0 < p < 1$

3.3.1 Haar Function

For this section, the closed sets E of Sect. 3.2 are restricted to be closed intervals, the weak $*$ convergence of Theorem 3.1 b is replaced by the L^p convergence of Definition 3.2, and the δ-harmonics are now the Haar distributional antiderivatives of δ measures.

The *Haar function* [21] $\psi : \mathbb{R} \to \mathbb{C}$ is defined as

$$\psi(t) := \begin{cases} 1, & \text{if } t \in [0, 1/2), \\ -1, & \text{if } t \in [1/2, 1), \\ 0, & \text{otherwise,} \end{cases}$$

i.e., $\psi = \chi_{[0,1/2)} - \chi_{[1/2,1)}$, where χ_X is the characteristic function of the set $X \subseteq \mathbb{R}$. Our results deal with the Haar wavelet system in the space $L^p(\mathbb{R})$, where $0 < p < 1$.

Definition 3.2 Let $f \in L^p(\mathbb{R})$, $0 < p < 1$. We denote the *p-quasinorm* of f by $\|f\|_{L^p(\mathbb{R})}$, where

$$\|f\|_{L^p(\mathbb{R})} := \left(\int_{\mathbb{R}} |f(t)|^p \, dt \right)^{1/p}.$$

If $p \geq 1$, the function $\|\cdot\|_{L^p(\mathbb{R})} : L^p(\mathbb{R}) \to [0, \infty)$ defines a norm. For $0 < p < 1$, $\|\cdot\|_{L^p(\mathbb{R})}$ is not a norm on $L^p(\mathbb{R})$. However, the function $\rho : L^p(\mathbb{R}) \times L^p(\mathbb{R}) \to [0, \infty)$, defined as

$$\forall f, g \in L^p(\mathbb{R}), \quad \rho(f, g) := \|f - g\|^p_{L^p(\mathbb{R})} = \int_{\mathbb{R}} |f(t) - g(t)|^p \, dt,$$

is a metric; and with this metric $L^p(\mathbb{R})$ is a complete metric space, e.g., [12, 15], cf. [34].

Definition 3.3 Let $p > 0$, and given the generating Haar function ψ as an element of $L^p(\mathbb{R})$. The *Haar dyadic wavelet system* for ψ is the sequence $\{\psi_{m,n} : (m, n) \in \mathbb{Z} \times \mathbb{Z}\} \subset L^p(\mathbb{R})$, where each $\psi_{m,n}$ is defined as

$$\psi_{m,n}(t) := 2^{m/p} \psi(2^m t - n) \quad \text{on } \mathbb{R}.$$

This system is fundamental in wavelet theory [14, 23, 33, 44] for $p = 2$.

3.3.2 *Haar Approximation Theorems*

Theorem 3.2 *Given* $N > 0$. *Let* $f \in L^p(\mathbb{R})$, *where* $0 < p < 1$, *and suppose* $\mathrm{supp}(f) \subseteq [-2^N, 2^N]$. *Assume that there exists* M *such that for* $n \in \{-2^{M+N}, \ldots, 2^{M+N} - 1\}$, f *is constant on* $I_{M,n} := [n/2^M, (n + 1)/2^M) = \mathrm{supp}(\psi_{M,n})$. *Then, for all* $\epsilon > 0$, *there is a constructible sequence of finite sums,*

$$f_{M,N,k} := \sum_{(m,n) \in S_{M,N,k} \subseteq \mathbb{Z}\times\mathbb{Z}} a_{m,n}(f) \psi_{m,n}, \quad a_{m,n}(f) \in \mathbb{C},$$

indexed by $k \geq 1$, *with the following properties:*

$$(m, n) \in S_{M,N,k} \quad \text{implies} \quad \mathrm{supp}(\psi_{m,n}) \subseteq \mathrm{supp}(f),$$

the finite index sequences $S_{M,N,k}$ *are the same for all such* f, *and*

$$\exists K = K(\epsilon) \text{ such that } \forall k > K, \quad \|f - f_{M,N,k}\|_{L^p(\mathbb{R})} < \epsilon/2.$$

The finite index sequences $S_{M,N,k}$ *are constructed in the proof.*

Proof Let us define $f_0 := f$ so that f_0 is constant on each $I_{M,n}$. We define f_{-1} as the *pairwise average* of the function f_0. By this we mean that if $n \in \{-2^{M+N-1}, \ldots, 2^{M+N-1} - 1\}$, then

$$\forall t \in I_{M-1,n}, \quad f_{-1}(t) := \frac{1}{2}\left(f_0\left(\frac{2n}{2^M}\right) + f_0\left(\frac{2n+1}{2^M}\right)\right),$$

i.e., f_{-1} is constant on $I_{M-1,n}$. We define f_{-2} as the pairwise average of the function f_{-1}, i.e., f_{-2} is constant on $I_{M-2,n}$, and so on. We also define the error function $e_{-1} := f_0 - f_{-1}$, $e_{-2} := f_{-1} - f_{-2}$, and so on. We compute

$$\begin{aligned} e_{-1}\left(\frac{2n}{2^M}\right) &= f_0\left(\frac{2n}{2^M}\right) - f_{-1}\left(\frac{2n}{2^M}\right) \\ &= f_0\left(\frac{2n}{2^M}\right) - \frac{1}{2}\left(f_0\left(\frac{2n}{2^M}\right) + f_0\left(\frac{2n+1}{2^M}\right)\right) \\ &= \frac{1}{2}\left(f_0\left(\frac{2n}{2^M}\right) - f_0\left(\frac{2n+1}{2^M}\right)\right) \\ &= -f_0\left(\frac{2n+1}{2^M}\right) + \frac{1}{2}\left(f_0\left(\frac{2n}{2^M}\right) + f_0\left(\frac{2n+1}{2^M}\right)\right) \\ &= -f_0\left(\frac{2n+1}{2^M}\right) + f_{-1}\left(\frac{2n+1}{2^M}\right), \end{aligned}$$

so that

$$e_{-1}\left(\frac{2n}{2^M}\right) = -e_{-1}\left(\frac{2n+1}{2^M}\right).$$

Since

$$\forall t \in I_{M-1,n}, \quad \psi_{m,n}(t) = \begin{cases} 2^{m/p}, & \text{if } t \in I_{m+1,2n} \\ -2^{m/p}, & \text{if } t \in I_{m+1,2n+1}, \end{cases}$$

where $n \in \{-2^{N+M-1}, \ldots, 2^{N+M-1} - 1\}$, we have

$$e_{-1}(t)\psi_{M-1,n}(t) = \begin{cases} 2^{(M-1)/p}\, e_{-1}(\frac{2n}{2^M}), & \text{if } t \in I_{M,2n} \\ -2^{(M-1)/p}\, e_{-1}(\frac{2n+1}{2^M}), & \text{if } t \in I_{M,2n+1}. \end{cases}$$

Therefore, we obtain

$$\forall t \in [-2^N, 2^N), \quad e_{-1}(t) = \sum_{n=-2^{N+M-1}}^{2^{N+M-1}-1} 2^{-(M-1)/p} e_{-1}\left(\frac{2n}{2^M}\right)\psi_{M-1,n}(t).$$

We now construct the sum to approximate $f = f_0$.

$$\begin{aligned} f_0 &= f_{-1} + e_{-1} \\ &= f_{-1} + \sum_{n=-2^{N+M-1}}^{2^{N+M-1}-1} a_{M-1,n}\psi_{M-1,n} \end{aligned}$$

$$
\begin{aligned}
&= f_{-2} + e_{-2} + \sum_{n=-2^{N+M-1}}^{2^{N+M-1}-1} a_{M-1,n}\psi_{M-1,n} \\
&= f_{-2} + \sum_{n=-2^{N+M-2}}^{2^{N+M-2}-1} a_{M-2,n}\psi_{M-2,n} + \sum_{n=-2^{N+M-1}}^{2^{N+M-1}-1} a_{M-1,n}\psi_{M-1,n} \\
&= f_{-2} + \sum_{m=1}^{2} \sum_{n=-2^{N+M-m}}^{2^{N+M-m}-1} a_{M-m,n}\psi_{M-m,n} \\
&\vdots \\
&= f_{+,-} + \sum_{m=1}^{M+N} \sum_{n=-2^{N+M-m}}^{2^{N+M-m}-1} a_{M-m,n}\psi_{M-m,n} \\
&:= f_{+,-} + s_{M,N},
\end{aligned}
$$

where

$$\forall m, n, \quad a_{M-m,n} := 2^{-(M-m)/p}\, e_{-m}\left(\frac{2n}{2^{M-m+1}}\right),$$

and

$$f_{+,-} := f_{+} + f_{-},$$

where

$$f_{+} := f^{\alpha} := \alpha\, \chi_{[0,2^N)}, \quad \text{with} \quad \alpha := \frac{1}{2^{M+N}} \sum_{n=0}^{2^{M+N}-1} f_0\left(\frac{n}{2^M}\right),$$

$$f_{-} := f^{\beta} := \beta\, \chi_{[-2^N,0)}, \quad \text{with} \quad \beta := \frac{1}{2^{M+N}} \sum_{n=-2^{M+N}}^{-1} f_0\left(\frac{n}{2^M}\right).$$

Note that

$$
\begin{aligned}
&1 \le m \le M+N \text{ and } -2^{M+N-m} \le n \\
&\le 2^{M+N-m} - 1 \text{ implies } \operatorname{supp}(\psi_{M-m,n}) \subseteq [-2^N, 2^N].
\end{aligned}
$$

Further, the first step in the construction of $S_{M,N,k}$ is to include all of these values of m and n. The second step is to include all those indices used in the constructions of s_k^{α} and s_k^{β}, which we now give. Combining these two sets defines $S_{M,N,k}$.

We construct sums s_k^α, and s_k^β to approximate f^α and f^β, respectively. For s_k^α, we proceed as follows. Define

$$f_1^\alpha(t) := s_1^\alpha(t) := \alpha 2^{N/p}\psi_{-N,0}(t) = \alpha\psi(2^{-N}t),$$
$$f_2^\alpha(t) := 2\alpha 2^{(N-1)/p}\psi_{-N+1,1}(t) = 2\alpha\psi(2^{-N+1}t - 1),$$

and

$$s_2^\alpha := s_1^\alpha + f_2^\alpha,$$

so that

$$f^\alpha - s_2^\alpha = 4\alpha\, \chi_{[2^N - 2^N/2^2, 2^N)}.$$

Thus,

$$f^\alpha - s_2^\alpha = \begin{cases} 0 & \text{if } t < 3\cdot 2^{N-2}, \\ 4\alpha & \text{if } 3\cdot 2^{N-2} \le t < 2^N, \\ 0 & \text{if } t \ge 2^N. \end{cases}$$

Next, define

$$f_3^\alpha(t) := 4\alpha 2^{(N-2)/p}\psi_{-N+2,3}(t) = 4\alpha\psi(2^{-N+2}t - 3)$$

and

$$s_3^\alpha := s_2^\alpha + f_3^\alpha,$$

so that

$$f^\alpha - s_3^\alpha = 8\alpha\, \chi_{[2^N - 2^N/2^3, 2^N)}.$$

In general, we set

$$f_k^\alpha(t) := 2^{k-1}\alpha 2^{(N-k+1)/p}\psi_{-N+k-1,2^{k-1}-1}(t) = 2^{k-1}\alpha\psi(2^{-N+k-1}t - (2^{k-1}-1))$$

and

$$s_k^\alpha := s_{k-1}^\alpha + f_k^\alpha,$$

so that

$$f^\alpha - s_k^\alpha = 2^k\alpha\, \chi_{[2^N - 2^N/2^k, 2^N)}.$$

Consequently,

$$\forall p > 0, \quad \|f^\alpha - s_k^\alpha\|_{L^p(\mathbb{R})} = ((2^k\alpha)^p 2^{N-k})^{1/p} = \alpha 2^{N/p} 2^{k(1-1/p)},$$

and, so,

$$\forall 0 < p < 1, \quad \lim_{k\to\infty} \|f^\alpha - s_k^\alpha\|_{L^p(\mathbb{R})} = 0.$$

For the given $\epsilon > 0$ in the statement of this theorem, let $\epsilon_0 := (\epsilon/2)(1/2^{1/p})$. Then, there exists $K_\alpha > 0$ such that

$$\forall k > K_\alpha, \quad \|f^\alpha - s_k^\alpha\|_{L^p(\mathbb{R})} < \epsilon_0,$$

and, therefore,

$$\forall k > K_\alpha, \quad \rho(f^\alpha, s_k^\alpha) = (2^k\alpha)^p 2^{N-k} < \frac{1}{2}\left(\frac{\epsilon}{2}\right)^p.$$

Similarly, for s_k^β, we define

$$\begin{aligned} f_1^\beta(t) := s_1^\beta(t) &:= -\beta 2^{N/p}\psi_{-N,-1}(t) = -\beta\psi(2^{-N}t + 1), \\ f_k^\beta(t) &:= -2^{k-1}\beta 2^{(N-k+1)/p}\psi_{-N+k-1,-2^{k-1}}(t) \\ &= -2^{k-1}\beta\psi(2^{-N+k-1}t + 2^{k-1}), \end{aligned}$$

and

$$s_k^\beta := s_{k-1}^\beta + f_k^\beta,$$

so that

$$f^\beta - s_k^\beta = 2^k\beta\, \chi_{[-2^N, -2^N + 2^N/2^k)}.$$

Consequently,

$$\forall p > 0, \quad \|f^\beta - s_k^\beta\|_{L^p(\mathbb{R})} = ((2^k\beta)^p 2^{N-k})^{1/p} = \beta 2^{N/p} 2^{k(1-1/p)},$$

and, so,

$$\forall 0 < p < 1, \quad \lim_{k\to\infty} \|f^\beta - s_k^\beta\|_{L^p(\mathbb{R})} = 0.$$

For the given $\epsilon > 0$ in the statement of this theorem, let $\epsilon_0 := (\epsilon/2)(1/2^{1/p})$. Then, there exists $K_\beta > 0$ such that

$$\forall k > K_\beta, \quad \|f^\beta - s_k^\beta\|_{L^p(\mathbb{R})} < \epsilon_0,$$

and, therefore,

$$\forall k > K_\beta, \quad \rho(f^\beta, s_k^\beta) = (2^k\beta)^p 2^{N-k} < \frac{1}{2}\left(\frac{\epsilon}{2}\right)^p.$$

We shall define $K > 0$ such that

$$\forall k > K, \quad (2^k\alpha)^p 2^{N-k} + (2^k\beta)^p 2^{N-k} < \left(\frac{\epsilon}{2}\right)^p. \tag{3.2}$$

In fact, we can set $K := \max\{K_\alpha, K_\beta\}$, or, letting $c := \max\{\alpha, \beta\}$, we can set

$$K := \lceil \frac{\log(\epsilon) - \log(2^{1+1/p}c2^{N/p})}{(1-1/p)\log(2)} \rceil.$$

In either case, (3.2) is obtained for all $k > K$.

Finally, define the finite sum,

$$f_{M,N,k} := s_k^\alpha + s_k^\beta + s_{M,N}.$$

Then, for all $k > K$ and for all $0 < p < 1$, we have the estimate

$$\begin{aligned}\|f - f_{M,N,k}\|_{L^p(\mathbb{R})}^p &= \rho(f, f_{M,N,k}) \\ &\leq \rho(f^\alpha, s_k^\alpha) + \rho(f^\beta, s_k^\beta) \\ &< \frac{1}{2}\left(\frac{\epsilon}{2}\right)^p + \frac{1}{2}\left(\frac{\epsilon}{2}\right)^p \\ &< \left(\frac{\epsilon}{2}\right)^p,\end{aligned}$$

and so

$$\forall k > K, \quad \|f - f_{M,N,k}\|_{L^p(\mathbb{R})} < \epsilon/2.$$

□

Theorem 3.3 *Given $N > 0$. Let $f \in L^p(\mathbb{R})$, where $p > 0$, and suppose f is a continuous function on $\mathbb{R}$, with* $\text{supp}(f) \subseteq [-2^N, 2^N]$. *Then, we have the assertion that*

$$\forall \epsilon > 0, \ \exists M = M(\epsilon) > 0 \text{ and } f_M, \quad \text{such that}$$

$$\forall n \in \{-2^{M+N}, \ldots, 2^{M+N} - 1\}, \ f_M \text{ is constant on } I_{M,n},$$

and

$$\|f - f_M\|_{L^p(\mathbb{R})} < \epsilon/2.$$

Proof Define f_M as

$$f_M := \sum_{n=-2^{M+N}}^{2^{M+N}-1} f(t_n)\, \chi_{I_{M,n}}, \quad \text{where } t_n \in I_{M,n}.$$

The sequence $\{f_M\}$, $M = 0, 1, 2, \ldots$ is uniformly convergent to f on $[-2^N, 2^N)$. Thus, for $\epsilon > 0$ as in the statement, we take $\epsilon_0 = (\epsilon/2)/(2^{(N+1)/p})$. Then, there is $M_0 > 0$ such that,

$$\forall M > M_0 \text{ and } \forall t \in [-2^N, 2^N), \quad |f(t) - f_M(t)| < \epsilon_0.$$

Hence, for $M > M_0$ we have

$$\begin{aligned}
\|f - f_M\|_{L^p(\mathbb{R})} &= \left(\int |f(t) - f_M(t)|^p\, dt\right)^{1/p} \\
&\leq \left(\int_{-2^N}^{2^N} \Big(\sup_{t\in[-2^N,2^N)} |f(t) - f_M(t)|\Big)^p\, dt\right)^{1/p} \\
&< \left(\int_{-2^N}^{2^N} (\epsilon_0)^p\, dt\right)^{1/p} \\
&= (2^{N+1}\epsilon_0^p)^{1/p} \\
&= 2^{(N+1)/p}\frac{\epsilon/2}{2^{(N+1)/p}} \\
&= \epsilon/2.
\end{aligned}$$

□

Theorem 3.4 *Let $f \in L^p(\mathbb{R})$, where $0 < p < 1$, and suppose f is a continuous function on $\mathbb{R}$, with $\mathrm{supp}(f) \subseteq [-2^N, 2^N]$ for some N. Then, for all $\epsilon > 0$, there is an $M = M(\epsilon)$, and there is a sequence of finite sums,*

$$f_{M,N,k} := \sum_{(m,n)\in S_{M,N,k}\subseteq \mathbb{Z}\times\mathbb{Z}} a_{m,n}(f)\psi_{m,n}, \quad a_{m,n}(f) \in \mathbb{C},$$

indexed by $k \geq 1$, with the following properties:

$$(m, n) \in S_{M,N,k} \quad \text{implies} \quad \text{supp}(\psi_{m,n}) \subseteq \text{supp}(f),$$

and

$$\exists K = K(\epsilon) \text{ such that } \forall k > K, \quad \|f - f_{M,N,k}\|_{L^p(\mathbb{R})} < \epsilon.$$

Proof Given $\epsilon > 0$. From Theorem 3.3 we know that there is an M and an f_M such that

$$\forall n \in \{-2^{M+N}, \ldots, 2^{M+N} - 1\}, \; f_M \text{ is constant on } I_{M,n},$$

and

$$\|f - f_M\|_{L^p(\mathbb{R})} < \frac{\epsilon}{2^{1/p}}.$$

Thus,

$$\rho(f, f_M) < \epsilon^p/2. \tag{3.3}$$

From Theorem 3.2, we know that there is a $K = K(\epsilon)$ and a sequence of sums,

$$f_{M,N,k} = \sum_{(m,n)\in S_{M,N,k}} a_{m,n}\psi_{m,n}, \quad a_{m,n} \in \mathbb{C},$$

such that

$$\forall k > K, \; \|f_M - f_{M,N,k}\|_{L^p(\mathbb{R})} < \frac{\epsilon}{2^{1/p}}.$$

Therefore,

$$\forall k > K, \; \rho(f_M, f_{M,N,k}) < \epsilon^p/2. \tag{3.4}$$

From (3.3) and (3.4) we have

$$\rho(f, f_{M,N,k}) \leq \rho(f, f_M) + \rho(f_M, f_{M,N,k}) = \epsilon^p,$$

and, hence,

$$\forall k > K, \; \|f - f_{M,N,k}\|_{L^p(\mathbb{R})} < \epsilon.$$

□

3.3.3 Adaptive Haar Approximation

Definition 3.4 Let $p > 0$. Take $A < B$, and set

$$C := (A+B)/2 \quad \text{and} \quad R := (B-A)/2$$

(C for center, R for radius). Given the generating Haar function ψ as an element of $L^p(\mathbb{R})$. The *adaptive Haar dyadic wavelet system* for $[A, B]$ is the sequence $\{\widetilde{\psi}_{m,n} : (m,n) \in \mathbb{Z} \times \mathbb{Z}\}$, where each $\widetilde{\psi}_{m,n}$ is defined as

$$\widetilde{\psi}_{m,n}(t; C, R) := \left(\frac{2^m}{R}\right)^{1/p} \psi\left(\frac{2^m(t-C)}{R} - n\right) \quad \text{on } \mathbb{R}.$$

The support of $\widetilde{\psi}_{m,n}$ is $\overline{\widetilde{I}_{m,n}}$, where

$$\widetilde{I}_{m,n} := \left[\frac{Rn}{2^m} + C, \frac{R(n+1)}{2^m} + C\right),$$

and

$$\|\widetilde{\psi}_{m,n}(\cdot; C, R)\|_{L^p(\mathbb{R})} = \left(\int |\widetilde{\psi}_{m,n}(t; C, R)|^p \, dt\right)^{1/p} = \left((2^m/R)|\widetilde{I}_{m,n}|\right)^{1/p}$$
$$= \left((2^m/R)(R/2^m)\right)^{1/p} = 1.$$

Theorem 3.5 *Given* $[A, B] \subseteq \mathbb{R}$. *Let* $f \in L^p(\mathbb{R})$, *where* $0 < p < 1$, *and suppose* $\operatorname{supp}(f) \subseteq [A, B]$. *Assume that there exists* M *such that for* $n \in \{-2^M, \ldots, 2^M - 1\}$, f *is constant on* $\widetilde{I}_{M,n}$. *Then, for all* $\epsilon > 0$, *there is a sequence of finite sums,*

$$f_{M,k} = \sum_{(i,j) \in S_{M,k} \subseteq \mathbb{Z} \times \mathbb{Z}} \tilde{a}_{i,j} \widetilde{\psi}_{i,j}, \quad \tilde{a}_{i,j} \in \mathbb{C},$$

indexed by $k \geq 1$, *with the following properties:*

$$(i,j) \in S_{M,k} \quad \text{implies} \quad \operatorname{supp}(\widetilde{\psi}_{i,j}) \subseteq \operatorname{supp}(f),$$

and

$$\exists K = K(\epsilon) \text{ such that } \forall k > K, \quad \|f - f_{M,k}\|_{L^p(\mathbb{R})} < \epsilon/2.$$

Proof The proof is analogous to the proof of Theorem 3.2. Define C and R from Definition 3.4. As in the proof of Theorem 3.2, we have:

$$\begin{aligned} f_0 &= f_{+,-} + \sum_{m=1}^{M} \sum_{n=-2^{M-m}}^{2^{M-m}-1} \tilde{a}_{M-m,n} \widetilde{\psi}_{M-m,n}, \\ &:= f_{+,-} + s_M, \end{aligned}$$

where

$$\forall m, n, \quad \tilde{a}_{M-m,n} := R^{1/p} 2^{-(M-m)/p} e_{-m}\left(\frac{2n}{2^{M-m+1}}\right),$$

and

$$f_{+,-} := f_+ + f_-,$$

where

$$f_+ := f^\alpha := \alpha\, \chi_{[C,B)}, \quad \text{with} \quad \alpha := \frac{1}{2^M} \sum_{n=0}^{2^M-1} f_0\left(\frac{n}{2^M}\right),$$

$$f_- := f^\beta := \beta\, \chi_{[A,C)}, \quad \text{with} \quad \beta := \frac{1}{2^M} \sum_{n=-2^M}^{-1} f_0\left(\frac{n}{2^M}\right).$$

We construct sums s^α, and s^β to approximate f^α and f^β, respectively. For s^α, we proceed as follows. Define

$$\begin{aligned} f_1^\alpha := s_1^\alpha &:= \alpha R^{1/p} \widetilde{\psi}_{0,0}, \\ f_k^\alpha &= 2^{k-1} \alpha R^{1/p} 2^{(-k+1)/p} \widetilde{\psi}_{k-1,2^{k-1}-1}, \end{aligned}$$

and

$$s_k^\alpha := s_{k-1}^\alpha + f_k^\alpha,$$

so that

$$f^\alpha - s_k^\alpha = 2^k \alpha\, \chi_{[B-R/2^k,B)}.$$

Consequently,

$$\forall p > 0, \quad \|f^\alpha - s_k^\alpha\|_{L^p(\mathbb{R})} = ((2^k\alpha)^p R 2^{-k})^{1/p} = \alpha R^{1/p} 2^{k(1-1/p)},$$

and, so,

$$\forall 0 < p < 1, \quad \lim_{k\to\infty} \|f^\alpha - s_k^\alpha\|_{L^p(\mathbb{R})} = 0.$$

Similarly, for s^β, we define

$$f_1^\beta := s_1^\beta := -\beta R^{1/p}\widetilde{\psi}_{0,-1},$$
$$f_k^\beta := -2^{k-1}\beta R^{1/p} 2^{(-k+1)/p}\widetilde{\psi}_{k-1,-2^{k-1}},$$

and

$$s_k^\beta := s_{k-1}^\beta + f_k^\beta,$$

so that

$$f^\beta - s_k^\beta = 2^k\beta\, \chi_{[A,A+R/2^k)}.$$

Consequently,

$$\forall p > 0, \quad \|f^\beta - s_k^\beta\|_{L^p(\mathbb{R})} = ((2^k\beta)^p R 2^{-k})^{1/p} = \beta R^{1/p} 2^{k(1-1/p)},$$

and, so,

$$\forall 0 < p < 1, \quad \lim_{k\to\infty} \|f^\beta - s_k^\beta\|_{L^p(\mathbb{R})} = 0.$$

We shall define $K > 0$ such that

$$\forall k > K, \quad (2^k\alpha)^p c 2^{-k} + (2^k\beta)^p c 2^{-k} < \left(\frac{\epsilon}{2}\right)^p. \tag{3.5}$$

In fact, let $c = \max\{\alpha, \beta\}$, and set

$$K := \lceil \frac{\log(\epsilon) - \log(2^{1+1/p} c R^{1/p})}{(1 - 1/p)\log(2)} \rceil,$$

so that (3.5) is obtained for all $k > K$.

Finally, define the finite sum,

$$f_{M,k} := s_k^\alpha + s_k^\beta + s_M.$$

Then, as in the proof of Theorem 3.2, we have that

$$\forall k > K, \quad \|f - f_{M,k}\|_{L^p(\mathbb{R})} < \epsilon/2.$$

□

Theorem 3.6 *Let* $f \in L^p(\mathbb{R})$, *where* $p > 0$, *and suppose* f *is a continuous function on* $\mathbb{R}$, *with* $\text{supp}(f) \subseteq [A, B]$. *Then,*

$$\forall \epsilon > 0, \ \exists M = M(\epsilon) > 0 \text{ and } f_M, \quad \text{such that}$$

$$\forall n \in \{-2^M, \ldots, 2^M - 1\}, \ f_M \text{ is constant on } \widetilde{I}_{M,n},$$

and

$$\|f - f_M\|_{L^p(\mathbb{R})} < \epsilon/2.$$

Proof The proof is analogous to the proof of Theorem 3.3. Define C and R from Definition 3.4, and then define f_M as

$$f_M := \sum_{n=-2^M}^{2^M-1} f(t_n)\, \chi_{\widetilde{I}_{M,n}}, \quad \text{where } t_n \in \widetilde{I}_{M,n}.$$

Since f is defined on $[A, B)$, we take $\epsilon_0 := (\epsilon/2)/(2R^{1/p})$. Then, as in the proof of Theorem 3.3, we proceed to assert that

$$\exists M_0, \text{ such that } \forall M > M_0, \quad \|f - f_M\|_{L^p(\mathbb{R})} < \epsilon/2.$$

□

Theorem 3.7 *Let* $f \in L^p(\mathbb{R})$, *where* $0 < p < 1$, *and suppose* f *is a continuous function on* $\mathbb{R}$, *with* $\text{supp}(f) \subseteq [A, B]$. *Then, for all* $\epsilon > 0$, *there is an* $M = M(\epsilon)$, *and there is a sequence of finite sums,*

$$f_{M,k} = \sum_{(i,j)\in S_{M,k}\subseteq \mathbb{Z}\times\mathbb{Z}} \tilde{a}_{i,j}\widetilde{\psi}_{i,j}, \quad \tilde{a}_{i,j} \in \mathbb{C},$$

indexed by $k \geq 1$, *with the following properties:*

$$(i, j) \in S_{M,k} \quad \text{implies} \quad \text{supp}(\widetilde{\psi}_{i,j}) \subseteq \text{supp}(f),$$

and

$$\exists K = K(\epsilon) \text{ such that } \forall k > K, \quad \|f - f_{M,k}\|_{L^p(\mathbb{R})} < \epsilon.$$

Proof The proof follows from the proof of Theorem 3.4. We use Theorems 3.6 and 3.5, and follow the steps as in the proof of Theorem 3.4. □

3.3.4 The Haar MRA of $L^p(\mathbb{R})$, $p \in [1, \infty)$

Theorem 3.8 (Haar MRA of $L^p(\mathbb{R})$, $p \in [1, \infty)$) *Let $\phi := \mathbb{1}_{[0,1)}$, take $p \in [1, \infty)$, define $\tau_n\phi(\cdot) := \phi(\cdot - n)$, and set*

$$V_0 := \left\{ \sum_{n\in\mathbb{Z}} a_n\, \tau_n\phi : \{a_n\} \in \ell^p(\mathbb{Z}) \right\},$$

where the sums converge in $L^p(\mathbb{R})$. For each $j \in \mathbb{Z}$, define

$$V_j = \{g(t) = 2^{j/p} f(2^j t) : f \in V_0\}.$$

Then, $\{V_j\}$, ϕ is an MRA of $L^p(\mathbb{R})$ with scaling function ϕ in the sense that the conditions defining an MRA of $L^2(\mathbb{R})$ are valid except that now $\{\tau_n\phi\}$ is a bounded unconditional basis for V_0, and

$$\overline{\bigcup_{j\in\mathbb{Z}} V_j} = L^p(\mathbb{R}) \quad \text{in} \quad L^p(\mathbb{R}).$$

It is straightforward to show that each V_j is a closed linear subspace of $L^p(\mathbb{R})$, and it is not difficult to prove that

$$\forall j \in \mathbb{Z}, \quad V_j \subseteq V_{j+1},$$

$$\bigcap_{j\in\mathbb{Z}} V_j = \{0\},$$

and

$$\overline{\bigcup_{j\in\mathbb{Z}} V_j} = L^p(\mathbb{R}). \tag{3.6}$$

Even though it is elementary, we shall verify that $\{\tau_n\phi\}$ is a bounded unconditional basis for V_0.

First, to prove that $\{\tau_n\phi\} = \{\tau_n\chi_{[0,1)}\}$ is a basis for the Banach space $V_0 \subseteq L^p(\mathbb{R})$, we let

$$f = \sum_{n\in\mathbb{Z}} a_n\tau_n\phi = \sum_{n\in\mathbb{Z}} b_n\tau_n\phi \text{ in } L^p(\mathbb{R}),$$

where $\{a_n\}, \{b_n\} \in \ell^p(\mathbb{Z})$. In this case,

$$0 = \int_{\mathbb{R}} |\sum_{n\in\mathbb{Z}} (a_n - b_n)\tau_n\phi|^p = \sum_{n\imath\mathbb{Z}} \int_n^{n+1} |a_n - b_n|^p = \sum_{n\imath\mathbb{Z}} |a_n - b_n|^p,$$

and this implies $a_n = b_n$ for each $n \in \mathbb{Z}$. Thus, we obtain the desired unique representation, and so $\{\tau_n\phi\}$ is a basis for V_0 in $L^p(\mathbb{R})$.

Next, for $\{c_n\} \in \ell^p(\mathbb{Z})$,

$$\int_{\mathbb{R}} |\sum_{n\in\mathbb{Z}} c_n\tau_n\phi|^p = \sum_{n\in\mathbb{Z}} |c_n|^p,$$

and so

$$\|f\|_{L^p(\mathbb{R})} = \|\{c_n\}\|_{\ell^p(\mathbb{Z})}, \tag{3.7}$$

where $f = \sum c_n\tau_n\phi$ in $L^p(\mathbb{R})$. Further, note that

$$\forall n \in \mathbb{Z}, \quad \|\tau_n\phi\|_{L^p(\mathbb{R})} = \left(\int_n^{n+1} 1^p dt\right)^{\frac{1}{p}} = 1.$$

Thus, $\{\tau_n\phi\}$ is a bounded basis for V_0 in $L^p(\mathbb{R})$-norm for each $p \geq 1$.

Finally, we prove this basis is unconditional. Given $\{b_n, c_n \in \mathbb{C} : n \in F \subseteq \mathbb{Z}$, a finite set$\}$, and assume $|b_n| \leq |c_n|$ for each $n \in F$. We shall prove that

$$\left\|\sum_{n\in\mathbb{Z}} b_n\tau_n\phi\right\|_{L^p(\mathbb{R})} \leq \left\|\sum_{n\in\mathbb{Z}} c_n\tau_n\phi\right\|_{L^p(\mathbb{R})}. \tag{3.8}$$

Because of (3.7), the left side of (3.8) is $\|\{b_n\}\|_{\ell^p(\mathbb{Z})}$ and the right side is $\|\{c_n\}\|_{\ell^p(\mathbb{Z})}$. Consequently, since $|b_n| \leq |c_n|$ we obtain (3.8).

Hence, for each $p \geq 1$, $\{\tau_n\phi\}$ is a bounded unconditional basis for V_0 considered s a closed subspace of $L^p(\mathbb{R})$.

Remark 3.3

(a) The hypotheses in Theorem 3.8 are the same as in the fundamental Meyer-Mallat MRA theorem for $L^2(\mathbb{R})$ in the special case of the Haar function, except that the hypothesis in the L^2 case that $\{\tau_n\phi\}$ is an ONB for V_0 is replaced by the hypothesis that $\{\tau_n\phi\}$ is a bounded unconditional basis for V_0. Bounded unconditional bases are equivalent to exact frames, which are equivalent to Riesz bases, i.e., topological isomorphisms of ONBs, see [16], [4, p. 456], [45]. Köthe (1936) proved the implication that bounded unconditional bases are Riesz bases.

(b) In light of the fact that $\{\tau_n\phi\}$ is a bounded unconditional basis for V_0 considered as a closed subspace of $L^1(\mathbb{R})$, we note that Pełczyński (1961) proved that $L^1(\mathbb{R})$ and $L^1(\mathbb{T})$ have no unconditional bases, see [36], [29, p. 24] for a short proof by V. D. Milman, and [37, pp. 256 ff.].

(c) If $0 < p < 1$, then, as noted in Sect. 3.3.1, $L^p(\mathbb{R})$ is not a normed vector space when convergence is defined by $\|\ldots\|_{L^p(\mathbb{R})}$, but it is a complete metric vector space, with metric ρ defined by

$$\forall f, g \in L^p(\mathbb{R}), \quad \rho(f, g) = \|f - g\|^p_{L^p(\mathbb{R})} = \int_{\mathbb{R}} |f(t) - g(t)|^p \, dt.$$

In this setting, all of the assertions of Theorem 3.8 remain valid except (3.6), which requires Minkowski's inequality for its verification in the cases $p \geq 1$.

3.4 Applications

3.4.1 External Beam Radiation Therapy

External beam radiation therapy (EBRT) for prostate cancer uses high-energy beams, such as X-rays or protons, to kill cancer cells. During the treatment, the high-energy beams are generated by a linear accelerator that aims the beams at the prostate gland. In particular, the disease is treated using an external source of radiation. EBRT kills cancer cells by damaging the genetic material that controls how cells grow and divide. Healthy cells in the beam's path have the ability to repair radiation damage as opposed to cancer cells that lack the ability to repair deoxyribonucleic acid (DNA) damage. Excess radiation to non-cancerous cells can result in short and/or long term adverse effects. The goal of treatment is to destroy the cancerous cells while sparing as much of the normal surrounding tissue as possible. The planning for EBRT involves intensity modulated radiation therapy (IMRT), which is a specific type of 3D treatment planning based on imaging, e.g., a CT scan, in order to target the radiation beam to the precise location and shape of the tumor. An important issue is to quantify the amount of radiation received by the body. In fact the EBRT technology makes this measurement by subtracting the dosage of radiation from the amount of radiation that exits the body. EBRT is a well-established methodology, and is part of our theme since analysis of the tumor is made in terms of processing within the tumor itself.

EBRT may also be used if one has prostate cancer that returns after surgery.

Internal radiation therapy (IRT) or brachytherapy is another form of radiation therapy used to treat prostate cancer. IRT involves placing radioactive sources in the prostate gland, where the radiation can kill the cancer cells while causing less damage to nearby healthy tissue. Medically, IRT minimizes the risk of side effects. Once gain, the methodology fits-in naturally with our theme since analysis of the tumor is made in terms of processing within the tumor itself.

Image based treatment planning is a technical and important tool used as a prequel for effective EBRT and IRT, e.g., see [40, 41].

3.4.2 Communications

Interference is one of the major challenging issues in cellular communications. What follows is a broad, brief outline and comparison of 1G, 2G, 3G, 4G, and 5G developments, e.g., see [24, 38], ending in the brief discussion of a natural 5G scenario formulated in terms of our *theme* of signal reconstruction by means of harmonics with supports within that of the signal.

Co-channel interference is essentially the scenario of more than one transmission on the same frequency. Optimally, the transmitters are far enough apart, e.g., as determined by Google maps, so that the interference created is small enough to ensure that the system meets acceptable (SINR) signal to interference plus noise ratio guidelines. A key aspect of cell network planning, where a cell is the geographic area that is covered by a single base station in a cellular network, is to assure that these guidelines are met.

Adjacent channel interference is the scenario when transmissions are occurring in adjacent channels, e.g., channel 1 and channel 2, and there is energy leakage from one channel into the other. This phenomenon occurs primarily because of equipment limitations (and it can be explained theoretically in terms of the uncertainty principle). For example, it is generally difficult to build effective band-pass filters for transmitters and receivers. In fact, ideally, band-pass filters are only supposed to allow only frequencies between, say, γ_1 and $\gamma_2 > \gamma_1$, and so if $\gamma > \gamma_2$ or if $\gamma < \gamma_1$ then frequency information of the signal containing γ would be zeroed-out. In reality this is never the case. Filters have roll-off into some band from $\gamma_1 - \epsilon$ to γ_1 and from γ_2 to $\gamma_2 + \epsilon$. This elementary set-up, in which we know ϵ can not be 0, allows us to compare 1G, 2G, 3G, and 4G systems, and to analyze the ramifications when the support for a carrier is entirely within the band $[\gamma_1, \gamma_2]$.

The problem of limiting band leakage has been an ongoing struggle for most of recent telecommunications history since the 1930s. It took from 1968, when cellular was first proposed, until about 1980, when the first commercial systems were launched, to make the technology viable, and part of that early solution was to construct filters that were close to being band-pass, and where essentially the only interference problems arose in co-adjacent, adjacent, and second-adjacent channels, this last being interference on channel 1 from channel 3 and/or vice-versa. In fact, in first generation communications systems, viz.,1G and first generation digital (1980s through to the early 1990s), the channels were about 30 kHz wide, and the equipment was vulnerable to leakage. Designing a cell network plan which avoided unacceptable interference was often not completely resolved.

2G systems brought with it equipment that was more advanced. Adjacent channel interference was still a challenge, but the filters could now be designed to eliminate second-adjacent channel problems. This was true for both GSM (Global System for Mobiles) with a channel bandwidth of 200 kHz and narrowband CDMA (Code-Division Multiple Access) with a channel bandwidth of about 1.25 MHz.

3G gave rise to wide band CDMA systems, which implement wide 5MHz carriers resulting in the elimination of adjacent channel issues. There is an engineering sleight of hand here. In fact, narrowband CDMA transmits 1.22 million chips per second through a 1.25 MHz channel, and so naturally there is some leakage. On the other hand, wide band CDMA sends 3.84 Mcps (million chips per second) through a 5 MHz channel. Thus, since the chip rate is comparatively low versus channel bandwidth it is much easier to avoid unacceptable leakage.

4G LTE (Long Term Evolution) cellular systems employ OFDMA (Orthogonal Frequency Division Multiple Access) technology for the downlink, i.e., transmissions from the base station to the mobile phone. LTE carriers are partitioned into orthogonal sub-carriers. In naive terms, the design ensures that specific local leakages should cancel each other and effectively eliminate interference between the subcarriers. Unfortunately, this technology must be modified for the uplink, i.e., transmissions from the mobile phone to the base station, in order to accommodate the power and engineering limitations of mobile devices. For the uplink, using OFDMA would often require large differences in transmit power for adjacent subcarriers, creating a signal with a high peak-to-average power ratio (PAR). Such signals would be difficult for the power-limited mobile to produce. SC-FDMA (Single Carrier-Frequency Division Multiple Access) is employed for the uplink.

In moving to 5G, one goal is to look beyond OFDMA with the purpose of constructing non-orthogonal sub-carriers that still have limited leakage. This is called NOMA (Non-Orthogonal Multiple Access). Many proposed 5G devices will operate at low power and, like 4G mobiles, may not always be able to create OFDMA signals because of physical device constraints and required power profiles. However, unlike 4G mobiles, 5G devices may be required to transmit significantly more data than SC-FDMA can accommodate. It is at this point that we see the emerging potential of the theory of frames in 5G [5, 11], and, more to the point for our *theme* in this paper, our point of view of signal approximation from within the signal's support. In particular, it becomes useful to construct signals that are supported entirely inside one sub-carrier. This would be a natural starting point for effective approximation having controllable leakage in order to construct high-data-rate signals having the desired properties.

Acknowledgments Although treated superficially and qualitatively herein, the author gratefully acknowledges the applicable expertise of the following: Dr. Vladimir Ioffe with regard to external beam radiation therapy; Dr. Alfredo Nava-Tudela with regard to the discrete cosine transform and image compression; and Dr. Michael Dellomo with regard to research in communication theory. Finally, the author thanks the two referees for their careful reading of the manuscript; and he has followed-through on all of their cogent suggestions.

References

1. D.R. Adams, L.I. Hedberg, *Function Spaces and Potential Theory* (Springer, Berlin, 1999)
2. D. Austin, What is ... JPEG. Not. Am. Math. Soc. **55**(2), 226–229 (2008)
3. J.J. Benedetto, *Spectral Synthesis* (Academic Press, New York, 1975)
4. J.J. Benedetto, Irregular sampling and frames, in *Wavelets: a Tutorial in Theory and Applications*, ed. by C.K. Chui (Academic Press Inc., San Diego, 1992), pp. 445–507
5. J.J. Benedetto, Frame decompositions, sampling, and uncertainty principle inequalities, in *Wavelets: Mathematics and Applications*, ed. by J.J. Benedetto, M.W. Frazier (CRC Press, Boca Raton, 1994), pp. 247–304
6. J.J. Benedetto, *Harmonic Analysis and Applications*. Studies in Advanced Mathematics (CRC Press, Boca Raton, 1997). MR1400886 (97m:42001)
7. J.J. Benedetto, W. Czaja, *Integration and Modern Analysis*. Birkhäuser Advanced Texts (Springer-Birkhäuser, New York, 2009)
8. J.J. Benedetto, A. Nava-Tudela, Sampling in image representation and compression, in *New Perspectives in Approximation and Sampling Theory*, ed. by G. Schmeisser, A. Zayed (Springer-Birkhäuser, New York, 2014). Chapter 7, invited
9. J.J. Benedetto, F. Njeunje, Haar approximation from within for $L^p(R^d)$, $0 < p < 1$. J. Sampl. Theory Signal Process. Data Anal. **19:2**(1), 1–40 (2021). https://doi.org/10.1007/s43670-020-00001-z
10. N. Bourbaki, *Intégration*, 2nd edn. (Hermann, Paris, 1965)
11. O. Christensen, *An Introduction to Frames and Riesz Bases*, 2nd edn. (Springer-Birkhäuser, New York, 2016 (2003))
12. K. Conrad, L^p space for $0 < p < 1$. Accessed on Conrad U Conn Website (2017)
13. A. Croisier, D. Esteban, C. Galand, Perfect channel splitting by use of interpolation, decimation, tree decomposition techniques, in *Int. Conf. Information Sciences/Systems*, Patras (1976), pp. 443–446
14. I. Daubechies, *Ten Lectures on Wavelets*. CBMS-NSF Regional Conference Series in Applied Mathematics (Society for Industrial and Applied Mathematics, Philadelphia, 1992)
15. M.M. Day, The spaces L^p with $0 < p < 1$. Bull. Am. Math. Soc. **46**(10), 816–823 (1940)
16. R.J. Duffin, A.C. Schaeffer, A class of nonharmonic Fourier series. Trans. Am. Math. Soc. **72**, 341–366 (1952)
17. D. Esteban, C. Galand, Applications of quadrature mirror filters to split-band voice coding schemes, in *Int Conf Acoust Speech Signal Proc*, Hartford (1977), pp. 191–195
18. V.I. Filippov, Subsystems of the Faber–Schauder system in function space. Izvestiya Vysshikh Uchebnykh Zavedenii. Matematika **2**, 78–85 (1991)
19. V.I. Filippov, On subsystems of the Haar system in the space (E)ϕ with $\lim_{t\to\infty} \phi(t) = 0$. Zametki **51**, 97–106 (1993)
20. V.I. Filippov, P. Oswald, Representation in L^p by series of translates and dilates of one function. J. Approx. Theory **82**(1), 15–29 (1995)
21. A. Haar, Zur Theorie der orthogonalen Funktionensysteme. Math. Ann. **69**(3), 331–371 (1910) MR1511592
22. L.I. Hedberg, Two approximation problems in function spaces. Ark. Mat. **16**(1–2), 51–81 (1978)
23. E. Hernández, G. Weiss, *A First Course on Wavelets*. Studies in Advanced Mathematics (CRC Press, Boca Raton, 1996). With a foreword by Yves Meyer. 97i:42015
24. H. Holma, A. Toskala (eds.), *WCDMA for UMTS: HSPA Evolution and LTE*, 5th edn. (John Wiley and Sons, New York, 2010)
25. J. Horváth, A remark on the representation theorems of Frederick Riesz and Laurent Schwartz, in *Essays Dedicated to Laurent Schwartz on the Occasion of his 65th Birthday*, ed. by L. Nachbin (Academic Press, New York, 1981)
26. I. Kaplansky, *Commutative Rings* (Allyn and Bacon, Boston, 1970)

27. Y. Katznelson, *An Introduction to Harmonic Analysis*, 3rd original 1968 edn. Cambridge Mathematical Library (Cambridge University Press, Cambridge, 2004, 1976, 1968)
28. R.S. Laugesen, Affine synthesis onto L^p when $0 < p \leq 1$. J. Fourier Anal. Appl. **14**(2), 235–266 (2008)
29. J. Lindenstrauss, L. Tzafriri, *Classical Banach Spaces I, Sequence Spaces*, vol. Ergebnisse der Mathematik 92 (Springer, Berlin, 1977)
30. S. Mallat, Multiresolution approximations and wavelet orthonormal bases of $L^2(R)$. Trans. Am. Math. Soc. **315**, 69–87 (1989)
31. S. Mallat, *A Wavelet Tour of Signal Processing, third edition: The Sparse Way* (Academic Press, New York, 2008)
32. Y. Meyer, Ondelettes, fonctions splines et analyses graduées. Rend. Sem. Mat. Univ. Politec. Torino **45**(1), 1–42 (1987, 1988). MR981153 (90e:42047)
33. Y. Meyer, *Ondelettes et Opérateurs* (Hermann, Paris, 1990)
34. A. Nava-Tudela, Predictive refinement methodology for compressed sensing imaging. arXiv e-prints, February 2020, arXiv:2002.09765 [cs.IT]
35. P. Oswald, L^p-approximation durch Reihen nach dem Haar-orthogonalsystem und dem Faber—Schauder-System. J. Approx. Theory **33**(1), 1–27 (1981)
36. A. Pełczyński, On the impossibility of embedding of the space L in certain Banach spaces. Coll. Math. **8**, 199–203 (1961)
37. A. Pietsch, *History of Banach Spaces and Linear Operators* (Birkhäuser, Boston, 2007)
38. T.S. Rappaport, *Wireless Communications: Principles and Practice*, 2nd edn. (Prentice-Hall, Upper Saddle River, 2002)
39. W. Rudin, *Fourier Analysis on Groups* (John Wiley and Sons, New York, 1962)
40. N. Shah, V. Ioffe, Frequency of Gleason score 7 to 10 in 5100 elderly prostate cancer patients. Nat. Rev. Urol. **18**(4), 181–187 (2016). PMID: 28127259 PMCID: PMC5260947. https://doi.org/10.3909/riu0732
41. N. Shah, V. Ioffe, A. Kapur, A comparative analysis of prostate cancer pre-treatment characteristics stratified by age. Can. J. Urol. **21**(2), 7213–7216 (2014). PMID: 24775574
42. M. Smith, T. Barnwell, Exact reconstruction for tree-structured subband codes. IEEE Trans. Acoust. Speech Signal Proc. **34**, 434–441 (1986)
43. D. Taubman, M. Marcellin, *JPEG2000 Image Compression Fundamentals, Standards and Practice: Image Compression Fundamentals, Standards and Practice* (Springer Science and Business Media, New York, 2012)
44. D.F. Walnut, *An Introduction to Wavelet Aanalysis* (Birkhäuser, New York, 2002)
45. R.M. Young, *An Introduction to Nonharmonic Fourier Series* (Academic Press, New York, 1980). Revised edition 2001

Chapter 4
An Update on the Compactness of Bilinear Commutators

Árpád Bényi and Rodolfo H. Torres

Dedicated to the memory of Guido L. Weiss

Abstract We survey some established results as well as some recent progresses in the study of the compactness of commutators of bilinear singular integrals with pointwise multiplication, and present a previously anticipated result related to the commutator with the Bilinear Hilbert Transform.

2010 Mathematics Subject Classification Primary: 42B20, 47B07; Secondary: 42B25, 47G99

4.1 Introduction

The study of commutators of singular integrals with pointwise multiplication is by now a well-developed area of harmonic analysis which has found over the years numerous applications and extensions. Much of the germ of this area of research is in the seminal work of Coifman, Rochberg, and Weiss, *Factorization theorems for Hardy spaces in several variables*, [13]. Aside from books and monographs, this is the most cited article of Professor Weiss (Guido) which, as of the writing of this note, has 682 citations in MathSciNet. The work in [13] is an important component of Guido's legacy which continues to have a tremendous impact in the field. We will forever fondly remember Guido and would like to dedicate this note to his memory.

Á. Bényi
Department of Mathematics, Western Washington University, Bellingham, WA, USA
e-mail: benyia@wwu.edu

R. H. Torres (✉)
Department of Mathematics, University of California, Riverside, CA, USA
e-mail: rodolfo.h.torres@ucr.edu

E. Hernández et al. (eds.), *The Mathematical Heritage of Guido Weiss*,
Applied and Numerical Harmonic Analysis,
https://doi.org/10.1007/978-3-031-76793-7_4

Recall that the commutator between an operator K and pointwise multiplication by a function b is defined by

$$[K, b](f) = K(bf) - bK(f).$$

Obviously, if b is a bounded function and K is bounded on L^p then so is $[K, b]$. It is the cancelation produced between the two terms in the definition of $[K, b]$ what makes sometimes possible the boundedness of the commutator for some appropriate b which may not be bounded. This cancellation can be clearly seen in the kernel of $[K, b]$ too. If $k(x, y)$ is the kernel of K, then the kernel of $[K, b]$ is

$$k(x, y)(b(y) - b(x)).$$

This also hints that additional smoothness on b may lead to further improvements on the properties of $[K, b]$ such as compactness, which is what we will revisit here.

Like numerous other contributions of Guido, the results presented in [13] are deeply rooted in fundamental results in analysis nowadays well-understood, but which were in their infancy in the 1970s. As stated by D. Sarason [28] in his Math Review of the aforementioned work:

> The result depends upon the duality between H^1 and BMO (the space of functions of bounded mean oscillation [see C. Fefferman and Stein, Acta Math. 129 (1972), no. 3–4, 137–193]) and upon a new result, very interesting in its own right, linking BMO and the L^p boundedness of certain commutator operators. The latter result is as follows. Let K be a Calderón-Zygmund singular integral operator with a smooth kernel, and let B be the operator of multiplication by the function b. If b is in BMO, then the commutator $[K, B]$ is bounded on L^p for $1 < p < \infty$, its norm being no larger than $\|b\|_{BMO}$ times a constant which depends only on K and p. Conversely, if, for some p in $(1, \infty)$, the commutator $[K, B]$ is bounded on L^p whenever K is one of the Riesz transforms, then b is in BMO. Two proofs are given for the first half (the difficult half) of this theorem, one involving direct estimates of certain maximal operators, the other involving B. Muckenhoupt's theory of weights...

In particular, one of the two proofs provided in [13] for the boundedness of the commutator relies on the so-called *Cauchy Integral Trick (CIT)*, which through the use of complex analysis allows to represent the commutator as a weighted average of the operator itself. Essentially, defining

$$\Psi(z)f(x) := e^{-z\,b(x)}K(e^{z\,b}\,f)(x),$$

one can then write

$$[K, b]f(x) = \Psi'(0)f(x) = \frac{1}{2\pi i}\int_{|z|=\delta} \frac{\Psi(z)f(x)}{z^2}\,dz.$$

This allows one to obtain, through Minkowski's inequality, the L^p-boundedness of the commutator through weighted estimates of the operator itself. This approach also uses the deep connection between BMO and weight theory taking advantage of the fact that, loosely speaking, the logarithms of Muckenhoupt weights are BMO functions.

After [13], the boundedness of the commutators with certain non-degenerate Calderón-Zygmund operators and the necessity of b to be in BMO were further studied by Janson [21] and Uchiyama [34] and, in this last article, Uchiyama actually showed the compactness of $[K, b]$ if the function b is in CMO. This CMO subspace of BMO is the closure in the BMO norm of C_c^∞, the space of infinitely differentiable functions with compact support. It also coincides with the space of vanishing mean oscillation studied by Coifman and Weiss in [12] and which they proved to be the pre-dual of H^1.

Many other results for commutators for other type of operators and in different contexts followed. In particular, the CIT has been by now extended repeatedly in the literature. It would be impossible to survey all of the extensions of the commutator results in this short article, but we refer the reader to the detailed presentations in Bényi et al. [5] (based around the CIT) and the survey by Wick [36] and the references therein.

In the rest of our presentation we will recall some of the first results on the study of compactness for commutators of several bilinear singular integrals to motivate and explain the most recent progress also presented. With the exception of a result about the separate-compactness of the commutator of the Bilinear Hilbert Transform in Sect. 4.5, all the other material presented in Sects. 4.2–4.4 has appeared already in other publications as indicated along the narrative. We will often only present the results in a simplified form, avoiding the most general hypothesis. We also emphasize that this area of research remains extremely active and new results on the subject are constantly produced. We do not intend this update on the topic to be exhaustive, but focus mostly on results originated by our own (and with other colleagues) contributions. We regret that we do not have space here to acknowledge several other not so closely related results by many other authors. A multitude of the compactness results, which for simplicity we will only state in the bilinear case, have not only extensions to higher multilinearity situations, but also to the study of iterated commutators, operators with different degrees of singularity, and other spaces of functions beyond the Lebesgue ones. The literature cited may guide the reader interested in such possible extensions.

4.2 Compact Bilinear Operators and Compact Sets of Lebesgue Spaces

We will follow the terminology in our work [1]. As stated there, we can trace back the natural notion of compact multilinear operator to the work of Calderón in [7] (the article that was also the birth place of important interpolation results and the famous Calderón Reproducing Formula). In particular, if X, Y, Z are normed or quasi-normed spaces, a bilinear operator $T : X \times Y \to Z$ is said to be *jointly compact*, or simply *compact*, if $\{T(x, y) : \|x\|, \|y\| \leq 1\}$ is a precompact set in Z.

Let $B_{1,X}$ and $B_{1,Y}$ be the closed unit balls in X and Y. As in the linear case, compactness is equivalent to the fact that for any sequence $\{(x_n, y_n)\} \subset B_{1,X} \times B_{1,Y}$, the sequence $\{T(x_n, y_n)\}$ has a convergent subsequence in Z. Other obvious alternative characterizations analogous to the linear situation also hold in the multilinear setting. It is also true that a compact bilinear operator is automatically continuous and that if Z is complete, then the subspace of compact bilinear operators is closed in the space of all bounded operators $T : X \times Y \to Z$ with respect to the operator norm. This is a very useful fact allowing in many cases to reduce the study of compactness to a regularized sequence of operators T_j converging to T in the operator norm.

There is also an intriguing notion of separate compactness. $T : X \times Y \to Z$ is *compact in the first variable* if $T_y = T(\cdot, y) : X \to Z$ is compact for all $y \in Y$ with a similar definition of *compact in the second variable*. If both such properties hold, we say that T is *separately compact*. Separate compactness implies a parallel notion of separate continuity and even boundedness of T if one of the spaces X or Y are Banach. However separate compactness does not in general imply compactness. See [1] for further properties and examples.

All the proofs about the compactness of bilinear commutators that we are aware of are done, after some regularization of the operators, using the á priori knowledge of the boundedness of the commutators. One then proceeds to show that the image of the product of unit balls is a precompact set in the target space. The most powerful and commonly used tool to establish the compactness of a subset of a Lebesgue space $L^r = L^r(\mathbb{R}^n)$ is the Fréchet-Kolmogorov-Riesz theorem, see e.g. [38, p.275], and its less well-know extension to the quasi-Banach case by Tsuji [33].

Theorem 4.1 *Let* $0 < r < \infty$. *A subset* $\mathcal{K} \subset L^r$ *is precompact if and only if the following three conditions are satisfied:*

(a) $\mathcal{K}$ *is bounded in* L^r*;*
(b) $\lim_{A\to\infty} \int_{|x|>A} |f(x)|^r\, dx = 0$ *uniformly for* $f \in \mathcal{K}$*;*
(c) $\lim_{t\to 0} \|f(\cdot + t) - f\|_{L^r} = 0$ *uniformly for* $f \in \mathcal{K}$.

There are versions of this theorem for weighted Lebesgue spaces, which we denote by $L^r(w) = L^r(\mathbb{R}^n, w dx)$. By a *weight* we mean a locally integrable function $w : \mathbb{R}^n \to [0, \infty]$ such that $0 < w < \infty$ a.e. Recall that for $1 \leq p < \infty$, a weight w is in the classical Muckenhoupt class A_p if

$$[w]_{A_p} = \sup_Q \left(\fint_Q w \right) \left(\fint_Q w^{1-p'} \right)^{\frac{p}{p'}} < \infty, \tag{4.1}$$

where the supremum is taken over all cubes Q with sides parallel to the axes in $\mathbb{R}^n$ (and the usual modification for the case $p = 1$).

One first extension of Theorem 4.1 to the weighted case is due to Clop and Cruz [11]. Using boundedness properties of the Hardy-Littlewood maximal function M, they were able to overcome a crucial use of the translation invariance property of the Lebesgue measure, absent in the weighted situation, and establish the following.

Theorem 4.2 *Let $1 < r < \infty$ and $w \in A_r$ and let $\mathcal{K} \subset L^r(w)$. If*

(a) $\mathcal{K}$ is bounded in $L^r(w)$;
(b) $\lim_{A\to\infty} \int_{|x|>A} |f(x)|^r\, w(x)\, dx = 0$ uniformly for $f \in \mathcal{K}$;
(c) $\lim_{t\to 0} \|f(\cdot + t) - f\|_{L^r(w)} = 0$ uniformly for $f \in \mathcal{K}$;

then $\mathcal{K}$ is precompact in $L^r(w)$.

Notice the restriction of w being an A_r weight in the target space and hence the limitation in the range of exponents. Nonetheless, a nice extension outside the Muckenhoupt classes was recently obtained by Xue et al. [37] eliminating these restrictions.

Theorem 4.3 *Assume that for some $1 < p_0 < \infty$, w and $w^{-1/(p_0-1)}$ are weights. Then $\mathcal{K} \subset L^r(w)$ is precompact for $0 < r < \infty$ if and only if*

(a) $\mathcal{K}$ is bounded in $L^r(w)$;
(b) $\lim_{A\to\infty} \int_{|x|>A} |f(x)|^r\, w(x)\, dx = 0$ uniformly for $f \in \mathcal{K}$;
(c) $\lim_{t\to 0} \|f(\cdot + t) - f\|_{L^r(w)} = 0$ uniformly for $f \in \mathcal{K}$.

We note that a version of the above theorem for even more general weights was obtained by Guo and Zhao [20] (after the arXiv preprint version of [37]), but it will not be needed for our purposes.

4.3 Bilinear Calderón-Zygmund Operators and Their Commutators

For the purpose of this presentation, a *bilinear Calderón-Zygmund operator (BCZO)* T is a bilinear bounded map from $L^{p_1} \times L^{p_2}$ to L^p, for all $1 < p_1, p_2 < \infty$ and

$$\frac{1}{p_1} + \frac{1}{p_2} = \frac{1}{p}, \tag{4.2}$$

associated to a kernel $K(x, y, z)$, which is defined away from the diagonal $x = y = z$ and has the properties[1]

$$|K(x, y, z)| \lesssim \frac{1}{(|x-y| + |x-z|)^{2n}}, \tag{4.3}$$

$$|\nabla K(x, y, z)| \lesssim \frac{1}{(|x-y| + |x-z|)^{2n+1}}. \tag{4.4}$$

[1] The regularity conditions (4.4) admit similar weaker versions as in the linear theory.

By this association, we mean that for all $f, g \in L_c^\infty$

$$T(f,g)(x) = \iint_{\mathbb{R}^{2n}} K(x,y,z) f(y) g(z)\, dy dz, \quad x \notin \operatorname{supp} f \cap \operatorname{supp} g.$$

A fully developed multilinear Calderón-Zygmund theory exists by now. We refer to the work of Grafakos and Torres [18] and the cited literature therein for a complete description. Note that a natural end point in this theory would be $L^1 \times L^1$ to $L^{1/2}$, but only a weak-type substitute of such boundedness is valid. The fact that $BCZO$s map into the quasi-Banach range creates often difficulties in their study in such ranges because of the lack of duality in L^r for $r < 1$, among other things.

A weighted multilinear Calderón-Zygmund theory exists too. A first natural approach using products of A_p weights was first developed by Grafakos and Torres in [19]. But there is, however, a surprisingly much larger set of weights that can be considered. They were discovered by Lerner et al. in [25]. For $\mathbf{p} = (p_1, p_2) \in (1,\infty) \times (1,\infty)$ and a pair of weights $\mathbf{w} = (w_1, w_2)$, let

$$\nu_{\mathbf{w}} = \nu_{\mathbf{w},\mathbf{p}} = w_1^{\frac{p}{p_1}} w_2^{\frac{p}{p_2}}.$$

Following [25], we say that $\mathbf{w}$ belongs to the class of vector weights $\mathbf{A_p}$ if

$$[\mathbf{w}]_{\mathbf{A_p}} = \sup_Q \left(\fint_Q \nu_{\mathbf{w}} \right) \left(\fint_Q w_1^{1-p_1'} \right)^{\frac{p}{p_1'}} \left(\fint_Q w_2^{1-p_2'} \right)^{\frac{p}{p_2'}} < \infty.$$

A useful characterization of these classes given in [25] is as follows,

$$\mathbf{w} \in \mathbf{A_p} \Leftrightarrow \begin{cases} \nu_{\mathbf{w}} \in A_{2p} \\ \sigma_1 = w_1^{1-p_1'} \in A_{2p_1'} \\ \sigma_2 = w_2^{1-p_2'} \in A_{2p_2'}. \end{cases} \tag{4.5}$$

It is possible to show that for p_1, p_2, p as in (4.2),

$$A_p \times A_p \subsetneq A_{\min(p_1,p_2)} \times A_{\min(p_1,p_2)} \subsetneq A_{p_1} \times A_{p_2} \subsetneq \mathbf{A_p}. \tag{4.6}$$

Note, in particular, that w_1 and w_2 may not be Muckenhoupt weights but σ_1 and σ_2 as defined above always are. This is a crucial and very useful property of the vector weights.

The notion that these are the appropriate classes of weights to be considered in the bilinear setting is supported by the fact that they characterize the boundedness

$$\mathcal{M} : L^{p_1}(w_1) \times L^{p_2}(w_2) \to L^p(\nu_{\mathbf{w}}),$$

of the *bi(sub)linear maximal function* defined by

$$\mathcal{M}(f,g)(x) = \sup_{Q \ni x} \left(\fint_Q |f(y)|\,dy \right) \left(\fint_Q |g(z)|\,dz \right).$$

These classes of weights are also the right ones for the boundedness of the form

$$T : L^{p_1}(w_1) \times L^{p_2}(w_2) \to L^p(\nu_{\mathbf{w}}),$$

of a $BCZO$, as was also proved in [25].

For a bilinear operator T and a function b, their *(first order) commutators* are defined by

$$[T,b]_1(f,g) = T(bf,g) - bT(f,g),$$

$$[T,b]_2(f,g) = T(f,bg) - bT(f,g).$$

It is very natural to expect commutators of a $BCZO$ with BMO functions to be bounded. This was first obtained in a limited range by Pérez and Torres [27, Proposition 3.1].

Theorem 4.4 *Let T be a $BCZO$ and $b \in BMO$. Then for $1 < p_1, p_2, p < \infty$ satisfying* (4.2) *and $j = 1, 2$,*

$$[T,b]_j : L^{p_1} \times L^{p_2} \to L^p.$$

The unnecessary restriction in the range $p > 1$ (recall the natural end-point is $p = 1/2$) in [27] was due to the use of the CIT and the needed weighted linear estimates to do so. This restriction was later removed and even weighted versions were obtained. For the product of classical weights this was done by Tang [29] while for the $\mathbf{A_p}$ case the following result was obtained in [25, Theorem 3.18].

Theorem 4.5 *Let T be a $BCZO$ and $b \in BMO$. Then for $1 < p_1, p_2 < \infty$ and p satisfying* (4.2), *$\mathbf{w} = (w_1, w_2) \in \mathbf{A_p}$ and $j = 1, 2$,*

$$[T,b]_j : L^{p_1}(w_1) \times L^{p_2}(w_2) \to L^p(\nu_{\mathbf{w}}).$$

Let us turn now our attention to the compactness results. The first result in the subject, which also opened this line of research, was obtained by the authors of this article in [1] relying on arguments from [34] and Theorem 4.1. Because only the classical version of Theorem 4.1 for $p \geq 1$ was used, a limited range for compactness was proved in [1]. However, in view of the extension of the characterization of compact sets in [33] to the quasi-Banach case, a simple generalization to the full range of exponents of the results in [1] was later presented by Torres et al. in [32]. Hence, the following holds.

Theorem 4.6 *Let T be a $BCZO$ and $b \in CMO$. Then for $1 < p_1, p_2 < \infty$ and p satisfying* (4.2), *and $j = 1, 2$,*

$$[T, b]_j : L^{p_1} \times L^{p_2} \to L^p$$

is compact.

The weighted case was much harder to obtain since a delicate regularization of the BCZO is first needed. The first result by Bényi et al. [4, Theorem 1.1] was the following.

Theorem 4.7 *Let T be a $BCZO$ and $b \in CMO$. Then for $1 < p_1, p_2, p < \infty$ satisfying* (4.2), *$\mathbf{w} = (w_1, w_2) \in A_p \times A_p$ and $j = 1, 2$,*

$$[T, b]_j : L^{p_1}(w_1) \times L^{p_2}(w_2) \to L^p(\nu_{\mathbf{w}}).$$

is compact.

It is not hard to see that if $\mathbf{w} \in A_p \times A_p$ then $\nu_{\mathbf{w}} \in A_p$, so Theorem 4.2, which, as already observed, requires the weight in the target L^p space of the operator to be in the corresponding Muckenhoupt class for the same p, can be used. However, the collection of weights is arbitrarily restricted by this technical condition. A minor generalization was also proved in [4], but it remained an open question whether more general weights could be used. This was finally answered in [37]. After the authors therein established Theorem 4.3, they could prove the complete and most general version in the weighted case; [37, Theorem 1.5 and Remark 3].

Theorem 4.8 *Let T be a $BCZO$ and $b \in CMO$. Then for $1 < p_1, p_2 < \infty$ and p satisfying* (4.2), *$\mathbf{w} = (w_1, w_2) \in \mathbf{A_p}$, and $j = 1, 2$,*

$$[T, b]_j : L^{p_1}(w_1) \times L^{p_2}(w_2) \to L^p(\nu_{\mathbf{w}})$$

is compact.

Before we conclude this section, we briefly mention what is known about the necessity of the condition $b \in CMO$ for compactness of the commutators. First of all, the necessity of $b \in BMO$ for the boundedness of certain $BCZO$ for $p > 1$ was obtained by Chaffee [8] and then revisited by different methods by Li and Wick [26]. For the full range $p > 1/2$ this was recently obtained by Wang et al. [35]. The necessity then of $b \in CMO$ for compactness was established by Chaffee et al. [9] for some BCZOs with homogeneous kernels and certain non-degeneracy condition. This last work was based on bilinear versions of modifications of the work in [34] for the linear case done by Chen et al. in [10].

4.4 Compactness of Commutators of Bilinear Pseudodifferential Operators and a New Subspace of *BMO*

While, as mentioned in the last section, the compactness of the commutators of $BCZO$s with certain homogeneous kernels implies that the multiplicative function b must be in CMO, it is possible that for operators with non-homogeneous kernels having a greater rate of decay, other subspaces of BMO could be considered. The study in the bilinear case is motivated by an article in the linear case by Cordes [14] back in the 70's. He looked at commutators of some multipliers and pseudodifferential operators. We note that this work appeared about the same time as the commutator result for singular integrals was obtained in [13]. The corresponding results for bilinear pseudodifferential operators were recently investigated by Torres and Xue [31].

For the purposes of this article, we say that an operator of the form

$$T(f,g)(x) = \int_{\mathbb{R}^n \times \mathbb{R}^n} \sigma(x,\xi,\eta)\widehat{f}(\xi)\widehat{g}(\eta)e^{ix(\xi+\eta)}d\xi d\eta$$

is a *bilinear pseudodifferential operator (BPSDO)*,[2] if its *symbol* σ satisfies the estimates

$$|D_x^\beta D_{\xi,\eta}^\alpha \sigma(x,\xi,\eta)| \lesssim_{\alpha,\beta} (1+|\xi|+|\eta|)^{-|\alpha|}, \tag{4.7}$$

for all multi-indices α, β. It is by now well-known that $BPSDO$s are $BCZO$s but their kernels satisfy better estimates than (4.3) and (4.4). Not only do they have a higher degree of smoothness

$$|\partial^\alpha K(x,y,z)| \lesssim_\alpha (|x-y|+|x-z|)^{-2n-|\alpha|} \tag{4.8}$$

for all multi-indices α whenever $x \neq y$ or $x \neq z$, but also a rapid decay away from the diagonal,

$$|\partial^\alpha K(x,y,z)| \lesssim_{\alpha,N} (|x-y|+|x-z|)^{-2n-N} \tag{4.9}$$

for all α and $N > 0$, if $|x-y|+|x-z| > 1$; see for example Bényi et al. [2]. It is this fast decay property (4.9) which essentially allows one to consider commutators with functions which are limits in BMO of functions whose derivatives go to zero at infinity rather than having compact support. To make this more precise, we recall several spaces using the same notation of [31] and refer also to the work of

[2] This class of operators is often denoted by $BS^0_{1,0}$ in the literature.

Bourdaud [6] for illuminating clarifications about different subspaces of BMO and inconsistencies found in the literature.

Let $C_u = C_u(\mathbb{R}^n)$ be the space of uniformly continuous functions and define

$$VMO = \overline{C_u \cap BMO}^{BMO},$$

$$A_\infty = \{b \in C^\infty \cap L^\infty : D^\alpha b(x) \to 0, |x| \to \infty, \alpha \neq 0\},$$

$$MMO = \overline{A_\infty}^{BMO},$$

$$B_\infty = \{b \in C^\infty \cap BMO : D^\alpha b(x) \to 0, |x| \to \infty, \alpha \neq 0\},$$

and finally

$$XMO = \overline{B_\infty}^{BMO}.$$

There are many other equivalent definitions of VMO in the literature but this space is not the space CMO we have been considering. The space A_∞ and its closure in L^∞ rather than BMO were used in [14] because the work in [13] was not available yet and the arguments in [14] needed that some commutators were á priori bounded.

It was observed in [31] that

$$CMO \subsetneq MMO \subsetneq XMO \subset VMO,$$

but proper containment in the last inclusion could not be established or disproved by those authors. Nonetheless, the following was obtained in [31, Theorem 1.1].

Theorem 4.9 *Let T be a $BPSDO$ and $b \in XMO$. Then for $1 < p_1, p_2 < \infty$ and p satisfying* (4.2), *and $j = 1, 2$,*

$$[T, b]_j : L^{p_1} \times L^{p_2} \to L^p$$

is compact.

An interesting ingredient in the proof of the above result, and not needed in the other results presented for commutators, is the use of boundedness properties for positive translation invariant bilinear operators studied by Grafakos and Soria [17].

The proper containment

$$XMO \subsetneq VMO$$

was finally resolved in a very recent work of Tao et al. [30], who also provided a nice characterization of the space along the lines of the one given in [34] for CMO. Finally, the same authors were also able to obtain the weighted version of Theorem 4.9. Namely, [30, Theorem 1.4],

Theorem 4.10 *Let T be a $BPSDO$ and $b \in XMO$. Then for* $1 < p_1, p_2 < \infty$ *and p satisfying* (4.2), $\mathbf{w} = (w_1, w_2) \in \mathbf{A_p}$, *and* $j = 1, 2$,

$$[T, b]_j : L^{p_1}(w_1) \times L^{p_2}(w_2) \to L^p(\nu_{\mathbf{w}})$$

is compact.

Certainly, the full power of the estimates (4.8) and (4.9) for all α and N is not needed to establish Theorem 4.9 or 4.10, but only a limited number of them. See [31] and [30] for some reduced regularity and decay assumptions.

4.5 Separate Compactness for the Commutators of the *BHT*

We conclude this article with a result about the commutator of perhaps the most famous bilinear operator, the *Bilinear Hilbert Transform (BHT)*, which is defined by

$$BHT(f, g)(x) = p.v. \int_{\mathbb{R}} \frac{f(x-y)g(x+y)}{y}\, dy.$$

This operator is much more singular than the $BCZO$s we have been considering so far and its boundedness properties (for which the full range of valid exponents is still unknown) cannot be obtained using the established methods of the Calderón-Zygmund theory. A breakthrough result using refined multiscale analysis techniques was obtained by Lacey and Thiele [23, 24].

Theorem 4.11 *For* $1 < p_1, p_2 < \infty$, $2/3 < p < \infty$ *satisfying* (4.2), *it holds that*

$$BHT : L^{p_1}(\mathbb{R}) \times L^{p_2}(\mathbb{R}) \to L^p(\mathbb{R}).$$

There is also a (sub)-bilinear maximal operator associated with the BHT defined by

$$MBHT(f, g)(x) = \sup_{t>0} \frac{1}{t} \int_{|y|<t} |f(x-y)||g(x+y)|\, dy.$$

It was proved by Lacey [22] that for the same range of indices as that of Theorem 4.11,

$$MBHT : L^{p_1}(\mathbb{R}) \times L^{p_2}(\mathbb{R}) \to L^p(\mathbb{R}).$$

The difficult part of the boundedness has to do with the case of $p \in (2/3, 1]$.The proof for the range $p > 1$ is rather straightforward and follows from Hölder's inequality and the boundedness of the Hardy-Littlewood Maximal function. The deep insight, however, given in [22] regarding the boundedness of the maximal operator $MBHT$ is that it follows from an appropriate pointwise control in terms of the following maximal operator

$$T^*(f,g)(x) = \sup_{0<\epsilon<\delta} \Big| \underbrace{\int_{\epsilon<|y|<\delta} \frac{f(x-y)g(x+y)}{y}\,dy}_{I_{\epsilon\delta}} \Big|.$$

In the calculations below, we will actually need a variation of this maximal operator, namely

$$T_*(f,g)(x) = \sup_{\delta>0} \Big| \underbrace{\int_{|y|>\delta} \frac{f(x-y)g(x+y)}{y}\,dy}_{I_{\delta}} \Big|.$$

Clearly, for fixed $0<\epsilon<\delta$, we have $I_{\epsilon\delta} = I_\delta - I_\epsilon$, which immediately implies the pointwise estimate $T^*(f,g)(x) \le 2T_*(f,g)(x)$. Moreover, letting $\epsilon \to 0$, we also get $T_*(f,g)(x) \le T^*(f,g)(x) + |BHT(f,g)(x)|$, and hence the $L^p \times L^q \to L^r$ boundedness of T^* and T_* are equivalent. Re-writing the content of [22, Theorem 1.5] so that it fits our purposes, we have the following.

Theorem 4.12 *Let* $1<p_1,p_2<\infty$ *and* $2/3<p<\infty$ *satisfy* (4.2). *Then,*

$$T_* : L^{p_1}(\mathbb{R}) \times L^{p_2}(\mathbb{R}) \to L^p(\mathbb{R}).$$

Weighted estimates for the BHT are far from being understood but some partial results were obtained by Culiuc et al. in [16]. They are still weaker than what one may expect but enough to run a CIT and Bényi et al. [5, Theorem 5.7] obtained.

Theorem 4.13 *Let* $b \in BMO$, $\mathbf{w} = (w_1, w_2)$ *be such that* $w_j^2 \in A_{p_j}$, *where* $1<p_j<\infty$, $j=1,2$ *and* $1<p<\infty$ *satisfy* (4.2). *Then, for* $j=1,2$,

$$[BHT,b]_j : L^{p_1}(w_1) \times L^{p_2}(w_2) \to L^p(\nu_{\mathbf{w}}).$$

Note the stronger assumption $w_j^2 \in A_{p_j}$, rather than $w_j \in A_{p_j}$ in the above theorem and the restriction of the range of p. Nonetheless, this is enough to use some general extrapolation results of Cruz-Uribe and Martell [15] and obtain the following unweighted version in the same range of exponents for which the BHT is bounded; [5, Corollary 5.8].

Theorem 4.14 *Let* $1<p_1,p_2<\infty$, $2/3<p<\infty$ *satisfy* (4.2) *and* $b \in BMO$. *Then, for* $j=1,2$,

$$[BHT,b]_j : L^{p_1} \times L^{p_2} \to L^p$$

and

$$\|[BHT, b]_j(f, g)\|_{L^p} \lesssim_{p_1,p_2} \|b\|_{BMO}\|f\|_{L^{p_1}}\|g\|_{L^{p_2}}. \tag{4.10}$$

Since we know now the boundedness of the commutators, we wish to revisit here the comments about separate compactness of commutators of BHT stated in the concluding remarks of [3]. We will follow the approach therein for bilinear fractional integral operators, which are also separately compact if $b \in CMO$. There are several differences, however, due to the non-positivity of the BHT. For convenience we switch now from the notation p_1, p_2, p to p, q, r.

Theorem 4.15 *Let* $1 < p, q, r < \infty$ *be such that* $\frac{1}{p} + \frac{1}{q} = \frac{1}{r}$, *and* $b \in CMO$. *Then, for* $j = 1, 2$,

$$[BHT, b]_j : L^p \times L^q \to L^r$$

is separately compact.

Proof By symmetry, we will only consider the commutator $[BHT, b]_1$. We aim to prove that the conditions *(a)*, *(b)*, and *(c)* of Theorem 4.1 hold for the set of values $[BHT, b]_1(f, g)$ in L^r, where $g \in L^q$ is fixed and $f \in L^p$ with $\|f\|_{L^p} \leq 1$. Using the estimate from Theorem 4.14 and general properties of compact bilinear operators, we may assume that $b \in C_c^\infty$ and observe also that condition *(a)* is trivially satisfied. A change of variables allows us to write the commutator in question as

$$[BHT, b]_1(f, g)(x) = \int_{\mathbb{R}} \frac{b(y) - b(x)}{x - y} f(y)g(2x - y)\,dy.$$

The proof of *(b)* is very much the same as in the proof of [3, Theorem 3.2]. For the sake of completeness we sketch it below. Assume $R > 1$ is large enough (to be chosen later) so that $|x| \geq R$ implies $x \notin 2\,\mathrm{supp}\, b$. Using Hölder's inequality twice, we get the following pointwise estimates:

$$\begin{aligned} |[BHT, b]_1(f, g)(x)| &\lesssim \|b\|_\infty |x|^{-1} \left(\int_{\mathrm{supp}\, b} |f(y)|^{q'}\,dy \right)^{1/q'} \|g\|_{L^q} \\ &\lesssim \|b\|_\infty |x|^{-1} |\mathrm{supp}\, b|^{1/r'} \|f\|_{L^p} \|g\|_{L^q}. \end{aligned}$$

Therefore, raising to the power $r > 1$ and integrating over the set $\{x \in \mathbb{R} : |x| > R\}$, we get for $\epsilon > 0$ that

$$\int_{|x|>R} |[BHT, b]_1(f, g)(x)|^r\,dx \lesssim \left(\int_{|x|>R} \frac{1}{|x|^r}\,dx \right) \|b\|_\infty^r |\mathrm{supp}\, b|^{r-1} \|g\|_{L^q}^r < \epsilon^r$$

by taking $R = R(\epsilon) > 0$ sufficiently large; this proves (b) (note that $r > 1$ is needed here).

We are left to show the continuity condition *(c)*, that is

$$\lim_{t\to 0} \|[BHT, b]_1(f, g)(\cdot + t) - [BHT, b]_1(f, g)\|_{L^r} = 0,$$

uniformly for $\|f\|_{L^p} \leq 1$ and $g \in L^q$ fixed. As in [3] we write

$$\begin{aligned}[BHT, b]_1(f, g)(x) &= \int_{\mathbb{R}} \frac{b(y) - b(x)}{x - y} f(y)g(2x - y)\,dy \\ &= \int_{\mathbb{R}} (b(y) - b(x))K_g(x, y)f(y)\,dy,\end{aligned}$$

where

$$K_g(x, y) = \frac{g(2x - y)}{x - y}.$$

Then, given $t \in \mathbb{R}$ small, we decompose

$$[BHT, b]_1(f, g)(x + t) - [BHT, b]_1(f, g)(x) = A(x) + B(x) + C(x) + D(x),$$

where

$$\begin{aligned}A(x) &= \int_{|x-y|>\delta} (b(x + t) - b(x))K_g(x, y)f(y)\,dy, \\ B(x) &= \int_{|x-y|>\delta} (b(x + t) - b(y))(K_g(x + t, y) - K_g(x, y))f(y)\,dy, \\ C(x) &= \int_{|x-y|\leq\delta} (b(y) - b(x))K_g(x, y)f(y)\,dy, \\ D(x) &= \int_{|x-y|\leq\delta} (b(x + t) - b(y))K_g(x + t, y)f(y)\,dy,\end{aligned}$$

with $\delta > 0$ to be chosen later.

First, notice that

$$|A(x)| \leq |t|\|\nabla b\|_\infty \left| \int_{|x-y|>\delta} \frac{f(y)g(2x - y)}{x - y}\,dy \right|.$$

Therefore,

$$|A(x)| \leq |t|\|\nabla b\|_\infty T_*(f, g)(x),$$

and based on Theorem 4.12 we immediately get

$$\|A\|_{L^r} \lesssim |t| \|\nabla b\|_\infty \|f\|_{L^p} \|g\|_{L^q}.$$

Next we concentrate on the C term and simply obtain

$$|C(x)| \leq \|\nabla b\|_\infty \int_{|y|\leq\delta} |f(x-y)||g(x+y)|\,dy.$$

Thus, raising to the power r, then integrating with respect to the x variable, using Minkowski's integral inequality (the values of r allow to do so) and then Hölder's inequality, we get

$$\begin{aligned}
\|C\|_{L^r} &\leq \|\nabla b\|_\infty \left(\int_{\mathbb{R}} \left(\int_{|y|\leq\delta} |f(x-y)||g(x+y)|\,dy\right)^r dx\right)^{\frac{1}{r}} \\
&\leq \|\nabla b\|_\infty \int_{|y|\leq\delta} \left(\int_{\mathbb{R}} |f(x-y)|^r |g(x+y)|^r\,dx\right)^{\frac{1}{r}} dy \\
&\leq \|\nabla b\|_\infty \int_{|y|\leq\delta} \|f\|_{L^p} \|g\|_{L^q}\,dy \\
&\leq \delta \|\nabla b\|_\infty \|f\|_{L^p} \|g\|_{L^q}.
\end{aligned}$$

The estimates for the D term are similar; we get

$$\|D\|_{L^r} \leq (\delta + |t|) \|\nabla b\|_\infty \|f\|_{L^p} \|g\|_{L^q}.$$

We are left to estimate appropriately the B term. We begin by further splitting $B(x) = E(x) + F(x)$, where

$$\begin{aligned}
E(x) &= \int_{|x-y|>\delta} (b(x+t) - b(y)) \frac{g(2x+2t-y) - g(2x-y)}{x-y} f(y)\,dy \\
&= \int_{|x-y|>\delta} (b(x+t) - b(y)) \frac{(\tau_{2t} g - g)(2x-y) f(y)}{x-y}\,dy
\end{aligned}$$

and

$$\begin{aligned}
F(x) &= \int_{|x-y|>\delta} \left(\frac{1}{x+t-y} - \frac{1}{x-y}\right) g(2x+2t-y) f(y)\,dy \\
&= -t \int_{|x-y|>\delta} \frac{b(x+t) - b(y)}{(x+t-y)(x-y)} \tau_{2t} g(2x-y) f(y)\,dy;
\end{aligned}$$

here, τ_a denotes the shift operator $\tau_a g(x) = g(x+a)$.

We estimate pointwise $E(x)$ as follows,

$$|E(x)| \le |b(x+t)|T_*(f, \tau_{2t}g - g)(x) + T_*(bf, \tau_{2t}g - g).$$

Thus, using again Theorem 4.12, we see that

$$\begin{aligned}\|E\|_{L^r} &\lesssim \|b\|_\infty \|f\|_{L^p}\|\tau_{2t}g - g\|_{L^q} + \|bf\|_{L^p}\|\tau_{2t}g - g\|_{L^q}\\ &\lesssim \|b\|_\infty \|f\|_{L^p}\|\tau_{2t}g - g\|_{L^q}.\end{aligned}$$

Next, we notice that

$$\begin{aligned}F(x) = -tb(x+t)&\int_{|x-y|>\delta} \frac{\tau_{2t}g(2x-y)f(y)}{(x+t-y)(x-y)}\,dy\\ &+t\int_{|x-y|>\delta} \frac{\tau_{2t}g(2x-y)(bf)(y)}{(x+t-y)(x-y)}\,dy\\ &= F_1(x) + F_2(x).\end{aligned}$$

Now, since

$$|F_1(x)| \lesssim |t||b(x+t)| \int_{|x-y|>\delta} \frac{|\tau_{2t}g(2x-y)||f(y)|}{(x-y)^2}\,dy,$$

using again the Minkowski integral inequality and then Hölder's inequality gives

$$\begin{aligned}\|F_1\|_{L^r} &\lesssim |t|\|b\|_\infty \left(\int\left(\int_{|y|>\delta} \frac{1}{y^2}|\tau_{2t}g(x+y)||f(x-y)|\,dy\right)^r dx\right)^{\frac{1}{r}}\\ &\lesssim |t|\|b\|_\infty \int_{|y|>\delta}\frac{1}{y^2}\left(\int |\tau_{2t}g(x+y)|^r|f(x-y)|^r\,dx\right)^{\frac{1}{r}} dy\\ &\lesssim |t|\|b\|_\infty\left(\int_{|y|>\delta}\frac{1}{y^2}\,dy\right)\|f\|_{L^p}\|g\|_{L^q} \lesssim \frac{|t|}{\delta}\|b\|_\infty\|f\|_{L^p}\|g\|_{L^q}.\end{aligned}$$

A similar argument works for the F_2 term and we get

$$\|F_2\|_{L^r} \lesssim \frac{|t|}{\delta}\|bf\|_{L^p}\|g\|_{L^q}.$$

Therefore, $\|F\|_{L^r} \lesssim \frac{|t|}{\delta}\|b\|_\infty\|f\|_{L^p}\|g\|_{L^q}$. Since $g \in L^q$, for a given $\epsilon > 0$ we can find $\gamma = \gamma(\epsilon, g) > 0$ such that $|t| < \gamma$ implies

$$\|\tau_{2t}g - g\|_{L^q} < \epsilon.$$

Finally, putting together all the estimates for A, B, C, D by choosing $|t| < \epsilon^2$ and $\delta = |t|/\epsilon$ we get that

$$\|[BHT, b]_1(f, g)(\cdot + t) - [BHT, b]_1(f, g)\|_{L^r} \lesssim \epsilon.$$

This shows (c) holds, thus finishing our proof of compactness in the first variable.

A moment of reflection should convince the reader that the computations could be repeated if we fixed f instead of g, proving compactness in the other variable. We omit the details. □

Perhaps a modification of the arguments could yield the full range $2/3 < r < \infty$. Also, the reader may be wondering why we cannot prove compactness instead of separate compactness. All the terms in the above proof but B produce estimates which are uniformly small in f and g. That will be needed to establish compactness. However, the term B can be made arbitrarily small only if $\|\tau_{2t}g - g\|_{L^q}$ is small, which cannot be achieved uniformly if g is not fixed. We encountered the same obstacle in the case of fractional bilinear versions of the BHT studied in [3]. We do not know if our computations could be modified to prove the (full notion of) compactness. Alternatively, compactness may not hold. These questions remain.

Acknowledgments We would like to thank the anonymous referees for their comments and for pointing out the reference [20], which was unknown to us.

References

1. Á. Bényi, R.H. Torres, Compact bilinear operators and commutators. Proc. Am. Math. Soc. **141**(10), 3609–3621 (2013)
2. Á. Bényi, D. Maldonado, V. Naibo, R.H. Torres, On the Hörmander classes of bilinear pseudodifferential operators. Integr. Equ. Oper. Theory **67**, 341–364 (2010)
3. Á. Bényi, W. Damián, K. Moen, R.H. Torres, Compactness properties of commutators of bilinear fractional integrals. Math. Z. **280**(1–2), 569–582 (2015)
4. Á. Bényi, W. Damián, K. Moen, R.H. Torres, Compact bilinear commutators: the weighted case. Michigan Math. J **64**(1), 39–51 (2015)
5. Á. Bényi, K. Moen, J.M. Martell, E. Stachura, R.H. Torres, Boundedness results for commutators with BMO functions via weighted estimates: a comprehensive approach. Math. Ann. **376**(1–2), 61–102 (2020)
6. G. Bourdaud, Remarques sur certains sous-espaces de $BMO(\mathbb{R}^n)$ et de $bmo(\mathbb{R}^n)$. Ann. Inst. Fourier (Grenoble) **52**, 1187–1218 (2002)
7. A.P. Calderón, Intermediate spaces and interpolation, the complex method. Studia Math. **24**, 113–190 (1964)
8. L. Chaffee, Characterization of bounded mean oscillation through commutators of bilinear singular integrals. Proc. R. Soc. Edin. Sect. A **146**(6), 1159–1166 (2016)
9. L. Chaffee, P. Chen, Y. Han, R.H. Torres, L.A. Ward, Characterization of compactness of commutators of bilinear singular integral operators. Proc. Am. Math. Soc. **146**(9), 3943–3953 (2018)
10. Y. Chen, Y. Ding, X. Wang, Compactness of commutators of Riesz potential on Morrey spaces. Potential Anal. **30**, 301–313 (2009)

11. A. Clop, V. Cruz, Weighted estimates for Beltrami equations. Ann. Acad. Sci. Fenn. Math. **38**, 91–113 (2013)
12. R. Coifman, G. Weiss, Extension of Hardy spaces and their use in analysis. Bull. Am. Math. Soc. **83**, 569–645 (1977)
13. R. Coifman, R. Rochberg, G. Weiss, Factorization theorems for Hardy spaces in several variables. Ann. Math. **103**, 611–635 (1976)
14. H.O. Cordes, On compactness of commutators of multiplications and convolutions, and boundedness of pseudodifferential operators. J. Funct. Anal. **18**, 115–131 (1975)
15. D.V. Cruz-Uribe, J.M. Martell, Limited range multilinear extrapolation with applications to the bilinear Hilbert transform. Math. Ann. **371**(1–2), 615–653 (2018)
16. A. Culiuc, F. Di Plinio, Y. Ou, Domination of multilinear singular integrals by positive sparse forms. J. Lond. Math. Soc. **98**(2), 369–392 (2018)
17. L. Grafakos, J. Soria, Translation-invariant bilinear operators with positive kernels. Integr. Equ. Oper. Theory **66**(2), 253–264 (2010)
18. L. Grafakos, R.H. Torres, Multilinear Calderón-Zygmund theory. Adv. Math. **165**, 124–164 (2002)
19. L. Grafakos, R.H. Torres, Maximal operator and weighted norm inequalities for multilinear singular integrals. Indiana Univ. Math. J. **51**(5), 1261–1276 (2002)
20. W. Guo, G. Zhao, On relatively compact sets in quasi-Banach function spaces. Proc. Am. Math. Soc. **148**(8), 3359–3373 (2020)
21. S. Janson, Mean oscillation and commutators of singular integral operators. Ark. Mat. **16**, 263–270 (1978)
22. M.T. Lacey, The bilinear maximal functions map into L^p for $2/3 < p \leq 1$. Ann. Math. **151**(1), 35–57 (2000)
23. M.T. Lacey, C.M. Thiele, L^p estimates on the bilinear Hilbert transform for $2 < p < \infty$. Ann. Math. **146**, 693–724 (1997)
24. M.T. Lacey, C.M. Thiele, On Calderón's conjecture. Ann. Math. **149**, 475–496 (1999)
25. A. Lerner, S. Ombrosi, C. Pérez, R.H. Torres, R. Trujillo-González, New maximal functions and multiple weights for the multilinear Calderón-Zygmund theory. Adv. Math. **220**, 1222–1264 (2009)
26. J. Li, B.D. Wick, Weak factorizations of the Hardy space $H^1(\mathbb{R}^n)$ in terms of multilinear Riesz transforms. Can. Math. Bull. **60**(3), 571–585 (2017)
27. C. Pérez, R.H. Torres, Sharp maximal function estimates for multilinear singular integrals. Contemp. Math. **320**, 323–331 (2003)
28. D. Sarason, Review of [13]. Mathematical Reviews, MR0412721 (54 # 843)
29. L. Tang, Weighted estimates for vector-valued commutators of multilinear operators. Proc. R. Soc. Edin. Sect. A **138**, 897–922 (2008)
30. J. Tao, Q. Xue, D. Yang, W. Yuan, XMO and weighted compact bilinear commutators. J. Fourier Anal. Appl. **27**(3), 34 pp. (2021), Paper No. 60
31. R.H. Torres, Q. Xue, On compactness of commutators of multiplication and bilinear pseudodifferential operators and a new subspace of BMO. Rev. Mat. Iberoam. **36**(3), 939–956 (2020)
32. R.H. Torres, Q. Xue, J. Yan, Compact bilinear commutators: the quasi-Banach space case. J. Anal. **26**, 227–234 (2018)
33. M. Tsuji, On the compactness of space L^p $(p > 0)$ and its application to integral operators. Kodai Math. Sem. Rep. **3**(101), 33–36 (1951)
34. A. Uchiyama, On the compactness of operators of Hankel type. Tôhoku Math. J. **30**(1), 163–171 (1978)
35. D. Wang, J. Zhou, Z. Teng, Characterizations of weighted BMO space and its application. Acta Math. Sin. English Ser. **37**(8), 1278–1292 (2021). arXiv:1707.01639
36. B.D. Wick, Commutators, BMO, Hardy spaces and factorization: a survey. Real Anal. Exchange **45**(1), 1–28 (2020)
37. Q. Xue, K. Yabuta, J. Yan, Weighted Fréchet–Kolmogorov theorem and compactness of vector-valued multilinear operators. J. Geom. Anal. **31**, 9891–9914 (2021) (Previously in arXiv:1806.06656v2)
38. K. Yosida, *Functional Analysis* (Springer, Berlin, 1995)

Chapter 5
Irregularities of Distribution on Two-Point Homogeneous Spaces

Luca Brandolini, Bianca Gariboldi, and Giacomo Gigante

Abstract We study the irregularities of distribution on two-point homogeneous spaces. Our main result is the following: let d be the real dimension of a two point homogeneous space $\mathcal{M}$, let $\left(\{a_j\}_{j=1}^N, \{x_j\}_{j=1}^N\right)$ be a system of positive weights and points on $\mathcal{M}$ and let

$$D_r(x) = \sum_{j=1}^{N} a_j \chi_{B_r(x)}(x_j) - \mu(B_r(x))$$

be the discrepancy associated with the ball $B_r(x)$. Then, if $d \not\equiv 1 (\mathrm{mod}\, 4)$, for any radius $0 < r < \pi/2$, we obtain the sharp estimate

$$\int_{\mathcal{M}} \left(|D_r(x)|^2 + |D_{2r}(x)|^2\right) d\mu(x) \geqslant cN^{-1-\frac{1}{d}}.$$

5.1 Introduction

A d-dimensional Riemannian manifold $\mathcal{M}$ with distance ρ is said to be a two-point homogeneous space if given four points $x_1, x_2, y_1, y_2 \in \mathcal{M}$ such that $\rho(x_1, y_1) = \rho(x_2, y_2)$, then there exists an isometry g of $\mathcal{M}$ such that $gx_1 = x_2$ and $gy_1 = y_2$. Wang in [19] has completely characterized compact connected two-point homogeneous spaces. More precisely if we assume that ρ is normalized so that $\mathrm{diam}(\mathcal{M}) = \pi$, then $\mathcal{M}$ is isometric to one of the following compact rank 1 symmetric spaces:

L. Brandolini (✉) · B. Gariboldi · G. Gigante
Dipartimento di Ingegneria Gestionale, dell'Informazione e della Produzione, Università degli Studi di Bergamo, Dalmine, BG, Italy
e-mail: luca.brandolini@unibg.it; biancamaria.gariboldi@unibg.it; giacomo.gigante@unibg.it

E. Hernández et al. (eds.), *The Mathematical Heritage of Guido Weiss*,
Applied and Numerical Harmonic Analysis,
https://doi.org/10.1007/978-3-031-76793-7_5

(i) the Euclidean sphere $S^d = SO(d+1)/SO(d) \times \{1\}, d \geqslant 1$;
(ii) the real projective space $P^n(\mathbb{R}) = O(n+1)/O(n) \times O(1), n \geqslant 2$;
(iii) the complex projective space $P^n(\mathbb{C}) = U(n+1)/U(n) \times U(1), n \geqslant 2$;
(iv) the quaternionic projective space $P^n(\mathbb{H}) = Sp(n+1)/Sp(n) \times Sp(1), n \geqslant 2$;
(v) the octonionic projective plane $P^2(\mathbb{O})$.

In the following we will assume that $\mathcal{M}$ is one of the above symmetric spaces and that d is its real dimension. In particular the real dimension of $P^n(\mathbb{F})$ is $d = nd_0$, where $d_0 = 1, 2, 4, 8$ according to the real dimension of $\mathbb{F} = \mathbb{R}$, $\mathbb{C}$, $\mathbb{H}$ and $\mathbb{O}$ respectively. In the case of S^d it will be convenient to set $d_0 = d$. See [9, pp. 176–178], see also [11, 16] and [21]. In the following, to keep notation simple, we will use

$$a = \frac{d-2}{2}, \qquad b = \frac{d_0-2}{2}.$$

Let μ be the Riemannian measure on $\mathcal{M}$ normalized so that $\mu(\mathcal{M}) = 1$ and let $B_r(x) = \{y \in \mathcal{M} : \rho(x, y) < r\}$.

For a given set of points $\{x_j\}_{j=1}^N \subset \mathcal{M}$ and positive weights $\{a_j\}_{j=1}^N$ such that $a_1 + a_2 + \cdots + a_N = 1$ we define the discrepancy of the ball $B_r(x)$ by

$$D_r(x) = \sum_{j=1}^N a_j \chi_{B_r(x)}(x_j) - \mu(B_r(x)) \tag{5.1}$$

This quantity compares the Riemannian measure μ with the discrete measure $\sum_{j=1}^N a_j \delta_{x_j}$ testing these two measures on the ball $B_r(x)$, and generalizes the classical spherical cap discrepancy defined when $\mathcal{M}$ is the Euclidean sphere S^d. It is known that, no matter how well distributed the points $\{x_j\}_{j=1}^N$ are, there are balls for which the discrepancy $D_r(x)$ cannot be too small. See [1] for an introduction to this subject and in particular page 181 and Corollary 24C. For example M. Skriganov in [16, Theorem 2.2] has proved, in the case of equal weights, that if η is a positive, locally integrable function on $(0, \pi)$ such that

$$\int_0^\pi \eta(r) \left(\sin \frac{1}{2} r\right)^{d-1} \left(\cos \frac{1}{2} r\right)^{d_0-1} dr < +\infty,$$

then there exists $c > 0$ such that for every distribution of N points

$$\int_0^\pi \int_{\mathcal{M}} |D_r(x)|^2 \, d\mu(x) \eta(r) dr \geqslant c N^{-1-\frac{1}{d}}.$$

A very general non-constructive result guarantees the existence of point distributions that exhibit the optimal decay $N^{-1-1/d}$ (see e.g. [4, Corollary 8.2]). In fact, Skriganov [16, Corollary 2.1] has proved that for well distributed optimal cubature formulas one has

$$\int_0^{\pi} \int_{\mathcal{M}} |D_r(x)|^2 \, d\mu(x) \eta(r) dr \leqslant c N^{-1-\frac{1}{d}}. \tag{5.2}$$

The existence of such cubature formulas for the sphere has been proved by Bondarenko, Radchenko and Vyazovska in [2]. These results have been recently extended to Riemannian manifolds in [6, 10]. See also the remarks that follow the statement of Theorem 5.4.

Our first result is in the same spirit of Theorem 2 in Chapter 6 in [13].

Theorem 5.1 *Let $d \not\equiv 1 \pmod 4$. Then, for all $0 < r < \pi/2$, there is a constant $C > 0$ such that for every set of points $\{x_j\}_{j=1}^N \subset \mathcal{M}$ and positive weights $\{a_j\}_{j=1}^N$ such that $a_1 + a_2 + \cdots + a_N = 1$ we have*

$$\int_{\mathcal{M}} |D_r(x)|^2 \, d\mu(x) + \int_{\mathcal{M}} |D_{2r}(x)|^2 \, d\mu(x) \geqslant C N^{-1-\frac{1}{d}}.$$

Again this result is optimal in view of Corollary 8.2 in [4]. Also, the same technique used by Skriganov to prove the estimate from above for cubature formulas (5.2) actually gives a stronger, uniform estimate in the radius r,

$$\int_{\mathcal{M}} |D_r(x)|^2 \, d\mu(x) \leqslant c N^{-1-\frac{1}{d}}. \tag{5.3}$$

See Theorem 5.4 for a precise statement.

When $d \equiv 1 \pmod 4$ our technique fails. In similar settings it is known that the discrepancy can be actually a bit smaller than the expected value $N^{-1-\frac{1}{d}}$. See [15, Theorem 3.1]. See also [3, 12] and [14].

Our final result is an estimate for the discrepancy of balls of a single fixed radius that holds in every dimension.

Theorem 5.2 *For every $\varepsilon > 0$ and for almost every $0 < r < \pi$ there is a constant $C > 0$ such that for every set of points $\{x_j\}_{j=1}^N \subset \mathcal{M}$ and positive weights $\{a_j\}_{j=1}^N$ such that $a_1 + a_2 + \cdots + a_N = 1$ we have*

$$\int_{\mathcal{M}} |D_r(x)|^2 \, d\mu(x) \geqslant C N^{-1-\frac{3}{d}-\varepsilon}.$$

The paper is organized as follows. In Sect. 5.2 we collect some basic facts about two-point homogeneous spaces. In Sect. 5.3 we prove Theorems 5.1 and 5.2. In Sect. 5.4 we show that suitable cubature formulas give the optimal discrepancy (5.3).

5.2 Two-Point Homogeneous Spaces

If o is a fixed point in $\mathcal{M}$, then $\mathcal{M}$ can be identified with the homogeneous space G/K, where G is the group of isometries of $\mathcal{M}$ and K is the stabilizer of o. We will also identify functions $F(x)$ on $\mathcal{M}$ with right K-invariant functions $f(g)$ on G by setting $f(g) = F(x)$ when $go = x$. If μ is the Riemannian measure on $\mathcal{M}$ normalized so that $\mu(\mathcal{M}) = 1$, then μ is invariant under the action of G, in other words, for every $g \in G$,

$$\int_{\mathcal{M}} F(gx)d\mu(x) = \int_{\mathcal{M}} F(x)d\mu(x)$$

Definition 5.1 A function F on $\mathcal{M}$ is a zonal function if for every $x \in \mathcal{M}$ and every $k \in K$ we have $F(kx) = F(x)$.

Lemma 5.1 *Let F be a zonal function. Then $F(x)$ depends only on $\rho(x, o)$. Furthermore, defining F_0 so that $F(x) = F_0(\rho(x, o))$ we have*

$$\int_{\mathcal{M}} F(x)d\mu(x) = \int_0^{\pi} F_0(r)A(r)dr, \tag{5.4}$$

where

$$A(r) = c(a, b)\left(\sin\frac{1}{2}r\right)^{2a+1}\left(\cos\frac{1}{2}r\right)^{2b+1}$$

and

$$c(a, b) = \left(\int_0^{\pi}\left(\sin\frac{1}{2}r\right)^{2a+1}\left(\cos\frac{1}{2}r\right)^{2b+1} dr\right)^{-1} = \frac{\Gamma(a+b+2)}{\Gamma(a+1)\Gamma(b+1)}.$$

In particular if

$$B_r(x) = \{y \in \mathcal{M} : \rho(x, y) < r\},$$

there exist two constants $c_1(d, d_0)$ and $c_2(d, d_0)$ such that for every $r \in [0, \pi]$

$$c_1(d, d_0)r^d \leqslant \mu(B_r(x)) \leqslant c_2(d, d_0)r^d. \tag{5.5}$$

Moreover if $k \geqslant 1$, then a ball of radius kr can be covered by $\frac{c_2(d,d_0)}{c_1(d,d_0)}(2(k+1))^d$ balls of radius r.

Proof Let $x, y \in \mathcal{M}$ such that $\rho(x, o) = \rho(y, o)$. Since $\mathcal{M}$ is two-point homogeneous there exists $g \in G$ such that $gx = y$ and $go = o$. Thus, $g \in K$ and $F(y) = F(gx) = F(x)$. Equation (5.4) follows from (4.17) in [9]. The estimate

(5.5) is an immediate consequence of (5.4). The last assertion is an estimate of the maximum number of disjoint balls of radius $r/2$ that can fit in a ball of radius $(k+1)r$. □

Let Δ be the Laplace-Beltrami operator on $\mathcal{M}$, let $\lambda_0, \lambda_1, \ldots,$ be the distinct eigenvalues of $-\Delta$ arranged in increasing order, let $\mathcal{H}_m$ be the eigenspace associated with the eigenvalue λ_m, and let d_m its dimension. It is well known that

$$L^2(\mathcal{M}) = \bigoplus_{m=0}^{+\infty} \mathcal{H}_m. \tag{5.6}$$

If $F(x) = F_0(\rho(x, o))$ is a zonal function on $\mathcal{M}$, then

$$\Delta F(x) = \frac{1}{A(t)} \frac{d}{dt} \left(A(t) \frac{d}{dt} F_0(t) \right) \bigg|_{t=\rho(x,o)} \tag{5.7}$$

(see (4.16) in [9]).

Definition 5.2 The zonal spherical function of degree $m \in \mathbb{N}$ with pole $x \in \mathcal{M}$ is the unique function $Z_x^m \in \mathcal{H}_m$, given by the Riesz representation theorem, such that for every $Y \in \mathcal{H}_m$

$$Y(x) = \int_{\mathcal{M}} Y(y) Z_x^m(y) d\mu(y).$$

The next lemma summarizes the main properties of zonal functions and it is essentially taken from [17] where the case $\mathcal{M} = S^d$ is discussed in detail.

Lemma 5.2

(i) *If $Y_m^1, \ldots, Y_m^{d_m}$ is an orthonormal basis of $\mathcal{H}_m \subset L^2(\mathcal{M})$, then*

$$Z_x^m(y) = \sum_{\ell=1}^{d_m} \overline{Y_m^\ell(x)} Y_m^\ell(y).$$

(ii) *Z_x^m is real valued and $Z_x^m(y) = Z_y^m(x)$.*
(iii) *If $g \in G$, then $Z_{gx}^m(gy) = Z_x^m(y)$.*
(iv) *$Z_x^m(x) = \|Z_x^m\|_2^2 = d_m$.*
(v) *$Z_o^m(x)$ is a zonal function and*

$$Z_o^m(x) = \frac{d_m}{P_m^{a,b}(1)} P_m^{a,b}(\cos(\rho(x, o))) \tag{5.8}$$

where $P_m^{a,b}$ are the Jacobi polynomials.

(vi) $\left\{d_m^{-1/2} Z_o^m\right\}_{m=0}^{+\infty}$ *is an orthonormal basis of the subspace of* $L^2(\mathcal{M})$ *of zonal functions.*

(vii) $\lambda_m = m(m+a+b+1)$.

(viii) $d_m \approx m^{d-1}$.

Proof

(i) By the defining property of Z_x^m

$$Z_x^m(y) = \sum_{\ell=1}^{d_m} \int_{\mathcal{M}} Z_x^m(z)\overline{Y_m^\ell(z)}d\mu(z)Y_m^\ell(y) = \sum_{\ell=1}^{d_m} \overline{Y_m^\ell(x)}Y_m^\ell(y).$$

(ii) Since the basis can be taken to be of real valued functions, then by point i) Z_x^m is real valued.

(iii) Let $Y \in \mathcal{H}_m(\mathcal{M})$. Since for every g, $Y(g^{-1}y) \in \mathcal{H}_m(\mathcal{M})$, then

$$\begin{aligned}\int_{\mathcal{M}} Y(y)Z_{gx}^m(gy)d\mu(y) &= \int_{\mathcal{M}} Y(g^{-1}y)Z_{gx}^m(y)d\mu(y)\\ &= Y(g^{-1}(gx)) = Y(x),\end{aligned}$$

and the uniqueness of Z_x^m shows that $Z_{gx}^m(gy) = Z_x^m(y)$.

(iv) Observe that

$$\|Z_x^m\|_2^2 = \int_{\mathcal{M}} Z_x^m(y)\overline{Z_x^m(y)}d\mu(y) = \overline{Z_x^m(x)} = \sum_{\ell=1}^{d_m} |Y_m^\ell(x)|^2.$$

Since by (iii) this quantity is constant in x, it equals its integral over $\mathcal{M}$, which is exactly d_m.

(v) By point (iii) we have

$$Z_o(kx) = Z_{k^{-1}o}(x) = Z_o(x).$$

Equation (5.8) follows from the fact that the radial part of the Laplace-Beltrami operator (5.7) coincides with the Jacobi operator.

(vi) follows from the fact that the Jacobi polynomials

$$\left\{P_m^{a,b}(\cos(t))\right\}_{m=0}^{+\infty}$$

form an orthogonal basis of $L^2((0,\pi), A(t)dt)$. See Chapter 4 in [18].

(vii) follows from [18, Theorem 4.2.2].

(viii) follows from (iv), (v) and (vi) computing explicitly the L^2 norm of $d_m^{-1/2} Z_o^m$. Using (4.1.1) in [18] to evaluate $P_m^{a,b}(1)$ and (4.3.3) in [18] to evaluate the L^2

norm of $P_m^{a,b}(x)$ one obtains

$$\begin{aligned} d_m &= (2m+a+b+1)\frac{\Gamma(b+1)}{\Gamma(a+1)\Gamma(a+b+2)} \\ &\times \frac{\Gamma(m+a+b+1)}{\Gamma(m+b+1)}\frac{\Gamma(m+a+1)}{\Gamma(m+1)} \\ &\approx m^{d-1}. \end{aligned}$$

□

5.3 Proofs of the Main Results

For the proofs of our main results we need some lemmas. In the next one we obtain an expression for the discrepancy that separates the contribution of the point distribution (and weights) from the contribution of the geometry of the set $B_r(x)$.

Lemma 5.3 *Let $D_r(x)$ be as in (5.1). Then*

$$\int_{\mathcal{M}} |D_r(x)|^2 \, d\mu(x) = \sum_{m=1}^{+\infty} \sum_{\ell=1}^{d_m} \left| \sum_{j=1}^{N} a_j Y_m^{\ell}(x_j) \right|^2 d_m^{-2} \left| \int_{B_r(o)} Z_o^m(y) d\mu(y) \right|^2 .$$

Proof By (5.6) we have

$$\int_{\mathcal{M}} |D_r(x)|^2 \, d\mu(x) = \sum_{m=0}^{+\infty} \int_{\mathcal{M}} |\mathbb{P}_m D_r(x)|^2 d\mu(x)$$

(here $\mathbb{P}_m$ denotes the orthogonal projection operator of $L^2(\mathcal{M})$ onto $\mathcal{H}_m$). Let $g \in G$ such that $gx_j = o$. Since $\chi_{B_r(o)}$ is a zonal function, then

$$\begin{aligned} \chi_{B_r(x)}(x_j) = \chi_{B_r(x_j)}(x) = \chi_{B_r(o)}(gx) &= \sum_{m=0}^{+\infty} d_m^{-1} \int_{B_r(o)} Z_o^m(y) d\mu(y) Z_o^m(gx) \\ &= \sum_{m=0}^{+\infty} d_m^{-1} \int_{B_r(o)} Z_o^m(y) d\mu(y) Z_{x_j}^m(x). \end{aligned}$$

Since $Z_{x_j}^m(x) \in \mathcal{H}_m$, we have

$$\mathbb{P}_m \left(\chi_{B_r(\cdot)}(x_j)\right)(x) = d_m^{-1} \int_{B_r(o)} Z_o^m(y) d\mu(y) Z_{x_j}^m(x).$$

Overall we obtain

$$\mathbb{P}_m D_r(x) = \sum_{j=1}^{N} a_j d_m^{-1} \int_{B_r(o)} Z_o^m(y) d\mu(y) Z_{x_j}^m(x) - \delta_0(m)\mu(B_r(o)).$$

In particular $\mathbb{P}_0 D_r(x) = 0$ and for $m > 0$

$$\begin{aligned}\mathbb{P}_m D_r(x) &= d_m^{-1} \sum_{j=1}^{N} a_j \int_{B_r(o)} Z_o^m(y) d\mu(y) Z_{x_j}^m(x) \\ &= d_m^{-1} \sum_{\ell=1}^{d_m} \left(\sum_{j=1}^{N} a_j \overline{Y_m^\ell(x_j)} \right) \int_{B_r(o)} Z_o^m(y) d\mu(y) Y_m^\ell(x)\end{aligned}$$

Finally

$$\int_{\mathcal{M}} |D_r(x)|^2 d\mu(x) = \sum_{m=1}^{+\infty} \sum_{\ell=1}^{d_m} \left| \sum_{j=1}^{N} a_j Y_m^\ell(x_j) \right|^2 d_m^{-2} \left| \int_{B_r(o)} Z_o^m(y) d\mu(y) \right|^2.$$

□

In the next results we estimate the quantities

$$\sum_{\ell=1}^{d_m} \left| \sum_{j=1}^{N} a_j Y_m^\ell(x_j) \right|^2 \quad \text{and} \quad d_m^{-2} \left| \int_{B_r(o)} Z_o^m(y) d\mu(y) \right|^2.$$

Proposition 5.1 *Let $m_0 > 0$, then there exist $C_0, C_1 > 0$ such that for every $X \geqslant m_0$ we have*

$$\sum_{m=m_0}^{X} \sum_{\ell=1}^{d_m} \left| \sum_{j=1}^{N} a_j Y_m^\ell(x_j) \right|^2 \geqslant C_1 \sum_{j=1}^{N} a_j^2 X^d - C_0.$$

Proof By the Cassels-Montgomery inequality for manifolds (see [5, Theorem 1] or [22, Theorem 9]) along with the fact that $\sum_{m=0}^{X} d_m \approx \sum_{m=0}^{X} m^{d-1} \approx X^d$, we have

$$\sum_{m=0}^{X} \sum_{\ell=1}^{d_m} \left| \sum_{j=1}^{N} a_j Y_m^\ell(x_j) \right|^2 \geqslant C_1 \sum_{j=1}^{N} a_j^2 X^d.$$

Since

$$\sum_{m=0}^{m_0-1}\sum_{\ell=1}^{d_m}\left|\sum_{j=1}^{N}a_jY_m^\ell(x_j)\right|^2 \leqslant \max_{\substack{0\leqslant m\leqslant m_0-1\\ \ell=1,\dots,d_m}}\|Y_m^\ell\|_\infty^2\sum_{m=0}^{m_0-1}\sum_{\ell=1}^{d_m}\left|\sum_{j=1}^{N}a_j\right|^2$$

$$= \max_{\substack{0\leqslant m\leqslant m_0-1\\ \ell=1,\dots,d_m}}\|Y_m^\ell\|_\infty^2\sum_{m=0}^{m_0-1}d_m = C_0,$$

we have

$$\sum_{m=m_0}^{X}\sum_{\ell=1}^{d_m}\left|\sum_{j=1}^{N}a_jY_m^\ell(x_j)\right|^2 \geqslant C_1\sum_{j=1}^{N}a_j^2X^d - C_0.$$

□

Lemma 5.4 *For any* $0\leqslant r\leqslant\pi$ *and for any* $m\geqslant 1$ *we have*

$$\int_{B_r(o)}Z_o^m(x)d\mu(x) = \frac{c(a,b)d_m}{mP_m^{a,b}(1)}P_{m-1}^{a+1,b+1}(\cos r)\left(\sin\left(\frac{r}{2}\right)\right)^{2a+2}\left(\cos\left(\frac{r}{2}\right)\right)^{2b+2}.$$

Proof By (5.4), (5.8), and a change of variable,

$$\begin{aligned}\int_{B_r(o)}Z_o^m(x)d\mu(x) &= \frac{d_m}{P_m^{a,b}(1)}\int_0^r P_m^{a,b}(\cos t)A(t)dt\\
&= \frac{c(a,b)d_m}{P_m^{a,b}(1)}\int_0^r P_m^{a,b}(\cos t)\left(\sin\frac{t}{2}\right)^{2a+1}\left(\cos\frac{t}{2}\right)^{2b+1}dt\\
&= \frac{c(a,b)d_m}{2P_m^{a,b}(1)}\int_0^r P_m^{a,b}(\cos t)\left(\frac{1-\cos t}{2}\right)^a\\
&\quad\times\left(\frac{1+\cos t}{2}\right)^b\sin t\,dt\\
&= \frac{c(a,b)d_m}{2^{a+b+1}P_m^{a,b}(1)}\int_{\cos r}^1 P_m^{a,b}(x)(1-x)^a(1+x)^b dx.\end{aligned}$$

By Rodrigues' formula (see [18, (4.3.1)])

$$P_m^{a,b}(x)(1-x)^a(1+x)^b = -\frac{1}{2m}\frac{d}{dx}\left(P_{m-1}^{a+1,b+1}(x)(1-x)^{a+1}(1+x)^{b+1}\right), \tag{5.9}$$

thus

$$\int_{B_r(o)} Z_o^m(x)d\mu(x) = \frac{c(a,b)d_m}{2m2^{a+b+1}P_m^{a,b}(1)} P_{m-1}^{a+1,b+1}(\cos r)$$
$$\times (1-\cos r)^{a+1}(1+\cos r)^{b+1}$$
$$= \frac{c(a,b)d_m}{mP_m^{a,b}(1)} P_{m-1}^{a+1,b+1}(\cos r)\left(\sin\left(\frac{r}{2}\right)\right)^{2a+2}\left(\cos\left(\frac{r}{2}\right)\right)^{2b+2}.$$

□

In the following, to keep notation simple we set

$$M = m + \frac{a+b+1}{2}.$$

Lemma 5.5 *For all $\varepsilon > 0$ and for all integers $m \geqslant 1$,*

$$\int_{B_r(o)} Z_o^m(y)d\mu(y)$$
$$= c(a,b)d_m\Gamma(a+1)\left(\sin\left(\frac{r}{2}\right)\right)^{a+1}\left(\cos\left(\frac{r}{2}\right)\right)^{b+1} M^{-a-1}\left(\frac{r}{\sin r}\right)^{\frac{1}{2}} J_{a+1}(Mr)$$
$$+ d_m O(m^{-\frac{5}{2}-a}),$$

uniformly in $r \in [0, \pi-\varepsilon]$. Here J_{a+1} is the Bessel function of first type and order $a+1$.

Proof The following asymptotic expansion of the Jacobi polynomials is well known, (see [18, Theorem 8.21.12]),

$$\left(\sin\frac{r}{2}\right)^{a+1}\left(\cos\frac{r}{2}\right)^{b+1} P_{m-1}^{a+1,b+1}(\cos r) \tag{5.10}$$
$$= M^{-a-1}\frac{\Gamma(m+a+1)}{(m-1)!}\left(\frac{r}{\sin r}\right)^{\frac{1}{2}} J_{a+1}(Mr) + r^{\frac{1}{2}}O(m^{-\frac{3}{2}}).$$

Thus, by the previous lemma, recalling that (see (4.1.1) in [18])

$$P_m^{a,b}(1) = \frac{\Gamma(m+a+1)}{\Gamma(m+1)\Gamma(a+1)},$$

for every $r \in [0, \pi - \varepsilon]$ and $m \geqslant 1$,

$$\begin{aligned}
&\int_{B_r(o)} Z_o^m(y)d\mu(y) \\
&= \frac{c(a,b)d_m}{m P_m^{a,b}(1)} P_{m-1}^{a+1,b+1}(\cos r)\left(\sin\left(\frac{r}{2}\right)\right)^{2a+2}\left(\cos\left(\frac{r}{2}\right)\right)^{2b+2} \\
&= \frac{c(a,b)d_m}{m}\frac{\Gamma(m+1)\Gamma(a+1)}{\Gamma(m+a+1)}\left(\sin\left(\frac{r}{2}\right)\right)^{a+1}\left(\cos\left(\frac{r}{2}\right)\right)^{b+1} \\
&\quad \times M^{-a-1}\frac{\Gamma(m+a+1)}{(m-1)!}\left(\frac{r}{\sin r}\right)^{1/2} J_{a+1}(Mr) \\
&\quad + \frac{c(a,b)}{m}\frac{d_m}{P_m^{a,b}(1)}\left(\sin\left(\frac{r}{2}\right)\right)^{a+1}\left(\cos\left(\frac{r}{2}\right)\right)^{b+1} r^{\frac{1}{2}} O(m^{-\frac{3}{2}}) \\
&= c(a,b)d_m\Gamma(a+1)\left(\sin\left(\frac{r}{2}\right)\right)^{a+1}\left(\cos\left(\frac{r}{2}\right)\right)^{b+1} M^{-a-1}\left(\frac{r}{\sin r}\right)^{\frac{1}{2}} J_{a+1}(Mr) \\
&\quad + d_m O(m^{-\frac{5}{2}-a}).
\end{aligned}$$

□

The expression obtained in Lemma 5.3 shows that to estimate the discrepancy from below requires an estimate from below for the quantity

$$\left|\int_{B_r(o)} Z_o^m(y)d\mu(y)\right|^2.$$

Lemma 5.5 shows that this quantity may vanish due to the zeroes of the Bessel functions. Our next lemma shows that at least when $d \not\equiv 1 \pmod 4$ one can overcome this obstruction using two balls of different radii. See the proof of Theorem 2 in Chapter 6 in [13].

Lemma 5.6 *Assume that* $d \not\equiv 1 \pmod 4$. *Then for all* $0 < r < \pi/2$, *there exist positive constants* C *and* m_0 *such that for* $m \geqslant m_0$,

$$\left|\int_{B_r(o)} Z_o^m(x)d\mu(x)\right|^2 + \left|\int_{B_{2r}(o)} Z_o^m(x)d\mu(x)\right|^2 \geqslant C d_m^2 m^{-2a-3}.$$

Proof By the previous lemma, for all $m \geqslant 1$

$$\begin{aligned}
&\left|\int_{B_r(o)} Z_o^m(x)d\mu(x)\right|^2 + \left|\int_{B_{2r}(o)} Z_o^m(x)d\mu(x)\right|^2 \\
&\geqslant \left|c(a,b)d_m\Gamma(a+1)M^{-a-1}\right|^2
\end{aligned}$$

$$
\begin{aligned}
&\times \left\{ \left| \left(\sin\left(\frac{r}{2}\right)\right)^{a+1} \left(\cos\left(\frac{r}{2}\right)\right)^{b+1} \left(\frac{r}{\sin r}\right)^{\frac{1}{2}} J_{a+1}(Mr) \right|^2 \right. \\
&\left. + \left| (\sin r)^{a+1} (\cos r)^{b+1} \left(\frac{2r}{\sin 2r}\right)^{\frac{1}{2}} J_{a+1}(M2r) \right|^2 \right\} - d_m^2 O(m^{-4-2a}) \\
&\geqslant C d_m^2 m^{-2a-2} \left\{ |J_{a+1}(Mr)|^2 + |J_{a+1}(2Mr)|^2 \right\} - d_m^2 O(m^{-4-2a}).
\end{aligned}
$$

By the asymptotic expansion of the Bessel functions (see e.g. [17, Lemma 3.11, Chapter 4] or [20, p. 199]),

$$
J_{a+1}(w) = \sqrt{\frac{2}{\pi w}} \cos\left(w - (a+1)\frac{\pi}{2} - \frac{\pi}{4}\right) + O(w^{-\frac{3}{2}}), \tag{5.11}
$$

we have

$$
\begin{aligned}
&|J_{a+1}(Mr)|^2 + |J_{a+1}(2Mr)|^2 \\
&= \frac{2}{\pi Mr} \cos^2\left(Mr - (a+1)\frac{\pi}{2} - \frac{\pi}{4}\right) + \frac{2}{2\pi Mr} \cos^2\left(2Mr - (a+1)\frac{\pi}{2} - \frac{\pi}{4}\right) \\
&+ O(m^{-2}) \\
&\geqslant \frac{C}{m} \left(\cos^2\left(Mr - (a+1)\frac{\pi}{2} - \frac{\pi}{4}\right) + \cos^2\left(2Mr - (a+1)\frac{\pi}{2} - \frac{\pi}{4}\right)\right) + O(m^{-2}).
\end{aligned}
$$

Calling

$$
\omega = Mr - (a+1)\frac{\pi}{2} - \frac{\pi}{4},
$$

we have that

$$
\begin{aligned}
&\cos^2\left(Mr - (a+1)\frac{\pi}{2} - \frac{\pi}{4}\right) + \cos^2\left(2Mr - (a+1)\frac{\pi}{2} - \frac{\pi}{4}\right) \\
&= \cos^2(\omega) + \cos^2\left(2\omega + (a+1)\frac{\pi}{2} + \frac{\pi}{4}\right).
\end{aligned}
$$

We claim that this expression never vanishes. Assume the contrary, then $\omega = \frac{\pi}{2} + k\pi$ for some integer k, and $2\omega + (a+1)\frac{\pi}{2} + \frac{\pi}{4} = \frac{\pi}{2} + h\pi$ for some integer h. This implies

$$
\pi + 2k\pi + (a+1)\frac{\pi}{2} + \frac{\pi}{4} = \frac{\pi}{2} + h\pi.
$$

Recalling that $a = \frac{d-2}{2}$, this identity implies that $\frac{d-1}{4}$ is an integer. This contradicts $d \not\equiv 1 \pmod 4$. Therefore for all $\omega \in \mathbb{R}$,

$$\cos^2(\omega) + \cos^2\left(2\omega + (a+1)\frac{\pi}{2} + \frac{\pi}{4}\right) \geqslant c,$$

so that for large m

$$|J_{a+1}(Mr)|^2 + |J_{a+1}(2Mr)|^2 \geqslant \frac{C}{m}.$$

Finally,

$$\begin{aligned}\left|\int_{B_r(o)} Z_o^m(x)d\mu(x)\right|^2 + \left|\int_{B_{2r}(o)} Z_o^m(x)d\mu(x)\right|^2 &\geqslant Cm^{-2a-3}d_m^2 + d_m^2 O(m^{-2a-4}) \\ &\geqslant Cd_m^2 m^{-2a-3}.\end{aligned}$$

□

We can now prove Theorem 5.1.

Proof (of Theorem 5.1) Let m_0 as in Lemma 5.6 and let $X \geqslant m_0$. Then by Lemma 5.3, Proposition 5.1 and Lemma 5.6, we have

$$\begin{aligned}&\|D_r\|_2^2 + \|D_{2r}\|_2^2 \\ &= \sum_{m=1}^{+\infty} d_m^{-2}\left(\left|\int_{B_r(o)} Z_o^m(x)d\mu(x)\right|^2 + \left|\int_{B_{2r}(o)} Z_o^m(x)d\mu(x)\right|^2\right)\sum_{\ell=1}^{d_m}\left|\sum_{j=1}^{N} a_j Y_m^\ell(x_j)\right|^2 \\ &\geqslant \min_{m_0 \leqslant m \leqslant X}\left(d_m^{-2}\left(\left|\int_{B_r(o)} Z_o^m(x)d\mu(x)\right|^2 + \left|\int_{B_{2r}(o)} Z_o^m(x)d\mu(x)\right|^2\right)\right) \\ &\quad \times \sum_{m=m_0}^{X}\sum_{\ell=1}^{d_m}\left|\sum_{j=1}^{N} a_j Y_m^\ell(x_j)\right|^2 \\ &\geqslant C\left(\min_{m_0 \leqslant m \leqslant X} m^{-2a-3}\right)\left(C_1 X^d \sum_{j=1}^{N} a_j^2 - C_0\right) \\ &\geqslant C X^{-2a-3}\left(C_1 X^d \sum_{j=1}^{N} a_j^2 - C_0\right).\end{aligned}$$

Applying the Cauchy-Schwartz inequality to $\sum_{j=1}^{N} a_j = 1$ gives

$$\sum_{j=1}^{N} a_j^2 \geqslant \frac{1}{N}.$$

Let $N \geqslant N_0 = m_0^d C_1/(2C_0)$. Then, setting $X = \left[(2C_0C_1^{-1}N)^{1/d}\right] + 1$, we have $X \geqslant m_0$, $C_1 X^d \sum_{j=1}^{N} a_j^2 - C_0 \geqslant C_0$ and

$$\|D_r\|_2^2 + \|D_{2r}\|_2^2 \geqslant CN^{-1-\frac{1}{d}}. \tag{5.12}$$

Let now $N < N_0$ and let us consider the points and weights

$$\widetilde{x}_j = \begin{cases} x_j & 1 \leqslant j \leqslant N-1, \\ x_N & N \leqslant j \leqslant N_0, \end{cases} \qquad \widetilde{a}_j = \begin{cases} a_j & 1 \leqslant j \leqslant N-1, \\ \dfrac{a_N}{N_0 - N + 1} & N \leqslant j \leqslant N_0. \end{cases}$$

Since the discrepancy $\widetilde{D}_r$ of the points $\{\widetilde{x}_j\}_{j=1}^{N_0}$ and weights $\{\widetilde{a}_j\}_{j=1}^{N_0}$ coincides with the discrepancy D_r of the points $\{x_j\}_{j=1}^{N}$ and weights $\{a_j\}_{j=1}^{N}$, applying (5.12) to $\widetilde{D}_r$ gives

$$\|D_r\|_2^2 + \|D_{2r}\|_2^2 \geqslant CN_0^{-1-\frac{1}{d}} = C\left(\frac{N}{N_0}\right)^{1+\frac{1}{d}} N^{-1-\frac{1}{d}} \geqslant C\left(\frac{1}{N_0}\right)^{1+\frac{1}{d}} N^{-1-\frac{1}{d}}$$

also when $1 \leqslant N < N_0$. □

We now begin the discussion leading to the proof of Theorem 5.2. In Lemma 5.6 we proved that if $d \not\equiv 1 \pmod 4$, then for all $0 < r < \pi/2$, there exist positive constants C and m_0 such that for $m \geqslant m_0$,

$$\left|\int_{B_r(o)} Z_o^m(x)d\mu(x)\right|^2 + \left|\int_{B_{2r}(o)} Z_o^m(x)d\mu(x)\right|^2 \geqslant Cd_m^2 m^{-2a-3}.$$

In order to do it, we showed that for every $M = m + \dfrac{a+b+1}{2}$ either $|J_{a+1}(Mr)|$ or $|J_{a+1}(2Mr)|$ is greater than $cm^{-\frac{1}{2}}$, that is either Mr or $2Mr$ is away from the zeroes of the Bessel function J_{a+1}. On the other hand, a precise analysis on the position of the zeroes of the Jacobi polynomials and a measure theoretic argument allow to obtain, in any dimension d and for almost every r, the worse estimate

$$\left|\int_{B_r(o)} Z_o^m(x)d\mu(x)\right|^2 \geq cd_m^2 m^{-2a-5-\delta}.$$

Notice also that not only this estimate is worse, but actual values of r for which it is satisfied are unknown. Recent results show that properly choosen values of r give the optimal estimate $m^{-1/2}$ if $d \not\equiv 1 \mod 4$. See [23, 24].

We need the following result of Frenzen and Wong (see [8, Corollary 2]) on the zeroes of the Jacobi polynomials.

Theorem 5.3 (Frenzen-Wong) *Let $a \geqslant -\frac{3}{2}$, $a+b \geqslant -3$, and let $0 < \theta_{m-1,1} < \theta_{m-1,2} < \ldots < \theta_{m-1,m-1} < \pi$ be the zeros of $P_{m-1}^{a+1,b+1}(\cos\theta)$. Then, as $m \to +\infty$, we have*

$$\theta_{m-1,\ell} = \frac{j_{a+1,\ell}}{M} + \frac{1}{M^2}\left\{\left((a+1)^2 - \frac{1}{4}\right)\frac{1 - t\cot t}{2t} - \frac{(a+1)^2 - (b+1)^2}{4}\tan\frac{t}{2}\right\} + t^2 O\left(\frac{1}{m^3}\right)$$

where $j_{a+1,\ell}$ is the ℓ-th positive zero of the Bessel function $J_{a+1}(x)$ and $t = j_{a+1,\ell}/M$. The O-term is uniformly bounded for all values of $\ell = 1, \ldots, [\gamma m]$, where $\gamma \in (0, 1)$.

Lemma 5.7 *Let $0 < \gamma_1 < \gamma_2 < 1$. For every $m \geqslant 1$ and every $[\gamma_1 m] \leqslant \ell \leqslant [\gamma_2 m]$ we have*

$$\theta_{m-1,\ell} = \frac{\ell\pi + a\frac{\pi}{2} + \frac{\pi}{4}}{M} + O\left(\frac{1}{m^2}\right).$$

Proof This follows directly from McMahon's expansion (see e.g. (1.5) in [7])

$$j_{a+1,\ell} = \ell\pi + a\frac{\pi}{2} + \frac{\pi}{4} + O\left(\frac{1}{\ell}\right)$$

and Theorem 5.3. □

Lemma 5.8 *There exists $m_0 > 0$ such that for every $\delta > 0$ and for almost every $r \in (0, \pi)$ there exists a constant $C > 0$ such that for $m \geqslant m_0$,*

$$\left|\sin^{2a+2}\left(\frac{r}{2}\right)\cos^{2b+2}\left(\frac{r}{2}\right)P_{m-1}^{a+1,b+1}(\cos r)\right| \geqslant Cm^{-\frac{3}{2}-\delta}.$$

Proof Let $\theta_\ell = \theta_{m-1,\ell}$, for $\ell = 1, \ldots, m-1$, be the positive zeros of $P_{m-1}^{a+1,b+1}(\cos\theta)$, and let

$$\mathcal{P}_{m-1}^{a+1,b+1}(r) = \sin^{2a+2}\left(\frac{r}{2}\right)\cos^{2b+2}\left(\frac{r}{2}\right)P_{m-1}^{a+1,b+1}(\cos r).$$

Let $\varepsilon > 0$. Let γ_1 and γ_2 be such that for m sufficiently large

$$\frac{\varepsilon}{2} < \frac{[\gamma_1 m]\pi + a\frac{\pi}{2} + \frac{\pi}{4}}{M} < \varepsilon$$

and

$$\pi - \varepsilon < \frac{[\gamma_2 m]\pi + a\frac{\pi}{2} + \frac{\pi}{4}}{M} < \pi - \frac{\varepsilon}{2}.$$

Then, if $\eta \in (0, 1)$,

$$\int_{\varepsilon}^{\pi-\varepsilon} \left|\mathcal{P}_{m-1}^{a+1,b+1}(r)\right|^{-1+\eta} dr \leqslant \sum_{\ell=[\gamma_1 m]}^{[\gamma_2 m]-1} \int_{\theta_\ell}^{\theta_{\ell+1}} \left|\mathcal{P}_{m-1}^{a+1,b+1}(r)\right|^{-1+\eta} dr.$$

By (5.9) we have

$$\frac{d}{dr}\mathcal{P}_{m-1}^{a+1,b+1}(r) = \frac{1}{2} m \mathcal{P}_m^{a,b}(r)\sin(r).$$

Hence, if $\varepsilon/2 \leqslant r \leqslant \pi - \varepsilon/2$, then by (5.10)

$$\begin{aligned}\frac{d}{dr}\mathcal{P}_{m-1}^{a+1,b+1}(r) &= m \sin^{2a+1}\left(\frac{r}{2}\right)\cos^{2b+1}\left(\frac{r}{2}\right) P_m^{a,b}(\cos r)\\ &= m \sin^{a+1}\left(\frac{r}{2}\right)\cos^{b+1}\left(\frac{r}{2}\right) M^{-a}\frac{\Gamma(m+a+1)}{m!}\left(\frac{r}{\sin r}\right)^{\frac{1}{2}} J_a(Mr)\\ &+ \sin^{a+1}\left(\frac{r}{2}\right)\cos^{b+1}\left(\frac{r}{2}\right) r^{\frac{1}{2}} O(m^{-\frac{1}{2}}).\end{aligned}$$

Therefore

$$\left|\frac{d}{dr}\mathcal{P}_{m-1}^{a+1,b+1}(r)\right| \geqslant cm\,|J_a(Mr)| - O(m^{-\frac{1}{2}}). \tag{5.13}$$

Let $\tau > 0$ and let $r \in \left(\theta_\ell - \frac{\tau}{M}, \theta_\ell + \frac{\tau}{M}\right)$. Then by Lemma 5.7

$$Mr - a\frac{\pi}{2} - \frac{\pi}{4} \in \left(\ell\pi + O\left(\frac{1}{m}\right) - \tau, \ell\pi + O\left(\frac{1}{m}\right) + \tau\right),$$

so that

$$\left|\cos\left(Mr - a\frac{\pi}{2} - \frac{\pi}{4}\right)\right| \geqslant c.$$

By the asymptotic expansion of the Bessel function (5.11) we therefore obtain

$$|J_a(Mr)| \geqslant \sqrt{\frac{2}{\pi Mr}} \left|\cos\left(Mr - a\frac{\pi}{2} - \frac{\pi}{4}\right)\right| - O(m^{-\frac{3}{2}}) \geqslant cm^{-\frac{1}{2}},$$

and by (5.13)

$$\left|\frac{d}{dr}\mathcal{P}_{m-1}^{a+1,b+1}(r)\right| \geqslant cm^{\frac{1}{2}}.$$

Thus,

$$\begin{aligned}
&\int_{\theta_\ell}^{\theta_{\ell+1}} \left|\mathcal{P}_{m-1}^{a+1,b+1}(r)\right|^{-1+\eta} dr \\
&= \int_{\theta_\ell}^{\theta_\ell+\frac{\tau}{M}} \left|\mathcal{P}_{m-1}^{a+1,b+1}(r)\right|^{-1+\eta} dr + \int_{\theta_\ell+\frac{\tau}{M}}^{\theta_{\ell+1}-\frac{\tau}{M}} \left|\mathcal{P}_{m-1}^{a+1,b+1}(r)\right|^{-1+\eta} dr \\
&\quad + \int_{\theta_{\ell+1}-\frac{\tau}{M}}^{\theta_{\ell+1}} \left|\mathcal{P}_{m-1}^{a+1,b+1}(r)\right|^{-1+\eta} dr \\
&\leqslant \int_{\theta_\ell}^{\theta_\ell+\frac{\tau}{M}} \left|cm^{\frac{1}{2}}(r-\theta_\ell)\right|^{-1+\eta} dr + \int_{\theta_\ell+\frac{\tau}{M}}^{\theta_{\ell+1}-\frac{\tau}{M}} \left|cm^{\frac{1}{2}}\left(\frac{\tau}{M}\right)\right|^{-1+\eta} dr \\
&\quad + \int_{\theta_{\ell+1}-\frac{\tau}{M}}^{\theta_{\ell+1}} \left|cm^{\frac{1}{2}}(\theta_{\ell+1}-r)\right|^{-1+\eta} dr \\
&\leqslant cm^{-\frac{1}{2}-\frac{\eta}{2}}.
\end{aligned}$$

Therefore

$$\begin{aligned}
\int_{\varepsilon}^{\pi-\varepsilon} \left|\mathcal{P}_{m-1}^{a+1,b+1}(r)\right|^{-1+\eta} dr &\leqslant \sum_{\ell=[\gamma_1 m]}^{[\gamma_2 m]-1} \int_{\theta_\ell}^{\theta_{\ell+1}} \left|\mathcal{P}_{m-1}^{a+1,b+1}(r)\right|^{-1+\eta} dr \\
&\leqslant cm \cdot m^{-\frac{1}{2}-\frac{\eta}{2}} = cm^{\frac{1}{2}-\frac{\eta}{2}}.
\end{aligned}$$

Finally

$$\begin{aligned}
\int_{\varepsilon}^{\pi-\varepsilon} \left(\sum_{m=1}^{+\infty} m^{-\sigma} \left|\mathcal{P}_{m-1}^{a+1,b+1}(r)\right|^{-1+\eta}\right) dr &= \sum_{m=1}^{+\infty} \int_{\varepsilon}^{\pi-\varepsilon} m^{-\sigma} \left|\mathcal{P}_{m-1}^{a+1,b+1}(r)\right|^{-1+\eta} dr \\
&\leqslant c \sum_{m=1}^{+\infty} m^{\frac{1}{2}-\sigma-\frac{\eta}{2}}.
\end{aligned}$$

If $\sigma > 3/2 - \eta/2$

$$\sum_{m=1}^{+\infty} m^{-\sigma} \left|\mathcal{P}_{m-1}^{a+1,b+1}(r)\right|^{-1+\eta}$$

converges for almost every $r \in (\varepsilon, \pi - \varepsilon)$ so that for almost every $r \in (\varepsilon, \pi - \varepsilon)$ we have

$$cm^{-\frac{\sigma}{1-\eta}} \leqslant \left|\mathcal{P}_{m-1}^{a+1,b+1}(r)\right|.$$

□

We are ready to prove Theorem 5.2.

***Proof** (of Theorem 5.2)* By Lemmas 5.4 and 5.8, there exists m_0 such that for every $\delta > 0$, for almost every $r \in (0, \pi)$ and for $m \geqslant m_0$ we obtain

$$\begin{aligned}\left|\int_{B_r(o)} Z_o^m(y)d\mu(y)\right|^2 &= \left|\frac{c(a,b)d_m}{m P_m^{a,b}(1)} P_{m-1}^{a+1,b+1}(\cos r)\right. \\ &\quad \left.\times \left(\sin\left(\frac{r}{2}\right)\right)^{2a+2} \left(\cos\left(\frac{r}{2}\right)\right)^{2b+2}\right|^2 \\ &\geqslant c\left|\frac{d_m}{m P_m^{a,b}(1)} m^{-\frac{3}{2}-\delta}\right|^2 \geqslant c d_m^2 m^{-2a-5-2\delta}.\end{aligned}$$

Then, by Lemma 5.3 and Proposition 5.1, for every $X \geqslant m_0$ we have

$$\begin{aligned}&\int_{\mathcal{M}} |D_r(x)|^2 \, d\mu(x) \\ &\geqslant \sum_{m=m_0}^{X} \sum_{\ell=1}^{d_m} \left|\sum_{j=1}^{N} a_j Y_m^\ell(x_j)\right|^2 \min_{m_0 \leqslant m \leqslant X} d_m^{-2} \left|\int_{B_r(o)} Z_o^m(y)d\mu(y)\right|^2 \\ &\geqslant c\left(C_1 X^d \sum_{j=1}^{N} a_j^2 - C_0\right) \min_{m_0 \leqslant m \leqslant X} \{m^{-2a-5-2\delta}\} \\ &\geqslant c\left(C_1 X^d \frac{1}{N} - C_0\right) X^{-3-2\delta-d}.\end{aligned}$$

Let $N \geqslant N_0 = m_0^d C_1/(2C_0)$. Then, setting $X = \left[(2C_0 C_1^{-1} N)^{1/d}\right] + 1$, we have $X \geqslant m_0$, $C_1 X^d N^{-1} - C_0 \geqslant C_0$ and

$$\int_{\mathcal{M}} |D_r(x)|^2 d\mu(x) \geqslant cN^{-1-\frac{3}{d}-\frac{2\delta}{d}}$$

for all $N \geqslant N_0$. The same argument used in the proof of Theorem 5.1 gives the result for every $N \geqslant 1$. □

5.4 Cubature Formulas

For a given $N \geqslant 1$, let $\{a_j\}_{j=1}^N$ be a set of weights and $\{x_j\}_{j=1}^N$ be a set of points in $\mathcal{M}$. We say that $\left(\{a_j\}_{j=1}^N, \{x_j\}_{j=1}^N\right)$ is a cubature of strength X on $\mathcal{M}$ if

$$\int_{\mathcal{M}} P(x)d\mu(x) = \sum_{j=1}^{N} a_j P(x_j)$$

for every

$$P \in \bigoplus_{0 \leqslant m \leqslant X} \mathcal{H}_m.$$

In the following theorem we will show that under suitable assumptions a cubature of strength $\kappa N^{1/d}$ has optimal discrepancy. This was already observed by M. Skriganov in [16, Corollary 2.1], although in his result the discrepancy contains an integration in the radius r of the balls with respect to an absolutely continuous measure in $[0, \pi]$. The same technique actually gives a stronger estimate, which is uniform in the radius r.

Theorem 5.4 *Let A, B, ε and κ be positive constants. There exists $c > 0$ such that if $N \geqslant 1$, and $\left(\{a_j\}_{j=1}^N, \{x_j\}_{j=1}^N\right)$ gives a cubature of strength $\kappa N^{1/d}$ satisfying*

$$0 \leqslant a_j \leqslant \frac{A}{N}, \tag{5.14}$$

and for every $x \in \mathcal{M}$

$$\#\left\{j : d(x_j, x) \leqslant N^{-\frac{1}{d}}\right\} \leqslant B, \tag{5.15}$$

then, for every $r \in [0, \pi - \varepsilon]$ we have

$$\int_{\mathcal{M}} |D_r(x)|^2 d\mu(x) \leqslant cN^{-1-\frac{1}{d}}. \tag{5.16}$$

Observe that cubatures as required by the above theorem do exist. Indeed, in [6] it is proved that for every $A \geqslant 1$ there exists a constant $C(A, \mathcal{M}) > 0$, depending only on A and $\mathcal{M}$, such that if $N \geqslant C(A, \mathcal{M})X^d$ and the weights $\{a_j\}_{j=1}^N$ satisfy the conditions (5.14) and

$$\sum_{j=1}^{N} a_j = 1,$$

then there is a choice of points $\{x_j\}_{j=1}^N$ such that $\left(\{a_j\}_{j=1}^N, \{x_j\}_{j=1}^N\right)$ is a cubature of strength X. The construction starts with a well separated set of points $\{y_j\}_{j=1}^N$ and ends with a new set of points $\{x_j\}_{j=1}^N$ in such a way that each x_j has distance at most $c_1(A, \mathcal{M})N^{-1/d}$ from y_j. This implies that the number of points x_j contained in any ball of radius $N^{-1/d}$ is uniformly bounded so that (5.15) is satisfied with a constant B depending on $\mathcal{M}$ and A. It suffices to apply the above construction with $X = \kappa N^{1/d}$ ($\kappa \leqslant C(A, \mathcal{M})^{-1/d}$). See also [10] for the case of equal weights, and [2] for the original work on the sphere by Bondarenko, Radchenko and Vyazovska.

Proof First of all we observe that it is enough to prove (5.16) for N sufficiently large. Let N_0 be a positive constant to be chosen later and assume $N \geqslant N_0$. Since $\left(\{a_j\}_{j=1}^N, \{x_j\}_{j=1}^N\right)$ gives a cubature of strength $\kappa N^{1/d}$, for every $m \leqslant \kappa N^{1/d}$ we have

$$\sum_{j=1}^{N} a_j Y_m^{\ell}(x_j) = 0.$$

Then by Lemmas 5.3 and 5.4 we have

$$\begin{aligned}
&\int_{\mathcal{M}} |D(x)|^2 d\mu(x) \\
&\quad = \sum_{m>\kappa N^{\frac{1}{d}}} \sum_{\ell=1}^{d_m} \left| \sum_{j=1}^{N} a_j Y_m^{\ell}(x_j) \right|^2 \\
&\quad \times \left| c(a,b) \frac{P_{m-1}^{a+1,b+1}(\cos r)}{m P_m^{a,b}(1)} \left(\sin\left(\frac{r}{2}\right)\right)^{2a+2} \left(\cos\left(\frac{r}{2}\right)\right)^{2b+2} \right|^2.
\end{aligned}$$

We want to show that there exist $c > 0$ and $\zeta > 0$ such that for every $r \in [0, \pi - \varepsilon]$, for every $N \geqslant 1$ and for every $m > \kappa N^{1/d}$

$$\begin{aligned}
&\left| P_{m-1}^{a+1,b+1}(\cos r) \left(\sin\left(\frac{r}{2}\right)\right)^{2a+2} \left(\cos\left(\frac{r}{2}\right)\right)^{2b+2} \right|^2 \qquad (5.17) \\
&\leqslant cN \int_0^{\zeta N^{-\frac{1}{d}}} \left| P_{m-1}^{a+1,b+1}(\cos u) \left(\sin\left(\frac{u}{2}\right)\right)^{2a+2} \left(\cos\left(\frac{u}{2}\right)\right)^{2b+2} \right|^2 du.
\end{aligned}$$

By (5.10) we have the expansion

$$\begin{aligned}
&\left(\sin\frac{r}{2}\right)^{a+\frac{3}{2}}\left(\cos\frac{r}{2}\right)^{b+\frac{3}{2}} P_{m-1}^{a+1,b+1}(\cos r)\\
&= M^{-a-1}\frac{\Gamma(m+a+1)}{\sqrt{2}(m-1)!}\sqrt{r}J_{a+1}(Mr) + O(rm^{-\frac{3}{2}})
\end{aligned}$$

uniformly in $r \in [0, \pi - \varepsilon]$ and $m > \kappa N^{\frac{1}{d}}$. Hence

$$\begin{aligned}
&\left|P_{m-1}^{a+1,b+1}(\cos r)\left(\sin\left(\frac{r}{2}\right)\right)^{2a+2}\left(\cos\left(\frac{r}{2}\right)\right)^{2b+2}\right|^2\\
&\leqslant \left|P_{m-1}^{a+1,b+1}(\cos r)\left(\sin\left(\frac{r}{2}\right)\right)^{a+\frac{3}{2}}\left(\cos\left(\frac{r}{2}\right)\right)^{b+\frac{3}{2}}\right|^2\\
&\leqslant \left|M^{-a-1}\frac{\Gamma(m+a+1)}{\sqrt{2}(m-1)!}\sqrt{r}J_{a+1}(Mr)\right|^2 + O(rm^{-2})\\
&\leqslant cm^{-1} + O(rm^{-2}) \leqslant cm^{-1}.
\end{aligned}$$

Let $0 < \zeta < \pi N_0^{\frac{1}{d}}$. For every $m > \kappa N^{\frac{1}{d}}$,

$$\begin{aligned}
&\int_0^{\zeta N^{-\frac{1}{d}}}\left|P_{m-1}^{a+1,b+1}(\cos u)\left(\sin\left(\frac{u}{2}\right)\right)^{2a+2}\left(\cos\left(\frac{u}{2}\right)\right)^{2b+2}\right|^2 du\\
&= \int_0^{\zeta N^{-\frac{1}{d}}}\left|P_{m-1}^{a+1,b+1}(\cos u)\left(\sin\left(\frac{u}{2}\right)\right)^{a+\frac{3}{2}}\left(\cos\left(\frac{u}{2}\right)\right)^{b+\frac{3}{2}}\right|^2\\
&\quad\times\left(\sin\left(\frac{u}{2}\right)\right)^{2a+1}\left(\cos\left(\frac{u}{2}\right)\right)^{2b+1} du\\
&\geqslant c\int_{\zeta N^{-\frac{1}{d}}/2}^{\zeta N^{-\frac{1}{d}}}\left|P_{m-1}^{a+1,b+1}(\cos u)\left(\sin\left(\frac{u}{2}\right)\right)^{a+\frac{3}{2}}\left(\cos\left(\frac{u}{2}\right)\right)^{b+\frac{3}{2}}\right|^2 u^{2a+1}du\\
&\geqslant c\int_{\zeta N^{-\frac{1}{d}}/2}^{\zeta N^{-\frac{1}{d}}}\left|\sqrt{u}J_{a+1}(Mu) + O(um^{-\frac{3}{2}})\right|^2 u^{2a+1}du\\
&\geqslant c\int_{\zeta N^{-\frac{1}{d}}/2}^{\zeta N^{-\frac{1}{d}}}\left|\sqrt{u}J_{a+1}(Mu)\right|^2 u^{2a+1}du\\
&\qquad - c\int_{\zeta N^{-\frac{1}{d}}/2}^{\zeta N^{-\frac{1}{d}}}\left|\sqrt{u}J_{a+1}(Mu)O(um^{-\frac{3}{2}})\right| u^{2a+1}du\\
&=\mathcal{A} - \mathcal{B}.
\end{aligned}$$

To estimate $\mathcal{A}$ we use (5.11). We have

$$\begin{aligned}
\mathcal{A} &= c\int_{\zeta N^{-\frac{1}{d}}/2}^{\zeta N^{-\frac{1}{d}}} \left|\sqrt{u}J_{a+1}(Mu)\right|^2 u^{2a+1}du \\
&= c\int_{\zeta N^{-\frac{1}{d}}/2}^{\zeta N^{-\frac{1}{d}}} \left|\left(M^{-\frac{1}{2}}\cos\left(Mu-(a+1)\frac{\pi}{2}-\frac{\pi}{4}\right)+O(m^{-\frac{3}{2}}u^{-1})\right)\right|^2 u^{2a+1}du \\
&\geqslant cm^{-1}\int_{\zeta N^{-\frac{1}{d}}/2}^{\zeta N^{-\frac{1}{d}}} \left|\cos\left(Mu-(a+1)\frac{\pi}{2}-\frac{\pi}{4}\right)\right|^2 u^{d-1}du \\
&\quad - c\int_{\zeta N^{-\frac{1}{d}}/2}^{\zeta N^{-\frac{1}{d}}} m^{-2}u^{-1}u^{2a+1}du \\
&\geqslant cm^{-1}(\zeta N^{-\frac{1}{d}})^{d-1}\int_{\zeta N^{-\frac{1}{d}}/2}^{\zeta N^{-\frac{1}{d}}} \left|\cos\left(Mu-(a+1)\frac{\pi}{2}-\frac{\pi}{4}\right)\right|^2 du \\
&\quad - cm^{-2}(\zeta N^{-\frac{1}{d}})^{d-1} \\
&\geqslant cm^{-2}(\zeta N^{-\frac{1}{d}})^{d-1}\int_{M\zeta N^{-\frac{1}{d}}/2}^{M\zeta N^{-\frac{1}{d}}} \left|\cos\left(v-(a+1)\frac{\pi}{2}-\frac{\pi}{4}\right)\right|^2 dv \\
&\quad - cm^{-2}(\zeta N^{-\frac{1}{d}})^{d-1} \\
&\geqslant cm^{-2}(\zeta N^{-\frac{1}{d}})^{d-1}M\zeta N^{-\frac{1}{d}} - cm^{-2}(\zeta N^{-\frac{1}{d}})^{d-1} \\
&\geqslant cm^{-1}\zeta^{d-1}N^{-1}(\zeta - cm^{-1}N^{\frac{1}{d}}) \geqslant cm^{-1}\zeta^{d-1}N^{-1}(\zeta - c\kappa^{-1}) \\
&\geqslant cm^{-1}\zeta^{d-1}N^{-1}
\end{aligned}$$

for $\zeta > 2\kappa^{-1}c$ (and therefore for N_0 large enough). Now we estimate $\mathcal{B}$:

$$\begin{aligned}
\mathcal{B} &= c\int_{\zeta N^{-\frac{1}{d}}/2}^{\zeta N^{-\frac{1}{d}}} \left|\sqrt{u}J_{a+1}(Mu)O(um^{-\frac{3}{2}})\right| u^{2a+1}du \\
&\leqslant c\int_{\zeta N^{-\frac{1}{d}}/2}^{\zeta N^{-\frac{1}{d}}} m^{-2}u^{d}du \leqslant cm^{-2}(\zeta N^{-\frac{1}{d}})^{d+1} = cm^{-2}\zeta^{d+1}N^{-1-\frac{1}{d}}.
\end{aligned}$$

It follows that

$$\int_0^{\zeta N^{-\frac{1}{d}}} \left| P_{m-1}^{a+1,b+1}(\cos u) \left(\sin\left(\frac{u}{2}\right)\right)^{2a+2} \left(\cos\left(\frac{u}{2}\right)\right)^{2b+2} \right|^2 du$$

$$\geqslant cm^{-1}\zeta^{d-1}N^{-1} - cm^{-2}\zeta^{d+1}N^{-1-\frac{1}{d}} = c\zeta^{d-1}m^{-1}N^{-1}\left(1 - cm^{-1}\zeta^2 N^{-\frac{1}{d}}\right)$$

$$\geqslant c\zeta^{d-1}m^{-1}N^{-1}\left(1 - cN^{-\frac{2}{d}}\right) \geqslant cm^{-1}N^{-1},$$

and (5.17) is proved. Thus,

$$\int_{\mathcal{M}} |D_r(x)|^2 d\mu(x)$$

$$= \sum_{m > \kappa N^{\frac{1}{d}}} \sum_{\ell=1}^{d_m} \left| \sum_{j=1}^{N} a_j Y_m^\ell(x_j) \right|^2$$

$$\times \left| c(a,b) \frac{P_{m-1}^{a+1,b+1}(\cos r)}{m P_m^{a,b}(1)} \left(\sin\left(\frac{r}{2}\right)\right)^{2a+2} \left(\cos\left(\frac{r}{2}\right)\right)^{2b+2} \right|^2$$

$$\leqslant cN \sum_{m > \kappa N^{\frac{1}{d}}} \sum_{\ell=1}^{d_m} \left| \sum_{j=1}^{N} a_j Y_m^\ell(x_j) \right|^2$$

$$\times \int_0^{\zeta N^{-\frac{1}{d}}} \left| c(a,b) \frac{P_{m-1}^{a+1,b+1}(\cos u)}{m P_m^{a,b}(1)} \left(\sin\left(\frac{u}{2}\right)\right)^{2a+2} \left(\cos\left(\frac{u}{2}\right)\right)^{2b+2} \right|^2 du$$

$$= cN \int_0^{\zeta N^{-\frac{1}{d}}} \int_{\mathcal{M}} |D_u(x)|^2 \, d\mu(x) du.$$

Since $u \leqslant \zeta N^{-1/d}$ using (5.15) and noticing that by Lemma 5.1 a ball of radius u can be covered by $\frac{c_2(d,d_0)}{c_1(d,d_0)}(2(\zeta+1))^d$ balls of radius $N^{-1/d}$ we have

$$|D_u(x)| = \left| \sum_{j=1}^{N} a_j \chi_{B_u(x)}(x_j) - \mu(B_u(x)) \right|$$

$$\leqslant \frac{A}{N} \sum_{j=1}^{N} \chi_{B_u(x)}(x_j) + \mu(B_u(x))$$

$$\leqslant \frac{A}{N} \frac{c_2(d,d_0)}{c_1(d,d_0)}(2(\zeta+1))^d B + \frac{c_2(d,d_0)\zeta^d}{N} \leqslant \frac{c}{N},$$

so that

$$\int_{\mathcal{M}} |D_u(x)|^2 d\mu(x) \leqslant \frac{C}{N^2}.$$

Finally,

$$\int_{\mathcal{M}} |D_r(x)|^2 \, d\mu(x) \leqslant cN \int_0^{\zeta N^{-\frac{1}{d}}} \int_{\mathcal{M}} |D_u(x)|^2 \, d\mu(x) du$$
$$\leqslant cN \int_0^{\zeta N^{-\frac{1}{d}}} \frac{C}{N^2} du = cN^{-1-\frac{1}{d}}.$$

□

Acknowledgments The authors wish to thank L. Colzani, A. Monguzzi and G. Travaglini for several helpful discussions on the subject of this paper.

We, the authors, belong to different mathematical generations, but we all stand on the shoulders of a giant, Guido Weiss. L. Brandolini visited Guido in St. Louis several times, B. Gariboldi was a student of Guido's first Italian Ph.D. student, Leonardo Colzani, and G. Gigante was Guido's Ph.D. student from 1996 to 1999. For us Guido was a teacher, a mentor and a friend. We miss him a lot. Grazie Guido.

References

1. J. Beck, W.W.L. Chen, *Irregularities of Distribution* (Cambridge University Press, Cambridge, 1987)
2. A. Bondarenko, D. Radchenko, M. Viazovska, Optimal asymptotic bounds for spherical designs. Ann. Math. **178**, 443–452 (2013)
3. L. Brandolini, L. Colzani, G. Gigante, G. Travaglini, L^p and weak-L^p estimates for the number of integer points in translated domains. Math. Proc. Camb. Philos. Soc. **159**, 471–480 (2015)
4. L. Brandolini, W.W.L. Chen, L. Colzani, G. Gigante, G. Travaglini, Discrepancy and numerical integration on metric measure spaces. J. Geom. Anal. **29**, 328–369 (2019)
5. L. Brandolini, B. Gariboldi, G. Gigante, On a sharp lemma of Cassels and Montgomery on manifolds. Math. Ann. **379**, 1807–1834 (2021)
6. M. Ehler, U. Etayo, B. Gariboldi, G. Gigante, T. Peter, Asymptotically optimal cubature formulas on manifolds for prefixed weights. J. Approx. Theory **271** (2021)
7. A. Elbert, Some recent results on the zeros of Bessel functions and orthogonal polynomials, Proceedings of the Fifth International Symposium on Orthogonal Polynomials, Special Functions and their Applications (Patras, 1999). J. Comput. Appl. Math. **133**, 65–83 (2001)
8. C.L. Frenzen, R. Wong, A uniform asymptotic expansion of the Jacobi polynomials with error bounds. Can. J. Math. **37**, 979–1007 (1985)
9. R. Gangolli, Positive definite kernels on homogeneous spaces and certain stochastic processes related to Lévy's Brownian motion of several parameters. Ann. Inst. H. Poincaré Sect. B (N.S.) **3**, 121–226 (1967)
10. B. Gariboldi, G. Gigante, Optimal asymptotic bounds for designs on manifolds. Anal. PDE **14**, 1701–1724 (2021)
11. S. Helgason, *Groups and Geometric Analysis* (AMS, Providence, 2000)

12. S.V. Konyagin, M.M. Skriganov, A.V. Sobolev, On a lattice point problem arising in the spectral analysis of periodic operators. Mathematika **50**, 87–98 (2003)
13. H.L. Montgomery, Ten lectures on the interface between analytic number theory and harmonic analysis, in *CBMS Regional Conference Series in Mathematics*, vol. 84 (AMS, Providence, 1994)
14. L. Parnovski, N. Sidorova, Critical dimensions for counting lattice points in Euclidean annuli. Math. Model. Nat. Phenom. **5**, 293–316 (2010)
15. L. Parnovski, A.V. Sobolev, On the Bethe-Sommerfeld conjecture for the polyharmonic operator. Duke Math. J. **107**, 209–238 (2001)
16. M.M. Skriganov, Point distributions in two-point homogeneous spaces. Mathematika **65**, 557–587 (2019)
17. E.M. Stein, G. Weiss, *Introduction to Fourier Analysis on Euclidean Spaces* (Princeton University Press, Princeton, 1971)
18. G. Szegő, *Orthogonal Polynomials* (American Mathematical Society, Providence, 1975)
19. H.C. Wang, Two-point homogeneous spaces. Ann. Math **55**, 177–191 (1952)
20. G.N. Watson, *A Treatise on the Theory of Bessel Functions* (Cambridge University Press, Cambridge, 1944)
21. J.A. Wolf, *Spaces of Constant Curvature* (University of California Berkley, Berkley, 1972)
22. B. Gariboldi, G. Gigante, Almost positive kernels on compact Riemannian manifolds. Math. Z. **302**(2), 783–801 (2022)
23. L. Brandolini, B. Gariboldi, G. Gigante, A. Monguzzi, Single radius spherical cap discrepancy on compact two-point homogeneous spaces. Math. Z. **308**(3), 52 (2024)
24. D. Bilyk, M. Mastrianni, S. Steinerberger, Single radius spherical cap discrepancy via gegenbadly approximable numbers. Adv. Math. **452**, Paper No. 109812, 18 (2024)

Chapter 6
Analytic Families of Operators in Extrapolation Theory with Application to Average Operators

María J. Carro

With love to Guido

Abstract Our goal is to prove weak type (1, 1) boundedness for an operator T_A which is given as an average of a family of operators $\{T_j\}_j$ satisfying certain estimates in the context of weighted Lebesgue spaces. In particular, we shall prove that, if there exists $\alpha > 0$, $s > 0$ and $C > 0$ such that, for every weight v in the Muckenhoupt class A_p and every measurable set E,

$$\sup_j \|T_j \chi_E\|_{L^{p,\infty}(v)} \leq \frac{C}{(p-1)^s} \|v\|_{A_p}^{\alpha} v(E).$$

and, for some $u_0 \in A_1$ fixed,

$$\sup_j \|T_j \chi_E\|_{L^{1,\infty}(u_0)} \leq C_{u_0} u_0(E),$$

then, for every $\beta > 0$, there exists a constant $C_\beta > 0$, so that

$$\sup_{y>0} \frac{y}{\left(1 + \log^+(1 + \log^+ \frac{1}{y})\right)^\beta} \int_{\{|T_A \chi_E| > y\}} u_0(x)dx \leq C_\beta \, u_0(E).$$

Partially supported by grants PID2020-113048GB-I00 funded by MCIN/AEI/10.13039/501100011033, CEX2019-000904-S funded by MCIN/AEI/ 10.13039/501100011033 and Grupo UCM-970966 (Spain).

M. J. Carro (✉)
Department of Analysis and Applied Mathematics, Complutense University of Madrid, Madrid, Spain
e-mail: mjcarro@ucm.es

E. Hernández et al. (eds.), *The Mathematical Heritage of Guido Weiss*,
Applied and Numerical Harmonic Analysis,
https://doi.org/10.1007/978-3-031-76793-7_6

Our main technique is inspired on the theory of analytic families of operators and it is closely related to the Rubio de Francia's extrapolation theory.

6.1 Introduction

Let $\{T_j\}_{j\in\mathbb{N}}$ be a collection of operators so that

$$T_j : L^1 \longrightarrow L^{1,\infty}, \qquad \sup_j ||T_j||_{L^1\longrightarrow L^{1,\infty}} < \infty.$$

Let $(c_j)_j \in \ell^1$ and set

$$S = \sum_j c_j T_j.$$

Since $L^{1,\infty}$ is not a Banach space, the weak type $(1, 1)$ boundedness of S may fail. However, (we refer to Sect. 6.2 for the definition of the Muckenhoupt class of weights A_p, $p \geq 1$) it has been very recently proved [1] that if, for every $u \in A_1$,

$$T_j : L^1(u) \longrightarrow L^{1,\infty}(u), \qquad \sup_j ||T_j||_{L^1(u)\longrightarrow L^{1,\infty}(u)} < \infty, \tag{6.1}$$

then the restricted weak type $(1, 1)$ boundedness of S holds, for every $u \in A_1$; that is, for every measurable set and every $u \in A_1$,

$$||S\chi_E||_{L^{1,\infty}(u)} \leq C_u ||c||_{\ell^1} u(E).$$

We observe that, by Rubio de Francia extrapolation theory [18], condition (6.1) implies that, for every $p > 1$ and every $v \in A_p$,

$$T_j : L^p(v) \longrightarrow L^{p,\infty}(v), \qquad \sup_j ||T_j||_{L^p(v)\longrightarrow L^{p,\infty}(v)} < \infty;$$

that is, condition (6.1) hides a weighted estimate at level $p > 1$. On the other hand, and this is the main motivation of this paper, there are examples of operators for which the hypothesis (6.1) is only known for the case $u = 1$ and not for every $u \in A_1$, but some weighted estimate at the level $p > 1$ is also true. The main goal of this paper concerns with this situation.

Now, the Rubio de Francia extrapolation theorem (see [7, 11–13, 18]), can be formulated, nowadays, as follows (see [9, 10]): Let T be an operator so that, for some $1 \leq p_0 < \infty$, and for every $w \in A_{p_0}$, we have

$$T : L^{p_0}(w) \longrightarrow L^{p_0}(w), \qquad ||T|| \leq N(\|w\|_{A_{p_0}}), \tag{6.2}$$

where $N(t), t > 0$, is an increasing function. Then, for every $1 < p < \infty$ and every $w \in A_p$,

$$T : L^p(w) \longrightarrow L^p(w), \qquad ||T|| \le K(w) \tag{6.3}$$

where

$$K(w) \le C_1 N\Big(\frac{1}{p-1}\|w\|_{A_p}^{\max\left(1, \frac{p_0-1}{p-1}\right)}\Big).$$

In particular, the above estimate (6.3) shows that if, T satisfies (6.2) with $N(t) = t^\alpha$, then

$$T : L^p \longrightarrow L^p, \qquad \frac{1}{(p-1)^\alpha}, \qquad 1 < p \le p_0,$$

and using Yano's extrapolation theorem [15, 19], we get that

$$T : L(\log L)^\alpha \longrightarrow L^1 + L^\infty.$$

Hence, if (6.2) holds for our family of operators $\{T_j\}_j$ uniformly in j, same estimate can be deduced for the sum operator S defined before, and consequently S would satisfy the above boundedness on the space $L(\log L)^\alpha$. Our goal is to see, that under some extra condition on the operators $\{T_j\}_j$, this estimate can be improved. We should also mention here paper [16] where it is shown that for a given operator the optimality of the weighted L^p bounds in term of the A_p constant of the weight is related to the unweighted behavior of the operator when $p \to 1$. Hence, the relation between Rubio de Francia's extrapolation and Yano's extrapolation was already in the literature (see also [5]).

Our general context will be the following: let $\{T_\theta\}_\theta$ be a family of operators indexed in a probability measure space $(\mathcal{M}, P)$ such that

$$T_\theta : L^1 \longrightarrow L^{1,\infty}, \qquad \sup_\theta ||T_\theta||_{L^1 \longrightarrow L^{1,\infty}} < \infty.$$

What can we say about the boundedness, near L^1, of the average operator

$$T_A f(x) = \int_{\mathcal{M}} T_\theta f(x) dP(\theta),$$

whenever it is well defined? Our main result is the following:

Theorem 6.1 *Let $\{T_\theta\}_\theta$ be a family of operators indexed in a probability measure space satisfying the following conditions:*

- *There exist $\alpha > 0$ and $s > 0$ so that, for every $1 < p < 2$ with $\alpha p(p-1) < 1$ and every $v \in A_p$,*

$$T_\theta : L^p(v) \longrightarrow L^{p,\infty}(v), \qquad \sup_\theta ||T_\theta|| \le \frac{C}{(p-1)^s} \|v\|_{A_p}^\alpha,$$

with C a universal constant.
- *For some $u_0 \in A_1$ and every measurable set E,*

$$\sup_\theta \|T_\theta \chi_E\|_{L^{1,\infty}(u_0)} \le C_{u_0} u_0(E).$$

Then, for every $\beta > 0$, the averaging operator T_A satisfies the following estimate

$$\sup_{y>0} \frac{y}{\left(1 + \log^+(1 + \log^+ \frac{1}{y})\right)^\beta} \int_{\{|T_A \chi_E| > y\}} u_0(x) dx \le C_{u_0, \beta}\, u_0(E).$$

In particular:

Corollary 6.1 *If T_j satisfies Condition-A uniformly in j and*

$$\sup_j \|T_j \chi_E\|_{L^{1,\infty}} \lesssim |E|,$$

then, for every $\beta > 0$ and every $(c_j)_j \in \ell^1$, the operator $S = \sum_j c_j T_j$, satisfies that

$$\sup_{y>0} \frac{y}{(1 + \log^+(1 + \log^+ \frac{1}{y}))^\beta} |\{x : |S\chi_E(x)| > y\}| \lesssim C_\beta ||c||_{\ell^1} |E|.$$

Remark 6.1 We have not succeeded in proving the restricted weak type (1, 1) of T_A for the weight u_0 neither to find a counterexample and hence this remains as an open question.

As usual, we shall use the symbol $A \lesssim B$ to indicate that there exists a universal positive constant C, independent of all relevant parameters, such that $A \le CB$. $A \approx B$ means that $A \lesssim B$ and $B \lesssim A$. Moreover, even though C may depend on $p > 1$, we shall only be concerned about the dependence on p if this blows up when $p \to 1$.

6.2 Preliminary Results and Some Technical Lemmas

For our purposes, let us recall that the Lorentz spaces $L^{p,q}(u)$ is defined as the set of measurable functions such that

$$\|f\|_{L^{p,q}(u)} = \left(\frac{q}{p} \int_0^\infty f_u^*(t)^q t^{\frac{q}{p}-1}\, dt\right)^{1/q} = \left(q \int_0^\infty y^{q-1} \lambda_f^u(y)^{q/p} dy\right)^{1/q} < \infty,$$

and $L^{p,\infty}(u)$ is defined by the condition

$$\|f\|_{L^{p,\infty}(u)} = \sup_{t>0} t^{1/p} f_u^*(t) = \sup_{y>0} y\lambda_f^u(y)^{1/p} < \infty,$$

where f_u^* is the decreasing rearrangement of f, with respect to the weight u, defined by

$$f_u^*(t) = \inf\left\{y > 0 : \lambda_f^u(y) \le t\right\},$$

and $\lambda_f^u(y) = u\big(\{x : |f(x)| > y\}\big)$ is the distribution function of f with respect to u. We shall also use the standard notation $u(E) = \int_E u(x)\,dx$ and, if $u = 1$, we shall write $\lambda_f(y)$, f^* and $|E|$ (see [2]). With the above definition, it holds that, for every $1 < p$ and $1 \le q \le \infty$,

$$\left|\int f(x)g(x)u(x)dx\right| \le \left(\frac{p}{q}\right)^{1/q} \left(\frac{p'}{q'}\right)^{1/q'} ||f||_{L^{p,q}(u)}||g||_{L^{p',q'}(u)},$$

and, for every q, $||\chi_E||_{L^{p,q}(u)} = u(E)^{1/p}$.

Let us also recall some well known facts of the class A_p. Let M be the Hardy-Littlewood maximal operator, defined by

$$Mf(x) = \sup_{x\in Q} \frac{1}{|Q|}\int_Q |f(y)|\,dy,$$

where Q denotes a cube in $\mathbb{R}^n$ and let v be a positive locally integrable function w (called weight) such that

$$\|v\|_{A_r} = \sup_Q \left(\frac{1}{|Q|}\int_Q v(x)\,dx\right)\left(\frac{1}{|Q|}\int_Q v^{-1/(r-1)}(x)\,dx\right)^{r-1} < \infty,$$

with $r > 1$. This class of weights is known as the Muckenhoupt class A_r $(r > 1)$. If $r = 1$, we say that $u \in A_1$, if $Mu(x) \le Cu(x)$, at almost every point $x \in \mathbb{R}^n$ and $\|w\|_{A_1}$ will be the least constant C satisfying such inequality.

In [3, 17], it was proved that, if $p > 1$, then

$$M : L^p(v) \longrightarrow L^p(v)$$

is bounded if, and only if, $v \in A_p$ and also, for every $1 \le p < \infty$, M is of weak-type (p, p) if and only if $v \in A_p$ and, in this case,

$$\|M\|_{L^p(v)\to L^{p,\infty}(v)} \le C\|v\|_{A_p}^{\frac{1}{p}}.$$

These classes of weights satisfy the following properties:

1. $u \in A_1$ if and only if there exists $h \in L^1_{loc}(\mathbb{R}^n)$ and k such that $k, k^{-1} \in L^\infty$ satisfying that, for some $0 < \delta < 1$,
$$u(x) = k(x)(Mh)^\delta.$$
2. Factorization: $v \in A_p$ if and only if there exist $u_0, u_1 \in A_1$ such that
$$v = u_0 u_1^{1-p}, \qquad \|u_0\|_{A_1} \le \|v\|_{A_p}, \|u_1\|_{A_1} \le \|v\|_{A_p}^{\frac{1}{p-1}}.$$
Moreover, if $v = u_0 u_1^{1-p}$, then $\|v\|_{A_p} \lesssim \|u_0\|_{A_1} \|u_1\|_{A_1}^{p-1}$.
3. For every $g \in L^1_{loc}$ with $Mg > 0$, every $u \in A_1$, and $0 < \delta < 1$, it holds that $v(x) = u(x)(Mg)^{\delta(1-p)} \in A_p$ with
$$\|v\|_{A_p} \lesssim \frac{\|u\|_{A_1}}{(1-\delta)^{(p-1)}}. \tag{6.4}$$
4. If $u \in A_1$ and $0 < \theta < 1$, then $(Mf)^{1-\theta} u^\theta \in A_1$ with
$$\left\| (Mf)^{1-\theta} u^\theta \right\|_{A_1} \lesssim \frac{\|u\|_{A_1}}{\theta}.$$

The following lemma in complex variable will be fundamental in our theory.

Lemma 6.1 *Let $S = \{0 < Re\, z < 1\}$ and set $H(S)$ the space of analytic functions in S and continuous in $\bar{S}$. If $F \in H(S)$ and* $\sup_{t \in \mathbb{R}} |F(j+it)| \le M_j$, *with $j = 0, 1$, then for every* $0 < \theta < 1$,
$$|F^{N)}(\theta)| \lesssim \left(\frac{1}{\theta(1-\theta)}\right)^N M_0^{1-\theta} M_1^\theta \left(1 + \left|\log \frac{M_1}{M_0}\right|\right)^N.$$

Using the dual operator of M as in [8], it was also proved in [6] the following result.

Lemma 6.2 ([6]) *For every $1 \le p < \infty$ and every $u \in A_1$, there exists $C_{p,u}$ such that*
$$\left\| \frac{M(fu(Mg)^{1-p})}{u(Mg)^{1-p}} \right\|_{L^{p',\infty}(u(Mg)^{1-p})} \le C_{p,u} \left\| f \right\|_{L^{p',1}(u(Mg)^{1-p})}, \tag{6.5}$$
where if $p = 1$ $L^{p',\infty}(u(Mg)^{1-p}) = L^{p',1}(u(Mg)^{1-p}) = L^\infty$

The constant $C_{p,u}$ was not explicitly computed in [6], but it can be proved that
$$C_{p,u} \lesssim \|u\|_{A_1}^2.$$

6.3 Main Results

In this paper, we shall start with an operator T satisfying the following condition:

Condition-A There exists $\alpha > 0$ and $s > 0$, so that, for every $1 < p < 2$ with $\alpha p(p-1) < 1$ and every $v \in A_p$,

$$\|T\chi_E\|_{L^{p,\infty}(v)} \lesssim \frac{\|v\|_{A_p}^{\alpha}}{(p-1)^s} v(E)^{1/p}.$$

This is the case of important operators in Harmonic Analysis such as the Hardy-Littlewood maximal operator or the maximal Calderón-Zygmund operators [14], among many others.

Proposition 6.1 *Let T be an operator satisfying Condition-A. Then, for every $\alpha p(p-1) < \beta \leq 1$, every measurable set E and every $u \in A_1$,*

$$\sup_{y>0} y^p \int_{\{|T\chi_E(x)|>y\}} u(x) \frac{(M\chi_E(x))^{(1-p)}}{\left(\log \frac{e}{M\chi_E(x)}\right)^{\beta}} dx \lesssim C(p,\beta,\alpha)\|u\|_{A_1}^{\alpha p} u(E), \tag{6.6}$$

with

$$C(p,\beta,\alpha) = \frac{(p-1)^{\beta-s}}{(\beta - \alpha p(p-1))}.$$

Proof Set $C_p = (p-1)^{-s}$. Let $0 < \mu < 1$, and let $K \subset \{|T\chi_E(x)| > y\}$ be a compact set. Set the analytic function on the unit strip $S = \{0 < Re\, z < 1\}$,

$$F(z) = \int_K u(x) \left(\frac{M\chi_E(x)}{e}\right)^{\left(\frac{1-\mu}{2}z+\mu\right)(1-p)} dx.$$

Then, by (6.4), we have that, for $j = 0, 1$,

$$|F(j+it)| \lesssim \frac{C_p \|u\|_{A_1}^{\alpha p} u(E)}{(1-\mu)^{\alpha p(p-1)} y^p}$$

uniform in $t \in \mathbb{R}$. Thus,

$$|F(1/2)| = \int_K u(x) \left(\frac{M\chi_E(x)}{e}\right)^{\frac{1+3\mu}{4}(1-p)} dx \lesssim \frac{C_p \|u\|_{A_1}^{\alpha p} u(E)}{(1-\mu)^{\alpha p(p-1)} y^p}, \tag{6.7}$$

and, by Lemma 6.1,

$$(1-\mu)(p-1)\int_K u(x)(M\chi_E(x))^{\frac{1+3\mu}{4}(1-p)}\log\frac{e}{M\chi_E(x)}dx \lesssim |F'(1/2)|$$
$$\lesssim \frac{C_p\|u\|_{A_1}^{\alpha p}u(E)}{y^p(1-\mu)^{\alpha p(p-1)}},$$

and therefore, letting K tend to $\{|T\chi_E(x)|>y\}$, we obtain that

$$y^p\int_{\{|T\chi_E(x)|>y\}} u(x)(M\chi_E(x))^{\frac{1+3\mu}{4}(1-p)}\log\frac{e}{M\chi_E(x)}dx$$
$$\lesssim \frac{C_p\|u\|_{A_1}^{\alpha p}u(E)}{(1-\mu)^{\alpha p(p-1)+1}(p-1)}. \tag{6.8}$$

By (6.7) and (6.8), we get that, for every $t>0$,

$$y^p\int_{\{|T\chi_E(x)|>y\}} u(x)(M\chi_E(x))^{\frac{1+3\mu}{4}(1-p)}\min\left(1, t\log\frac{e}{M\chi_E(x)}\right)dx$$
$$\lesssim \frac{C_p\|u\|_{A_1}^{\alpha p}}{(1-\mu)^{\alpha p(p-1)}}\min\left(1,\frac{t}{(1-\mu)(p-1)}\right)u(E),$$

and integrating in $t\in(0,\infty)$ against $\frac{1}{t^{\gamma+1}}$, we obtain that, for every $0\le\gamma<1$,

$$\int_{\{|T\chi_E(x)|>y\}} u(x)\left(\frac{e}{M\chi_E(x)}\right)^{\frac{1+3\mu}{4}(p-1)}\left(\log\frac{e}{M\chi_E(x)}\right)^{\gamma}dx$$
$$\lesssim \frac{C_p\|u\|_{A_1}^{\alpha p}u(E)}{(1-\mu)^{\alpha p(p-1)+\gamma}(p-1)^{\gamma}}.$$

Hence, if $\alpha p(p-1)+\gamma<1$, we can integrate in the variable $\mu\in(0,1)$ and we obtain that if $\beta=1-\gamma$,

$$y^p\int_{\{|T\chi_E(x)|>y\}} u(x)(M\chi_E(x))^{(1-p)}\frac{1}{\left(\log\frac{e}{M\chi_E(x)}\right)^{\beta}}dx$$
$$\lesssim y^p\int_{\{|T\chi_E(x)|>y\}} u(x)(M\chi_E(x))^{\frac{1-p}{4}}dx + C_p\frac{(p-1)^{\beta}\|u\|_{A_1}^{\alpha p}}{(\beta-\alpha p(p-1))}u(E)$$
$$\lesssim C_p\left(\frac{(p-1)^{\beta}}{(\beta-\alpha p(p-1))}+1\right)\|u\|_{A_1}^{\alpha p}u(E),$$

as we wanted to see. □

Our next step is to prove that an estimate of the form (6.6) is also true for other values of p. To this end, the following lemma is needed.

Lemma 6.3 *Let $\alpha > 0$ and $0 < \varepsilon \leq 1$. Let $q > 1$ and $1 < p \leq \min(2, q)$ such that*

$$(\alpha q + 1)(p - 1) = \frac{\varepsilon}{2}.$$

Let us define $\gamma > 0$ and $\nu > 0$ such that

$$\frac{\varepsilon\gamma}{q-1} = \alpha p + \frac{1}{q}, \qquad \gamma\frac{p-1}{q-1} + \nu\frac{q-p}{q-1} = 1 \tag{6.9}$$

Then, it holds that:

$$\frac{1}{q'} < \nu, \qquad \frac{1}{1 - \frac{1}{\nu q'}} \leq 2\Big(1 + \frac{1}{\alpha p}\Big)q.$$

Proof We observe that,

$$\nu = \frac{q-1}{q-p}\Big(1 - \gamma\frac{p-1}{q-1}\Big) = \frac{q-1-\gamma(p-1)}{q-p} = \frac{q - 1 - \frac{(q-1)(\alpha p + \frac{1}{q})}{\varepsilon}(p-1)}{q-p}$$
$$= (q-1)\frac{\varepsilon - (\alpha p + \frac{1}{q})(p-1)}{\varepsilon(q-p)} > \frac{q-1}{q}$$

if and only if

$$\frac{\varepsilon - (\alpha p + \frac{1}{q})(p-1)}{\varepsilon(q-p)} > \frac{1}{q}.$$

That is,

$$\varepsilon - \Big(\alpha p + \frac{1}{q}\Big)(p-1) > \varepsilon\Big(1 - \frac{p}{q}\Big),$$

or equivalently

$$\varepsilon\frac{p}{q} > \Big(\alpha p + \frac{1}{q}\Big)(p-1),$$

$$\varepsilon > (\alpha pq + 1)\frac{p-1}{p} = \Big(\alpha q + \frac{1}{p}\Big)(p-1).$$

On the other hand,

$$\frac{\nu q'}{\nu q'-1} = \frac{q^{\frac{\varepsilon-(\alpha p+\frac{1}{q})(p-1)}{\varepsilon(q-p)}}}{q^{\frac{\varepsilon-(\alpha p+\frac{1}{q})(p-1)}{\varepsilon(q-p)}}-1} = \frac{q(\varepsilon-(\alpha p+\frac{1}{q})(p-1))}{q(\varepsilon-(\alpha p+\frac{1}{q})(p-1))-\varepsilon(q-p)}$$

$$= \frac{q(2(\alpha q+1)(p-1)-(\alpha p+\frac{1}{q})(p-1))}{2p(\alpha q+1)(p-1)-q(\alpha p+\frac{1}{q})(p-1))}$$

$$= \frac{q(2(\alpha q+1)-(\alpha p+\frac{1}{q}))}{2p(\alpha q+1)-q(\alpha p+\frac{1}{q})}$$

$$= \frac{2\alpha q^2+2q-\alpha pq-1}{\alpha pq+2p-1} \le \frac{2\alpha q^2+2q}{\alpha pq} \le 2\Big(1+\frac{1}{\alpha p}\Big)q,$$

and the result is proved. □

Proposition 6.2 *Let T be an operator satisfying Condition-A. Then, for every $0 < \varepsilon \le 1$, every $q > 1$, every measurable set E and every $u \in A_1$,*

$$\sup_{y>0} y^q \int_{\{|T\chi_E(x)|>y\}} u(x)\frac{(M\chi_E(x))^{(1-q)}}{\left(\log \frac{e}{M\chi_E(x)}\right)^{\varepsilon}}dx \lesssim \frac{1}{\varepsilon^{q(s+1+\alpha)}}\|u\|_{A_1}^{(\alpha+2)q}u(E). \tag{6.10}$$

Proof Let $0 \le \varepsilon < 1$ be fixed. Let K be a compact set such that $K \subset \{|T\chi_E(x)| > y\}$ and let us define

$$A_q := \int_K u(x)\frac{(M\chi_E(x))^{(1-q)}}{\left(\log \frac{e}{M\chi_E(x)}\right)^{\varepsilon}}dx.$$

Set p, γ and ν as in Lemma 6.3 and set $C_p = (p-1)^{-s}$. Then, we have that

$$A_q \le \int_K (M\chi_E)^{(1-p)}\frac{u^{\frac{p-1}{q-1}}\left[M\left(u\chi_K\frac{(M\chi_E)^{1-q}}{\left(\log\frac{e}{M\chi_E}\right)^{\varepsilon\nu}}\right)\right]^{\frac{q-p}{q-1}}}{\left(\log\frac{e}{M\chi_E}\right)^{\varepsilon\gamma\frac{p-1}{q-1}}}dx,$$

and since,

$$v = u^{\frac{p-1}{q-1}}\left[M\left(u\chi_K\frac{(M\chi_E)^{1-q}}{\left(\log\frac{e}{M\chi_E}\right)^{\varepsilon\nu}}\right)\right]^{\frac{q-p}{q-1}} \in A_1,$$

with $\|v\|_{A_1} \lesssim \frac{q-1}{p-1}\|u\|_{A_1}$ and, by (6.9), $\varepsilon\gamma\frac{p-1}{q-1} > \alpha p(p-1)$, we have by Proposition 6.1,

$$A_q \le \frac{B}{y^p}\int_E u^{\frac{p-1}{q-1}}(x)\left[M\left(u\chi_K\frac{(M\chi_E)^{1-q}}{(\log\frac{e}{M\chi_E(x)})^{\varepsilon v}}\right)\right]^{\frac{q-p}{q-1}}dx$$

$$= \frac{B}{y^p}\int_E\left[\frac{M\left(u\chi_K\frac{(M\chi_E)^{1-q}}{(\log\frac{e}{M\chi_E(x)})^{\varepsilon v}}\right)}{u(x)}\right]^{\frac{q-p}{q-1}}u(x)dx$$

$$= \frac{B}{y^p}\int_E\left[\frac{M\left(u\chi_K\frac{(M\chi_E)^{1-q}}{(\log\frac{e}{M\chi_E(x)})^{\varepsilon v}}\right)}{u(x)(M\chi_E(x))^{1-q}}\right]^{\frac{q-p}{q-1}}u(x)(M\chi_E(x))^{1-q}dx$$

with

$$B \lesssim C_p\frac{\|v\|_{A_1}^{\alpha p}(p-1)^{(\alpha p+\frac{1}{q})(p-1)}}{(\varepsilon\gamma\frac{p-1}{q-1}-\alpha p(p-1))} \lesssim C_p\frac{q\|v\|_{A_1}^{\alpha p}}{(p-1)}$$

$$\lesssim C_p\frac{q^{1+\alpha p}}{\varepsilon^{1+\alpha p}}\|u\|_{A_1}^{\alpha p} \approx \frac{q^{2+\alpha p}}{\varepsilon^{s+1+\alpha p}}\|u\|_{A_1}^{\alpha p}.$$

Then, by duality,

$$A_q \lesssim \frac{Bu(E)^{p/q}}{y^p}\left\|\frac{M\left(u\chi_K\frac{(M\chi_E)^{1-q}}{(\log\frac{e}{M\chi_E})^{\varepsilon v}}\right)}{u(M\chi_E)^{1-q}}\right\|_{L^{q',\infty}(u(M\chi_E)^{1-q})}^{\frac{q-p}{q-1}}.$$

Now, using (6.5) with the behavior of the constant $C_{p,u}$ we have that

$$\left\|\frac{M\left(u\chi_K\frac{(M\chi_E)^{1-q}}{(\log\frac{e}{M\chi_E})^{\varepsilon v}}\right)}{u(M\chi_E)^{1-q}}\right\|_{L^{q',\infty}(u(M\chi_E)^{1-q})}$$

$$\lesssim \|u\|_{A_1}^2\left\|\frac{\chi_K}{(\log\frac{e}{M\chi_E})^{\varepsilon v}}\right\|_{L^{q',1}(u(M\chi_E)^{1-q})}$$

$$= \|u\|_{A_1}^2\int_0^\infty\left(\int_{\{x\in K:\frac{1}{(\log\frac{e}{M\chi_E})^{\varepsilon v}}>z\}}u(x)(M\chi_E(x))^{1-q}dx\right)^{1/q'}dz$$

$$= t\|u\|_{A_1}^2\int_0^1\left(\int_{\{x\in K:\frac{1}{(\log\frac{e}{M\chi_E})^{\varepsilon v}}>z\}}u(x)(M\chi_E(x))^{1-q}dx\right)^{1/q'}dz$$

$$\lesssim \|u\|_{A_1}^2 \left(\int_K u(x) \frac{(M\chi_E)^{1-q}(x)}{\left(\log \frac{e}{M\chi_E(x)}\right)^{\varepsilon}} dx \right)^{1/q'} \int_0^1 z^{-\frac{1}{\nu q'}} dz$$

$$\lesssim \frac{\|u\|_{A_1}^2}{1 - \frac{1}{\nu q'}} A_q^{1/q'} \lesssim \|u\|_{A_1}^2 A_q^{1/q'}$$

Consequently, we have that

$$A_q \lesssim \frac{\|u\|_{A_1}^{\alpha p}}{\varepsilon^{s+1+\alpha p} y^p} u(E)^{p/q} \left(\|u\|_{A_1}^2 \right)^{\frac{q-p}{q-1}} A_q^{\frac{q-p}{q'(q-1)}} \lesssim \frac{\|u\|_{A_1}^{(\alpha+2)p}}{\varepsilon^{s+1+\alpha p} y^p} u(E)^{p/q} A_q^{\frac{q-p}{q}},$$

and hence,

$$A_q \lesssim \frac{\|u\|_{A_1}^{(\alpha+2)q}}{\varepsilon^{(s+1+\alpha p)\frac{q}{p}} y^q} u(E) \leq \frac{\|u\|_{A_1}^{(\alpha+2)q}}{\varepsilon^{q(s+1+\alpha)} y^q} u(E),$$

as we wanted to see. □

Remark 6.2 We observe that, in general, the constant has to blow up when $\varepsilon \to 0$, since on the contrary it can be proved that T will be of weak type $(1, 1)$ for every weight $u \in A_1$, which is false in general.

We shall now fix a weight $u_0 \in A_1$ and let us consider the condition:

$$||T\chi_E||_{L^{1,\infty}(u_0)} \lesssim u_0(E), \qquad \forall E. \tag{6.11}$$

Corollary 6.2 *Let T be an operator satisfying Condition-A and* (6.11). *Then, for every $q > 1$, every measurable set E, and every $\beta > (q-1)(s+1+\alpha)$,*

$$\sup_{y>0} \frac{y^q}{(1 + \log(1 + \log_+ \frac{1}{y}))^\beta} \int_{\{|T\chi_E(x)|>y\}} u_0(x)(M\chi_E(x))^{(1-q)} dx \lesssim C_{u_0,q} u_0(E).$$

Proof First of all we observe that,

$$\frac{y^q}{(1 + \log(1 + \log_+ \frac{1}{y}))^\beta} \int_{\{|T\chi_E(x)|>y, M\chi_E(x)>y\}} u_0(x)(M\chi_E(x))^{(1-q)} dx$$

$$\leq y \int_{\{|T\chi_E(x)|>y\, M\chi_E(x)>y\}} u_0(x) dx \leq C_{u_0} u(E)$$

Now, let $0 < \theta < 1$ and let $p = 1 + \frac{q-1}{\theta}$. Let us define, in the case $y < 1$, the following analytic function

$$F(z) = \int_K u_0(x) \frac{(\frac{M\chi_E(x)}{e})^{z(1-p)}}{\left(\log \frac{e}{M\chi_E(x)}\right)^{\varepsilon}} dx,$$

where $K \subset \{|T\chi_E(x)| > y, M\chi_E(x) \le y\}$ is an arbitrary compact set. Then, by (6.11),

$$|F(it)| \lesssim \frac{1}{y} u_0(E)$$

and, by (6.10),

$$|F(1+it)| \lesssim \frac{1}{\varepsilon^{p(s+1+\alpha)} y^p} u_0(E),$$

and thus, since $1 - q = \theta(1-p)$, we have by Lemma 6.1, that

$$\begin{aligned} |F'(\theta)| &\approx \left| (p-1) \int_K u_0(x) \log\left(\frac{e}{M\chi_E(x)}\right)^{1-\varepsilon} \left(\frac{M\chi_E(x)}{y}\right)^{1-q} dx \right| \\ &\lesssim \frac{C_q}{\varepsilon^{(q-1+\theta)(s+1+\alpha)} y^q} \left(1 + \log_+ \frac{1}{y}\right)\left(1 + \log_+ \frac{1}{\varepsilon}\right), \end{aligned}$$

from which it follows that, for every $\beta > (q-1)(s+1+\alpha)$, we can take θ so that

$$\frac{1}{\varepsilon^{(q-1+\theta)(s+1+\alpha)}} \left(1 + \log_+ \frac{1}{\varepsilon}\right) \le \frac{1}{\varepsilon^{\beta}},$$

and hence

$$\left(\log \frac{e}{y}\right)^{1-\varepsilon} \left| \int_K u_0(x) \left(\frac{M\chi_E(x)}{y}\right)^{1-q} dx \right| \lesssim \frac{C_q}{\varepsilon^{\beta} y^q} \left(1 + \log_+ \frac{1}{y}\right).$$

Therefore,

$$\sup_{\varepsilon} \frac{\varepsilon^{\beta}}{\left(\log \frac{e}{y}\right)^{\varepsilon}} \left| \int_K u_0(x) \left(\frac{M\chi_E(x)}{y}\right)^{1-q} dx \right| \lesssim \frac{C_q}{y^q},$$

from which the result follows in the case $y < 1$.

Now, when $y > 1$, we consider the analytic function

$$F(z) = \int_K u_0(x) \frac{(\frac{M\chi_E(x)}{y})^{z(1-p)}}{\left(\log \frac{ey}{M\chi_E(x)}\right)} dx.$$

Then, by (6.11) and (6.10),

$$|F(j+it)| \lesssim \frac{u_0(E)}{y}, \qquad j = 0, 1,$$

and thus,

$$|F'(1/2)| \approx \left|(q-1) \int_K u_0(x) \Big(\frac{M\chi_E(x)}{y}\Big)^{1-q} dx\right| \lesssim \frac{C_q}{y} u_0(E),$$

and the result follows, letting K tends to $\{|T\chi_E(x)| > y, M\chi_E(x) \le y\}$. □

6.4 Applications to Average Operators

The following result was proved in [4], but we include the proof for the sake of completeness.

Proposition 6.3 *If there exists $C > 0$ so that $\sup_{y>0} W(y)\lambda_{T_\theta f}(y) \le C$, then $\sup_{y>0} \widetilde{W}(y)\lambda_{T_A f}(y) \le C$, where*

$$\widetilde{W}(R) = \sup_{x \le R} \frac{R - x}{\int_x^\infty \frac{1}{W(u)} du}.$$

Proof Let $\phi(t) = \int_0^t h(s)\, ds$, with h a positive and increasing function. Then ϕ is a convex function and, by Jensen's inequality,

$$\phi(|T_A f(x)|) \le \int_M \phi(T_\theta f(x)|)\, dP(\theta).$$

Hence, for every $R > 0$, $\phi(R)\chi_{\{x;|T_A f(x)|>R\}}(x) \le \int_M \phi(T_\theta f(x))\, dP(\theta)$ and integrating over $\mathbb{R}^n$, we obtain

$$\phi(R)\lambda_{T_A f}(R) \le \int_M \int_{\mathbb{R}^n} \phi(T_\theta f(x))\, dx\, dP(\theta).$$

Now,

$$\int_{\mathbb{R}^n} \phi(T_\theta f(x))\,dx = \int_0^\infty \lambda_{T_\theta f}(y)d\phi(y) = \int_0^\infty \lambda_{T_\theta f}(y)h(y)\,dy$$

and hence, we deduce that

$$\phi(R)\lambda_{T_A f}(R) \le \int_M \int_0^\infty \lambda_{T_\theta f}(y)h(y)\,dy\,dP(\theta) \le C\int_0^\infty \frac{h(y)}{W(y)}\,dy,$$

and therefore, if $h \uparrow$ indicates that h is an increasing function, we get that

$$\left(\sup_{h\uparrow} \frac{\int_0^R h}{\int_0^\infty \frac{h(y)}{W(y)}\,dy}\right)\lambda_{T_A f}(R) \le C.$$

The result now follows by computing the exact expression of the function between parenthesis by a simple change of variable and the well known fact that

$$\sup_{f\downarrow} \frac{\int_0^\infty f(t)v(t)dt}{\int_0^\infty f(t)u(t)dt} = \sup_{r>0} \frac{\int_0^r v(t)dt}{\int_0^r u(t)dt}.$$

where $f \downarrow$ indicates that f is a decreasing function. □

Theorem 6.2 *If T_θ satisfies Condition-A uniformly in θ and, for some u_0,*

$$\sup_\theta \|T_\theta \chi_E\|_{L^{1,\infty}(u_0)} \lesssim u_0(E),$$

then, for every $\beta > 0$,

$$\sup_{y>0} \frac{y}{(1+\log^+(1+\log^+\frac{1}{y}))^\beta} \int_{\{|T_A\chi_E|>y\}} u_0(x)dx \lesssim C_\beta\, u_0(E).$$

Proof Let us fix $q > 1$ so that $\beta > (q-1)(s+1+\alpha)$, and let us choose $W(y) = \frac{y^q}{(1+\log^+(1+\log^+\frac{1}{y}))^\beta}$. Then a simple computation shows that, in this case,

$$(q-1)W(y) \lesssim \widetilde{W}(y),$$

and hence, by Proposition 6.3,

$$\sup_{y>0} \frac{y^q}{\left(1+\log(1+\log_+\frac{1}{y})\right)^\beta} \int_{\{|T_A\chi_E(x)|>y\}} u_0(x)(M\chi_E(x))^{(1-q)}dx \lesssim C_\beta\, u_0(E).$$

Therefore,

$$\begin{aligned}
&\frac{y}{(1+\log^+(1+\log^+\frac{1}{y}))^\beta}\int_{\{|T_A\chi_E|>y\}} u_0(x)dx \\
&\le \|u_0\|_{A_1} u_0(E) + \frac{y}{(1+\log^+(1+\log^+\frac{1}{y}))^\beta}\int_{\{|T_A\chi_E|>y, M\chi_E(x)\le y\}} u_0(x)dx \\
&\le \|u_0\|_{A_1} u_0(E) + \frac{y^q}{(1+\log^+(1+\log^+\frac{1}{y}))^\beta}\int_{\{|T_A\chi_E|>y\}} u_0(x)(M\chi_E(x))^{1-q}dx \\
&\lesssim C_\beta\, u_0(E),
\end{aligned}$$

and the result follows. □

Finally, the proof of Corollary 6.1 follows immediately.

References

1. S. Baena-Miret, M.J. Carro, On weak-type (1, 1) for averaging type operators. https://arxiv.org/abs/2301.05201
2. C. Bennett, R. Sharpley, *Interpolation of Operators*. Pure and Applied Mathematics, vol. 129 (Academic Press, Inc., Boston, 1988)
3. S. Buckley, Estimates for operator norms on weighted spaces and reverse Jensen inequalities. Trans. Am. Math. Soc. **340**, 253–272 (1993)
4. M.J. Carro, Weak type estimates for averaging operators, in *Function Spaces and Applications* (Delhi, 1997) (Narosa, New Delhi, 2000), pp. 36–43
5. M.J. Carro, C. Domingo-Salazar, Endpoint estimates for Rubio de Francia operators. Trans. Am. Math. Soc. **371**(3), 1621–1648 (2019)
6. M.J. Carro, J. Soria, Restricted weak-type Rubio de Francia extrapolation for $p > p_0$ with applications to exponential integrability estimates. Adv. Math. **290**, 888–918 (2016)
7. D. Cruz-Uribe, J.M. Martell, C. Pérez, *Weights, Extrapolation and the Theory of Rubio de Francia*, No. 223. Operator Theory: Advances and Applications, vol. 215 (Springer, Berlin, 2011)
8. A. de la Torre, On the adjoint of the maximal function, in *Function Spaces, Differential Operators and Nonlinear Analysis* (Paseky nad Jizerou, 1995), Prometheus, Prague (1996), pp. 189–194
9. O. Dragičević, L. Grafakos, M.C. Pereyra, S. Petermichl, Extrapolation and sharp norm estimates for classical operators on weighted Lebesgue spaces. Publ. Mat. **49**, 73–91 (2005)
10. J. Duoandikoetxea, Extrapolation of weights revisited: new proofs and sharp bounds. J. Funct. Anal. **260**(6), 1886–1901 (2011)
11. J. García-Cuerva, An extrapolation theorem in the theory of A_p weights. Proc. Am. Math. Soc. **87**, 422–426 (1983)
12. J. García-Cuerva, J.L. Rubio de Francia, *Weighted Norm Inequalities and Related Topics*, North-Holland Mathematics Studies, vol. 116, Notas de Matemática [Mathematical Notes], 104 (North-Holland Publishing Co., Amsterdam, 1985)
13. L. Grafakos, *Modern Fourier Analysis*, 2nd edn. Graduate Texts in Mathematics, vol. 250 (Springer, New York, 2009)

14. P.T. Hytönen, M.T. Lacey, H. Martikainen, T. Orponen, M.C. Reguera, E. Sawyer, I. Uriarte-Tuero, Weak and strong type estimates for maximal truncations of Calderón-Zygmund operators on Ap weighted spaces. J. Anal. Math. **118**(1), 177–220 (2012)
15. B. Jawerth, M. Milman, Extrapolation theory with applications. Mem. Am. Math. Soc. **89**(440), iv+82 pp. (1991)
16. T. Luque, C. Pérez, E. Rela, Optimal exponents in weighted estimates without examples. Math. Res. Lett. **22**(1), 183–201 (2015)
17. B. Muckenhoupt, Weighted norm inequalities for the Hardy maximal function. Trans. Am. Math. Soc. **165**, 207–226 (1972)
18. J.L. Rubio de Francia, Factorization theory and A_p weights. Am. J. Math. **106**(3), 533–547 (1984)
19. S. Yano, Notes on Fourier analysis. XXIX. An extrapolation theorem. J. Math. Soc. Jpn. **3**, 296–305 (1951)

Chapter 7
On Complex Analytic Tools, and the Holomorphic Rotation Methods

Ronald R. Coifman, Jacques Peyrière, and Guido Weiss

Abstract An extended holomorphic Radon transform enables a "method of rotations" to lift one dimensional adapted nonlinear holomorphic analysis tools to higher dimensions. In particular the various multiscale orthogonal bases of holomorphic functions in Hardy spaces in one dimension give rise to orthogonal bases of Harmonic functions on higher dimensional half spaces. We provide an overview of recent progress in nonlinear approximation using holomorphy, and discuss their extension to high dimension through the Radon transform.

7.1 Introduction

This paper in honor of Guido Weiss was written posthumously, jointly with him, as we had all of his initial notes and ideas related to the program described below.

Our task, here, is to recount ideas, explorations, and visions that Guido his collaborators and students, developed over the last 60 years. To point out the connection of ideas between the original views of the interplay between complex and real analysis as envisioned by Zygmund and his students Calderón, Guido Weiss, Eli Stein,and many others, 70 years ago, and the current approaches

Guido Weiss died before publication of this work was completed.

R. R. Coifman (✉)
Department of Mathematics, Program in Applied Mathematics, Yale University, New Haven, CT, USA
e-mail: coifman-ronald@yale.edu

J. Peyrière
Institut de Mathématiques d'Orsay, CNRS, Université Paris-Saclay, Orsay, France
e-mail: jacques.peyriere@universite-paris-saclay.fr

G. Weiss
1928–2021, St. Louis, MO, USA

E. Hernández et al. (eds.), *The Mathematical Heritage of Guido Weiss*,
Applied and Numerical Harmonic Analysis,
https://doi.org/10.1007/978-3-031-76793-7_7

introducing nonlinear multi resolution analysis for the organization and processing of complicated functions.

It was Zygmund's view that harmonic analysis provides the infrastructure linking all areas of analysis, from complex analysis to partial differential equations, to probability, number theory, and geometry.

In particular he pushed forward the idea that the remarkable tools of complex analysis, which include; contour integration, conformal mappings, factorization, tools which were used to provide miraculous proofs in real analysis, should be deciphered and converted to real variable tools. Together with Calderón, they bucked the trend for abstraction, prevalent at the time, and formed a school pushing forward this interplay between real and complex analysis. A principal bridge was provided by real variable methods, multiscale analysis, Littlewood Paley theory, and related Calderon representation formulas. Our aim, here, is to elaborate on the magic of complex analysis and indicate potential applications in Higher dimensions. An old idea of Calderón and Zygmund, the so called "rotation method", enabled the reductions of the study of L^p estimates for multi dimensional singular integrals to a superposition,over all directions,of Hilbert transforms. Thereby allowing the use of one complex variable methods. A related idea was the invention of systems of Harmonic functions satisfying generalised Cauchy Riemann equations, such as the Riesz systems exploiting their special properties [3].

Our goal is to extend these ideas to enable remarkable nonlinear complex analytic tools for the adapted analysis of functions in one variable, to apply in higher dimensions.

Guido has been pushing the idea that factorization theorems like Blaschke products are a key to a variety of nonlinear analytic methods. Our goal here is to demonstrate this point, deriving amazing approximation theorems, in one variable, and opening doors to higher dimensional applications. Application in which each harmonic function is the average of special holomorphic functions in planes and constant in orthogonal directions.

We start by describing recent developments in nonlinear complex analysis, exploiting the tools of factorization and composition. In particular we will sketch methods extending conventional Fourier analysis, exploiting both phase and amplitudes of holomorphic functions. The "miracles of nonlinear complex analysis", such as factorization and composition of functions lead to new versions of holomorphic wavelets, and relate them to multiscale dynamical systems.

Our story interlaces the role of the phase of signals with their analytic/geometric properties. The Blaschke factors are a key ingredient, in building analytic tools, starting with the Malmquist-Takenaka orthonormal bases of the Hardy space $\mathrm{H}^2(\mathbb{T})$, continuing with "best" adapted bases obtained through phase unwinding, and describing relations to composition of Blaschke products and their dynamics (on the disc and upper half plane). Specifically we construct multiscale orthonormal holomorphic wavelet bases, generalized scaled holomorphic orthogonal bases, to dynamical systems, obtained by composing Blaschke products.

We also, remark, that the phase of a Blaschke product is a one layer neural net with (arctan as an activation sigmoid) and that the composition is a "Deep Neural Net" whose "depth" is the number of compositions. Our results provide a wealth of related libraries of orthogonal bases.

We sketch these ideas in various "vignette" subsections and refer for more details on analytic methods [1], related to the Blaschke based nonlinear phase unwinding decompositions [2, 4, 9]. We also consider orthogonal decompositions of invariant subspaces of Hardy spaces. In particular we constructed a multiscale decomposition, described below, of the Hardy space of the upper half-plane.

Such a decomposition can be carried in the unit disk by conformal mapping. A somewhat different multiscale decomposition of the space $\mathrm{H}^2(\mathbb{T})$ has been constructed by using Malmquist-Takenaka bases associated with Blaschke products whose zeroes are $(1-2^{-n})\mathrm{e}^{2\mathrm{i}\pi j/2^n}$ where $n \geq 1$ and $0 \leq j < 2^n$ [6]. Here we provide a variety of multiscale decompositions by considering iterations of Blaschke products.

In the next chapter we will show how with help of an extended Radon transform we can introduce a method of rotations to enable us to lift the one dimensional tools to higher dimensions. In particular the various orthogonal bases of holomorphic functions in one dimension, give rise to orthogonal bases of Harmonic functions in the higher dimensional upper half space.

7.2 Preliminaries and Notation

For $p \geq 1$, $\mathrm{H}^p(\mathbb{T})$ stands for the space of analytic functions f on the unit disk $\mathbb{D}$ such that

$$\sup_{0<r<1} \int_0^{2\pi} |f(r\mathrm{e}^{\mathrm{i}\theta})|^p \frac{\mathrm{d}\theta}{2\pi} < +\infty.$$

Such functions have boundary values almost everywhere, and the Hardy space $\mathrm{H}^p(\mathbb{T})$ can be identified with the set of L^p functions on the torus $\mathbb{T} = \partial\mathbb{D}$ whose Fourier coefficients of negative order vanish. We will alternate between analysis on the disk, and the parallel theory for analytic functions on the upper half plane $\mathbb{H} = \{x + \mathrm{i}y \ : \ y > 0\}$. The space of analytic functions f on $\mathbb{H}$ such that

$$\sup_{y>0} \|f(\cdot + \mathrm{i}y)\|_{L^p(\mathbb{R})} < +\infty$$

is denoted by $\mathrm{H}^p(\mathbb{R})$. These functions have boundary values in $L^p(\mathbb{R})$ when $p \geq 1$. The space $\mathrm{H}^p(\mathbb{R})$ is identified to the space of L^p functions whose Fourier transform vanishes on the negative half line $(-\infty, 0)$.

7.3 Analysis on the Upper Half Plane

We present some known results [1], without proof. In this section one simply writes H^2 instead of $\mathsf{H}^2(\mathbb{R})$.

7.3.1 Malmquist-Takenaka Bases

Let $(a_j)_{1\leq j}$ be a sequence (finite or not) of complex numbers with positive imaginary parts and such that

$$\sum_{j\geq 0} \frac{\Im a_j}{1+|a_j|^2} < +\infty. \tag{7.1}$$

The corresponding Blaschke product is

$$\mathsf{B}(x) = \prod_{j\geq 0} \frac{\left|1+a_j^2\right|}{1+a_j^2} \frac{x-a_j}{x-\overline{a}_j},$$

where, 0/0, which appears if $a_j = \mathrm{i}$, should be understood as 1. The factors $\dfrac{\left|1+a_j^2\right|}{1+a_j^2}$ insure the convergence of this product when there are infinitely many zeroes. But, in some situations, it is more convenient to use other convergence factors as we shall see below.

Whether the series (7.1) is convergent or not, one defines (for $n \geq 0$) the functions

$$\phi_n(x) = \frac{1}{\sqrt{\pi}} \left(\prod_{0\leq j<n} \frac{x-a_j}{x-\overline{a}_j} \right) \frac{1}{x-\overline{a}_n}.$$

Then these functions form an orthonormal system in H^2. If the series (7.1) diverges, it is a Malmquist-Takenaka orthonormal basis of H^2, otherwise it is a basis of the orthogonal complement of $\mathsf{B}\,\mathsf{H}^2$ in H^2.

We remark that roughly a hundred years ago these bases were constructed [7, 11] through a Gram Schmidt orthogonalization of the list of rational functions with poles in the lower half plane.

Observe that for a rational function with a pole of order M at a the corresponding M basis functions have the form

$$\phi_n(x) = \mathrm{e}^{\mathrm{i}n\theta(x)} \frac{1}{x - \overline{a}_n} \qquad (n = 1..M).$$

These are localized "Fourier like" basis functions around the real part of a scaled by the imaginary part.

7.3.2 *Example of a Multiscale Wavelet Decomposition*

The infinite Blaschke products

$$G_n(x) = \prod_{j \le n} \frac{j - \mathrm{i}}{j + \mathrm{i}} \frac{x - j - \mathrm{i}}{x - j + \mathrm{i}} \quad \text{and} \quad G(x) = \prod_{j \in \mathbb{Z}} \frac{j - \mathrm{i}}{j + \mathrm{i}} \frac{x - j - \mathrm{i}}{x - j + \mathrm{i}}$$

can be expressed in terms of known functions:

$$G_n(x) = \frac{\Gamma(-\mathrm{i} - n)}{\Gamma(\mathrm{i} - n)} \frac{\Gamma(x - n + \mathrm{i})}{\Gamma(x - n - \mathrm{i})} \quad \text{and} \quad G(x) = \frac{\sin \pi(\mathrm{i} - x)}{\sin \pi(\mathrm{i} + x)}. \tag{7.2}$$

7.3.3 *An Orthonormal System*

Consider the function $\phi(x) = \dfrac{\Gamma(x - 1 + \mathrm{i})}{\sqrt{\pi}\,\Gamma(x - \mathrm{i})}$. It is easily checked that

$$\phi(x - n) = \frac{\Gamma(\mathrm{i} - n)}{\Gamma(-\mathrm{i} - n)} \frac{G_n(x)}{\sqrt{\pi}\big(x - (n + 1) + \mathrm{i}\big)}.$$

Set $\phi_n(x) = \phi(x - n)$. For fixed m, the functions ϕ_n / G_m, for $n \ge m$, form a Malmquist-Takenaka basis of $(G/G_m)\mathrm{H}^2$. In other terms, the functions ϕ_n, for $n \ge m$, form an orthonormal basis of $G_m\mathrm{H}^2 \ominus G\mathrm{H}^2$. This means that the functions ϕ_n (for $n \in \mathbb{Z}$) form a Malmquist-Takenaka basis of the orthogonal complement of $G\mathrm{H}^2$ in H^2.

7.3.3.1 Multiscale Decomposition

As $|1 - G(2^n x)| \le C 2^n$ all the products $\mathscr{B}_n(x) = \prod_{j<n} G(2^j x)$ are convergent and $\lim_{n \to -\infty} \mathscr{B}_n = 1$ uniformly.

Let $\mathscr{B} = \mathscr{B}_0$. Obviously, $\mathscr{B}_n(x) = \mathscr{B}(2^n x)$. Consider the following subspaces of H^2: $\mathsf{E}_n = \mathscr{B}_n \mathsf{H}^2$. This is a decreasing sequence. The space $\mathsf{E}_{+\infty} = \bigcap_{n\in\mathbb{Z}} \mathsf{E}_n$ is equal to $\{0\}$ since a function orthogonal to this space would have too many zeros, and the space $\mathsf{E}_{-\infty}$ = closure of $\bigcup_{n\in\mathbb{Z}} \mathsf{E}_n$ is equal to H^2 since $\mathscr{B}_n$ converges uniformly to 1 when n goes to $-\infty$.

For all n and j, let

$$\phi_{n,j}(x) = 2^{n/2}\phi(2^n x - j)\mathscr{B}(2^n x).$$

Then, for all n, $(\phi_{n,j})_{j\in\mathbb{Z}}$ is an orthonormal basis of $\mathsf{E}_n \ominus \mathsf{E}_{n+1}$. We conclude that $(\phi_{n,j})_{n,j\in\mathbb{Z}}$ is an orthonormal basis of H^2.

7.4 Adapted MT Bases, "Phase Unwinding"

We now find a "best" adapted Malmquist Takenaka basis to analyze or unwind the oscillations of a given function.

The idea is to peel off the oscillation of a function by dividing by its Blaschke product defined by the zeroes of the function, this procedure is iterated to yield an expansion in an orthogonal collection of functions or Blaschke products which of course are naturally embedded in a MT basis, once the zeroes are ordered.

7.4.1 The Unwinding Series

There is a natural way to iterate the Blaschke factorization, it is inspired by the power series expansion of a holomorphic function on the disk. If G has no zeroes inside $\mathbb{D}$, its Blaschke factorization is the trivial one $G = 1 \cdot G$, however, the function $G(z) - G(0)$ certainly has at least one root inside the unit disk $\mathbb{D}$ and will therefore yield some nontrivial Blaschke factorization $G(z) - G(0) = \mathsf{B}_1 G_1$. We write

$$\begin{aligned} F(z) &= \mathsf{B}(z) \cdot G(z) = \mathsf{B}(z) \cdot \big(G(0) + (G(z) - G(0)\big) \\ &= \mathsf{B}(z) \cdot \big(G(0) + \mathsf{B}_1(z)G_1(z)\big) = G(0)\mathsf{B}z + \mathsf{B}(z)\mathsf{B}_1(z)G_1(z). \end{aligned}$$

An iterative application gives rise to the *unwinding series*

$$F = a_1\mathsf{B}_1 + a_2\mathsf{B}_1\mathsf{B}_2 + a_3\mathsf{B}_1\mathsf{B}_2\mathsf{B}_3 + a_4\mathsf{B}_1\mathsf{B}_2\mathsf{B}_3\mathsf{B}_4 + \ldots$$

This orthogonal expansion first appeared in the PhD thesis of Michel Nahon [9] and independently by T. Qian in [8, 10] Detailed approximations in smoothness spaces were derived by S. Steinerberger in [2]. Given a general function F it is not numerically feasible to actually compute the roots of the function; a crucial insight in [9] is that this is not necessary—one can numerically obtain the Blaschke product in a stable way by using a method that was first mentioned in a paper of Guido and Mary Weiss [12] and has been investigated with respect to stability by Nahon [9] Using the boundedness of the Hilbert transform one can prove easily convergence in $L^p, 1 < p < \infty$.

7.4.2 *The Fast Algorithm of Guido and Mary Weiss [12]*

Our starting point is the theorem that any Hardy function can be decomposed as

$$F = \mathsf{B} \cdot G,$$

where B is a Blaschke product, that is a function of the form

$$\mathsf{B}(z) = z^m \prod_{i \in I} \frac{\overline{a_i}}{|a_i|} \frac{z - a_i}{1 - \overline{a_i} z},$$

where $m \in \mathbb{N}_0$ and $a_1, a_2, \cdots \in \mathbb{D}$ are zeroes inside the unit disk $\mathbb{D}$ and G has no roots in $\mathbb{D}$. For $|z| = 1$ we have $|\mathsf{B}(z)| = 1$ which motivates the analogy

$$\mathsf{B} \sim \text{frequency and } G \sim \text{amplitude}$$

for the function restricted to the boundary. However, the function G need not be constant: it can be any function that never vanishes inside the unit disk. If F has roots inside the unit disk, then the Blaschke factorization $F = \mathsf{B} \cdot G$ is going to be nontrivial (meaning $\mathsf{B} \not\equiv 1$ and $G \not\equiv F$). G should be 'simpler' than F because the winding number around the origin decreases.

In fact since $|F| = |G|$ and $\ln(G)$ is analytic in the disk we have formally that $G = \exp(\ln|F| + \mathrm{i}(\ln|F|)^{\sim})) = \exp(\mathscr{H}(\ln|F|))$ where $\mathscr{H}$ is the projection onto the Hardy space. and $\mathsf{B} = F/G$. G can be computed easily using the FFT [9].

7.4.3 *A Remarkable Unwinding*

The following is an explicit unwinding of a singular inner function in the upper half plane illustrating this amazing approximation of $\exp \frac{2\mathrm{i}\pi}{x}$:

$$\exp \frac{2\mathrm{i}\pi}{x} = \mathrm{e}^{-2\pi} + \left(1 - \mathrm{e}^{-4\pi}\right) \sum_{n \geq 0} (-1)^n \mathrm{e}^{-2n\pi} B(x)^{n+1},$$

where B is a Blaschke product whose zeros are $\{1/(j + \mathrm{i})\}_{j \in \mathbb{Z}}$.

7.5 Geometric Function Theory: The Role of Compositions of Blaschke Products

7.5.1 *Iteration of Blaschke Products*

We claim that by building Blaschke product through composition we open up rich dynamical structures, and libraries of corresponding Malmquist Takenaka bases.

We are interested in iteration of finite Blaschke products

$$\mathsf{B}(z) = \mathrm{e}^{\mathrm{i}\theta} z^{\nu} \prod_{j=1}^{\mu} \frac{z + a_j}{1 + \overline{a}_j z},$$

where μ and ν are nonnegative integers and the a_j are complex numbers of modulus less than 1.

It is well known that $\mathbb{T}$ and $\mathbb{D}$ are globally invariant under B, as well as the complement of $\overline{\mathbb{D}}$ in the Riemann sphere.

A careful discussion can be found in [1]. Here is the main result.

Theorem 7.1 *Let* B *be a finite Blaschke product with a fixed point* α *inside the unit disk. Then there exists a sequence* $\alpha, a_1, a_2, \ldots, a_j, \ldots$ *of complex numbers in the unit disk and an increasing sequence* $(\nu_j)_{j\geq 1}$ *of positive integers such that* $a_1, a_2, \ldots, a_{\nu_n}$ *are the zeros, counted according to their multiplicity, of* B_n *(the nth iterate of* B*). Moreover* $\sum_{j\geq 1}(1 - |a_j|) = +\infty$*. Also,* B_n *converges towards* α *uniformly on compact subsets of the open unit disk.*

7.5.2 *Dynamic Multiscale Analysis Through Composition of Blaschke Products*

Each Blaschke product B defines invariant subspaces of H^p. The projection on this space is given by the kernel $\dfrac{\mathsf{B}(z)\overline{\mathsf{B}(w)}}{z - w}$. This projection is continuous for $1 < p < +\infty$.

Let F be a Blaschke product of degree at least 2 with a fixed point inside the unit disk. Its iterates define a hierarchy of nested invariant subspaces $\mathsf{E}_n = F_n \mathsf{H}^2$.

Due to Theorem 7.1, $\bigcap_{n\geq 1} \mathsf{E}_n = \{0\}$.

The Takenaka construction provides orthonormal bases of $E_n \ominus E_{n+1}$. But this is not canonical as it depends on an ordering of the zeros of F_{n+1}/F_n.

Figure 7.1 shows 1st, 3rd, and 5th iterates of $F(z) = z(z - 2^{-1})/(1 - 2^{-1}z)$. Figure 7.2 displays the phase for the fourth iterate of $F(z) = z^2(z-2^{-2})/(1-2^{-2}z)$. The upper pictures display the phases modulo 2π (values in the interval $(-\pi, \pi]$) of theses Blaschke products while the lower pictures display minus the logarithms of their absolute value. The coordinates (x, y) correspond to the point e^{-y+ix}. On these figures it is easy to locate the zeros, specially by looking at the phase which then has an abrupt jump.

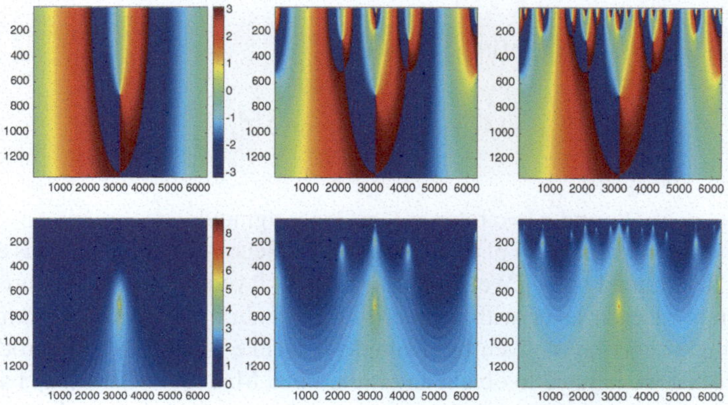

Fig. 7.1 The argument and the absolute value of $F(z)$, $F^{(3)}(z)$, and $F^{(5}(z)$, with $F(z) = \frac{z(z-2^{-1})}{1-2^{-1}z}$ and $z = \exp(-y + ix)$

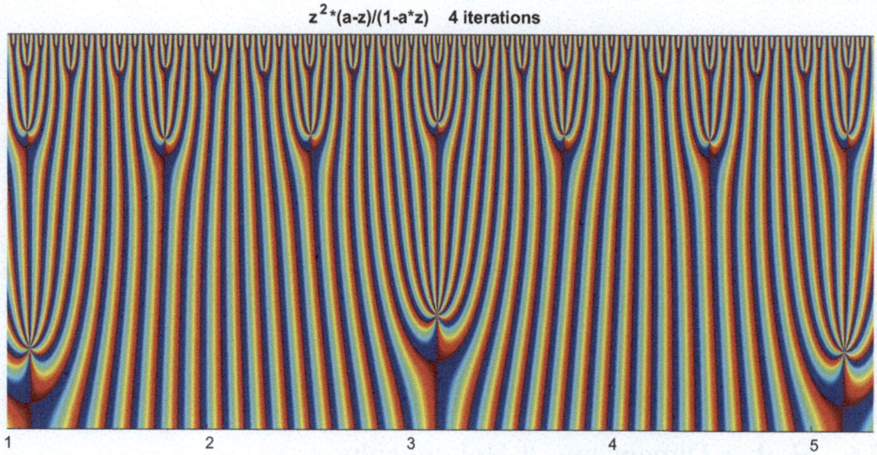

Fig. 7.2 The multiscale view of the argument of $F^{(4)}(z)$, with $F(z) = \frac{z^2(z-2^{-2})}{1-2^{-2}z}$ and $z = \exp(-y + ix)$

7.5.3 Remarks on Iteration of Blaschke Products as a "Deep Neural Net"

In the upper half plane let $(a_j)_{1\leq j}$ be a finite sequence of complex numbers with positive imaginary parts. The corresponding Blaschke product on the line is

$$\mathsf{B}(x) = \prod_{j\geq 0} \frac{x - a_j}{x - \overline{a}_j}.$$

We can write $\mathsf{B}(x) = \exp\big(\mathrm{i}\theta(x)\big)$, where

$$\theta(x) = \sum_{j\geq 0} \sigma\big((x-\alpha_j)/\beta_j\big)$$

with $a_j = \alpha_j + \mathrm{i}\beta_j$ and $\sigma = \arctan x + \pi/2$ is a sigmoid.

This is precisely the form of a single layer in a Neural Net, each unit has a weight and bias determined by a_j. We obtain the various layers of a deep net through the composition of each layer with a preceding layer. In our preceding examples we took a single short layer given by a simple Blaschke term with two zeroes in the first layer that we iterated to obtain an orthonormal Malmquist Takenaka basis (we could have composed different elementary products at each layer), demonstrating the versatility of the method to generate highly complex functional representations.

As an example let $F(z)$ be mapped from G, (7.2) in the section on wavelet construction.

$$F(z) = G(w) = \frac{\sin(\pi(\mathrm{i} - w))}{\sin(\pi(\mathrm{i} + w))} \quad \text{with} \quad w = \frac{\mathrm{i}(1-z)}{(1+z)}.$$

We can view the phase of F as a neural layer which when composed with itself results in a phase which is a two layer neural net represented graphically in Fig. 7.3.

Where each end of a color droplet corresponds to one zero or unit of the two layer net.

We refer to Daubechies et al. [5] for a description of a similar iteration for piecewise affine functions in which simple affine functions play the role of a Blaschke product.

7.6 Higher Dimensions, θ-Holomorphy

Our goal is to explore methodologies to use the remarkable analytic approximation theorems described above to enable deeper understanding of real analysis, in higher dimensions. We know that Blaschke factorization do not exist, nevertheless there are remarkable bases that can be lifted.

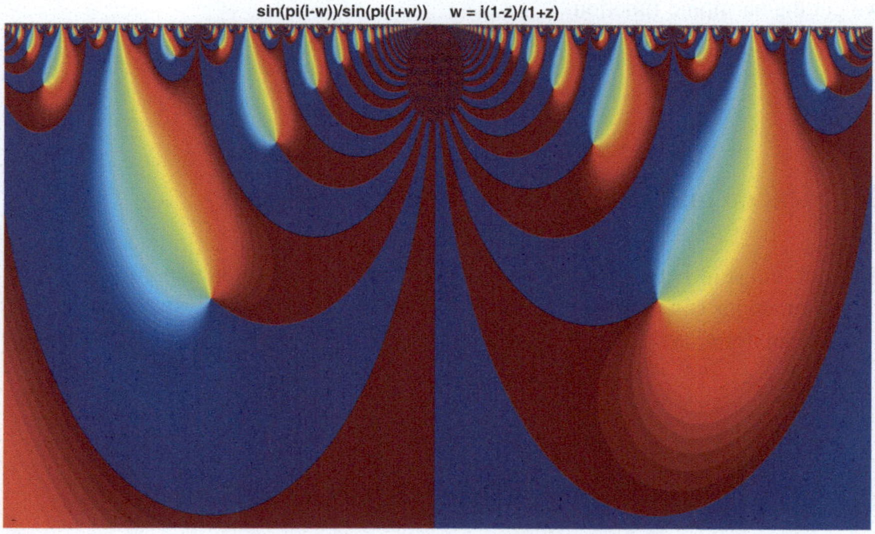

Fig. 7.3 Two iterations of $F(z) = G(w) = \frac{\sin(\pi(\mathrm{i}-w))}{\sin(\pi(\mathrm{i}+w))}$ with $w = \frac{\mathrm{i}(1-z)}{(1+z)}$

We start by observing that $Z_\theta = (x \cdot \theta + \mathrm{i}y) = t + \mathrm{i}y$ is harmonic in (x, y),(in 3 dimensions) and so is Z_θ^k This is a harmonic homogeneous polynomial of degree k in (x_1, x_2, y) that is constant in the direction perpendicular to θ. here we identified θ with $(\cos\theta, \sin\theta)$. It is well known [3] that

$$\frac{1}{2\pi}\int_0^{2\pi} \mathrm{e}^{-\mathrm{i}l\theta} Z_\theta^k \,\mathrm{d}\theta = Y_l^k(x_1, x_2, y) \quad (|l| \le k)$$

is the standard orthogonal basis of spherical Harmonics in 3 dimensions.

As a consequence we see that any Harmonic function $U(x, y)$ is a superposition over θ of holomorphic functions in Z_θ, more explicitly a Power series in Z_θ with coefficients depending on θ.

$$U(x, y) = \frac{1}{2\pi}\int_0^{2\pi} F_\theta(Z_\theta)\,\mathrm{d}\theta = \frac{1}{2\pi}\int_0^{2\pi} \sum_{k\ge 0} a_k(\theta) Z_\theta^k, \mathrm{d}\theta$$

where $a_k(\theta)$ is a trigonometric polynomial of degree k.

Another example, taking

$$F_\theta(Z_\theta) = \mathrm{e}^{-2\mathrm{i}\pi(rZ_\theta)}\mathrm{e}^{-2\mathrm{i}\pi(k\theta)}.$$

we get the harmonic function

$$J_k(r\sqrt{x_1^2 + x_2^2})\mathrm{e}^{-2\mathrm{i}\pi(k\phi)}\mathrm{e}^{-yr}$$

7.7 Radon and Fourier in the Upper Half Space

This relationship between holomorphic functions in planes as generating all harmonic functions can most easily be explored through Fourier analysis. We define the Radon transform, and relate it to the Fourier transform to lead to the θ-holomorphic version.

Let

$$\mathsf{R}_\theta f(t) = \int_{x\in\theta^\perp} f(t\theta + x)\,\mathrm{d}x. \tag{7.3}$$

Obviously $\mathsf{R}_{-\theta} f(t) = \mathsf{R}_\theta f(-t)$. The formula $\widehat{\mathsf{R}_\theta f}(t) = \hat{f}(u\theta)$ for Fourier transforms is well known.

For $f \in L^1(\mathbb{R}^n)$, consider its harmonic extension u to $\mathbb{R}_+^{n+1}$. For $x \in \mathbb{R}^n$ and $y > 0$ we have

$$\begin{aligned} u(x, y) = f \star \mathsf{P}_y(x) &= \int \mathrm{e}^{2\mathrm{i}\pi x\cdot\xi}\mathrm{e}^{-2\pi|\xi|y}\hat{f}(\xi)\,\mathrm{d}\xi \\ &= \int_{S_{n-1}} \left(\int_0^\infty \mathrm{e}^{2\mathrm{i}\pi r(x\cdot\theta+\mathrm{i}y)}\hat{f}(r\theta)r^{n-1}\mathrm{d}r\right)\mathrm{d}\theta \\ &= \int_{S_{n-1}} F_\theta(x\cdot\theta + \mathrm{i}y)\,\mathrm{d}\theta, \end{aligned}$$

where

$$F_\theta(z) = \int_0^\infty \mathrm{e}^{2\mathrm{i}\pi rz}\hat{f}(r\theta)r^{n-1}\mathrm{d}r = \int_0^\infty \mathrm{e}^{2\mathrm{i}\pi rz}\widehat{\mathsf{R}_\theta f}(r)\,r^{n-1}\mathrm{d}r. \tag{7.4}$$

When $n = 2$, we have

$$\widehat{F_\theta}(t) = \widehat{\mathsf{R}_\theta f}(t)\,t\,\mathbf{1}_{(0,+\infty)}(t) = \frac{1}{2\mathrm{i}\pi}\widehat{\mathsf{D}\mathsf{R}_\theta f}(t)\,\mathbf{1}_{(0,+\infty)}(t).$$

So, for $\Im z > 0$,

$$F_\theta(z) = -\frac{1}{4\pi^2}\int_{-\infty}^{\infty}\frac{\mathrm{d}(\mathsf{R}_\theta f(t))}{t - z} = -\frac{1}{4\pi^2}\int_{-\infty}^{+\infty}\frac{\mathsf{R}_\theta f(t)}{(t - z)^2}\,\mathrm{d}t.$$

For general n we get

$$F_\theta(z) = \frac{(n-1)!}{(2\mathrm{i}\pi)^n} \int_{-\infty}^{+\infty} \frac{\mathrm{R}_\theta f(t)}{(t-z)^n}\,\mathrm{d}t.$$

7.7.1 *Some Isometries*

We describe some computations in the case when $n = 2$ and mention the case of higher dimension at the end of this section.

In view of (7.4)

$$\widehat{F_\theta(\cdot + \mathrm{i}y)}(r) = \hat{f}(r\theta)\,\mathrm{e}^{-2\pi r y} r \mathbf{1}_{(0,\infty)}(r). \tag{7.5}$$

Hence, the Plancherel identity yields

$$\begin{aligned}\int_0^\infty \mathrm{d}y \int_{-\infty}^{+\infty} |F_\theta(t+\mathrm{i}y)|^2 \mathrm{d}t &= \int_0^\infty \int_0^\infty |\hat{f}(r\theta)|^2 r^2 \mathrm{e}^{-4\pi r y} \mathrm{d}r\mathrm{d}y \\ &= \frac{1}{4\pi} \int_0^\infty |\hat{f}(r\theta)|^2 r\mathrm{d}r\end{aligned}$$

Let $\|F\|_B^2 = \int_0^\infty \int_{-\infty}^{+\infty} |F_\theta(t+\mathrm{i}y)|^2 \mathrm{d}y\mathrm{d}t$ (this is the norm of the Bergman space on the upper half plane). Then

$$4\pi \int_0^{2\pi} \|F_\theta\|_B^2 \mathrm{d}\theta = \iint_{(0,+\infty)\times(0.2\pi)} \left|\hat{f}(r\theta)\right|^2 r\,\mathrm{d}r\mathrm{d}\theta = \|f\|^2_{L^2(\mathbb{R}^n)}. \tag{7.6}$$

We have

$$\frac{\partial u(x,y)}{\partial y} = -2\pi \int_{\mathbb{R}^2} \mathrm{e}^{2\mathrm{i}\pi\xi\cdot x} |\xi| \mathrm{e}^{-2\pi|\xi|y} \hat{f}(\xi)\,\mathrm{d}\xi.$$

$$\begin{aligned}\iint_{\mathbb{R}^3_+} \left|\frac{\partial u(x,y)}{\partial y}\right|^2 \mathrm{d}x\mathrm{d}y &= (2\pi)^2 \int_{\mathbb{R}^2} \left(\int_0^\infty \mathrm{e}^{-4\pi|\xi|y} \mathrm{d}y\right) |\hat{f}(\xi)|^2 |\xi|^2 \mathrm{d}\xi \\ &= \pi \int_{\mathbb{R}^n} |\hat{f}(\xi)|^2 |\xi| \mathrm{d}\xi\end{aligned} \tag{7.7}$$

Equation (7.5) yields $\int_{-\infty}^{+\infty} |F_\theta(t)|^2\,\mathrm{d}t = \int_0^{\infty} |\hat{f}(r\theta)|^2 r^2 \mathrm{d}r$, and

$$\int_0^{2\pi} \|F_\theta\|^2_{\mathrm{H}^2(\mathbb{R})} \mathrm{d}\theta = \int_{\mathbb{R}^2} |\hat{f}(\xi)|^2 |\xi| \mathrm{d}\xi. \tag{7.8}$$

Formulas (7.7) and (7.8) together give

$$\int_0^{2\pi} \|F_\theta\|^2_{\mathrm{H}^2(\mathbb{R})} \mathrm{d}\theta = \frac{1}{\pi} \iint_{\mathbb{R}^3_+} \left| \frac{\partial u(x,y)}{\partial y} \right|^2 \mathrm{d}x\mathrm{d}y, \tag{7.9}$$

In higher dimension formulas (7.6) and (7.9) become

$$\int_{S_{n-1}} \mathrm{d}\theta \iint_{\mathbb{R}_+2n+1} |F_\theta(t+\mathrm{i}y)|^2 y^{n-2} \mathrm{d}t\mathrm{d}y = \frac{1}{(4\pi)^{n-1}} |f\|^2_{L^2(\mathbb{R}^n)}$$

and

$$\iint_{\mathbb{R}^{n+1}_+} \left| \frac{\partial^k u(x,y)}{\partial y^k} \right| y^{2k-n} \mathrm{d}x\mathrm{d}y$$

$$= \frac{(4\pi)^{n-1}\Gamma(2k-n+1)}{2^{2k}} \int_{S_{n-1}} \|F_\theta\|^2_{H^2(\mathbb{R})} \mathrm{d}\theta. \tag{7.10}$$

7.7.2 *Remarks; "Lifted Analysis" of Harmonic Functions*

One of our goals is to enable the application of some of the one dimensional analytic approximation tools to higher dimensions. We refer to Michel Nahon's thesis [9] in which he decomposes a function in the plane as a sum of functions whose Fourier transform live in thin wedges, as a tool to extract features (gradients of Phase) from an image of a fingerprint. This illustrates potential variants of our current approach.

We envision a function in two variables represented as a superposition of $F_\theta(t+iy)$, each of which is approximated to error ϵ leading to a harmonic function approximation of error ϵ in the Dirichlet space. Similar estimations with different mix of Hilbert spaces can be easily derived as in [2], leading to faster rates of convergence (when more regularity is present).

Another obvious application is the representation of a Calderón Zygmund operator given by a Fourier multiplier homogeneous of degree 0, $\Omega(\theta)$, simply by averaging $\Omega(\theta)F_\theta(t+iy)$.

The representation of these operators is a version of the rotation method (no parity required on Ω). Also it provides a local representation method for generalized conjugate functions or C-Z operators, just by using the local spherical Harmonics version of the θ-holomorphic representation. In particular Riesz transforms correspond to $\Omega(\theta) = (\cos\theta, \sin\theta)$.

7.7.3 A Natural Ortho-Basis in the Dirichlet Space

We now use identity (7.9) to transfer an orthonormal basis of the Hardy space H^2 to an orthonormal system in the Dirichlet space in $\mathbb{R}^3_+$.

Start from the basis $\frac{i}{\sqrt{\pi}}\left(\frac{z-i}{z+i}\right)^n \frac{1}{z+i}$ of H^2 (this corresponds to the Fourier basis in the disc, mapped to the upper half plane Hardy space). We consider the generating function

$$F(z) = \frac{\mathrm{i}}{\sqrt{2\pi^3}} \sum_{n\geq 0} (z-i)^n t^n / (z+i)^{n+1}$$

and compute $G(x, y) = \int_0^{2\pi} F(z_\theta)\, \mathrm{d}\theta$. We get

$$G(x, y) = \sqrt{2/\pi} / \sqrt{(\rho^2 + 2y + 1)t^2 - 2(1-\rho^2)t + \rho^2 - 2y + 1},$$

where $\rho = \sqrt{x_1^2 + x_2^2 + y^2}$.

This also can be written as

$$G(x, y) = \frac{\sqrt{2/\pi}}{\sqrt{\rho^2 + 2y + 1}\,\sqrt{a^2t^2 - 2bat + 1}},$$

if one sets $a = \sqrt{\frac{\rho^2-2y+1}{\rho^2+2y+1}}$ and $b = \frac{\rho^2-1}{\sqrt{(\rho^2+1)^2-4y^2}}$.

It results that the functions $\dfrac{\sqrt{2/\pi}\, a^n P_n(b)}{\sqrt{\rho^2+2y+1}}$, where the P_n are the Legendre polynomials, form an orthonormal system in the Dirichlet space.

To get an orthonormal basis of the Dirichlet space in 3 dimensions, it suffices to take $\dfrac{\sqrt{2/\pi}\, 2a^n P_n(b)\, \mathrm{e}^{\mathrm{i}k\theta}}{\sqrt{\rho^2+2y+1}}$, with $k \in \mathbb{Z}$ and $n \geq 0$.

Of course such computation can be done in higher dimension: isometry (7.10) allows to transfer orthonormal bases of H^2 to orthonormal systems in a suitable Dirichlet space.

7.7.4 Concluding Remarks and Potential Applications

As we all know complex methods, such as interpolation of operators, or the remarkable proofs by Calderón of the boundedness in L^2 of commutators with the Hilbert transform, or the Cauchy integral on Lipschitz curves are powerful tools. Over the years the goal has been to convert them into real variable methods. In parallel the quest for higher dimensional complex tools is continuing, see the examples [3] in which various systems generalizing holomorphic functions to higher dimension are studied. The point here, is that the infinite dimensional θ-holomorphic functions generate all of these systems through the choice of appropriate multipliers (as described for the Riesz system).

Our goal here was to describe recent nonlinear analytic tools in the classical setting. and transfer them to the higher dimensional real setting. Together with Guido Weiss we had observed [3] that all harmonic functions in higher dimensions are combinations of holomorphic functions on subplanes which are constant in normal directions. The recent developments in one dimension as well as the isometries described here, and the corresponding efficient approximation methods, open the door for applications in higher dimensions, such as image denoising. See [4] for the impact of unwinding on precision Doppler analysis in 1 dimension, which we expect to carry over to 2 or 3 dimensions.

Observe also that, for simplicity, we restricted our attention to 2 dimensions in cylindrical coordinates. We could have defined more generally power series in the variable $Z_\epsilon = (x \cdot \epsilon)$ where ϵ satisfying; $(\epsilon \cdot \epsilon) = 0$, represents a point on the complex quadric with $|\Re\epsilon| = 1, |\Im\epsilon| = 1.$, or a two dimensional plane spanned by $\Re\epsilon, \Im\epsilon.$

Clearly we can extend the preceding discussion to this setting. Where; $Z_\epsilon = (x \cdot \epsilon)$ is the point $t + is$ in the complex plane with coordinate $t\Re\epsilon + is\Im\epsilon.$

References

1. R.R. Coifman, J. Peyrière, Phase unwinding, or invariant subspace decompositions of Hardy Spaces. J. Fourier Anal. Appl. **25**, 684–695 (2019)
2. R.R. Coifman, S. Steinerberger, Nonlinear phase unwinding of functions. J. Fourier Anal. Appl. **23**, 778–809 (2016)
3. R.R. Coifman, G. Weiss, *Analyse Harmonique Non-Commutative sur Certains Espaces Homogenes*. Lecture Notes i Mathematics, vol. 242 (Springer, Berlin, 1971)
4. R.R. Coifman, S. Steinerberger, H.T. Wu, Carrier frequencies, holomorphy and unwinding (2016). arXiv preprint arXiv:1606.06475, arxiv.org
5. I. Daubechies, R. DeVore, S. Foucart, B. Hanin, G. Petrova, Nonlinear approximation and (deep) ReLU networks. arXiv:1905.02199v1 [cs.LG] (5 May 2019)
6. H.G. Feichtinger, M. Pap, Hyperbolic wavelets and multiresolution in the Hardy space of the upper half plane, in *Blaschke Products and Their Applications* (Springer, Berlin, 2013)
7. F. Malmquist, Sur la determination d'une classe de fonctions analytiques par leurs valeurs dans un ensemble donne de poits. C.R. 6ieme Cong. Math. Scand. (Kopenhagen, 1925), Copenhagen, Gjellerups (1926), pp. 253–259

8. W. Mi, T. Qian, F. Wan, A fast adaptive model reduction method based on Takenaka-Malmquist systems. Syst. Control Lett. **61**(1), 223–230 (2012)
9. M. Nahon, Dissertation, Yale University (2000)
10. T. Qian, I.T. Ho, I.T. Leong, Y.B. Wang, Adaptive decomposition of functions into pieces of non-negative instantaneous frequencies. Int. J. Wavelets Multiresolut. Inf. Process. **8**(5), 813–833 (2010)
11. S. Takenaka, On the orthogonal functions and a new formula of interpolation. Jpn. J. Math. II, **2**, 129–145 (1925)
12. G. Weiss, M. Weiss, A derivation of the main results of the theory of H^p-spaces. Rev. Un. Mat. Argentina **20**, 63–71 (1962)

Chapter 8
Speed of Convergence in an Ergodic Theorem

Leonardo Colzani

Dedicated to Guido Weiss, teacher and friend. Grazie Guido.

Abstract If α is a vector in $\mathbb{R}^d$ and $f(x)$ a function on the torus $\mathbb{T}^d$, define

$$\mathcal{A}_N f(x) = \frac{1}{2N+1} \sum_{n=-N}^{N} f(x + n\alpha),$$

$$\mathcal{B}_N f(x) = \frac{1}{N} \sum_{n=1-N}^{N-1} \left(1 - \frac{|n|}{N}\right) f(x + n\alpha).$$

More generally, for suitable weights $\left\{\Phi\left(N^{-1}n\right)\right\}_{n=-\infty}^{+\infty}$ and sequences of integers $\{a(n)\}_{n=-\infty}^{+\infty}$, define

$$C_N f(x) = \left(\sum_{n=-\infty}^{+\infty} \Phi\left(N^{-1}n\right)\right)^{-1} \sum_{n=-\infty}^{+\infty} \Phi\left(N^{-1}n\right) f(x + a(n)\alpha).$$

We estimate the speed of convergence to $\int_{\mathbb{T}^d} f(x)dx$ of these averages.

The mutual configurations of celestial bodies, conjunctions, quadratures, oppositions, have always aroused great interest in astronomy and astrology. Nicole Oresme in *Tractatus de commensurabilitate vel incommensurabilitate motuum celi*, written around 1360, states that two bodies in commensurable circular motions have a finite

L. Colzani (✉)
Dipartimento di Matematica, Università degli Studi di Milano Bicocca, Milano, Italy
e-mail: leonardo.colzani@unimib.it

E. Hernández et al. (eds.), *The Mathematical Heritage of Guido Weiss*,
Applied and Numerical Harmonic Analysis,
https://doi.org/10.1007/978-3-031-76793-7_8

number of points of conjunction, while in the case of incommensurable circular motions have infinitely many points of conjunction, and *No sector of a circle is so small that two such mobiles could not conjunct in it at some future time, and in which they had not previously conjuncted.* See [27] for an exposition and analysis of his proof. The result of Oresme is a special case of a classical theorem of Leopold Kronecker. If the number 1 together with the entries of a vector α in $\mathbb{R}^d$ are linearly independent over the rationals, then the sequence $\{n\alpha\}_{n=1}^{+\infty}$ is dense in the torus $\mathbb{T}^d = \mathbb{R}^d/\mathbb{Z}^d$. See Chapter XXIII in [10]. This implies that for every function $f(x)$ continuous and non constant on the torus the sequence $\{f(n\alpha)\}_{n=1}^{+\infty}$ does not have a limit as $n \to +\infty$. Another classical result of Piers Bohl, Waclaw Sierpinski, and Hermann Weyl, states that the sequence $\{n\alpha\}_{n=1}^{+\infty}$ is uniformly distributed in the torus, and the arithmetic means of the sequence $\{f(n\alpha)\}_{n=1}^{+\infty}$ converge to the integral of the function,

$$\lim_{N\to+\infty}\left\{\frac{1}{N}\sum_{n=1}^{N} f(n\alpha)\right\} = \int_{\mathbb{T}^d} f(x)dx.$$

For the one dimensional case see Definition 1.1, Theorem 1.1, Theorem 2.1, and Example 2.1 in Chapter 1 of [17], and for the multidimensional case see Section 6 in Chapter 1 of [17], or Section 6.3 in Chapter 6 of [26]. More generally, Weyl has proved that if $\{a(n)\}_{n=-\infty}^{+\infty}$ is an arbitrary increasing sequence of integers, then for almost every α in $\mathbb{R}^d$, or which is the same in $\mathbb{T}^d$, the sequence $\{a(n)\alpha\}_{n=-\infty}^{+\infty}$ is equidistributed in the torus. See Theorem 4.1 in Chapter 1 of [17], or Theorem 6.21 in [26]. A quantitative version of these results is the Koksma Hlawka inequality, which in one variable and for an arbitrary sequence of points states that

$$\left|\frac{1}{N}\sum_{n=1}^{N} f(\alpha_n) - \int_0^1 f(x)dx\right| \le \mathcal{V}(f)\cdot\mathcal{D}\left(\{\alpha_n\}_{n=1}^{N}\right).$$

$\mathcal{V}(f)$ is the total variation of the function $f(x)$ and $\mathcal{D}(\{\alpha_n\}_{n=1}^{N})$ is the discrepancy of the sequence of points $\{\alpha_n\}_{n=1}^{N}$,

$$\mathcal{V}(f) = \sup_{0=x_0<x_1<\ldots<x_n=1}\left\{\sum_{j=0}^{n-1}\left|f(x_{j+1})-f(x_j)\right|\right\},$$

$$\mathcal{D}\left(\{\alpha_n\}_{n=1}^{N}\right) = \sup_{0\le x<y\le 1}\left\{\left|\frac{|\{n : x\le\alpha_n\le y\}|}{N} - (y-x)\right|\right\}.$$

See Theorem 5.1 in Chapter 2 of [17], and Section 7.3 in [26]. In several variables the discrepancy with respect to intervals has to be replaced by the discrepancy with respect to parallelepipeds with sides parallel to the axes, and the total variation has to be replaced by the Hardy Krause variation. See Definition 1.13 and Theorem 1.14

in [7] and Section 5 in Chapter 2 in [17]. See also [3] and Theorem 7.14 in [26] for an extension to piecewise smooth functions. In dimension $d = 1$ it is known that there exists an absolute constant c such that for any infinite sequence $\{\alpha_n\}_{n=1}^{+\infty}$ one has $\mathcal{D}(\{\alpha_n\}_{n=1}^{N}) \geq c \log(1+N)/N$ for infinitely many values of N, and this bound is best possible, there are sequences for which $\mathcal{D}(\{\alpha_n\}_{n=1}^{N}) \leq C \log(1+N)/N$ for every N. Examples of these low discrepancy sequences are the van der Corput sequence and Kronecker sequences $\{n\alpha\}_{n=1}^{+\infty}$ with the continued fraction expansions of the parameters α with bounded partial quotients. See Section 3 in Chapter 2 of [17]. In dimension $d > 1$ the discrepancy is larger than in dimension 1. It is a classical result in metric theory of Diophantine approximation that for almost every vector α the discrepancy of the Kronecker sequence $\{n\alpha\}_{n=1}^{N}$ is between $cN^{-1}\log^d(1+N)\log(\log(2+N))$ and $CN^{-1}\log^d(1+N)\log^{1+\varepsilon}(\log(2+N))$, for every $\varepsilon > 0$. The case $d = 1$ is due to Khintchin, while the generalization to $d > 1$ has been proved by Beck. See [2], and Theorem 1.72 and Theorem 1.91 in [7]. Hence, if the function $f(x)$ has bounded Hardy Krause variation, then for every $\varepsilon > 0$ and for almost every α one has

$$\left| \frac{1}{N}\sum_{n=1}^{N} f(n\alpha) - \int_{\mathbb{T}^d} f(x)dx \right| \leq C\frac{\log^d(1+N)\log^{1+\varepsilon}(\log(2+N))}{N}.$$

There have been many different attempts to improve the rate of convergence in the ergodic theorem. For a brief and limited survey, see the last remark at the end of the paper. This work is part of this line of research. Our purpose is not only to remove some of these logarithms, but also to improve the power N^{-1} by introducing appropriate summation methods for the sequence $\{f(n\alpha)\}_{n=-\infty}^{+\infty}$. For the definition and elementary properties of summation methods see Section 2.2 of [7] and Section 1 in Chapter III of [28]. More precisely, we shall study the speed of convergence to $\int_{\mathbb{T}^d} f(x)dx$ of the means

$$\left(\sum_{n=-\infty}^{+\infty} \Phi(N^{-1}n)\right)^{-1} \sum_{n=-\infty}^{+\infty} \Phi(N^{-1}n) f(x + a(n)\alpha).$$

$\Phi(s)$ is a bounded function with compact support in the real line, $\{a(n)\}_{n=-\infty}^{+\infty}$ is an increasing sequence of relative integers, and α is a vector in $\mathbb{T}^d$. For example, when $\Phi(s) = \chi_{\{0\leq s<1\}}(s)$ is the characteristic function of the interval $\{0 \leq s < 1\}$ one obtains the classical arithmetic means. When $\Phi(s) = (1-|s|)\,\chi_{\{-1<s<1\}}(s)$ one obtains the Cesaro means. And when $\Phi(s) = (1-|s|^2)^{\gamma}\chi_{\{-1<s<1\}}(s)$ one obtains the Bochner Riesz means of order γ. We shall show that for arbitrary functions $\Phi(s)$ with bounded variation, arbitrary increasing sequences $\{a(n)\}_{n=-\infty}^{+\infty}$, and for functions with absolutely convergent Fourier expansions, for almost every α in the torus the expected speed of convergence of the above means is of the order of $N^{-1/2}$ up to a small logarithmic transgression, and this result is essentially best possible. Moreover, we shall show that for suitable $\Phi(s)$ and $\{a(n)\}_{n=-\infty}^{+\infty}$, suitable $f(x)$, and for almost every α the convergence may be faster. For example, for arithmetic means of functions $f(x)$ with Fourier transforms $\{\widehat{f}(m)\}_{m\in\mathbb{Z}^d}$ in the Orlicz sequence space $L\log L(\mathbb{Z}^d)$ the speed of convergence is the order of N^{-1}, for Cesaro means of functions with Fourier transforms in $L^{1/2}(\mathbb{Z}^d)$ the speed of convergence is of the order of N^{-2}, and for Bochner Riesz means of order $\gamma > 0$ of functions with Fourier transforms in $L^{1/(1+\gamma)}(\mathbb{Z}^d)$ the speed of convergence is of the order of $N^{-1-\gamma}$. For connections between our results and other results already in the literature, see the remarks at the end of the paper.

Let us fix some notations in what follows. Denote by $\|t\|$ the distance of a real number t to the nearest integer. Functions on the torus $\mathbb{T}^d = \mathbb{R}^d/\mathbb{Z}^d$ are identified with periodic functions on $\mathbb{R}^d$ with periods $\mathbb{Z}^d$. The Fourier transform and Fourier expansion of an integrable function on the torus are defined by

$$\widehat{f}(m) = \int_{\mathbb{T}^d} f(x)\exp(-2\pi i m\cdot x)dx,$$

$$f(x) = \sum_{m\in\mathbb{Z}^d} \widehat{f}(m)\exp(2\pi i m\cdot x).$$

In what follows, α is a point in the torus $\mathbb{T}^d$, $\{a(n)\}_{n=-\infty}^{+\infty}$ is a strictly increasing sequence of relative integers, and $\Phi(s)$ is a bounded function with compact support in the real line $\{-\infty < s < +\infty\}$, with the property that there exist positive constants $c < C$ such that for every integer $N \geq 1$ one has

$$cN \leq \left|\sum_{n=-\infty}^{+\infty} \Phi(N^{-1}n)\right| \leq CN. \tag{8.1}$$

For example, if $\Phi(s)$ is bounded with compact support, and it is non negative and larger than $\varepsilon > 0$ in an interval larger than $\delta > 0$, then the above lower and upper bounds are satisfied for every $N > 1/\delta$. The discrepancy between Weyl sums and integral of a continuous function on the torus is defined by

$$\mathcal{D}f(N,\alpha) = \sup_{x\in\mathbb{T}^d}\left\{\left|\left(\sum_{n=-\infty}^{+\infty}\Phi(N^{-1}n)\right)^{-1}\sum_{n=-\infty}^{+\infty}\Phi(N^{-1}n)f(x+a(n)\alpha) - \int_{\mathbb{T}^d} f(y)dy\right|\right\}.$$

Our first result is quite obvious. If the function $\Phi(s)$ has too many ups and downs, then the weights $\left\{\Phi(N^{-1}n)\right\}_{n=-\infty}^{+\infty}$ can be essentially arbitrary, and the discrepancy $\mathcal{D}f(N,\alpha)$ does not necessarily converge to zero as $N \to +\infty$.

Theorem 8.1 *There exist a bounded compactly supported function $\Phi(s)$ with unbounded variation, an increasing sequence of relative integers $\{a(n)\}_{n=-\infty}^{+\infty}$, and a positive constant c, such that for every integrable function $f(x)$, every*

$m \in Z^d \setminus \{0\}$, and every α one has

$$\limsup_{N \to +\infty} \{\mathcal{D}f(N, \alpha)\} \geq c|\widehat{f}(m)|.$$

In the above theorem one can choose $\{a(n)\}_{n=-\infty}^{+\infty} = \{n\}_{n=-\infty}^{+\infty}$, the sequence of relative integers. The bad behaviour of the discrepancy is due to the unbounded variation of the function $\Phi(s)$. With bounded variation this pathology disappears.

Theorem 8.2 *Assume that $\Phi(s)$ is a function with compact support and bounded variation and with property* (8.1). *Assume that $\{a(n)\}_{n=-\infty}^{+\infty}$ is an increasing sequence of integers. Also assume that $\{\eta(N)\}_{N=1}^{+\infty}$ is a positive increasing sequence diverging to $+\infty$, in such a way that*

$$\sum_{j=0}^{+\infty} \frac{1}{\eta(2^j)} < +\infty.$$

Finally, assume that $f(x)$ has an absolutely convergent Fourier expansion,

$$\sum_{m \in \mathbb{Z}^d} |\widehat{f}(m)| < +\infty.$$

Then for almost all α, with a possible exception of a set of Lebesgue measure zero, there exist constants $c(f, \alpha)$ such that for every positive integer N one has

$$\mathcal{D}f(N, \alpha) \leq c(f, \alpha) N^{-1/2} \eta(N).$$

For example, in the above theorem one can choose $\eta(N) = \log^{1+\varepsilon}(1 + N)$, or $\eta(N) = \log(1 + N) \log^{1+\varepsilon}(\log(2 + N))$, with $\varepsilon > 0$. Without further restrictions on the weights $\Phi(s)$ and sequences $\{a(n)\}_{n=-\infty}^{+\infty}$ the above Theorem 8.2 is close to best possible. The rate of convergence $N^{-1/2}$ cannot be accelerated.

Theorem 8.3 *Assume that $\Phi(s)$ is a bounded function with compact support and with property* (8.1). *Assume that the sequence $\{a(n)\}_{n=-\infty}^{+\infty}$ is increasing and lacunary, $a(n-1)/a(n) > K > 1$ if $a(n) < 0$ and $a(n+1)/a(n) > K > 1$ if $a(n) > 0$. Assume that $\{\eta(N)\}_{N=1}^{+\infty}$ is a positive sequence diverging to $+\infty$. Finally, assume that $f(x)$ is a non constant integrable function on the torus. Then for almost every α one has*

$$\limsup_{N \to +\infty} \left\{ \sqrt{N} \eta(N) \mathcal{D}f(N, \alpha) \right\} = +\infty.$$

The above results show that for generic weights $\Phi(s)$ and sequences $\{a(n)\}_{n=-\infty}^{+\infty}$ the rate of convergence $N^{-1/2}$ cannot be accelerated, but under suitable assumptions on these weights and sequences and for suitable classes of functions $f(x)$ the rate of convergence may be faster than $N^{-1/2}$.

Theorem 8.4 *Assume that $\{a(n)\}_{n=-\infty}^{+\infty}$ is an increasing sequence of integers. Also assume that $\Phi(s)$ is a bounded function with compact support and with property (8.1). Assume that there exist positive constants H and ϑ such that for every positive integer N and every real number t one has*

$$\left| \left(\sum_{n=-\infty}^{+\infty} \Phi(N^{-1}n) \right)^{-1} \sum_{n=-\infty}^{+\infty} \Phi(N^{-1}n) \exp(2\pi i a(n)t) \right| \leq H(N\|t\|)^{-\vartheta}.$$

Finally assume that the function $f(x)$ has a Fourier transform $\{\widehat{f}(m)\}_{m\in\mathbb{Z}^d}$ with

$$\begin{cases} \sum_{m\in\mathbb{Z}^d} |\widehat{f}(m)| < +\infty & \text{if } \vartheta < 1, \\ \sum_{m\in\mathbb{Z}^d} |\widehat{f}(m)| \log(1 + 1/|\widehat{f}(m)|) < +\infty & \text{if } \vartheta = 1, \\ \sum_{m\in\mathbb{Z}^d} |\widehat{f}(m)|^{1/\vartheta} < +\infty & \text{if } \vartheta > 1. \end{cases}$$

Then for almost all α, with a possible exception of a set of Lebesgue measure zero, there exist constants $c(f,\alpha)$ such that for every positive integer N one has

$$\mathcal{D}f(N,\alpha) \leq c(f,\alpha)N^{-\vartheta}.$$

The speed of convergence $cN^{-\vartheta}$ cannot be accelerated in the following sense. Assume that there exists $K > 0$ such that for infinitely many positive integers N one has

$$\int_{\mathbb{T}} \left| \left(\sum_{n=-\infty}^{+\infty} \Phi(N^{-1}n) \right)^{-1} \sum_{n=-\infty}^{+\infty} \Phi(N^{-1}n) \exp(2\pi i a(n)t) \right| dt \geq KN^{-\vartheta}.$$

Then for every integrable function $f(x)$ and every m in $\mathbb{Z}^d \setminus \{0\}$, for infinitely many positive integers N one also has

$$\int_{\mathbb{T}^d} \mathcal{D}f(N,\alpha)d\alpha \geq KN^{-\vartheta}|\widehat{f}(m)|.$$

The above theorem is non void. The following are explicit examples of weights $\Phi(s)$ and sequences $\{a(n)\}_{n=-\infty}^{+\infty}$ to which the theorem applies.

Corollary 8.1 *If* $\{a(n)\}_{n=-\infty}^{+\infty} = \{n\}_{n=-\infty}^{+\infty}$ *is the sequence of relative integers, and if*

$$\Phi(s) = \begin{cases} 1 & if \ |s| \leq 1, \\ 0 & if \ |s| > 1, \end{cases}$$

then Theorem 8.4 applies with $\vartheta = 1$.

Corollary 8.2 *If* $\{a(n)\}_{n=-\infty}^{+\infty} = \{n\}_{n=-\infty}^{+\infty}$, *and if*

$$\Phi(s) = \begin{cases} 1 - |s| & if \ |s| \leq 1, \\ 0 & if \ |s| > 1, \end{cases}$$

then Theorem 8.4 applies with $\vartheta = 2$.

Corollary 8.3 *If* $\{a(n)\}_{n=-\infty}^{+\infty} = \{n\}_{n=-\infty}^{+\infty}$, *if* $\gamma > 0$, *and if*

$$\Phi(s) = \begin{cases} (1 - |s|^2)^\gamma & if \ |s| \leq 1, \\ 0 & if \ |s| > 1, \end{cases}$$

then Theorem 8.4 applies with $\vartheta = \gamma + 1$.

Corollary 8.4 *If* $\{a(n)\}_{n=-\infty}^{+\infty} = \{n\}_{n=-\infty}^{+\infty}$, *and if* $\Phi(s)$ *is a smooth function with compact support and with property* (8.1), *then Theorem 8.4 applies with every* $\vartheta > 0$.

For the curse of dimensionality it may be of some interest to remark that the implicit constants in Theorem 8.4 and corollaries depend on the quasi norms of the Fourier transforms $\{\widehat{f}(m)\}_{m\in\mathbb{Z}^d\setminus\{0\}}$, but they are otherwise independent of the dimension d. It is also interesting to observe that in Theorem 8.4 and corollaries the exceptional sets of parameters α that do not give an optimal rate of convergence, which are of measure 0, from the topological point of view for some functions are small, and for some other functions are not so small.

Theorem 8.5

1. *Let* $\{a(n)\}_{n=-\infty}^{+\infty}$ *and* $\Phi(s)$ *be as in Theorem 8.4, and let* $g(x) = \exp(2\pi i m \cdot x)$, *with* $m \in \mathbb{Z}^d \setminus \{0\}$. *If* $m \cdot \alpha \in \mathbb{Z}$ *then* $\mathcal{D}g(N, \alpha) = 1$, *and if* $m \cdot \alpha \notin \mathbb{Z}$ *then* $\mathcal{D}g(N, \alpha) \leq H \left\|m \cdot \alpha\right\|^{-\vartheta} N^{-\vartheta}$.
2. *In the one dimensional torus* $\mathbb{T}$, *if* $\nu > 1$ *define*

$$f(x) = \sum_{m=1}^{+\infty} m^{-\nu} \exp(2\pi i m x).$$

Also, with the above $f(x)$ define

$$\Omega = \left\{ \alpha \in \mathbb{T} : \limsup_{N \to +\infty} \left\{ N^{\nu/(\nu+1)} \left| \frac{1}{N} \sum_{n=0}^{N-1} f(n\alpha) \right| \right\} > 0 \right\}.$$

This set Ω has Lebesgue measure 0, but it contains a countable intersection of open dense sets.

Observe that the set $\{\alpha \in \mathbb{T}^d : m \cdot \alpha \in \mathbb{Z}\}$ in part 1 of the above theorem when $d = 1$ is a finite number of points, and when $d > 1$ it is contained in a finite number of $d - 1$ dimensional hyperplanes. Hence this set is meagre, it is of Baire first category. On the other hand, the set Ω in part 2 of the theorem is nonmeagre, it is of Baire second category. The first set is small, the second large.

Finally, the above Theorem 8.4 and corollaries have analogues with discrete means replaced by continuous means,

$$\left(\int_{-\infty}^{+\infty} \Phi(N^{-1}s) \, ds \right)^{-1} \int_{-\infty}^{+\infty} \Phi(N^{-1}s) f(x + s\alpha) ds.$$

We hope to return to this point in a sequel of this paper.

Let us now present the proofs of the above results.

A basic tool in the proofs of theorems and corollaries is the Fourier expansion of the discrepancy.

Lemma 8.1 *Assume that $\Phi(s)$ is a bounded function with compact support in the real line, and that $\{a(n)\}_{n=-\infty}^{+\infty}$ is a strictly increasing sequence of relative integers. If $f(x)$ is an integrable function on the torus, then*

$$\left(\sum_{n=-\infty}^{+\infty} \Phi(N^{-1}n) \right)^{-1} \sum_{n=-\infty}^{+\infty} \Phi(N^{-1}n) f(x + a(n)\alpha) - \int_{\mathbb{T}^d} f(y) dy$$

$$= \sum_{m \in \mathbb{Z}^d \setminus \{0\}} \left(\left(\sum_{n=-\infty}^{+\infty} \Phi(N^{-1}n) \right)^{-1} \sum_{n=-\infty}^{+\infty} \Phi(N^{-1}n) \exp(2\pi i a(n) m \cdot \alpha) \right)$$

$$\times \widehat{f}(m) \exp(2\pi i m \cdot x).$$

Proof A simple and straight computation. The lemma claims that the second term of the identity is the Fourier expansion of the first term, but if the function $f(x)$ has an absolutely convergent Fourier expansion, then the above equality holds also pointwise. □

Proof of Theorem 8.1 Let $\mathcal{P}$ be the sequence of prime numbers. To every prime p associate the interval $I(p) = \{0 \leq t \leq 1/(2p)\}$. Since $\sum_{p \in \mathcal{P}} 1/p = +\infty$, it is possible to choose points $\{r(p)\}_{p \in \mathcal{P}}$ such that the collection of intervals

$\{r(p)+I(p)\}_{p\in\mathcal{P}}$ covers every point of the torus infinitely many times. Define $\Phi(s)=0$ if $s<0$ or $s\geq 1$, and if N is a prime and $0\leq n\leq N-1$ define $\Phi(N^{-1}n)=2+\exp(-2\pi inr(N))$. Finally, when $0\leq s<1$ is not a rational $N^{-1}n$ with a prime denominator N, define $\Phi(s)$ in an arbitrary way, for example set $\Phi(s)=2$. Observe that since the rationals n/N with $1\leq n\leq N-1$ and a prime denominator N are all different, this function $\Phi(s)$ is well defined. Since the supremum norm dominates the integral norm, and since the integral norm dominates the size of the Fourier coefficients, by Lemma 8.1 for every $m\in\mathbb{Z}^d\setminus\{0\}$ one has

$$\sup_{x\in\mathbb{T}^d}\left\{\left|\left(\sum_{n=-\infty}^{+\infty}\Phi(N^{-1}n)\right)^{-1}\sum_{n=-\infty}^{+\infty}\Phi(N^{-1}n)f(x+n\alpha)-\int_{\mathbb{T}^d}f(y)dy\right|\right\}$$

$$\geq\int_{\mathbb{T}^d}\left|\left(\sum_{n=-\infty}^{+\infty}\Phi(N^{-1}n)\right)^{-1}\sum_{n=-\infty}^{+\infty}\Phi(N^{-1}n)f(x+n\alpha)-\int_{\mathbb{T}^d}f(y)dy\right|dx$$

$$\geq\left|\left(\sum_{n=-\infty}^{+\infty}\Phi(N^{-1}n)\right)^{-1}\sum_{n=-\infty}^{+\infty}\Phi(N^{-1}n)\exp(2\pi inm\cdot\alpha)\right||\widehat{f}(m)|.$$

In order to prove the theorem, it suffices to show that the superior limit as $N\to+\infty$ of the above expression can be bounded below by a positive constant independent of α. If $m\cdot\alpha$ is an integer, then

$$\left|\left(\sum_{n=-\infty}^{+\infty}\Phi(N^{-1}n)\right)^{-1}\sum_{n=-\infty}^{+\infty}\Phi(N^{-1}n)\exp(2\pi inm\cdot\alpha)\right|=1.$$

If $m\cdot\alpha$ is not an integer and if N is a prime, the above choice $\Phi(N^{-1}n)=2+\exp(-2\pi inr(N))$ if $0\leq n\leq N-1$ and $\Phi(N^{-1}n)=0$ otherwise gives

$$\left|\left(\sum_{n=-\infty}^{+\infty}\Phi(N^{-1}n)\right)^{-1}\sum_{n=-\infty}^{+\infty}\Phi(N^{-1}n)\exp(2\pi inm\cdot\alpha)\right|\geq$$

$$\left|\sum_{n=-\infty}^{+\infty}\Phi(N^{-1}n)\right|^{-1}\left(\left|\sum_{n=0}^{N-1}\exp(2\pi in(m\cdot\alpha-r(N)))\right|-2\left|\sum_{n=0}^{N-1}\exp(2\pi inm\cdot\alpha)\right|\right).$$

By the above definition of $\Phi(s)$ one has $N\leq\left|\sum_{n=-\infty}^{+\infty}\Phi(N^{-1}n)\right|\leq 3N$. Hence,

$$\left|\sum_{n=-\infty}^{+\infty}\Phi(N^{-1}n)\right|^{-1}\left|\sum_{n=0}^{N-1}\exp(2\pi inm\cdot\alpha)\right|$$

$$= \left| \sum_{n=-\infty}^{+\infty} \Phi(N^{-1}n) \right|^{-1} \left| \frac{\sin(N\pi m \cdot \alpha)}{\sin(\pi m \cdot \alpha)} \right| \leq \frac{1}{N\,|\sin(\pi m \cdot \alpha)|}.$$

This term is bounded by $c(\alpha)/N$, and it vanishes as $N \to +\infty$. Finally, if $m \cdot \alpha$ is in the interval $r(N) + I(N)$, then $0 \leq N(m \cdot \alpha - r(N)) \leq 1/2$, and

$$\left| \sum_{n=-\infty}^{+\infty} \Phi(N^{-1}n) \right|^{-1} \left| \sum_{n=0}^{N-1} \exp\left(2\pi i n(m \cdot \alpha - r(N))\right) \right|$$
$$= \left| \sum_{n=-\infty}^{+\infty} \Phi(N^{-1}n) \right|^{-1} \left| \frac{\sin\left(\pi N(m \cdot \alpha - r(N))\right)}{\sin\left(\pi(m \cdot \alpha - r(N))\right)} \right| \geq \frac{2}{3\pi}.$$

The last inequality follows from the inequalities $2s \leq \sin(\pi s) \leq \pi s$ if $0 \leq s \leq 1/2$. Since $m \cdot \alpha$ belongs to infinitely many intervals $r(p) + I(p)$, the theorem follows. □

The following elementary lemma is a link between functions in $\mathbb{T}$ and functions in $\mathbb{T}^d$, and it will be used repeatedly.

Lemma 8.2 *If $g(t)$ is a locally integrable function on $\{-\infty < t < +\infty\}$ periodic with period 1, then for every $m \in \mathbb{Z}^d \setminus \{0\}$ one has*

$$\int_{\mathbb{T}^d} g(m \cdot \alpha) d\alpha = \int_{\mathbb{T}} g(t) dt.$$

Proof For every real number s and non zero integer h the map $t \to ht + s$ is a measure preserving transformation of the one dimensional torus. If $g(t)$ is periodic with period 1,

$$\int_{\mathbb{T}} g(ht + s) dt = \int_0^1 g(ht + s) dt = h^{-1} \int_s^{s+h} g(t) dt = \int_0^1 g(t) dt = \int_{\mathbb{T}} g(t) dt.$$

This implies that for every $m = (h, k)$, with $h \in \mathbb{Z} \setminus \{0\}$ and $k \in \mathbb{Z}^{d-1}$,

$$\int_{\mathbb{T}^d} g(m \cdot \alpha) d\alpha = \int_{\mathbb{T}^{d-1}} \left(\int_{\mathbb{T}} g(h\beta + k \cdot \gamma) d\beta \right) d\gamma$$
$$= \int_{\mathbb{T}^{d-1}} \left(\int_{\mathbb{T}} g(t) dt \right) d\beta = \int_{\mathbb{T}} g(t) dt.$$

□

The following maximal function is the key to the proof of Theorem 8.2.

Lemma 8.3 *Assume that $\Phi(s)$ is a function in the real line with compact support and bounded variation, and that $\{a(n)\}_{n=-\infty}^{+\infty}$ is a strictly increasing sequence of relative integers. Also assume that $\{\eta(N)\}_{N=1}^{+\infty}$ is a positive increasing sequence diverging to $+\infty$, with $\sum_{j=0}^{+\infty}\left(\eta(2^j)\right)^{-1} < +\infty$. Then one has*

$$\left\{\int_{\mathbb{T}}\left(\sup_{N\geq 1}\left\{N^{-1/2}\left(\eta(N)\right)^{-1}\left|\sum_{n=-\infty}^{+\infty}\Phi(N^{-1}n)\exp(2\pi ia(n)t)\right|\right\}\right)^2 dt\right\}^{1/2} < +\infty.$$

Proof Assume that $\Phi(s)$ has support contained in $\{-u \leq s \leq u\}$, so that $\Phi(N^{-1}n) = 0$ when $n < -Nu$ or $n > Nu$. A summation by parts gives

$$\sum_{n=-\infty}^{+\infty}\Phi(N^{-1}n)\exp(2\pi ia(n)t) =$$

$$\sum_{-Nu-1\leq n\leq Nu}\left(\Phi(N^{-1}n) - \Phi(N^{-1}(n+1))\right)\left(\sum_{-Nu-1\leq k\leq n}\exp(2\pi ia(k)t)\right).$$

Hence, if $\mathcal{V}(\Phi)$ denotes the total variation of $\Phi(s)$,

$$\left|\sum_{n=-\infty}^{+\infty}\Phi(N^{-1}n)\exp(2\pi ia(n)t)\right|$$

$$\leq \mathcal{V}(\Phi)\sup_{-Nu-1\leq n\leq Nu}\left\{\left|\sum_{-Nu-1\leq k\leq n}\exp(2\pi ia(k)t)\right|\right\}.$$

Given $N \geq 1$, choose $j \geq 0$ with $2^j \leq N < 2^{j+1}$. Then,

$$N^{-1/2}\left(\eta(N)\right)^{-1}\left|\sum_{n=-\infty}^{+\infty}\Phi(N^{-1}n)\exp(2\pi ia(n)t)\right|$$

$$\leq c2^{-j/2}\left(\eta(2^j)\right)^{-1}\sup_{-2^{j+1}u-1\leq h\leq k\leq 2^{j+1}u}\left\{\left|\sum_{n=h}^{k}\exp(2\pi ia(n)t)\right|\right\}$$

$$\leq c\sum_{j=0}^{+\infty}2^{-j/2}\left(\eta(2^j)\right)^{-1}\sup_{-2^{j+1}u-1\leq h\leq k\leq 2^{j+1}u}\left\{\left|\sum_{n=h}^{k}\exp(2\pi ia(n)t)\right|\right\}.$$

Observe that the last sum does not depend on N. Hence,

$$\left\{\int_{\mathbb{T}}\left(\sup_{N\geq 1}\left\{N^{-1/2}(\eta(N))^{-1}\left|\sum_{n=-\infty}^{+\infty}\Phi(N^{-1}n)\exp(2\pi ia(n)t)\right|\right\}\right)^2 dt\right\}^{1/2}$$

$$\leq c\sum_{j=0}^{+\infty}2^{-j/2}\left(\eta(2^j)\right)^{-1}\left\{\int_{\mathbb{T}}\left(\sup_{-2^{j+1}u-1\leq h\leq k\leq 2^{j+1}u}\left\{\left|\sum_{n=h}^{k}\exp(2\pi ia(n)t)\right|\right\}\right)^2 dt\right\}^{1/2}.$$

Finally, by the theorem of Carleson Hunt on the almost everywhere convergence of Fourier series, the maximal operator associated to the partial sums of Fourier series is bounded on $L^p(\mathbb{T})$ for every $1 < p < +\infty$. See [13]. Hence there exist absolute constants independent of j such that

$$\left\{\int_{\mathbb{T}}\left(\sup_{-2^{j+1}u-1\leq h\leq k\leq 2^{j+1}u}\left\{\left|\sum_{n=h}^{k}\exp(2\pi ia(n)t)\right|\right\}\right)^2 dt\right\}^{1/2}$$

$$\leq c\left\{\int_{\mathbb{T}}\left|\sum_{-2^{j+1}u-1\leq n\leq 2^{j+1}u}\exp(2\pi ia(n)t)\right|^2 dt\right\}^{1/2}$$

$$= c\left\{\sum_{-2^{j+1}u-1\leq n\leq 2^{j+1}u}1\right\}^{1/2}\leq c2^{j/2}.$$

By the way, with the theorem of Rademacher Menchov (see [28, Theorem 10.21, Chapter XIII]), instead of the more sophisticated theorem of Carleson Hunt, one obtains a similar result with only a small logarithmic transgression. □

Proof of Theorem 8.2 It suffices to show that if $f(x)$ has an absolutely convergent Fourier expansion, then the maximal function of the discrepancy is square integrable on the torus,

$$\left\{\int_{\mathbb{T}^d}\left(\sup_{N\geq 1}\left\{N^{1/2}(\eta(N))^{-1}\mathcal{D}f(N,\alpha)\right\}\right)^2 d\alpha\right\}^{1/2}\leq c\sum_{m\in\mathbb{Z}^d\setminus\{0\}}|\widehat{f}(m)|.$$

By the Fourier expansion of the discrepancy in Lemma 8.1,

$$\mathcal{D}f(N,\alpha)$$

$$\le \sum_{m\in\mathbb{Z}^d\setminus\{0\}} |\widehat{f}(m)| \left(\left| \sum_{n=-\infty}^{+\infty} \Phi(N^{-1}n) \right|^{-1} \left| \sum_{n=-\infty}^{+\infty} \Phi(N^{-1}n) \exp(2\pi i a(n) m\cdot\alpha) \right| \right).$$

With the assumption $cN \le \left|\sum_{n=-\infty}^{+\infty} \Phi(N^{-1}n)\right| \le CN$, one has

$$\left\{ \int_{\mathbb{T}^d} \left(\sup_{N\ge 1} \left\{ N^{1/2} (\eta(N))^{-1} \mathcal{D}f(N,\alpha) \right\} \right)^2 d\alpha \right\}^{1/2} \le c \sum_{m\in\mathbb{Z}^d\setminus\{0\}} |\widehat{f}(m)| \times$$

$$\left\{ \int_{\mathbb{T}^d} \left(\sup_{N\ge 1} \left\{ N^{-1/2} (\eta(N))^{-1} \left| \sum_{n=-\infty}^{+\infty} \Phi(N^{-1}n) \exp(2\pi i a(n) m\cdot\alpha) \right| \right\} \right)^2 d\alpha \right\}^{1/2}.$$

Finally, by Lemma 8.2 the last integral is independent of $m \in \mathbb{Z}^d \setminus \{0\}$, and by Lemma 8.3 it is finite. □

Proof of Theorem 8.3 If $f(x)$ is a non constant integrable function on the torus, then there exists $m \in \mathbb{Z}^d \setminus \{0\}$ with $|\widehat{f}(m)| \neq 0$. Since the supremum norm of a function dominates the integral norm, and since the integral norm dominates the sizes of the Fourier coefficients, by Lemma 8.1 one has

$$\sup_{x\in\mathbb{T}^d} \left\{ \left| \left(\sum_{n=-\infty}^{+\infty} \Phi(N^{-1}n) \right)^{-1} \sum_{n=-\infty}^{+\infty} \Phi(N^{-1}n) f(x+a(n)\alpha) - \int_{\mathbb{T}^d} f(y)dy \right| \right\}$$

$$\ge \left| \left(\sum_{n=-\infty}^{+\infty} \Phi(N^{-1}n) \right)^{-1} \sum_{n=-\infty}^{+\infty} \Phi(N^{-1}n) \exp(2\pi i a(n) m\cdot\alpha) \right| |\widehat{f}(m)|.$$

Hence, since $cN \le \left|\sum_{n=-\infty}^{+\infty} \Phi(N^{-1}n)\right| \le CN$,

$$\sup_{N\ge 1,\, x\in\mathbb{T}^d} \left\{ \sqrt{N}\eta(N) \left| \left(\sum_{n=-\infty}^{+\infty} \Phi(N^{-1}n) \right)^{-1} \sum_{n=-\infty}^{+\infty} \Phi(N^{-1}n) f(x+n\alpha) \right.\right.$$

$$\left.\left. - \int_{\mathbb{T}^d} f(y)dy \right| \right\}$$

$$\ge c \sup_{N\ge 1} \left\{ N^{-1/2}\eta(N) \left| \sum_{n=-\infty}^{+\infty} \Phi(N^{-1}n) \exp(2\pi i a(n) m\cdot\alpha) \right| \right\} |\widehat{f}(m)|.$$

Set

$$\mathcal{M}(t) = \sup_{N\geq 1}\left\{N^{-1/2}\eta(N)\left|\sum_{n=-\infty}^{+\infty}\Phi(N^{-1}n)\exp(2\pi i a(n)t)\right|\right\}.$$

It suffices to prove that if $\{a(n)\}_{n=-\infty}^{+\infty}$ is a lacunary sequence, then for every m in $\mathbb{Z}^d\setminus\{0\}$ and for almost every α one has $\mathcal{M}(m\cdot\alpha)=+\infty$, which is the same as proving that for every $s>0$ the set of α with $\mathcal{M}(m\cdot\alpha)\leq s$ has measure 0. Define

$$\mathcal{A}(s) = \{t\in\mathbb{T} : \mathcal{M}(t)\leq s\},$$
$$\mathcal{B}(s) = \left\{\alpha\in\mathbb{T}^d : \mathcal{M}(m\cdot\alpha)\leq s\right\}.$$

By Lemma 8.2, $|\mathcal{A}(s)| = |\mathcal{B}(s)|$. By the definitions of $\mathcal{A}(s)$ and $\mathcal{M}(t)$,

$$\left\{\int_{\mathcal{A}(s)}\left|\sum_{n=-\infty}^{+\infty}\Phi(N^{-1}n)\exp(2\pi i a(n))\right|^2 dt\right\}^{1/2} \leq s\,|\mathcal{A}(s)|^{1/2}N^{1/2}(\eta(N))^{-1}.$$

Moreover for every $H\geq 0$, by the orthogonality of the exponentials on the torus,

$$\left\{\int_{\mathcal{A}(s)}\left|\sum_{n=-\infty}^{+\infty}\Phi(N^{-1}n)\exp(2\pi i a(n)t)\right|^2 dt\right\}^{1/2} \geq$$

$$\left\{\int_{\mathcal{A}(s)}\left|\sum_{|n|>H}\Phi(N^{-1}n)\exp(2\pi i a(n)t)\right|^2 dt\right\}^{1/2} - \left\{\sum_{|n|\leq H}\left|\Phi(N^{-1}n)\right|^2\right\}^{1/2}.$$

By [28, Lemma 6.5, Chapter V], if $\{a(n)\}_{n=-\infty}^{+\infty}$ is a lacunary sequence, then there exist positive constants c and H such that if N is sufficiently large,

$$\left\{\int_{\mathcal{A}(s)}\left|\sum_{|n|>H}\Phi(N^{-1}n)\exp(2\pi i a(n)t)\right|^2 dt\right\}^{1/2}$$

$$\geq c\,|\mathcal{A}(s)|^{1/2}\left\{\sum_{|n|>H}\left|\Phi(N^{-1}n)\right|^2\right\}^{1/2}$$

$$\geq c\,|\mathcal{A}(s)|^{1/2}\left\{\sum_{n=-\infty}^{+\infty}\left|\Phi(N^{-1}n)\right|^2\right\}^{1/2} - c\,|\mathcal{A}(s)|^{1/2}\left\{\sum_{|n|\leq H}\left|\Phi(N^{-1}n)\right|^2\right\}^{1/2}$$

$$\geq c\,|\mathcal{A}(s)|^{1/2}\,N^{1/2} - c\,|\mathcal{A}(s)|^{1/2}\left\{\sum_{|n|\leq H}\left|\Phi(N^{-1}n)\right|^2\right\}^{1/2}.$$

The last inequality $\sum_{n=-\infty}^{+\infty}\left|\Phi(N^{-1}n)\right|^2 \geq cN$ follows from the assumptions that $\Phi(s)$ is bounded with bounded support and that $\left|\sum_{n=-\infty}^{+\infty}\Phi(N^{-1}n)\right| \geq cN$. Hence,

$$c\,|\mathcal{A}(s)|^{1/2}\,N^{1/2} - (1 + c\,|\mathcal{A}(s)|^{1/2})\left\{\sum_{|n|\leq H}\left|\Phi(N^{-1}n)\right|^2\right\}^{1/2}$$
$$\leq s\,|\mathcal{A}(s)|^{1/2}\,N^{1/2}\,(\eta(N))^{-1}.$$

Since $\sum_{|n|\leq H}\left|\Phi(N^{-1}n)\right|^2$ is bounded while $\{\eta(N)\}_{N=1}^{+\infty}$ diverges as $N \to +\infty$, it follows that $|\mathcal{A}(s)| = 0$. □

The key tool in the proof of Theorem 8.4 are the function spaces $Weak-L^p(\mathbb{T}^d)$. Let us recall their definition. Denote by $|\Omega|$ the Lebesgue measure of a measurable set in the torus, and if $0 < p < +\infty$ denote by $Weak-L^p(\mathbb{T}^d)$ the function spaces defined by the quasi norms

$$\|g(\alpha)\|_{Weak-L^p(\mathbb{T}^d)} = \sup_{s>0}\left\{s^p\left|\left\{\alpha \in \mathbb{T}^d : |g(\alpha)| > s\right\}\right|\right\}^{1/p}.$$

When $1 < p < +\infty$ these spaces are Banach, and the above defined quasi norms are equivalent to norms. When $0 < p \leq 1$ these spaces are quasi Banach, and the following lemma describes the numerical sequences that allow to sum sequences of functions.

Lemma 8.4 *Assume that $0 < p < +\infty$, and that $\{g_m(\alpha)\}_{m\in\mathbb{Z}^d}$ is a sequence of functions in $Weak-L^p(\mathbb{T}^d)$ with uniformly bounded quasi norms, $\|g_m(\alpha)\|_{Weak-L^p(\mathbb{T}^d)} \leq c$ for every m. Also assume that $\{s(m)\}_{m\in\mathbb{Z}^d}$ is a numerical sequence with*

$$\begin{cases} \sum_{m\in\mathbb{Z}^d}|s(m)| < +\infty & \text{if } 1 < p < +\infty, \\ \sum_{m\in\mathbb{Z}^d,\ s(m)\neq 0}|s(m)|\log(1 + 1/|s(m)|) < +\infty & \text{if } p = 1, \\ \sum_{m\in\mathbb{Z}^d}|s(m)|^p < +\infty & \text{if } 0 < p < 1. \end{cases}$$

Then the series $\sum_{m\in\mathbb{Z}^d} s(m)g_m(\alpha)$ *converges absolutely almost everywhere and in the quasi norm of* $Weak - L^p(\mathbb{T}^d)$ *to a function in* $Weak - L^p(\mathbb{T}^d)$*, and*

$$\left\| \sum_{m\in\mathbb{Z}^d} s(m)g_m(\alpha) \right\|_{Weak-L^p(\mathbb{T}^d)}$$

$$\leq \begin{cases} c \sum_{m\in\mathbb{Z}^d} |s(m)| & \text{if } 1 < p < +\infty, \\ c \sum_{m\in\mathbb{Z}^d,\ s(m)\neq 0} |s(m)| \log(1 + 1/|s(m)|) & \text{if } p = 1, \\ c \left\{ \sum_{m\in\mathbb{Z}^d} |s(m)|^p \right\}^{1/p} & \text{if } 0 < p < 1. \end{cases}$$

Proof All of this is well known. For the case $1 < p < +\infty$ see [24, Section 3, Chapter V]. For the case $p = 1$ see [23, Lemma 2.3]. For the case $0 < p < 1$ see [25, Lemma 1.8]. □

Lemma 8.5 *Denote by* $\|t\|$ *the distance of a real number* t *to the nearest integer. If* $m \in \mathbb{Z}^d \setminus \{0\}$ *and* $\vartheta > 0$*, the functions* $\alpha \to \|m \cdot \alpha\|^{-\vartheta}$ *are in the space* $Weak - L^{1/\vartheta}(\mathbb{T}^d)$*, with quasi norms independent of* m*,*

$$\sup_{s>0} \left\{ s^{1/\vartheta} \left| \{\alpha \in \mathbb{T}^d : \|m \cdot \alpha\|^{-\vartheta} > s\} \right| \right\}^{\vartheta} = 2^{\vartheta}.$$

Proof This follows from Lemma 8.2, with $g(t)$ the characteristic function of the set $\{t \in \mathbb{T} : \|t\|^{-\vartheta} > s\}$,

$$\left| \{\alpha \in \mathbb{T}^d : \|m \cdot \alpha\|^{-\vartheta} > s\} \right| = \left| \{t \in \mathbb{T} : \|t\|^{-\vartheta} > s\} \right| = \begin{cases} 1 & \text{if } s \leq 2^{\vartheta}, \\ 2s^{-1/\vartheta} & \text{if } s \geq 2^{\vartheta}. \end{cases}$$

□

Proof of Theorem 8.4 Set

$$c(f, \alpha) = \sup_{N\geq 0} \left\{ N^{\vartheta} \mathcal{D}f(N, \alpha) \right\}.$$

In order to prove that $c(f, \alpha)$ is finite almost everywhere it suffices to prove the maximal inequality

$$\sup_{s>0} \left\{ s^{1/\vartheta} \left| \left\{ \alpha \in \mathbb{T}^d : c(f, \alpha) > s \right\} \right| \right\}^{\theta}$$

$$\leq \begin{cases} \sum_{m\in\mathbb{Z}^d} |\widehat{f}(m)| < +\infty & if\ \vartheta < 1, \\ \sum_{m\in\mathbb{Z}^d} |\widehat{f}(m)| \log(1 + 1/|\widehat{f}(m)|) < +\infty & if\ \vartheta = 1, \\ \left\{ \sum_{m\in\mathbb{Z}^d} |\widehat{f}(m)|^{1/\vartheta} \right\}^{\vartheta} & if\ \vartheta > 1. \end{cases}$$

By the assumptions on $\Phi(s)$ and the Fourier expansion of the discrepancy in Lemma 8.1,

$$\left| \left(\sum_{n=-\infty}^{+\infty} \Phi(N^{-1}n) \right)^{-1} \sum_{n=-\infty}^{+\infty} \Phi(N^{-1}n) f(x + a(n)\alpha) - \int_{\mathbb{T}^d} f(y)dy \right|$$

$$\leq \sum_{m\in\mathbb{Z}^d\setminus\{0\}} |\widehat{f}(m)| \left| \left(\sum_{n=-\infty}^{+\infty} \Phi(N^{-1}n) \right)^{-1} \sum_{n=-\infty}^{+\infty} \Phi(N^{-1}n) \exp(2\pi i a(n) m \cdot \alpha) \right|$$

$$\leq HN^{-\vartheta} \sum_{m\in\mathbb{Z}^d\setminus\{0\}} |\widehat{f}(m)| \, \|m \cdot \alpha\|^{-\vartheta}.$$

By Lemmas 8.4 and 8.5, and the assumptions on the Fourier transform $\{\widehat{f}(m)\}_{m\in\mathbb{Z}^d}$, the last series converges absolutely almost everywhere and in the quasi norm of $Weak - L^{1/\vartheta}(\mathbb{T}^d)$.

In order to see that the rate of convergence $cN^{-\vartheta}$ cannot be improved in general, observe that for every $m \in \mathbb{Z}^d \setminus \{0\}$ one has

$$\sup_{x\in\mathbb{T}^d} \left\{ \left| \left(\sum_{n=-\infty}^{+\infty} \Phi(N^{-1}n) \right)^{-1} \sum_{n=-\infty}^{+\infty} \Phi(N^{-1}n) f(x + a(n)\alpha) - \int_{\mathbb{T}^d} f(y)dy \right| \right\}$$

$$\geq \left| \left(\sum_{n=-\infty}^{+\infty} \Phi(N^{-1}n) \right)^{-1} \sum_{n=-\infty}^{+\infty} \Phi(N^{-1}n) \exp(2\pi i a(n) m \cdot \alpha) \right| |\widehat{f}(m)|.$$

By the assumption of the theorem and Lemma 8.2, for infinitely many N one has

$$\int_{\mathbb{T}^d} \left| \left(\sum_{n=-\infty}^{+\infty} \Phi(N^{-1}n) \right)^{-1} \sum_{n=-\infty}^{+\infty} \Phi(N^{-1}n) \exp(2\pi i a(n) m \cdot \alpha) \right| d\alpha \geq KN^{-\vartheta}.$$

□

Proof of Corollary 8.1 The Dirichlet kernel is

$$\sum_{n=-N}^{N} \exp(2\pi int) = \frac{\sin((2N+1)\pi t)}{\sin(\pi t)}.$$

When $t = 0$ the sum is $2N + 1$. By the inequality $2\,\|t\| \le |\sin(\pi t)| \le \pi\,\|t\|$, it follows that

$$\left|\frac{1}{2N+1}\sum_{n=-N}^{N} \exp(2\pi int)\right| \le \min\left\{1, ((4N+2)\,\|t\|)^{-1}\right\}.$$

Hence Theorem 8.4 applies with $\vartheta = 1$. □

Proof of Corollary 8.2 The Fejér kernel is

$$\sum_{n=-N}^{N}\left(1-\frac{|n|}{N}\right)\exp(2\pi int) = \frac{1}{N}\left(\frac{\sin(\pi Nt)}{\sin(\pi t)}\right)^2.$$

When $t = 0$ the sum is N. It follows that

$$\left|\frac{1}{N}\sum_{n=-N}^{N}\left(1-\frac{|n|}{N}\right)\exp(2\pi int)\right| \le \min\left\{1, (2N\,\|t\|)^{-2}\right\}.$$

Hence Theorem 8.4 applies with $\vartheta = 2$. □

Proof of Corollary 8.3 The Bessel functions have the integral representation

$$\int_{-N}^{+N}\left(1-\frac{|t|^2}{N^2}\right)^{\gamma}\exp(2\pi ist)\,dt$$
$$= \pi^{-\gamma}\Gamma(\gamma+1)\,N^{-\gamma+1/2}\,|s|^{-\gamma-1/2}\,J_{\gamma+1/2}(2\pi N\,|s|).$$

Hence, the Poisson summation formula gives a series expansion of the Bochner Riesz kernel,

$$\sum_{n=-N}^{N}\left(1-\frac{|n|^2}{N^2}\right)^{\gamma}\exp(2\pi int)$$
$$= \pi^{-\gamma}\Gamma(\gamma+1)\,N^{-\gamma+1/2}\sum_{k=-\infty}^{+\infty}|t+k|^{-\gamma-1/2}\,J_{\gamma+1/2}(2\pi N\,|t+k|).$$

See [24, Theorem 4.15, Chapter IV and Theorem 2.4, Chapter VII]. Recall that $\left|J_{\gamma+1/2}(z)\right| \leq c\,|z|^{\gamma+1/2}$ when $|z| \to 0$ and $\left|J_{\gamma+1/2}(z)\right| \leq c\,|z|^{-1/2}$ when $|z| \to +\infty$. See [24, Lemma 3.11, Chapter IV]. It follows that the above series are absolutely and uniformly convergent, and the use of the Poisson summation formula is legitimate. Assume for simplicity that $t \neq 1/2 + k$, with k integer. In the above expansion the leading term is the one with $|t+k| < 1/2$, and it has the form

$$\pi^{-\gamma}\Gamma(\gamma+1)\,N^{-\gamma+1/2}\,\|t\|^{-\gamma-1/2}\,J_{\gamma+1/2}(2\pi N\,\|t\|)\,.$$

This term is bounded by $cN\min\{1, (N\,\|t\|)^{-\gamma-1}\}$. The sum of the remaining terms, those with $|t+k| > 1/2$, can be bounded independently of t by

$$cN^{-\gamma}\sum_{|t+k|>1/2}|t+k|^{-\gamma-1} \leq cN^{-\gamma}.$$

Finally observe that when $t = 0$,

$$cN \leq \sum_{|n|\leq N/2}\left(1-\frac{1}{4}\right)^{\gamma} \leq \sum_{n=-N}^{N}\left(1-\frac{|n|^2}{N^2}\right)^{\gamma} \leq \sum_{|n|\leq N} 1 \leq CN.$$

It follows that

$$\left|\left(\sum_{n=-N}^{N}\left(1-\frac{|n|^2}{N^2}\right)^{\gamma}\right)^{-1}\sum_{n=-N}^{N}\left(1-\frac{|n|^2}{N^2}\right)^{\gamma}\exp\left(2\pi i n t\right)\right| \leq c\min\{1, (N\,\|t\|)^{-\gamma-1}\}.$$

Hence Theorem 8.4 applies with $\vartheta = \gamma + 1$. □

Proof of Corollary 8.4 If $\Phi(s)$ is smooth with compact support, the Poisson summation formula and repeated integrations by parts give

$$\begin{aligned}
&\sum_{n=-\infty}^{+\infty}\Phi(N^{-1}n)\exp\left(2\pi i n t\right)\\
&=\sum_{n=-\infty}^{+\infty}\int_{-\infty}^{+\infty}\Phi(N^{-1}s)\exp\left(2\pi i(t+n)s\right)ds\\
&=(-1)^k N\sum_{n=-\infty}^{+\infty}\left(2\pi i N(t+n)\right)^{-k}\int_{-\infty}^{+\infty}\exp\left(2\pi i N(t+n)s\right)\frac{d^k}{dt^k}\Phi(s)ds.
\end{aligned}$$

Hence,

$$\left| \left(\sum_{n=-\infty}^{+\infty} \Phi(N^{-1}n) \right)^{-1} \sum_{n=-\infty}^{+\infty} \Phi(N^{-1}n) \exp(2\pi int) \right|$$

$$\leq (2\pi)^{-k} \left(\int_{-\infty}^{+\infty} \left| \frac{d^k}{dt^k} \Phi(s) \right| ds \right) \left(N \left| \sum_{n=-\infty}^{+\infty} \Phi(N^{-1}n) \right|^{-1} \right)$$

$$\times \left(\sum_{n=-\infty}^{+\infty} (N|t+n|)^{-k} \right)$$

$$\leq c \, (N \|t\|)^{-k} .$$

□

The following lemma is the key tool in the proof of Theorem 8.5.

Lemma 8.6 *In the one dimensional torus* $\mathbb{T}$, *if* $\nu > 1$ *define*

$$f(x) = \sum_{m=1}^{+\infty} m^{-\nu} \exp(2\pi imx) .$$

If p, q, N *are positive integers, with* p *and* q *coprime, and if* $N \geq 2q^{\nu+1}$, *then*

$$\left| \frac{1}{N} \sum_{n=0}^{N-1} f(np/q) \right| > \left(\frac{1}{2} \sum_{m=1}^{+\infty} m^{-\nu} \right) q^{-\nu} .$$

Proof The series defining $f(x)$ is absolutely convergent, and a rearrangement gives

$$\frac{1}{N} \sum_{n=0}^{N-1} f(np/q) = \sum_{m=1}^{+\infty} m^{-\nu} \left(\frac{1}{N} \sum_{n=0}^{N-1} \exp(2\pi inmp/q) \right) .$$

Dividing the integers m into multiples of q and non multiples of q, one gets

$$\left| \sum_{m=1}^{+\infty} m^{-\nu} \left(\frac{1}{N} \sum_{n=0}^{N-1} \exp(2\pi inmp/q) \right) \right|$$

$$\geq \left| \sum_{m=jq} m^{-\nu} \left(\frac{1}{N} \sum_{n=0}^{N-1} \exp(2\pi inmp/q) \right) \right| - \sum_{m \neq jq} m^{-\nu} \left| \frac{1}{N} \sum_{n=0}^{N-1} \exp(2\pi inmp/q) \right| .$$

If $m = jq$ is a multiple of q, then $\exp(2\pi i n m p/q) = 1$ for every n, and

$$\left|\sum_{m=jq} m^{-\nu}\left(\frac{1}{N}\sum_{n=0}^{N-1}\exp(2\pi i n m p/q)\right)\right| = q^{-\nu}\sum_{j=1}^{+\infty} j^{-\nu}.$$

If $m \neq jq$ is not a multiple of q, write $N = H + Kq$, with $0 \leq H < q$. Then

$$\left|\frac{1}{N}\sum_{n=0}^{N-1}\exp(2\pi i n m p/q)\right| = \left|\frac{1}{N}\frac{\exp(2\pi i N m p/q) - 1}{\exp(2\pi i m p/q) - 1}\right|$$

$$= \left|\frac{1}{N}\frac{\exp(2\pi i H m p/q) - 1}{\exp(2\pi i m p/q) - 1}\right| = \left|\frac{1}{N}\sum_{n=0}^{H-1}\exp(2\pi i n m p/q)\right| \leq \frac{H}{N} < \frac{q}{N}.$$

Hence,

$$\sum_{m\neq jq} m^{-\nu}\left|\frac{1}{N}\sum_{n=0}^{N-1}\exp(2\pi i n m p/q)\right| < \frac{q}{N}\sum_{m=1}^{+\infty} m^{-\nu}.$$

In conclusion,

$$\left|\frac{1}{N}\sum_{n=0}^{N-1} f(np/q)\right| > q^{-\nu}\sum_{j=1}^{+\infty} j^{-\nu} - \frac{q}{N}\sum_{m=1}^{+\infty} m^{-\nu} = \left(1 - \frac{q^{\nu+1}}{N}\right)\left(\sum_{m=1}^{+\infty} m^{-\nu}\right) q^{-\nu}.$$

□

Proof of Theorem 8.5 The proof of 1 is elementary. If $g(x) = \exp(2\pi i m \cdot x)$, with $m \in \mathbb{Z}^d \setminus \{0\}$, then

$$\left(\sum_{n=-\infty}^{+\infty}\Phi(N^{-1}n)\right)^{-1}\sum_{n=-\infty}^{+\infty}\Phi(N^{-1}n) g(x + a(n)\alpha) - \int_{\mathbb{T}^d} g(y)dy$$

$$= \left(\left(\sum_{n=-\infty}^{+\infty}\Phi(N^{-1}n)\right)^{-1}\sum_{n=-\infty}^{+\infty}\Phi(N^{-1}n)\exp(2\pi i a(n) m \cdot \alpha)\right)\exp(2\pi i m \cdot x).$$

If $m \cdot \alpha \in \mathbb{Z}$, then $\exp(2\pi i a(n) m \cdot \alpha) = 1$, and

$$\left(\sum_{n=-\infty}^{+\infty}\Phi\left(N^{-1}n\right)\right)^{-1}\sum_{n=-\infty}^{+\infty}\Phi\left(N^{-1}n\right)\exp(2\pi i a(n) m \cdot \alpha) = 1.$$

Hence if $m \cdot \alpha \in \mathbb{Z}$ then $\mathcal{D}g(N, \alpha) = 1$. If $m \cdot \alpha \notin \mathbb{Z}$ then, by the assumptions in Theorem 8.4,

$$\left| \left(\sum_{n=-\infty}^{+\infty} \Phi\left(N^{-1}n\right) \right)^{-1} \sum_{n=-\infty}^{+\infty} \Phi\left(N^{-1}n\right) \exp\left(2\pi i a(n)\, m \cdot \alpha\right) \right| \leq H\left(N \|m \cdot \alpha\|\right)^{-\vartheta}.$$

Hence if $m \cdot \alpha \notin \mathbb{Z}$ then $\mathcal{D}g(N, \alpha) \leq H \|m \cdot \alpha\|^{-\vartheta} N^{-\vartheta}$.

Now let us prove 2. By Theorem 8.4 and Corollary 8.1 the set Ω has Lebesgue measure 0. Let us prove that it contains a countable intersection of open dense sets. Observe that $N \geq 2q^{\nu+1}$ if and only if $q^{-\nu} \geq 2^{\nu/(\nu+1)} N^{-\nu/(\nu+1)}$. Hence, by Lemma 8.6, if $N \geq 2q^{\nu+1}$ then

$$\left| \frac{1}{N} \sum_{n=0}^{N-1} f(np/q) \right| > 2^{-1} \zeta(\nu)\, q^{-\nu} \geq 2^{-1/(\nu+1)} \zeta(\nu)\, N^{-\nu/(\nu+1)}.$$

Define

$$A(N) = \left\{ \alpha \in \mathbb{T} : \left| \frac{1}{N} \sum_{n=0}^{N-1} f(n\alpha) \right| > 2^{-1/(\nu+1)} \zeta(\nu)\, N^{-\nu/(\nu+1)} \right\}.$$

Each set $A(N)$ is open, hence also $\Omega(M) = \cup_{N \geq M} A(N)$ is open. Moreover, $\Omega(M)$ contains every fraction p/q, hence it is dense. Finally, the set Ω in the statement of the theorem contains $\cap_{M \geq 1} \Omega(M)$. □

Remark 8.1 It is curious to compare the above ergodic theorems and corollaries with the approximation properties of Fourier series. While Theorem 8.4 and corollaries seem to suggest that for ergodic means stronger summation methods also give faster convergence, the approximation properties of Fourier series seem to go in the opposite direction. Stronger summation methods of Fourier series may enlarge the class of functions for which convergence holds true, but do not necessarily accelerate the rate of convergence. Assume $d = 1$, and denote by $\mathcal{S}_N f(x)$ and $\mathcal{F}_N f(x)$ the partial sums and arithmetic means of the Fourier expansion of a function $f(x)$,

$$\mathcal{S}_N f(x) = \sum_{m=-N}^{+N} \widehat{f}(m) \exp(2\pi i m x),$$

$$\mathcal{F}_N f(x) = \sum_{m=-N}^{+N} \left(1 - \frac{|m|}{N+1}\right) \widehat{f}(m) \exp(2\pi i m x),$$

The partial sums of the Fourier expansions of continuous functions may not converge pointwise, but when they converge the approximation is close to optimal up to a logarithmic transgression. If $P_N(x)$ is a trigonometric polynomial of degree at most N, then

$$\begin{aligned}
&\sup_{x\in\mathbb{T}}\{|\mathcal{S}_N f(x)) - f(x)|\} \le \sup_{x\in\mathbb{T}}\{|f(x) - P_N(x)|\} \\
&+ \sup_{x\in\mathbb{T}}\left\{\left|\int_{\mathbb{T}}\left(\sum_{m=-N}^{+N} \exp(2\pi i m(x-y))\right)(f(y) - P_N(y))\,dy\right|\right\} \\
&\le \left(1 + \int_{\mathbb{T}}\left|\sum_{m=-N}^{+N} \exp(2\pi i m y)\right| dy\right)\sup_{x\in\mathbb{T}}\{|f(x) - P_N(x)|\}.
\end{aligned}$$

Since the last integral is asymptotic to $4\pi^{-2}\log(N)$ as $N \to +\infty$, if $\mathcal{E}_N(f)$ denotes the best approximation in the supremum norm of $f(x)$ with trigonometric polynomials of degree at most N, then

$$\sup_{x\in\mathbb{T}}\{|\mathcal{S}_N f(x)) - f(x)|\} \le c\,\mathcal{E}_N(f)\log(1+N).$$

The arithmetic means of the Fourier expansions of continuous functions always converge, but the approximation is never better than c/N. For every $|m| \le N$,

$$\sup_{x\in\mathbb{T}}\{|\mathcal{F}_N f(x)) - f(x)|\} \ge \int_{\mathbb{T}} |\mathcal{F}_N f(x)) - f(x)|\,dx \ge \frac{|m|}{N+1}\left|\widehat{f}(m)\right|.$$

Similarly, the approximation with Bochner Riesz means of order $\gamma > 0$ is never better than c/N^2. In conclusion, partial sums of Fourier series may convergence faster than arithmetic means, or Bochner Riesz means.

Remark 8.2 The classical Birkhoff ergodic theorem states that if T is an ergodic endomorphism of a probability space $\mathbb{X}$ and if $f(x)$ is an integrable function on $\mathbb{X}$, then for almost every x in $\mathbb{X}$ one has

$$\lim_{N\to+\infty}\left\{\frac{1}{N}\sum_{n=1}^{N} f\left(T^n(x)\right)\right\} = \int_{\mathbb{X}^d} f(y)dy.$$

See [21, Chapter 2]. In this paper $\mathbb{X} = \mathbb{T}^d$ and $T(x) = x + \alpha$. The assumption in the theorems that $f(x)$ has an absolutely convergent Fourier expansion implies that $f(x)$ is continuous and bounded. This boundedness is a necessary assumption in order to have a result for every x. If $f(x)$ is unbounded, then the sum $\sum_{n=-\infty}^{+\infty}\Phi(N^{-1}n) f(x + a(n)\alpha)$ may be unbounded for some x's.

Remark 8.3 The intuition behind the estimates in Theorem 8.2 is quite simple. If $f(x)$ is square integrable then a straightforward double application of Parseval's equality to Lemma 8.1 gives

$$\left\{\int_{\mathbb{T}^d}\int_{\mathbb{T}^d}\left|\left(\sum_{n=-\infty}^{+\infty}\Phi(N^{-1}n)\right)^{-1}\sum_{n=-\infty}^{+\infty}\Phi(N^{-1}n)f(x+a(n)\alpha)\right.\right.$$

$$\left.\left.-\int_{\mathbb{T}^d}f(y)dy\right|^2 dxd\alpha\right\}^{1/2}$$

$$=\left|\sum_{n=-\infty}^{+\infty}\Phi(N^{-1}n)\right|^{-1}\left\{\sum_{n=-\infty}^{+\infty}\left|\Phi(N^{-1}n)\right|^2\right\}^{1/2}\left\{\sum_{m\in\mathbb{Z}^d\setminus\{0\}}|\widehat{f}(m)|^2\right\}^{1/2}.$$

Under the assumption that $\Phi(s)$ is bounded with compact support, and that $cN \leq \left|\sum_{n=-\infty}^{+\infty}\Phi(N^{-1}n)\right| \leq CN$, then

$$\left|\sum_{n=-\infty}^{+\infty}\Phi(N^{-1}n)\right|^{-1}\left\{\sum_{n=-\infty}^{+\infty}\left|\Phi(N^{-1}n)\right|^2\right\}^{1/2} \leq cN^{-1/2}.$$

Hence the above equality suggests a speed of convergence $cN^{-1/2}$ for almost every (α, x) in $\mathbb{T}^d \times \mathbb{T}^d$. On the other hand, observe that in order to prove that this speed of convergence $cN^{-1/2}$ holds true for almost every (α, x), the above equality does not suffice, but rather one has to estimate a maximal function,

$$\mathcal{M}f(\alpha, x) =$$

$$\sup_{N\geq 1}\left\{N^{1/2}\left|\left(\sum_{n=-\infty}^{+\infty}\Phi(N^{-1}n)\right)^{-1}\sum_{n=-\infty}^{+\infty}\Phi(N^{-1}n)f(x+a(n)\alpha)-\int_{\mathbb{T}^d}f(y)dy\right|\right\}.$$

Bayart et al. in [1] proved that this maximal function is not bounded on square integrable functions, but it becomes bounded if $N^{1/2}$ is replaced by N^{ν} with $\nu < 1/2$. In Theorem 8.2 the assumption that $f(x)$ is square integrable is replaced by the more restricted assumption of an absolutely convergent Fourier expansion, however one obtains the stronger result that the speed of convergence is bounded by $N^{-1/2}\log^{1+\varepsilon}(1+N)$ for every x and almost every α.

Remark 8.4 The assumptions in Theorems 8.2 and 8.4 and corollaries concern the summability properties of the Fourier transforms $\{\widehat{f}(m)\}_{m\in\mathbb{Z}^d}$. If $f(x)$ is Hölder continuous with exponent $\delta > 0$, then there exists $c > 0$ such that for every $R \geq 1$,

$$\left\{\sum_{|m|\geq R}|\widehat{f}(m)|^2\right\}^{1/2} \leq cR^{-\delta}.$$

Hence, for every $0 < p \leq 1$,

$$\sum_{m\in\mathbb{Z}^d\setminus\{0\}} |\widehat{f}(m)|^p \leq \sum_{k=0}^{+\infty} \left\{ \sum_{2^k\leq|m|<2^{k+1}} 1 \right\}^{1-p/2} \left\{ \sum_{|m|\geq 2^k} |\widehat{f}(m)|^2 \right\}^{p/2}$$

$$\leq c \sum_{k=0}^{+\infty} 2^{-kp(\delta-d(1/p-1/2))}.$$

Hence, if $f(x)$ is Hölder continuous with exponent $\delta > d/2$, then $f(x)$ has an absolutely convergent Fourier expansion, and also in $L\log L(\mathbb{Z})$, or in $L^p(\mathbb{Z})$ with $p > 2d/(2\delta + d)$. See [28, Section 3, Chapter VI] for the case $d = 1$, and [19, Chapter 4 and Section 3, Chapter 6] for the case $d \geq 1$. However, it should be emphasized that the assumptions in Theorems 8.2 and 8.4 are not smoothness assumptions. There are functions with Fourier transforms in $L\log L(\mathbb{Z})$, or in $L^p(\mathbb{Z})$ with $0 < p < 1$, which do not have bounded variation, and which are not Hölder continuous of any positive order. An example is a variation on the theme of the Weierstrass function,

$$f(x) = \sum_{n=1}^{+\infty} 2^{-n} \cos(2\pi n! x).$$

Since the Fourier coefficients of a function $g(x)$ Hölder continuous of an order $\delta > 0$ has Fourier coefficients $|\widehat{g}(m)| \leq c|m|^{-\delta}$, see [28, Section 4, Chapter II], the above Weierstrass function is not Hölder continuous of any positive order.

Remark 8.5 As said in the introduction, the motivation for this work comes from an attempt to improve the speed of convergence in the ergodic theorem. In the literature there are several results. We quote only a few.

Krengel in [16] and Kakutani and Petersen in [14] proved that no general statements can be made about the rate of convergence in ergodic theorems. See also Section 3.2 in Chapter 3 of [21]. In particular, if T is an ergodic measure preserving transformation of the interval $\{0 \leq x \leq 1\}$ and if $\{\varepsilon(N)\}_{N=1}^{+\infty}$ is a positive sequence converging to 0, then there exists a continuous function $f(x)$ such that for almost every x one has

$$\limsup_{N\to+\infty} \left\{ \varepsilon(N)^{-1} \left| \frac{1}{N} \sum_{n=1}^{N} f(T^n(x)) - \int_0^1 f(y)dy \right| \right\} = +\infty.$$

Moreover, for every $1 \leq p \leq +\infty$,

$$\limsup_{N\to+\infty} \left\{ \varepsilon(N)^{-1} \left\{ \int_0^1 \left| \frac{1}{N} \sum_{n=1}^{N} f(T^n(x)) - \int_0^1 f(y)dy \right|^p dx \right\}^{1/p} \right\} = +\infty.$$

Bayart et al. [1] proved that the expected speed of convergence of Weyl sums of continuous, or more generally square integrable functions, is slightly less than $N^{-1/2}$. More precisely, they proved that if $f(x)$ is square integrable and if $\nu < 1/2$ then for almost every $(\alpha, x) \in \mathbb{T}^d \times \mathbb{T}^d$ one has

$$\limsup_{N\to+\infty}\left\{N^{\nu}\left|\frac{1}{N}\sum_{n=1}^{N} f(x+n\alpha) - \int_{\mathbb{T}^d} f(y)dy\right|\right\} = 0.$$

They also proved that if $\nu = 1/2$ then there exist continuous functions $f(x)$ such that for almost every $(\alpha, x) \in \mathbb{T}^d \times \mathbb{T}^d$ one has

$$\limsup_{N\to+\infty}\left\{N^{1/2}\left|\frac{1}{N}\sum_{n=1}^{N} f(x+n\alpha) - \int_{\mathbb{T}^d} f(y)dy\right|\right\} = +\infty.$$

Confirming a conjecture of Erdös and Szüs, Kesten in [15] and Petersen in [20] proved that if $f(x)$ is the characteristic function of an interval $\{a \le x \le b\}$, with $0 < b - a < 1$, then there exists $c(f, \alpha)$ such that for every N

$$\frac{1}{N}\sum_{n=1}^{N} f(x+n\alpha) - \int_0^1 f(y)dy \le \frac{c(f,\alpha)}{N}$$

if and only if $b - a = n\alpha - k$ for some integers n and k. Hence for a characteristic function the speed of convergence cN^{-1} is the exception not the rule. Sets of bounded remainders for continuous irrational rotations on multidimensional tori are studied in Grepstad and Larcher in [8] and Grepstad and Lev in [9].

Hellekalek and Larcher in [11] and [12] proved that if $f(x)$ is a continuously differentiable function on $\{0 \le x \le 1\}$ with $df(x)/dx$ Lipschitz continuous and with $f(0) \ne f(1)$, hence $f(x)$ has a jump discontinuity as a function on the torus $\mathbb{T}$, then for every α and every x one has

$$\limsup_{N\to+\infty}\left\{N\left|\frac{1}{N}\sum_{n=1}^{N} f(x+n\alpha) - \int_0^1 f(y)dy\right|\right\} = +\infty.$$

It is also proved that if $f(0) = f(1)$, hence $f(x)$ is continuous as a function on the torus $\mathbb{T}$ and the derivative may have a jump discontinuity, then for almost every α with respect to Lebesgue measure one has

$$\limsup_{N\to+\infty}\left\{N\sup_{0\le x\le 1}\left\{\left|\frac{1}{N}\sum_{n=1}^{N} f(x+n\alpha) - \int_0^1 f(y)dy\right|\right\}\right\} < +\infty.$$

Observe that a discontinuous function cannot have a Fourier expansion absolutely convergent. On the other hand, the assumption that $f(0) = f(1)$ and

$df(x)/dx$ Lipschitz continuous in $\{0 \leq x \leq 1\}$ implies that $|\widehat{f}(m)| \leq cm^{-2}$. More generally, if $df(x)/dx$ Hölder continuous with exponent $\varepsilon > 0$, then $|\widehat{f}(m)| \leq cm^{-1-\varepsilon}$.

Dick and Pillichshammer in [6] proved that if $\{\alpha_n\}_{n=0}^{+\infty}$ is a van der Corput sequence on the interval $\{0 \leq x \leq 1\}$, and if the Fourier coefficients of the function $f(x)$ have decay $|\widehat{f}(m)| \leq c|m|^{-1-\varepsilon}$ for some $\varepsilon > 0$, then

$$\left|\frac{1}{N}\sum_{n=1}^{N} f(\alpha_n) - \int_0^1 f(x)dx\right| \leq \frac{c}{N}.$$

It should be pointed out that the assumption $f(x)$ of bounded variation, or the assumption $|\widehat{f}(m)| \leq c|m|^{-1-\varepsilon}$ in the above results, are a sort of regularity assumption on the function and it is quite different from the assumptions in Theorems 8.2 and 8.4 that the functions have absolutely convergent Fourier expansions, or Fourier transforms in $L\log L(\mathbb{Z})$, or in $L^p(\mathbb{Z})$ with $0 < p < 1$. The above quoted results and our results apply to quite different classes of functions. In particular, Theorems 8.2 and 8.4 can be applied also to some functions with fractal graphs, such as the Weierstrass functions mentioned in Remark 8.4.

Set

$$\sup_{N\geq 0}\left\{\left|\sum_{n=1}^{N} f(x+n\alpha) - N\int_{\mathbb{T}} f(y)dy\right|\right\} = c(f,\alpha,x).$$

In [18] Liardet, and again in [22] Schoissengeier, proved that if $c(f,\alpha,x)$ is finite at a particular point x_0, $c(f,\alpha,x_0) < +\infty$, then $c(f,\alpha,x)$ is finite at every other point x, and $c(f,\alpha,x) \leq c(f,\alpha,x_0)$. Indeed the proof holds true also in several variables. In [22] there is also a study of properties of the set of $B(f) = \{\alpha \in \mathbb{T},\ c(f,\alpha,x) < +\infty\}$. In particular, it is proved that if the complement of $B(f)$ in the torus has a cardinality smaller than the continuum, then $f(x)$ is a trigonometric polynomial. Hence, if $f(x)$ is not a trigonometric polynomial, in Corollary 8.1 the exceptional set Ω has measure zero but the cardinality of continuum.

In Theorems 8.2 and 8.4 above the exceptional sets of α which give a slow speed of convergence have been measured with the Lebesgue measure. In [4] there is a study of the Hausdorff dimension of these exceptional sets, in the particular case of the of weight $\Phi(s) = \chi_{\{|s|\leq 1\}}(s)$.

Acknowledgments I was the first Italian doctoral student of Guido Weiss, the first of a long series. I owe much gratitude to Guido, and to Mitchell Taibleson, the supervisors of my doctoral dissertation. I learned a lot from them, and above all we have been friends. The content of this paper is related to [4] and to a forthcoming paper in collaboration with Bianca Gariboldi and Alessandro Monguzzi (see [5]). I would like to thank them, and also I would like to thank Luca Brandolini, Giacomo Gigante, Giancarlo Travaglini, for several useful discussions.

References

1. F. Bayart, Z. Buczolich, Y. Heurteaux, Fast and slow points of Birkhoff sums. Ergodic Theory Dyn. Syst. **40**(12), 3236–3256 (2020)
2. J. Beck, Probabilistic diophantine approximation. I. Kronecker sequences. Ann. Math. **140**, 449–502 (1994)
3. L. Brandolini, L. Colzani, G. Gigante, G. Travaglini, On the Koksma-Hlawka inequality. J. Complex. **29**, 158–172 (2013)
4. L. Colzani, Speed of convergence of Weyl sums over Kronecker sequences. Monatsh. Math. **200**(2), 209–228 (2023)
5. L. Colzani, B. Gariboldi, A. Monguzzi, Summability and speed of convergence in an ergodic theorem. J. Math. Anal. Appl. **536** (2024)
6. J. Dick, F. Pillichshammer, Periodic functions with bounded remainder. Arch. Math. **87**(6), 554–563 (2006)
7. M. Drmota, R.F. Tichy, *Sequences, Discrepancies and Applications*, Lecture Notes in Mathematics 1651 (Springer, 1997)
8. S. Grepstad, G. Larcher, Sets of bounded remainder for a continuous irrational rotation on $[0, 1]^2$. Acta Arith. **176**, 365–395 (2016)
9. S. Grepstad, N. Lev, Sets of bounded discrepancy for multi-dimensional irrational rotation. Geom. Funct. Anal. **25**, 87–133 (2015)
10. G.H. Hardy, E.M. Wright, *An Introduction to the Theory of Numbers* (Oxford University Press, 2008)
11. P. Hellekalek, G. Larcher, On functions with bounded remainder. Ann. Inst. Fourier (Grenoble) **39**, 17–26 (1989)
12. P. Hellekalek, G. Larcher, On Weyl sums and skew products over irrational rotations. Theoret. Comput. Sci. **65**, 189–196 (1989)
13. R.A. Hunt, On the convergence of Fourier series, in *Proceedings of the Conference on Orthogonal Expansions and their Continuous Analogues held at Southern Illinois University, Edwardsville* (Southern Illinois University Press, Carbondale, 1967), pp. 235–255
14. S. Kakutani, K. Petersen, The speed of convergence in the ergodic theorem. Monatsh. Math. **91**, 11–18 (1981)
15. H. Kesten, On a conjecture of Erdös and Szüsz related to uniform distribution mod 1. Acta Arithmetica **12**, 193–212 (1966/1967)
16. U. Krengel, On the speed of convergence in the ergodic theorem. Monatsh. Math. **86**, 3–6 (1978/1979)
17. L. Kuipers, H. Niederreiter, *Uniform Distribution of Sequences*, Wiley Interscience (Wiley, 1974)
18. P. Liardet, Regularities of distribution. Compos. Math. **61**, 267–293 (1987)
19. G.G. Lorentz, *Approximation of Functions* (Chelsea Publishing Company, 1986)
20. K. Petersen, On a series of cosecants related to a problem in ergodic theory. Compos. Math. **26**, 313–317 (1973)
21. K. Petersen, *Ergodic Theory* (Cambridge University Press, 1989)
22. J. Schoissengeier, Regularity of distribution of $(n\alpha)$-sequences. Acta Arithm. **133**, 127–157 (2008)
23. E.M. Stein, N.J. Weiss, On the convergence of Poisson integrals. Trans. Am. Math. Soc. **140**, 35–54 (1969)
24. E.M. Stein, G. Weiss, *Introduction to Fourier Analysis on Euclidean Spaces* (Princeton University Press, 1971)

25. E.M. Stein, M.H. Taibleson, G. Weiss, Weak type estimates for maximal operators on certain H(p) classes, in *Proceedings of the Seminar on Harmonic Analysis (Pisa 1980)*, Rendiconti del Circolo Matematico di Palermo, Supplemento 1 (1981), pp. 81–87
26. G. Travaglini, *Number Theory, Fourier Analysis and Geometric Discrepancy*, London Mathematical Society Student Texts 81 (Cambridge University Press, 2014)
27. J. von Plato, Oresme's proof of the density of rotations of a circle through an irrational angle. Hist. Math. **20**, 428–433 (1993)
28. A. Zygmund, *Trigonometric Series* (Cambridge University Press, 1968)

Chapter 9
Characterizations of Product Hardy Spaces on Stratified Groups by Singular Integrals and Maximal Functions

Michael G. Cowling, Zhijie Fan, Ji Li, and Lixin Yan

Abstract A large part of the theory of Hardy spaces on products of Euclidean spaces has been extended to the setting of products of stratified Lie groups. This includes characterisation of H^1 by square functions and by atomic decompositions, proof of the duality of H^1 with BMO, and description of many interpolation spaces. Until now, however, two aspects of the classical theory have been conspicuously absent: the characterisation of H^1 by singular integrals (of Christ–Geller type) or by (vertical or nontangential) maximal functions. In this paper we fill in these gaps by developing new techniques on products of stratified groups, using the ideas in Chen et al. (Flag Hardy space theory on Heisenberg groups and applications. arXiv:2102.07371, 2021) on the Heisenberg group with flag structure.

9.1 Introduction and Statement of Main Results

Hardy spaces first appeared in the study of boundary behaviour of holomorphic functions on the disc and upper half plane. The modern theory of Hardy spaces began in 1960, when E. M. Stein and G. Weiss [30] considered functions defined

M. G. Cowling (✉)
School of Mathematics and Statistics, University of New South Wales, Sydney, NSW, Australia
e-mail: m.cowling@unsw.edu.au

Z. Fan
Department of Mathematics, Wuhan University, Wuhan, P. R. China
e-mail: ZhijieFan@whu.edu.cn

J. Li
Department of Mathematics, Macquarie University, Sydney, NSW, Australia
e-mail: ji.li@mq.edu.au

L. Yan
Department of Mathematics, Macquarie University, Sydney, NSW, Australia

Department of Mathematics, Sun Yat-sen (Zhongshan) University, Guangzhou, P. R. China
e-mail: mcsylx@mail.sysu.edu.cn

E. Hernández et al. (eds.), *The Mathematical Heritage of Guido Weiss*,
Applied and Numerical Harmonic Analysis,
https://doi.org/10.1007/978-3-031-76793-7_9

on $\mathbb{R}^n \times \mathbb{R}^+$, and it took off in the early 1970s, with the remarkable work of C. Fefferman and Stein [9] and then R. R. Coifman and Weiss [7]. Much of this theory has been extended to more general spaces of homogeneous type, in the sense of Coifman and Weiss [6, 7]. In the late 1970s, G. B. Folland and Stein [13] characterised the Hardy space $\mathsf{H}^1(G)$ on a stratified group G in terms of atomic decompositions, square functions, area functions, and maximal functions. The area integrals and maximal functions involve taking integrals or suprema over cones in $G \times \mathbb{R}_+$. Soon after, M. Christ and D. Geller [5] showed that there are singular integral operators $\mathcal{R}_0, \dots, \mathcal{R}_n$ on a stratified Lie group such that $f \in \mathsf{H}^1(G)$ if and only if all $\mathcal{R}_j f \in \mathsf{L}^1(G)$. Here $\mathcal{R}_0$ is the identity operator and the other $\mathcal{R}_j$ are Riesz transformations, that is, convolutions with derivatives of a potential.

Harmonic analysis on product spaces $\mathbb{R}^m \times \mathbb{R}^n$ was born in the late 1970s and studied extensively in the 1980s, in particular by S.-Y. A. Chang, R. Fefferman, R. F. Gundy, J.-L. Journé, J. Pipher, and Stein (see [1, 11, 15, 23, 25, 28]), motivated by problems on the boundary behavior of holomorphic functions in several complex variables, which require consideration of approach regions that behave differently in different variables. Harmonic analysis on product spaces is influenced by classical harmonic analysis, but is different in that the different factors in the product may be dilated independently. The terms one-parameter and multiparameter are often used to highlight the different structures of the dilations considered. As in classical harmonic analysis, an important part of the theory is the development of Hardy and BMO spaces, their duality and the connections to atomic decompositions. A key ingredient is Journé's covering lemma, which provides a tool to replace general open sets by rectangles with controlled geometry.

Since the 1980s, the development of multiparameter harmonic analysis proceeded apace; recent contributions in the area include [12, 22, 24, 26, 27]. Much of the product space theory on $\mathbb{R}^m \times \mathbb{R}^n$ has been extended to more general product spaces, including the duality of H^1 with BMO, characterisation of H^1 by square functions and atomic decompositions, and description of various interpolation spaces. In [2, 16–18], the theory of Hardy spaces H^p, for p less than and close to 1, has been developed on products $X_1 \times X_2$ of spaces of homogeneous type. Hence on products $G_1 \times G_2$ of stratified Lie groups, there is already a well-defined Hardy space $\mathsf{H}^1(G_1 \times G_2)$ that may be characterised by atomic decompositions and by square or area functions.

Two aspects of the classical theory that have been conspicuous by their absence until now are a singular integral characterisation of Christ–Geller type and a maximal function characterisation. The main difficulty is that the geometrical structure is harder to handle than in the one-parameter case. For example, one may obtain the atomic decomposition from the nontangential maximal function in the one-parameter case by using the classical Calderón–Zygmund and Whitney decompositions involving cubes, but these decompositions are absent in the multiparameter case.

This paper fills these gaps for products of stratified Lie groups, with Theorems 9.1 and 9.2 below. For simplicity, and because new methods would otherwise be needed, we consider products of only two groups. Our new techniques come from [3], where similar results are proved on the Heisenberg group with its flag structure.

Unexplained definitions may be found below.

Theorem 9.1 *The double Riesz transformations $\mathcal{R}^{[1]}_{j_1} \otimes \mathcal{R}^{[2]}_{j_2}$ characterise the Hardy space $\mathsf{H}^1(G_1 \times G_2)$. That is, $f \in \mathsf{H}^1(G_1 \times G_2)$ if and only if each $\mathcal{R}^{[1]}_{j_1} \otimes \mathcal{R}^{[2]}_{j_2} f$ is in $\mathsf{L}^1(G_1 \times G_2)$, and moreover*

$$\|f\|_{\mathsf{H}^1(G_1\times G_2)} \eqsim \sum_{j_1=0}^{d_1} \sum_{j_2=0}^{d_2} \left\| \mathcal{R}^{[1]}_{j_1} \otimes \mathcal{R}^{[2]}_{j_2} f \right\|_{\mathsf{L}^1(G_1\times G_2)}.$$

Using Theorem 9.1 and the H^1-BMO duality (see for example [16]), we obtain a decomposition of functions in the product space $\mathsf{BMO}(G_1 \times G_2)$.

Corollary 9.1 *For a function u on $G_1 \times G_2$, the following are equivalent:*

(a) $u \in \mathsf{BMO}(G_1 \times G_2)$;
(b) there exist $g_{j_1,j_2} \in \mathsf{L}^\infty(G_1 \times G_2)$ such that

$$u = \sum_{j_1=0}^{d_1} \sum_{j_2=0}^{d_2} \mathcal{R}^{[1]}_{j_1} \otimes \mathcal{R}^{[2]}_{j_2} (g_{j_1,j_2}).$$

Write $\Gamma(g_1, g_2)$ for the product $\Gamma_1(g_1) \times \Gamma_2(g_2)$ of the cones treated by Folland and Stein [11], and for suitable functions $\psi^{[i]}$ on G_i, define the nontangential maximal function:

$$\mathcal{N}_\psi(f)(g_1, g_2) := \sup\left\{ \left| f * (\psi^{[1]}_{t_1} \otimes \psi^{[2]}_{t_2})(h_1, h_2) \right| : (h_1, h_2) \in \Gamma(g_1, g_2), t_1, t_2 \in \mathbb{R}_+ \right\},$$

where $\psi^{[i]}_{t_i}$ is a normalised dilate of $\psi^{[i]}$.

Theorem 9.2 *The nontangential maximal operator $\mathcal{N}_\psi$ characterises the Hardy space $\mathsf{H}^1(G_1 \times G_2)$. That is, $f \in \mathsf{H}^1(G_1 \times G_2)$ if and only if $\mathcal{N}_\psi f$ is in $\mathsf{L}^1(G_1 \times G_2)$; moreover*

$$\|f\|_{\mathsf{H}^1(G_1\times G_2)} \eqsim \left\| \mathcal{N}_\psi(f) \right\|_{\mathsf{L}^1(G_1\times G_2)}.$$

This paper is organised as follows. In Sect. 9.2, we remind the reader of some background on stratified Lie groups and analysis thereupon, and introduce some notation to simplify the formulae in the case of products of such groups. In Sect. 9.3, we review some of the main results on Hardy spaces on products of stratified groups; many of these are valid in the more general context of products of spaces of homogeneous type. Then we prove our main theorems on the characterisations of $\mathsf{H}^1(G_1 \times G_2)$, by Riesz transforms in Sect. 9.4 and by maximal functions in Sect. 9.5. The results proved are actually somewhat more general than stated in Theorems 9.1

and 9.2, but precise statements require more notation than we have established at this point.

"Constants" are always positive real numbers; we write $A \lesssim B$ when there is a constant C such that $A \leq CB$, and $A \simeq B$ when $A \lesssim B$ and $B \lesssim A$. We denote the identity of a group by o, and the indicator function of a set E by χ_E.

9.2 Preliminaries

9.2.1 Stratified Nilpotent Lie Groups

Let G be a (real and finite dimensional) stratified nilpotent Lie group of step k with Lie algebra $\mathfrak{g}$. This means that we may write $\mathfrak{g}$ as a vector space direct sum $\mathfrak{v}_1 \oplus \cdots \oplus \mathfrak{v}_k$, where $[\mathfrak{v}_1, \mathfrak{v}_j] = \mathfrak{v}_{j+1}$ when $1 \leq j \leq k$; here $\mathfrak{v}_{k+1} = \{0\}$. Let Q denote the homogeneous dimension $\sum_{j=1}^{k} j \dim \mathfrak{v}_j$ of G.

There is a one-parameter family of automorphic dilations δ_t on $\mathfrak{g}$, given by

$$\delta_t(X_1 + X_2 + \cdots + X_k) = tX_1 + t^2 X_2 + \cdots + t^k X_k;$$

here each $X_j \in \mathfrak{v}_j$ and $t > 0$. The exponential mapping $\exp : \mathfrak{g} \to G$ is a diffeomorphism, and we identify $\mathfrak{g}$ and G. The dilations extend to automorphic dilations of G, also denoted by δ_t, by conjugation with exp. The natural bi-invariant Haar measure on G is the Lebesgue measure on $\mathfrak{g}$, lifted to G using exp.

By [19], the group G may be equipped with a smooth subadditive homogeneous norm ρ, a continuous function from G to $[0, \infty)$ that is smooth on $G \setminus \{o\}$ and satisfies

(a) $\rho(g^{-1}) = \rho(g)$;
(b) $\rho(xy) \leq \rho(x) + \rho(y)$
(c) $\rho(\delta_t(g)) = t\rho(g)$ for all $g \in G$ and $t > 0$;
(d) $\rho(g) = 0$ if and only if $g = o$,

Abusing notation, we define $\rho(g, g') = \rho(g^{-1}g')$ for all $g, g' \in G$; this defines a metric on G. We write $B(g, r)$ for the open ball with centre g and radius r with respect to ρ:

$$B(g, r) = gB(o, r) = g\{h \in G : \rho(h) < 1\}.$$

The metric space (G, ρ) is *geometrically doubling*; that is, there exists $N \in \mathbb{N}$ such that every metric ball $B(x, 2r)$ may be covered by at most N balls of radius r.

We remind the reader that a stratified Lie group is a space of homogenous type in the sense of Coifman and Weiss [6, 7], and analysis on stratified Lie groups uses much from the theory of such spaces. In particular, we frequently deal with *molecules*, that is, functions ψ that satisfy *standard decay and smoothness conditions*, meaning that there is a parameter $\varepsilon \in (0, 1]$, which we fix once and for all, and a constant C such that

$$
\begin{aligned}
|\psi(g)| &\le C \frac{1}{(1+\rho(g))^{Q+\varepsilon}} \\
\left|\psi(g) - \psi(g')\right| &\le C \frac{\rho(g^{-1}g')^{\varepsilon}}{(1+\rho(g)+\rho(g'))^{Q+2\varepsilon}}
\end{aligned}
\tag{9.1}
$$

for all $g, g' \in G$. We often impose an additional *cancellation* condition, namely

$$
\int_G \psi(g)\, \mathrm{d}g = 0. \tag{9.2}
$$

We write $\|\psi\|_{\mathsf{M}(G)}$ for the least constant C such that the conditions (9.1) hold, $\mathsf{M}(G)$ for the Banach space of all such functions ψ, and $\mathsf{M}_0(G)$ for the subspace of $\mathsf{M}(G)$ of all ψ that also satisfy condition (9.2).

The *normalised dilate* f_t of a function f on G by $t > 0$ is given by $f_t := t^{-Q} f \circ \delta_{1/t}$, and the *convolution* $f * f'$ of measurable functions f and f' on G is defined by

$$
f * f'(g) = \int_G f(h) f'(h^{-1}g)\, \mathrm{d}h = \int_G f(gh^{-1}) f'(h)\, \mathrm{d}h.
$$

Take left-invariant vector fields $\mathcal{X}_1, \dots, \mathcal{X}_n$ on G that form a basis of $\mathfrak{v}_1$, and define the *sub-Laplacian* $\mathcal{L} = -\sum_{j=1}^{n} (\mathcal{X}_j)^2$. Observe that each $\mathcal{X}_j$ is homogeneous of degree 1 and $\mathcal{L}$ is homogeneous of degree 2, in the sense that

$$
\mathcal{X}_j (f \circ \delta_t) = t \left(\mathcal{X}_j f\right) \circ \delta_t \qquad \text{and} \qquad \mathcal{L}(f \circ \delta_t) = t^2 \left(\mathcal{L} f\right) \circ \delta_t
$$

for all $t > 0$ and all $f \in C^2(G)$.

Associated to the sub-Laplacian, there are various Riesz potential operators $\mathcal{L}^{-\alpha}$, where $\alpha > 0$; these are convolution operators with homogeneous kernels—see Folland [10]. The *Riesz transformation* $\mathcal{R}_j := \mathcal{X}_j \mathcal{L}^{-1/2}$ is a singular integral operator, and is bounded on $\mathsf{L}^p(G)$ when $1 < p < \infty$ as well as from the Folland–Stein Hardy space $\mathsf{H}^1(G)$ to $\mathsf{L}^1(G)$. We define $\mathcal{R}_0$ to be the identity operator $\mathcal{I}$.

The Hardy–Littlewood maximal operator $\mathcal{M}$ on G is defined using the metric balls:

$$
\mathcal{M} f(g) := \sup \left\{ \frac{1}{|B(g', r)|} \int_{B(g',r)} \left|f(g'')\right| \mathrm{d}g'' : g \in B(g', r) \right\}.
$$

For future use, we note that the layer cake formula implies that, if μ is a radial decreasing function on G (that is, $\mu(g)$ depends only on $\rho(g)$ and decreases as $\rho(g)$ increases), then

$$|f| * \mu_\varepsilon(g) \leq \|\mu\|_{\mathsf{L}^1(G)} \mathcal{M}f(g) \qquad \forall g \in G. \tag{9.3}$$

9.2.2 Functional Calculus for the Sub-Laplacian

The sub-Laplacian $\mathcal{L}$ has a spectral resolution:

$$\mathcal{L}(f) = \int_0^\infty \lambda \, \mathrm{d}\mathcal{E}_{\mathcal{L}}(\lambda) f \qquad \forall f \in \mathsf{L}^2(G),$$

where $\mathcal{E}_{\mathcal{L}}(\lambda)$ is a projection-valued measure supported on $[0, \infty)$, the spectrum of $\mathcal{L}$. For a bounded Borel function $\eta : [0, \infty) \to \mathbb{C}$, we define the operator $F(\mathcal{L})$ spectrally:

$$\eta(\mathcal{L}) f = \int_0^\infty \eta(\lambda) \, \mathrm{d}\mathcal{E}_{\mathcal{L}}(\lambda) f \qquad \forall f \in \mathsf{L}^2(G).$$

This operator is a convolution with a Schwartz distribution on G.

Take a smooth function $\eta : \mathbb{R}_+ \to \mathbb{R}$, supported in $[1/2, 2]$, such that $\sum_{n\in\mathbb{Z}} \eta(2^{-n}s) = 1$ for all $s \in \mathbb{R}_+$. The convolution kernels $k_{\eta(\mathcal{L}_i)}$ of the operators $\eta(\mathcal{L}_i)$ on G are Schwartz functions, by [20]. Moreover, we may write $\eta(t\mathcal{L}_i) = t\mathcal{L}_i \psi(t\mathcal{L}_i)$, where $\psi(t) := t^{-1}\eta(t)$ for all $t \in \mathbb{R}_+$ and $\operatorname{supp} \psi \subset [1/2, 2]$, and deduce that

$$k_{\eta(t\mathcal{L}_i)} = t\mathcal{L}_i k_{\psi(t\mathcal{L}_i)}.$$

Integration by parts now implies that

$$\int_G k_{\eta(t\mathcal{L}_i)}(g) \, \mathrm{d}g = \int_G t\mathcal{L}_i k_{\Psi(t\mathcal{L}_i)}(g) \, \mathrm{d}g = 0.$$

9.2.3 The Heat and Poisson Kernels

Let p_t and P_t, where $t > 0$, be the heat and Poisson kernels associated to the sub-Laplacian operator $\mathcal{L}$, that is, the convolution kernels of the operators $\mathrm{e}^{t\mathcal{L}}$ and $\mathrm{e}^{t\sqrt{\mathcal{L}}}$ on G. We write Q_t for $t\partial_t P_t$, and to simplify notation later, we often write P instead of P_1. We warn the reader that P_t and Q_t are the normalised dilates of P_1 and Q_1

by the factor t, but p_t is the normalised dilate of p_1 by a factor of $t^{1/2}$. Let ∇ denote the subgradient on G and $\bar{\nabla}$ denote the gradient (∇, ∂_t) on $G \times \mathbb{R}_+$.

Lemma 9.1 *The kernels p_t and P_t are $\mathbb{R}_+$-valued. Further, p_t and P_t have integral 1, while Q_t has integral 0 for all $t \in \mathbb{R}_+$. Finally, there exists a constant c such that*

$$p_t(g) \lesssim t^{-Q/2} \exp\left(-\rho^2(g)/ct\right)$$

$$\left|\bar{\nabla} p_t(g)\right| \lesssim t^{-(Q+1)/2} \exp\left(-\rho^2(g)/ct\right)$$

$$P_t(g) \eqsim \frac{t}{(t^2 + \rho(g)^2)^{(Q+1)/2}}$$

$$\left|\bar{\nabla} P_t(g)\right| \lesssim \frac{t}{(t^2 + \rho(g)^2)^{(Q+2)/2}}$$

for all $g \in G$ and $t \in \mathbb{R}_+$.

Proof For the heat kernel estimates, see [31, Theorem IV.4.2]. Note that there is a version of the first estimate with the opposite inequality and a different constant c.

The estimates for P_t and Q_t follow from the subordination formula

$$\mathrm{e}^{-t\sqrt{\mathcal{L}}} = \frac{1}{2\sqrt{\pi}} \int_0^\infty \frac{t\mathrm{e}^{-t^2/4v}}{\sqrt{v}} \mathrm{e}^{-v\mathcal{L}} \frac{dv}{v}.$$

For the case of the Heisenberg group, much of this is worked out in detail in [3]. □

This lemma implies that the heat kernel p_1 and the Poisson kernel P_1 (and their derivatives) both satisfy the standard decay and smoothness conditions (9.1); the derivatives also satisfy the cancellation condition (9.2).

Lemma 9.1 also implies the following standard corollary, whose proof we omit.

Corollary 9.2 *Suppose that $f \in \mathsf{L}^p(G)$, where $1 \le p \le \infty$. Then $\|f * P_t\|_{\mathsf{L}^p(G)}$ and $\left\|f * t\bar{\nabla} P_t\right\|_{\mathsf{L}^p(G)}$ are uniformly bounded as t runs over $\mathbb{R}_+$. Further,*

$$\lim_{t \to 0} f * P_t = f;$$

the convergence is both pointwise almost everywhere, and in the $\mathsf{L}^p(G)$ norm if $1 \le p < \infty$ and in the weak-star topology if $p = \infty$. Finally,

$$\lim_{t \to 0} f * t\bar{\nabla} P_t = 0;$$

the convergence is both pointwise almost everywhere, and in the strong operator topology if $f \in \mathsf{L}^1(G)$, in the $\mathsf{L}^p(G)$ norm if $1 < p < \infty$ and in the weak-star topology if $p = \infty$.

9.2.4 Systems of Pseudodyadic Cubes

We use the Hytönen–Kairema [21] families of "dyadic cubes" in geometrically doubling metric spaces. We state a version of [21, Theorem 2.2] that is simpler, in that we work on well-behaved metric spaces rather than general pseudometric spaces. The Hytönen–Kairema construction builds on seminal work of Christ [4] and of Sawyer and Wheeden [29].

Theorem 9.3 ([21]) *Let* (G, ρ) *be a metric stratified group and* c_0, C_0 *and* κ *constants such that* $0 < c_0 \le C_0 < \infty$ *and* $12C_0\kappa \le c_0$. *Then for all* $k \in \mathbb{Z}$, *there exist families* $\mathscr{Q}^k(G)$ *of* pseudodyadic cubes Q *with* centres $z(Q)$, *such that:*

(a) G *is the disjoint union of all* $Q \in \mathscr{Q}^k(G)$, *for each* $k \in \mathbb{Z}$;
(b) $B(z(Q), c_0\kappa^k/3) \subseteq Q \subseteq B(z(Q), 2C_0\kappa^k)$ *for all* $Q \in \mathscr{Q}^k(G)$;
(c) if $Q \in \mathscr{Q}^k(G)$ *and* $Q' \in \mathscr{Q}^{k'}(G)$ *where* $k \le k'$, *then either* $Q \cap Q' = \emptyset$ *or* $Q \subseteq Q'$; *in the second case,* $B(z(Q), 2C_0\kappa^k) \subseteq B(z(Q'), 2C_0\kappa^{k'})$;

The family of pseudodyadic cubes Q in $\mathscr{Q}^k(G)$, where $k \in \mathbb{Z}$, of Theorem 9.3 will be called a *Hytönen–Kairema set of cubes* on G. We write $\mathscr{Q}(G)$ for the union of all $\mathscr{Q}^k(G)$. Given a cube $Q \in \mathscr{Q}^k(G)$, we denote the quantity κ^k by $\ell(Q)$, by analogy with the side-length of a Euclidean cube.

9.2.5 Products of Stratified Groups

We equip products of stratified groups G_1 and G_2 with a product structure: the basic geometric objects are rectangles, which are products of balls, and pseudodyadic rectangles, which are products of pseudodyadic cubes. We write $\mathscr{P}^{\boldsymbol{j}}(\boldsymbol{G})$ for the collection of all pseudodyadic rectangles that are products of cubes in $\mathscr{Q}^{j_1}(G_1)$ and in $\mathscr{Q}^{j_2}(G_2)$; $\mathscr{P}(\boldsymbol{G})$ for the collection of all pseudodyadic rectangles, and $\mathscr{R}(\boldsymbol{G})$ for the collection of all rectangles. We let $\boldsymbol{\ell} : \mathscr{P}(\boldsymbol{G}) \to \boldsymbol{T}$ be the function such that $\ell_i(Q_1 \times Q_2) = \ell(Q_i)$, the "side-length" of Q_i.

We carry forward the notation from Sect. 9.2.1, modified by adding a subscript i or superscript $[i]$ to clarify that we are dealing with G_i. To shorten the formulae, we often use bold face type to indicate a product object: thus we write $\boldsymbol{G}$, $\boldsymbol{g}$, $\boldsymbol{r}$ and $\boldsymbol{t}$ in place of $G_1 \times G_2$, (g_1, g_2), (r_1, r_2) and (t_1, t_2). For example, $B_i(g_i, r_i)$ denotes the open ball on G_i with centre g_i and radius r_i, with respect to the homogeneous norm ρ_i, and a typical rectangle $R(\boldsymbol{g}, \boldsymbol{r})$ is then a product $B_1(g_1, r_1) \times B_2(g_2, r_2)$. We also write $\boldsymbol{t}\,\mathrm{d}\boldsymbol{t}$ in place of $t_1 t_2\,\mathrm{d}t_1\,\mathrm{d}t_2$, and $\boldsymbol{T}$ for the product parameter space $\mathbb{R}_+ \times \mathbb{R}_+$.

The element of Haar measure on $\boldsymbol{G}$ is denoted $\mathrm{d}\boldsymbol{g}$, but may be written as $\mathrm{d}g_1\,\mathrm{d}g_2$ for calculations. The convolution $f * f'$ of functions f and f' on $\boldsymbol{G}$ is defined by

$$(f * f')(\boldsymbol{g}) := \int_{\boldsymbol{G}} f(\boldsymbol{h}) f'(\boldsymbol{h}^{-1}\boldsymbol{g})\,\mathrm{d}\boldsymbol{h}.$$

We define the strong maximal operator $\mathcal{M}_S$ by

$$\mathcal{M}_S(f)(\boldsymbol{g}) := \sup\left\{\frac{1}{|R|}\int_R |f(\boldsymbol{h})|\, \mathrm{d}\boldsymbol{h} : R \ni \boldsymbol{g},\, R \in \mathscr{R}(\boldsymbol{G})\right\}.$$

It is a straightforward exercise to show that $\mathcal{M}_S$ is dominated by the iterated Hardy–Littlewood maximal operators in the factors:

$$\mathcal{M}_S f \le \mathcal{M}_1\mathcal{M}_2(f) \qquad \text{and} \qquad \mathcal{M}_S f \le \mathcal{M}_2\mathcal{M}_1(f) \qquad \forall f \in \mathrm{L}^1_{\mathrm{loc}}(\boldsymbol{G}).$$

When $1 < p \le \infty$, the operators $\mathcal{M}_1$ and $\mathcal{M}_2$ in the factors are L^p-bounded, so the iterated maximal operators and the strong maximal operator are also L^p-bounded.

Given functions $\psi^{[1]}$ on G_1 and $\psi^{[2]}$ on G_2, we often deal with the product of their normalised dilates on $G_1 \times G_2$, and we abbreviate this to $\psi_{\boldsymbol{t}}$:

$$\psi_{\boldsymbol{t}} := \psi^{[1]}_{t_1} \otimes \psi^{[2]}_{t_2}.$$

If $\psi^{[1]} \in \mathsf{M}(G_1)$ and $\psi^{[2]} \in \mathsf{M}(G_2)$, then

$$|f * \psi_{\boldsymbol{t}}(\boldsymbol{g})| \lesssim \mathcal{M}_S(f)(\boldsymbol{g}) \qquad \forall \boldsymbol{g} \in \boldsymbol{G} \quad \forall f \in \mathrm{L}^1(\boldsymbol{G}),$$

much as argued to prove (9.3), but with "biradial" in place of "radial".

Given an open subset U of $\boldsymbol{G}$ with finite measure $|U|$, we define the enlargement $\widetilde{U}$ of U using the strong maximal operator $\mathcal{M}_S$:

$$\widetilde{U} := \left\{\boldsymbol{g} \in \boldsymbol{G} : \mathcal{M}_S\chi_U(\boldsymbol{g}) > \frac{1}{4}\right\}.$$

We write $\mathscr{M}(U)$ for the family of maximal pseudodyadic rectangles contained in U.

We let $P_{\boldsymbol{t}} := P^{[1]}_{t_1} \otimes P^{[2]}_{t_2}$; when $t_1 = 0$ or $t_2 = 0$, we interpret this as a distribution supported in G_2 or in G_1 in the obvious way. We write $Q^{[i]}_{t_i}$ for the convolution kernel of the operator $t_i\,\partial_{t_i} \mathrm{e}^{-t_i\sqrt{\mathcal{L}_i}}$; then $Q^{[i]}_{t_i} = t_i\,\partial_{t_i} P^{[i]}_{t_i}$. By arguing as in Corollary 9.2, it is easy to see that for any measurable subset V of $\boldsymbol{G}$,

$$\lim_{t_1 \to 0} \chi_V * (Q^{[1]}_{t_1} \otimes P^{[2]}_{t_2})(\boldsymbol{g}) = 0 \tag{9.4}$$

for almost all $\boldsymbol{g}$ in $\boldsymbol{G}$ and in the weak-star topology of $\mathrm{L}^\infty(\boldsymbol{G})$.

The double Riesz transforms $\mathcal{R}^{[1]}_{j_1} \otimes \mathcal{R}^{[2]}_{j_2} f$, where $0 \le j_i \le d_i$, of a suitable function f on $\boldsymbol{G}$ are defined in the obvious way: when j_1 and j_2 are nonzero,

$$\mathcal{R}^{[1]}_{j_1} \otimes \mathcal{R}^{[2]}_{j_2} f := X^{[1]}_{j_1}\mathcal{L}_1^{-1/2} X^{[2]}_{j_2}\mathcal{L}_2^{-1/2} f, \tag{9.5}$$

and if $j_i = 0$ we replace $X^{[i]}_{j_i}\mathcal{L}_i^{-1/2}$ by the identity operator $\mathcal{I}_i$.

9.3 The Known Product Hardy Spaces

9.3.1 The Atomic Hardy Space

Fix a constant C and Hytönen–Kairema sets of pseudodyadic cubes in G_1 and G_2. A pseudodyadic rectangle R is a product $Q_1 \times Q_2$ of pseudodyadic cubes in the factors G_1 and G_2.

An integrable function a_R is said to be a *particle* associated to the pseudodyadic rectangle R if the following support and product cancellation conditions hold:

$$\operatorname{supp} a_R \subseteq CR \tag{9.6}$$

and

$$\int_{G_1} a_R(g_1, \cdot)\, \mathrm{d}g_1 = 0 \qquad \text{and} \qquad \int_{G_2} a_R(\cdot, g_2)\, \mathrm{d}g_2 = 0 \tag{9.7}$$

(almost everywhere).

A function a on $\boldsymbol{G}$ is said to be a *product atom* associated to an open subset U of $\boldsymbol{G}$ of finite measure if a satisfies the following support and size conditions:

$$\operatorname{supp} a \subset \widetilde{U}$$

$$\|a\|_{\mathsf{L}^2(\boldsymbol{G})} \le \left|\widetilde{U}\right|^{-1/2},$$

and we may decompose a as a sum $\sum_{R \in \mathscr{M}(U)} a_R$ of particles a_R associated to the pseudodyadic rectangles $R \in \mathscr{M}(U)$ in such a way that

$$\left(\sum_{R \in \mathscr{M}(U)} \|a_R\|^2_{\mathsf{L}^2(\boldsymbol{G})} \right)^{1/2} \le |U|^{-1/2}.$$

Definition 9.1 We say that $f \in \mathsf{L}^1(\boldsymbol{G})$ belongs to the atomic Hardy space $\mathsf{H}^1_{\mathrm{atom}}(\boldsymbol{G})$ if and only if it is possible to represent f as a sum

$$f = \sum_{n \in \mathbb{N}} \lambda_n a_n,$$

where a_n is an atom and $\lambda_n \in \mathbb{R}_+$ for all n, and $\sum_{n \in \mathbb{N}} \lambda_n < \infty$. We define the norm $\|f\|_{\mathsf{H}^1_{\mathrm{atom}}(\boldsymbol{G})}$ to be the infimum of the sums $\sum_{n \in \mathbb{N}} \lambda_n$ over all such representations of f.

It is often more convenient to impose a stronger requirement on particles, namely, that $a_R = \mathcal{L}_1^{N_1} \mathcal{L}_1^{N_2} b_R$ for some $\mathsf{L}^2(\boldsymbol{G})$ function b_R in the domain of $\mathcal{L}_1^{N_1} \mathcal{L}_1^{N_2}$ and for large integers N_1 and N_2; this means that a_R has many vanishing moments, which may make calculations easier. We may show that this stronger requirement on particles gives the same atomic Hardy space, using telescopic series arguments to make moments vanish.

9.3.2 Square Function and Area Function Hardy Spaces

For $\boldsymbol{g} \in \boldsymbol{G}$ and $\beta \in [0, \infty)$, we write $\Gamma^{\beta}(\boldsymbol{g})$ for the product cone $\Gamma_1^{\beta}(g_1) \times \Gamma_2^{\beta}(g_2)$, where

$$\Gamma_i^{\beta}(g_i) := \{(h_i, t_i) \in G_i \times \mathbb{R}_+ : \rho_i(g_i, h_i) \leq \beta t_i\}.$$

We work on the domain $G_1 \times G_2 \times \mathbb{R}_+ \times \mathbb{R}_+$.

Take functions $\psi^{[i]}$ on G_i that satisfy the standard decay, smoothness and cancellation conditions (9.1) and (9.2). Recall that $\psi_{\boldsymbol{t}}$ denotes the product function $\psi_{t_1}^{[1]} \otimes \psi_{t_2}^{[2]}$.

Definition 9.2 For $\psi^{[i]}$ as above and $\beta > 0$, we define $\mathcal{S}_{\psi,\beta}(f)(\boldsymbol{g})$ to be

$$\left(\iint_{\Gamma^{\beta}(\boldsymbol{g})} \frac{|(f * \psi_{\boldsymbol{t}}(\boldsymbol{h})|^2}{|R(\boldsymbol{o}, \beta \boldsymbol{t})|} \, \mathrm{d}\boldsymbol{h} \, \frac{\mathrm{d}\boldsymbol{g}}{\boldsymbol{t}} \right)^{1/2}$$

for all $\boldsymbol{g} \in \boldsymbol{G}$ and $f \in \mathsf{L}^1(\boldsymbol{G})$. We also define

$$\mathcal{S}_{\psi,0}(f)(\boldsymbol{g}) := \left(\int_{\boldsymbol{T}} |f * \psi_{\boldsymbol{t}}(\boldsymbol{g})|^2 \, \frac{\mathrm{d}\boldsymbol{g}}{\boldsymbol{t}} \right)^{1/2}$$

for all $\boldsymbol{g} \in \boldsymbol{G}$ and $f \in \mathsf{L}^1(\boldsymbol{G})$. The Hardy space $\mathsf{H}^1_{\mathrm{sq},\psi,\beta}(\boldsymbol{G})$ is defined to be the space

$$\{f \in \mathsf{L}^1(\boldsymbol{G}) : \left\| \mathcal{S}_{\psi,\beta}(f) \right\|_{\mathsf{L}^1(\boldsymbol{G})} < \infty\},$$

equipped with the norm

$$\|f\|_{\mathsf{H}^1_{\mathrm{sq},\psi,\beta}(\boldsymbol{G})} := \left\| \mathcal{S}_{\psi,\beta}(f) \right\|_{\mathsf{L}^1(\boldsymbol{G})}.$$

Note that $\mathcal{S}_{\psi,\beta}(f)$ tends to $\mathcal{S}_{\psi,0}(f)$ as $\beta \to 0$, at least pointwise. There are also discrete versions of this definition, where the integrals over $\mathbb{R}_+$ are replaced by sums over powers of 2 (or some other base). We usually call $\mathcal{S}_{\psi,\beta}(f)$ an area function when $\beta > 0$ and a square function when $\beta = 0$, but it is more efficient to treat these together.

As mentioned earlier, much is known about Hardy spaces defined as above, and we summarise some of the main results. From [16], the space $\mathsf{H}^1_{\mathrm{sq},\psi,0}(\boldsymbol{G})$ is independent of the choice of the functions $\psi^{[i]}$, provided that they satisfy the decay, smoothness and cancellation conditions (9.1) and (9.2); discrete square functions and area operators $\mathcal{S}_{\psi,1}$ also characterise the same space, which we write simply as $\mathsf{H}^1(\boldsymbol{G})$. The key technique to prove these equivalences is a Plancherel–Pólya inequality. From [17] and [2], we see also that $\mathsf{H}^1(\boldsymbol{G})$ may be characterised using wavelet and atomic decompositions; more precisely, $\mathsf{H}^1_{\mathrm{atom}}(\boldsymbol{G}) = \mathsf{H}^1(\boldsymbol{G})$. Further, the double Riesz transformations $\mathcal{R}^{[1]}_{j_1} \otimes \mathcal{R}^{[1]}_{j_1}$ (see Definition 9.5) and similar singular integral operators are all bounded from $\mathsf{H}^1(\boldsymbol{G})$ to $\mathsf{L}^1(\boldsymbol{G})$. Finally, from [16], the dual of $\mathsf{H}^1(\boldsymbol{G})$ is the space BMO defined in terms of (suitable product) Carleson measures on $\boldsymbol{G}$.

In Sect. 9.3.3 below, we show that the space $\mathsf{H}^1_{\mathrm{sq},\psi,\beta}(\boldsymbol{G})$ is also independent of β.

Let ∇_i and $\mathcal{L}_i$ denote the subgradient and the sub-Laplacian on G_i; recall that $\bar{\nabla}_i$ denotes the gradient (∇_i, ∂_t) on $G_i \times \mathbb{R}_+$. The (vector-valued) convolution kernels of the operators $t_i \mathcal{L}_i \mathrm{e}^{-t_i \mathcal{L}_i}$ and $t_i \bar{\nabla}_i \mathrm{e}^{-t_i \sqrt{\mathcal{L}_i}}$ satisfy the decay, smoothness and cancellation conditions (9.1) and (9.2). Hence $\mathsf{H}^1(\boldsymbol{G})$ may also be characterised via the Littlewood–Paley area functions and square functions defined using the heat and Poisson kernels.

9.3.3 *Independence of Cone Angle*

Recall that $R(\boldsymbol{g}, \boldsymbol{t}) := B_1(g_1, t_1) \times B_2(g_2, t_2)$. Fix a parameter θ in $(0, 1)$.

If V is a closed subset of $\boldsymbol{G}$, then we say that $\boldsymbol{g} \in \boldsymbol{G}$ has global θ-density with respect to V if

$$\frac{|V \cap R(\boldsymbol{g}, \boldsymbol{t})|}{|R(\boldsymbol{g}, \boldsymbol{t})|} \geq \theta$$

for all $\boldsymbol{t} \in \boldsymbol{T}$. Let V^* be the set containing all points of global θ-density of V, then V^* is closed and $V^* \subseteq V$. Equivalently,

$$(V^*)^c = \{\boldsymbol{g} \in \boldsymbol{G} : \mathcal{M}_S(\chi_{V^c})(\boldsymbol{g}) > 1 - \theta\}.$$

It follows from the $\mathsf{L} \log \mathsf{L} \to \mathsf{L}^{1,\infty}$ estimate for the strong maximal function (see, for example, [8]) that $|(V^*)^c| \leq c_\theta \, |V^c|$, where

$$c_\theta = \frac{C}{1-\theta} \left(1 + \log_2^+ \left(\frac{1}{1-\theta} \right) \right).$$

For a closed subset V of $\boldsymbol{G}$, write

$$W^\beta(V) := \bigcup_{g \in V} \Gamma^\beta(\boldsymbol{g}).$$

Lemma 9.2 *Suppose that V is a closed set in $\boldsymbol{G}$ such that $|V^c| < \infty$. Then there exist constants $c_0 \le 1/4$ and C such that if $\beta > 1$ and $\theta = 1 - c_0\beta^{-Q_1-Q_2}$, then*

$$\iint_{W^\beta(V^*)} F(\boldsymbol{g},\boldsymbol{t})\,|R(\boldsymbol{o},\boldsymbol{t})|\,\mathrm{d}\boldsymbol{g}\,\mathrm{d}\boldsymbol{t} \lesssim \int_V \iint_{\Gamma(\boldsymbol{g})} F(\boldsymbol{h},\boldsymbol{t})\,\mathrm{d}\boldsymbol{h}\,\mathrm{d}\boldsymbol{t}\,\mathrm{d}\boldsymbol{g}$$

for all measurable nonnegative-real-valued functions F on $\boldsymbol{G} \times \boldsymbol{T}$.

Proof First, if $(\boldsymbol{h},\boldsymbol{t}) \in W^\beta(V^*)$, then there exists $\tilde{\boldsymbol{g}} \in V^* \cap R(\boldsymbol{h},\beta\boldsymbol{t})$. We see easily that

$$\left|R(\tilde{\boldsymbol{g}},\beta\boldsymbol{t}) \cap R(\boldsymbol{h},\boldsymbol{t})^c\right| \le \left(1 - 2c_0\beta^{-Q_1-Q_2}\right)|R(\tilde{\boldsymbol{g}},\beta\boldsymbol{t})|,$$

for some constant $c_0 \le 1/4$. Hence

$$\begin{aligned}
|V \cap R(\boldsymbol{h},\boldsymbol{t})| &\ge |V \cap R(\tilde{\boldsymbol{g}},\beta\boldsymbol{t})| - \left|R(\tilde{\boldsymbol{g}},\beta\boldsymbol{t}) \cap R(\boldsymbol{h},\boldsymbol{t})^c\right| \\
&\ge \left(\theta - 1 + 2c_0\beta^{-Q_1-Q_2}\right)|R(\tilde{\boldsymbol{g}},\beta\boldsymbol{t})| \\
&= c_0\beta^{-Q_1-Q_2}\,|R(\tilde{\boldsymbol{g}},\beta\boldsymbol{t})| \ge C\,|R(\boldsymbol{g},\boldsymbol{t})|.
\end{aligned}$$

Now, by Fubini's Theorem,

$$\begin{aligned}
&\int_V \iint_{\Gamma(\boldsymbol{g})} F(\boldsymbol{h},\boldsymbol{t})\,\mathrm{d}\boldsymbol{h}\,\mathrm{d}\boldsymbol{t}\,\mathrm{d}\boldsymbol{g} \\
&= \iint_{\boldsymbol{T}\times\boldsymbol{G}} \int_V \chi_{R(\boldsymbol{o},\boldsymbol{t})}\left(\boldsymbol{h}^{-1}\boldsymbol{g}\right) F(\boldsymbol{h},\boldsymbol{t})\,\mathrm{d}\boldsymbol{h}\,\mathrm{d}\boldsymbol{g}\,\mathrm{d}\boldsymbol{t} \\
&\ge \iint_{W^\beta(V^*)} \int_{\boldsymbol{G}} \chi_{R(\boldsymbol{o},\boldsymbol{t})}\left(\boldsymbol{h}^{-1}\boldsymbol{g}\right) F(\boldsymbol{h},\boldsymbol{t})\,\mathrm{d}\boldsymbol{g}\,\mathrm{d}\boldsymbol{h}\,\mathrm{d}\boldsymbol{t} \\
&= \iint_{W^\beta(V^*)} F(\boldsymbol{h},\boldsymbol{t})\,|R(\boldsymbol{o},\boldsymbol{t})|\,\mathrm{d}\boldsymbol{h}\,\mathrm{d}\boldsymbol{t},
\end{aligned}$$

as required. □

Proposition 9.1 *With the notation of Definition 9.2,*

$$\mathsf{H}^1_{\mathrm{sq},\psi,\beta}(\boldsymbol{G}) = \mathsf{H}^1_{\mathrm{sq},\psi,1}(\boldsymbol{G}),$$

and these spaces have equivalent norms for all $\beta > 0$.

Proof It suffices to suppose that $\beta > 1$ and show that

$$\left\| \mathcal{S}_{\psi,\beta}(f) \right\|_{\mathsf{L}^1(\boldsymbol{G})} \lesssim \beta^{Q_1+Q_2}(1+\log_2^+ \beta) \left\| \mathcal{S}_{\psi,1}(f) \right\|_{\mathsf{L}^1(\boldsymbol{G})} .$$

For all $\lambda > 0$, set

$$V = \{\boldsymbol{g} \in \boldsymbol{G} : \mathcal{S}_{\psi,1}(f)(\boldsymbol{g}) \leq \lambda\},$$

and $\theta = 1 - \beta^{-Q_1-Q_2}/4$. Then, from Lemma 9.2 and Fubini's theorem,

$$\begin{aligned}
\int_{V^*} \mathcal{S}_{\psi,\beta}(f)(\boldsymbol{g})^2 \,\mathrm{d}\boldsymbol{g} &= \int_{V^*} \iint_{\Gamma^\beta(\boldsymbol{g})} \frac{|f * \psi_t(\boldsymbol{h})|^2}{|R(\boldsymbol{o}, \beta t)|} \,\mathrm{d}\boldsymbol{h} \,\frac{\mathrm{d}g}{t} \,\mathrm{d}\boldsymbol{g} \\
&\lesssim \beta^{Q_1+Q_2} \iint_{W^\beta(V^*)} |f * \psi_t(\boldsymbol{h})|^2 \,\mathrm{d}\boldsymbol{h} \,\frac{\mathrm{d}g}{t} \\
&\lesssim \beta^{Q_1+Q_2} \int_V \iint_{\Gamma(\boldsymbol{g})} \frac{|f * \psi_t(\boldsymbol{h})|^2}{|R(\boldsymbol{o}, \beta t)|} \,\mathrm{d}\boldsymbol{h} \,\frac{\mathrm{d}g}{t} \,\mathrm{d}\boldsymbol{g} \\
&\eqsim \beta^{Q_1+Q_2} \int_V \mathcal{S}_{\psi,1}(f)^2 \,\mathrm{d}\boldsymbol{g}.
\end{aligned}$$

Therefore

$$\begin{aligned}
&\left|\{\boldsymbol{g} \in \boldsymbol{G} : \mathcal{S}_{\psi,\beta}(f)(\boldsymbol{g}) > \lambda\}\right| \\
&\quad \leq \left|(V^*)^c\right| + \frac{C}{\lambda^2} \int_{V^*} \mathcal{S}_{\psi,\beta}(f)(\boldsymbol{g})^2 \,\mathrm{d}\boldsymbol{g} \\
&\quad \leq C\beta^{Q_1+Q_2}(1+\log_2^+ \beta) \left(\left|(V^*)^c\right| + \frac{1}{\lambda^2} \int_V \mathcal{S}_{\psi,1}(f)(\boldsymbol{g})^2 \,\mathrm{d}\boldsymbol{g} \right).
\end{aligned}$$

Integrating with respect to λ yields

$$\left\| \mathcal{S}_{\psi,\beta}(f) \right\|_{\mathsf{L}^1(\boldsymbol{G})} \lesssim \beta^{Q_1+Q_2}(1+\log_2^+ \beta) \left\| \mathcal{S}_{\psi,1}(f) \right\|_{\mathsf{L}^1(\boldsymbol{G})},$$

which completes the proof of Proposition 9.1. □

9.3.4 Summary

The known results cited in Sect. 9.3.2 and our additional material here may be summarised in the following proposition.

Proposition 9.2 *The atomic Hardy space* $\mathsf{H}^1_{\mathrm{atom}}(\boldsymbol{G})$ *and the square function and area function Hardy spaces* $\mathsf{H}^1_{\mathrm{sq},\psi,\beta}$ *for different* ψ *and* β *coincide and have equivalent norms.*

9.4 The Singular Integral Characterisation

We consider a stratified Lie group G. Recall that $\mathcal{R}_0$ is the identity operator I, and when $1 \le j \le d_i$, the jth Riesz operator $\mathcal{R}_j$ on G is defined by

$$\mathcal{R}_j := X_j(\mathcal{L})^{-1/2};$$

its convolution kernel, k_j say, is smooth away from the identity of G, and homogeneous of degree $-Q$. According to Christ and Geller [5], $f \in \mathsf{H}^1(G)$ if and only if all $\mathcal{R}_j f \in \mathsf{L}^1(G)$, and there is a corresponding norm equivalence. We say that the singular integral operators $\mathcal{R}_j$, where $0 \le j \le d_j$, characterise $\mathsf{H}^1(G)$.

Definition 9.3 Suppose that the singular integral operators $\mathcal{K}^{[i]}_j$, where $0 \le j \le n_i$, characterise $\mathsf{H}^1(G_i)$, in the sense above. The space $\mathsf{H}^1_{\mathrm{SIO}}(\boldsymbol{G})$ is defined to be the set of all $f \in \mathsf{L}^1(\boldsymbol{G})$ such that

$$\sum_{j_1=0}^{n_1}\sum_{j_2=0}^{n_2} \left\| \mathcal{K}^{[1]}_{j_1} \otimes \mathcal{K}^{[2]}_{j_2} f \right\|_{\mathsf{L}^1(\boldsymbol{G})} < \infty,$$

with norm

$$\|f\|_{\mathsf{H}^1_{\mathrm{SIO}}(\boldsymbol{G})} := \sum_{j_1=0}^{n_1}\sum_{j_2=0}^{n_2} \left\| \mathcal{K}^{[1]}_{j_1} \otimes \mathcal{K}^{[2]}_{j_2} f \right\|_{\mathsf{L}^1(\boldsymbol{G})}.$$

In this section, we generalise Theorem 9.1, which states that the spaces $\mathsf{H}^1_{\mathrm{Riesz}}(\boldsymbol{G})$ and $\mathsf{H}^1(\boldsymbol{G})$ coincide and have equivalent norms.

Theorem 9.4 *Suppose that the singular integral operators* $\mathcal{K}^{[i]}_j$*, where* $0 \le j \le n_i$*, characterise* $\mathsf{H}^1(G_i)$*. Then the double singular integral operators* $\mathcal{K}^{[1]}_{j_1} \otimes \mathcal{K}^{[2]}_{j_2}$ *characterise the Hardy space* $\mathsf{H}^1(\boldsymbol{G})$*. That is,* $f \in \mathsf{H}^1(\boldsymbol{G})$ *if and only if each* $\mathcal{K}^{[1]}_{j_1} \otimes \mathcal{K}^{[2]}_{j_2} f$ *is in* $\mathsf{L}^1(\boldsymbol{G})$ *and moreover*

$$\|f\|_{\mathsf{H}^1(\boldsymbol{G})} \eqsim \sum_{j_1=0}^{n_1}\sum_{j_2=0}^{n_2} \left\| \mathcal{K}^{[1]}_{j_1} \otimes \mathcal{K}^{[2]}_{j_2} f \right\|_{\mathsf{L}^1(\boldsymbol{G})}.$$

It is known (see [17] and [2]) that singular integral operators associated to homogeneous kernels are bounded from $\mathsf{H}^1(\boldsymbol{G})$ to $\mathsf{L}^1(\boldsymbol{G})$, so it suffices to show that if all the double singular integral transforms of a function f are in $\mathsf{L}^1(\boldsymbol{G})$ then $f \in \mathsf{H}^1(\boldsymbol{G})$. Our proof of Proposition 9.3 below extends [3], which introduced a new method, using randomisation, to characterise flag Hardy space on Heisenberg groups by products of singular integrals.

9.4.1 The Square Function and Singular Integral Transforms

Proposition 9.3 *Suppose that* $f \in \mathsf{L}^2(\boldsymbol{G})$, *and* $\mathcal{K}_{j_1}^{[1]} \otimes \mathcal{K}_{j_2}^{[2]} f \in \mathsf{L}^1(\boldsymbol{G})$ *when* $j_i = 0, \dots, n_i$. *Then* $f \in \mathsf{H}^1(\boldsymbol{G})$, *and*

$$\|f\|_{\mathsf{H}^1(\boldsymbol{G})} \leq C \|f\|_{\mathsf{H}^1_{\mathrm{SIO}}(\boldsymbol{G})}.$$

Proof We use a randomisation argument coupled with the analogous one-parameter result of Christ and Geller [5]. Fix a smooth function η on $\mathbb{R}_+$, supported in $[1/2, 2]$, such that $\sum_{m\in\mathbb{Z}} \eta(2^{-m}t) = 1$ for all $t \in \mathbb{R}_+$. By Sect. 9.2.2, the convolution kernels $k_{\eta(\mathcal{L}_i)}$ of the operators $\eta(\mathcal{L}_i)$ on G_i are Schwartz functions of mean 0.

Let $r_m : [0, 1] \to \mathbb{R}$ be a collection of independent Rademacher random variables (see [14]). Fix i, take η as above, and define

$$T_s(f) = \sum_{m\in\mathbb{Z}} r_m(s)\eta(2^{-m}\mathcal{L}_i)f$$

for all $f \in \mathsf{H}^1(G_i)$ and all $s \in [0, 1]$. Straightforward calculation shows that

$$\left|\xi^k \partial_\xi^k \left(\sum_{m\in\mathbb{Z}} r_m(s)\eta(2^{-m}\xi)\right)\right| \leq C_k \qquad \forall \xi \in \mathbb{R}_+ \quad \forall k \in \mathbb{N},$$

and from the multiplier theorem (see for example, [13, Theorem 6.25]), the operator T_s is bounded from $\mathsf{H}^1(G_i)$ to $\mathsf{L}^1(G_i)$ with norm uniformly bounded for $s \in [0, 1]$. Together with the Christ–Geller characterisation [5, Theorem A], this implies that

$$\left\|\sum_{m\in\mathbb{Z}} r_m(s)\eta(2^{-m}\mathcal{L}_i)f\right\|_{\mathsf{L}^1(G_i)} \lesssim \left(\sum_{j_i=0}^{n_i} \left\|\mathcal{K}_{j_i}^{[i]} f\right\|_{\mathsf{L}^1(G_i)}\right)$$

for all $f \in \mathsf{L}^1(G_i)$ such that $\mathcal{K}^{[i]}_{j_i} f \in \mathsf{L}^1(G_i)$. Iteration of the argument shows that

$$\begin{aligned}
&\Big\| \sum_{m\in\mathbb{Z}} r_m(s_1)\eta(2^{-m}\mathcal{L}_1)\Big(\sum_{n\in\mathbb{Z}} r_n(s_2)\eta(2^{-n}\mathcal{L}_2) f\Big)\Big\|_{\mathsf{L}^1(\boldsymbol{G})} \\
&\qquad \lesssim \sum_{j_1=0}^{n_1} \Big\| \mathcal{K}^{[1]}_{j_1} \sum_{n\in\mathbb{Z}} r_n(s_2)\eta(2^{-n}\mathcal{L}_2) f\Big\|_{\mathsf{L}^1(\boldsymbol{G})} \\
&\qquad = \sum_{j_1=0}^{n_1} \Big\| \sum_{n\in\mathbb{Z}} r_n(s_2)\eta(2^{-n}\mathcal{L}_2)\mathcal{K}^{[1]}_{j_1} f\Big\|_{\mathsf{L}^1(\boldsymbol{G})} \\
&\qquad \lesssim \sum_{j_1=0}^{n_1}\sum_{j_2=0}^{n_2} \Big\| \mathcal{K}^{[1]}_{j_1} \otimes \mathcal{K}^{[2]}_{j_2} f\Big\|_{\mathsf{L}^1(\boldsymbol{G})},
\end{aligned}$$

because operators involving convolutions (even with distributions) on G_1 and operators involving convolutions (even with distributions) on G_2 commute. By Khinchin's inequality (see, for example, [14, Appendix C.5]), this implies that

$$\begin{aligned}
&\Big\| \Big(\sum_{n\in\mathbb{Z}}\sum_{m\in\mathbb{Z}} |\eta(2^{-m}\mathcal{L}_1)\eta(2^{-n}\mathcal{L}_2) f|^2\Big)^{1/2}\Big\|_{\mathsf{L}^1(\boldsymbol{G})} \\
&\quad \lesssim \iint_{[0,1]\times[0,1]} \Big\| \sum_{m\in\mathbb{Z}} r_m(s_1)\eta(2^{-m}\mathcal{L}_1) \sum_{n\in\mathbb{Z}} r_n(s_2)\eta(2^{-n}\mathcal{L}_2) f\Big\|_{\mathsf{L}^1(\boldsymbol{G})} \,\mathrm{d}s_1\,\mathrm{d}s_2 \\
&\quad \lesssim \sum_{j_1=0}^{n_1}\sum_{j_2=0}^{n_2} \Big\| \mathcal{K}^{[1]}_{j_1} \otimes \mathcal{K}^{[2]}_{j_2} f\Big\|_{\mathsf{L}^1(\boldsymbol{G})}.
\end{aligned}$$

This ends the proof of Proposition 9.3. □

Remark 9.1 It is straightforward to extend this result to products of more than two factors. It is just a matter of repeating the randomisation argument more times.

9.5 Maximal Function Characterisation

In this section, we prove Theorem 9.2. As we have already noted, in the one-parameter setting, a common strategy for showing that maximal functions characterise the Hardy space is to use atoms; this strategy does not work in the multi-parameter case. Merryfield [25] managed to extend the one-parameter result to the product space $\mathbb{R}^m \times \mathbb{R}^n$; his new tool is the solution of a particular Cauchy–Riemann type equation. However, it is not clear whether there is a version of his lemma on spaces of homogeneous type, or even just on homogeneous groups. In

[3], a new method, using Poisson kernels and harmonic functions, was introduced to characterise flag Hardy space on Heisenberg groups by maximal functions. Here we extend this method to product groups.

9.5.1 *The Maximal Function Hardy Spaces*

Recall that $\Gamma^\beta(\boldsymbol{g})$ denotes the cone with vertex $\boldsymbol{g}$ and angle β:

$$\Gamma^\beta(\boldsymbol{g}) := \{(\boldsymbol{h}, \boldsymbol{t}) \in \boldsymbol{G} \times \boldsymbol{T} : \rho_i(g_i, h_i) \le \beta t_i \text{ when } i = 1, 2\}.$$

Definition 9.4 Take functions $\zeta^{[i]} \in \mathsf{M}(G_i)$, and define the maximal operator $\mathcal{M}_{\zeta,\beta}$ by

$$\mathcal{M}_{\zeta,\beta}(f)(\boldsymbol{g}) := \sup_{\boldsymbol{h}\in\Gamma^\beta(\boldsymbol{g})} |f * \zeta_t(\boldsymbol{h})| \qquad \forall \boldsymbol{g} \in \boldsymbol{G}, \quad \forall f \in \mathsf{L}^1(\boldsymbol{G}).$$

The Hardy space $\mathsf{H}^1_{\max,\zeta,\beta}(\boldsymbol{G})$ is defined to be the space

$$\{f \in \mathsf{L}^1(\boldsymbol{G}) : \left\|\mathcal{M}_{\zeta,\beta} f\right\|_{\mathsf{L}^1(\boldsymbol{G})} < \infty\}$$

equipped with the norm

$$\|f\|_{\mathsf{H}^1_{\max,\zeta,\beta}(\boldsymbol{G})} := \left\|\mathcal{M}_{\zeta,\beta} f\right\|_{\mathsf{L}^1(\boldsymbol{G})},$$

for the Hardy space, we require that the integrals of the $\zeta^{[i]}$ are nonzero.

It is obvious that $\left\|\mathcal{M}_{\zeta,\gamma} f\right\|_{\mathsf{L}^1(\boldsymbol{G})} \le \left\|\mathcal{M}_{\zeta,\beta} f\right\|_{\mathsf{L}^1(\boldsymbol{G})}$ when $\gamma \le \beta$.

We treat the cases when $\beta > 0$ and when $\beta = 0$ together. In the important special cases when the $\zeta^{[i]}$ coincide with the Poisson or heat kernels, we have additional tools, such as Harnack or Moser inequalities. The possibly less well known Plancherel–Pólya inequality provides similar results for more general ζ.

To characterise $\mathsf{H}^1(\boldsymbol{G})$ by maximal functions, we are going to show two results.

Proposition 9.4 *If β is large enough, then the spaces $\mathsf{H}^1_{\mathrm{sq},P,1}(\boldsymbol{G})$ and $\mathsf{H}^1_{\max,P,\beta}(\boldsymbol{G})$ coincide and have equivalent norms.*

Proposition 9.5 *For different choices of φ and ζ and different choices of β and γ, the spaces $\mathsf{H}^1_{\max,\varphi,\beta}(\boldsymbol{G})$ and $\mathsf{H}^1_{\max,\zeta,\gamma}(\boldsymbol{G})$ coincide and have equivalent norms.*

Combining the above two results with Proposition 9.1 proves Theorem 9.2.

9.5.2 Part 1 of the Proof of Proposition 9.4

Evidently $\mathsf{H}^1_{\mathrm{sq},P,1}(G) \subseteq \mathsf{H}^1_{\max,P,\beta}(G)$, and

$$\left\| \mathcal{M}_{P,\beta} f \right\|_{\mathsf{L}^1(\boldsymbol{G})} \lesssim \left\| \mathcal{S}_{P,1} f \right\|_{\mathsf{L}^1(\boldsymbol{G})}. \tag{9.8}$$

Indeed, when $\left\| \mathcal{S}_{P,\beta} f \right\|_{\mathsf{L}^1(\boldsymbol{G})} < \infty$, then by [2], we may write $f = \sum_j \lambda_j a_j$, where each a_j is an atom, and $\sum_j |\lambda_j| \lesssim \left\| \mathcal{S}_{P,\beta} f \right\|_{\mathsf{L}^1(\boldsymbol{G})}$. Thus, to prove (9.8), it suffices to verify that

$$\left\| \mathcal{M}_{P,\beta}(a) \right\|_{\mathsf{L}^1(\boldsymbol{G})} \lesssim 1$$

for each product atom a as in Sect. 9.3.1. The Poisson kernels $P_1^{[1]}$ and $P_1^{[2]}$ satisfy the standard decay and smoothness conditions (9.1), and the atom a satisfies the standard product cancellation condition (9.7). Then the desired estimate of $\left\| \mathcal{M}_{P,\beta}(a) \right\|_{\mathsf{L}^1(\boldsymbol{G})}$ follows from standard product arguments and Journé's covering lemma.

We provide a brief outline of the proof here for completeness and for the reader's convenience. It suffices to show that, for each product atom a,

$$\left\| \mathcal{M}_{P,\beta}(a) \right\|_{\mathsf{L}^1(\boldsymbol{G})} \lesssim 1.$$

From Sect. 9.3.1, we may write a on $\boldsymbol{G}$ as a sum $\sum_{R \in \mathcal{M}(U)} a_R$, where U is an open subset of $\boldsymbol{G}$ of finite measure, and the particles a_R satisfy support and size conditions.

There are two steps to the proof: first, we find some small positive ϵ, such that for any pseudodyadic rectangle S containing the pseudodyadic rectangle R,

$$\int_{\boldsymbol{G} \setminus S} |\mathcal{M}_{P,\beta}(a_R)(\boldsymbol{g})| \, \mathrm{d}\boldsymbol{g} \lesssim \left(\frac{\ell_1(R)}{\ell_1(S)} + \frac{\ell_2(R)}{\ell_2(S)} \right)^{\epsilon} |R|^{1/2} \, \|a_R\|_{\mathsf{L}^2(\boldsymbol{G})}, \tag{9.9}$$

for all particles a_R associated to R. The estimation of this expression may be achieved by writing S as $Q_1 \times Q_2$ and breaking up $\boldsymbol{G} \setminus S$ into the three regions $(Q_1)^c \times (Q_2)^c$, $(Q_1)^c \times Q_2$ and $Q_1 \times (Q_2)^c$. In the first region we use the support and cancellation conditions on a_R to estimate $a_R * P_t(\boldsymbol{g})$, and show that

$$\int_{(Q_1)^c \times (Q_2)^c} |\mathcal{M}_{P,\beta}(a_R)(\boldsymbol{g})| \, \mathrm{d}\boldsymbol{g} \lesssim \left(\frac{\ell_1(R)}{\ell_1(S)} \cdot \frac{\ell_2(R)}{\ell_2(S)} \right)^{\epsilon} |R|^{1/2} \, \|a_R\|_{\mathsf{L}^2(\boldsymbol{G})};$$

in the second region we use the support and cancellation conditions to control the g_1 variable and Hölder's inequality and a Littlewood–Paley argument to control the g_2 variable, and show that

$$\int_{(Q_1)^c \times Q_2} |\mathcal{M}_{P,\beta}(a_R)(\boldsymbol{g})| \, \mathrm{d}\boldsymbol{g} \lesssim \left(\frac{\ell_1(R)}{\ell_1(S)} \right)^{\epsilon} |R|^{1/2} \, \|a_R\|_{\mathrm{L}^2(\boldsymbol{G})} \, ;$$

in the third region, we argue similarly, but with the roles of the variables reversed.

Once (9.9) is proved, we apply Journé's lemma to obtain an estimate for atoms rather than particles. Let a be an atom associated to the open set U, and write $a = \sum_{R \in m(U)} a_R$. Define

$$V = \{\boldsymbol{g} \in \boldsymbol{G} : \mathcal{M}_S \chi_U(\boldsymbol{g}) > 1/4\} \qquad \text{and} \qquad W = \{\boldsymbol{g} \in \boldsymbol{G} : \mathcal{M}_S \chi_V(\boldsymbol{g}) > 1/4\} \, .$$

Then $|W| \lesssim |V| \lesssim |U|$. Further,

$$\begin{aligned} &\int_{\boldsymbol{G}} |\mathcal{M}_{P,\beta}(a)(\boldsymbol{g})| \, \mathrm{d}\boldsymbol{g} \\ &\qquad = \int_W |\mathcal{M}_{P,\beta}(a)(\boldsymbol{g})| \, \mathrm{d}\boldsymbol{g} + \int_{\boldsymbol{G} \setminus W} |\mathcal{M}_{P,\beta}(a)(\boldsymbol{g})| \, \mathrm{d}\boldsymbol{g} \\ &\qquad \le |W|^{1/2} \left(\int_W |\mathcal{M}_{P,\beta}(a)(\boldsymbol{g})| \, \mathrm{d}\boldsymbol{g} \right)^{1/2} + \sum_{R \in m(U)} \int_{\boldsymbol{G} \setminus W} |\mathcal{M}_{P,\beta}(a_R)(\boldsymbol{g})| \, \mathrm{d}\boldsymbol{g}. \end{aligned}$$

The first term is estimated using the L^2-boundedness of $\mathcal{M}_{P,\beta}$:

$$|W|^{1/2} \left(\int_W |\mathcal{M}_{P,\beta}(a)(\boldsymbol{g})| \, \mathrm{d}\boldsymbol{g} \right)^{1/2} \lesssim |W|^{1/2} \, \|a\|_{\mathrm{L}^2(\boldsymbol{G})} \lesssim |U|^{1/2} |U|^{-1/2} = 1.$$

The second term is estimates using Journé's lemma, namely, for each $R \in m(U)$, we can find $S \in m(W)$ (depending on R) such that $R \subseteq S$ and

$$\sum_{r \in m(R)} \left(\frac{\ell_1(R)}{\ell_1(S)} + \frac{\ell_2(R)}{\ell_2(S)} \right)^{\epsilon} |R| \lesssim |U|.$$

Then from (9.9), Hölder's inequality, Journé's lemma, and the definitions,

$$\begin{aligned} &\sum_{R \in m(U)} \int_{\boldsymbol{G} \setminus W} |\mathcal{M}_{P,\beta}(a_R)(\boldsymbol{g})| \, \mathrm{d}\boldsymbol{g} \\ &\qquad \le \sum_{R \in m(U)} \int_{\boldsymbol{G} \setminus S} |\mathcal{M}_{P,\beta}(a_R)(\boldsymbol{g})| \, \mathrm{d}\boldsymbol{g} \end{aligned}$$

$$
\begin{aligned}
&\lesssim \sum_{R\in m(U)} \left(\frac{\ell_1(R)}{\ell_1(S)} + \frac{\ell_2(R)}{\ell_2(S)}\right)^{\epsilon} |R|^{1/2} \|a_R\|_{\mathsf{L}^2(\boldsymbol{G})} \\
&\lesssim \left(\sum_{R\in m(U)} \left(\frac{\ell_1(R)}{\ell_1(S)} + \frac{\ell_2(R)}{\ell_2(S)}\right)^{2\epsilon} |R|\right)^{1/2} \left(\sum_{R\in m(U)} \|a_R\|^2_{\mathsf{L}^2(\boldsymbol{G})}\right)^{1/2} \\
&\lesssim |U|^{1/2}|U|^{-1/2} = 1.
\end{aligned}
$$

Note that proving this result of products of more than two factors seems to be nontrivial; Journé's lemma requires us to have more than one "improving factor" $\ell_j(R)/\ell_j(S)$.

It remains to prove the opposite inclusion: $\mathsf{H}^1_{\max,P,\beta}(\boldsymbol{G}) \subseteq \mathsf{H}^1_{\mathrm{sq},P,1}(\boldsymbol{G})$, and

$$
\left\|\mathcal{S}_{P,1} f\right\|_{\mathsf{L}^1(\boldsymbol{G})} \lesssim \left\|\mathcal{M}_{P,\beta} f\right\|_{\mathsf{L}^1(\boldsymbol{G})}.
$$

We first treat a stratified group, and then a product of stratified groups.

9.5.3 Part 2 of the Proof of Proposition 9.4

In this part of the proof, we prove (9.10) for a stratified group G with no product structure. This simplifies the notation. Later the group G will be one of the factors of the product group $\boldsymbol{G}$ that we wish to consider.

We are going to use integration by parts, and need to know about the behaviour of certain harmonic functions on $G \times \mathbb{R}_+$ at the boundaries of this region. Suppose that $f \in \mathsf{L}^p(G)$, where $1 \le p \le \infty$, and consider the Poisson integral $f * P_t(g)$ and $f * Q_t(g)$, where $g \in G$ and $t \in \mathbb{R}_+$, whose behaviour as $t \to 0$ is discussed in Corollary 9.2.

From Lemma 9.1, if $f \in \mathsf{L}^\infty(G)$, then $f * P_t$ and $f * Q_t$ are bounded in $\mathsf{L}^\infty(G)$ as $t \to \infty$. If $f \in \mathsf{H}^1_{\max,P,\gamma}(G)$, then $\left\|f_{1/s} * P_1\right\|_{\mathsf{H}^1_{\max,P,\gamma}(G)}$ is bounded for all $s > 0$. Since

$$
f_{1/s} * P_1 \to \left(\int_G f(g)\,\mathrm{d}g\right) P_1 \qquad \text{as } s \to \infty
$$

in $\mathsf{L}^1(G)$ and $\sup_{t>1} P_t(\cdot) \notin \mathsf{L}^1(G)$ so $P_1 \notin \mathsf{H}^1_{\max,P,\gamma}(G)$, we see that f has mean 0. Thus

$$
\begin{aligned}
&\|f * P_t\|_{\mathsf{L}^1(G)} + \|f * Q_t\|_{\mathsf{L}^1(G)} \\
&\quad = \left\|f_{1/t} * P_1\right\|_{\mathsf{L}^1(G)} + \left\|f_{1/t} * Q_1\right\|_{\mathsf{L}^1(G)} \to 0
\end{aligned}
$$

and

$$\begin{aligned}\|f * P_t\|_{\mathsf{L}^\infty(G)} &+ \|f * Q_t\|_{\mathsf{L}^\infty(G)} \\ &= t^{-Q} \left\| f_{1/t} * P_1 \right\|_{\mathsf{L}^\infty(G)} + t^{-Q} \left\| f_{1/t} * Q_1 \right\|_{\mathsf{L}^\infty(G)} \to 0\end{aligned}$$

as $t \to \infty$. This convergence is also pointwise almost everywhere.

Proposition 9.6 *Suppose that G is a stratified Lie group and $\gamma > 0$. If β is large enough, then*

$$\mathsf{H}^1_{\max, P,\beta}(G) \subseteq \mathsf{H}^1_{\mathrm{sq}, P,\gamma}(G),$$

and there is a corresponding norm inequality:

$$\|f\|_{\mathsf{H}^1_{\mathrm{sq}, P,\gamma}(G)} \lesssim \|f\|_{\mathsf{H}^1_{\max, P,\beta}(G)} \qquad \forall f \in \mathsf{H}^1_{\max, P,\beta}(G). \tag{9.10}$$

Proof Take $f \in \mathsf{L}^1(G)$ such that $\mathcal{M}_{P,\beta}(f) \in \mathsf{L}^1(G)$. We assume that f is real-valued, for otherwise we may treat the real and imaginary parts separately. We may also suppose that f is smooth, by a simple mollification argument.

Fix $\alpha > 0$, and define

$$L_\beta(\alpha) := \left\{g \in G : \mathcal{M}_{P,\beta}(f)(g) \le \alpha\right\},$$

$$A_\beta(\alpha) := \left\{g \in G : \mathcal{M}_S(1 - \chi_{L_\beta(\alpha)})(g) < \frac{1}{4}\right\},$$

where $\mathcal{M}_S$ is the strong maximal operator, which is L^2 bounded. Then

$$A_\beta(\alpha) \subseteq L_\beta(\alpha) \qquad \text{and} \qquad \left|(L_\beta(\alpha))^c\right| \le \left|A_\beta(\alpha)^c\right| \lesssim \left|(L_\beta(\alpha))^c\right|. \tag{9.11}$$

Define also

$$W_\beta := \bigcup_{g \in A_\beta(\alpha)} \Gamma^\beta(g) \qquad \text{and} \qquad \widetilde{W}_\beta := \bigcup_{h \in L_\beta(\alpha)(f)} \Gamma^\beta(h).$$

We claim that there exists $C_0 \in (0, 1)$ such that

$$\chi_{L_\gamma(\alpha)} * P_t(g) \ge C_0 \qquad \forall (g, t) \in W_\gamma.$$

Indeed, by definition, for such (g, t),

$$(1 - \chi_{L_\gamma(\alpha)}) * \chi_{B(o,\gamma t)} < \frac{1}{4}\, |B(o, \gamma t)|,$$

that is,

$$\chi_{L_\gamma(\alpha)} * \chi_{B(o,\gamma t)} \ge \frac{3}{4}\, |B(o, \gamma t)|,$$

and the claim follows from Lemma 9.1. We also claim that if β is large enough, then there is a constant $C_1 \in (0, C_0)$, such that if $(g, t) \notin \widetilde{W}_\beta$, then

$$\chi_{L_\beta(\alpha)} * P_t(g) \le C_1.$$

Indeed, if $(g, t) \notin \widetilde{W}_\beta$ then $\rho(h^{-1}g) \ge \beta t$ for all $h \in L_\beta(\alpha)$. Hence,

$$\begin{aligned}\chi_{L_\beta(\alpha)} * P_t(g) &= \int_G \chi_{L_\beta(\alpha)}(h) P_t(h^{-1}g)\, \mathrm{d}h \le \int_{B(g,\beta t)^c} P_t(h^{-1}g)\, \mathrm{d}h \\ &= \int_{B(g,\beta t)^c} P_1(h^{-1}g)\, \mathrm{d}h \to 0\end{aligned}$$

as $\beta \to \infty$, proving our claim.

Take a smooth function $\eta : \mathbb{R} \to \mathbb{R}$ such that $\eta(s) = 1$ when $s \ge C_0$ and $\eta(s) = 0$ when $s \le C_1$. Define $\mathrm{H}_t := \chi_{L_\beta(\alpha)} * P_t$. Then

$$t\partial_t \mathrm{H}_t(g) = \chi_{L_\beta(\alpha)} * Q_t(g),$$

which is uniformly bounded for all $g \in G$ and $t \in \mathbb{R}_+$ and

$$t\partial_t \mathrm{H}_t \to 0 \qquad \text{as } t \to 0$$

pointwise almost everywhere, by Corollary 9.2,

It will suffice to show that

$$\int_{A_\gamma(\alpha)} \mathcal{S}_{P,\gamma}(f)(g)^2\, \mathrm{d}g \lesssim \int_{L_\beta(\alpha)} \mathcal{M}_{P,\beta}(f)(g)^2\, \mathrm{d}g + \alpha^2 \left|L_\gamma(\alpha)^c\right|. \tag{9.12}$$

Indeed, coupled with (9.11), this implies that

$$\begin{aligned}&\left|\{g \in G : \mathcal{S}_{P,\gamma}(f)(g) > \alpha\}\right| \\ &\quad\le \left|\{g \in A_\gamma(\alpha)^c : \mathcal{S}_{P,\gamma}(f)(g) > \alpha\}\right| + \left|\{g \in A_\gamma(\alpha) : \mathcal{S}_{P,\gamma}(f)(g) > \alpha\}\right| \\ &\quad\le \left|A_\gamma(\alpha)^c\right| + \frac{1}{\alpha^2} \int_{A_\gamma(\alpha)} \mathcal{S}_{P,\gamma}(f)(g)^2\, \mathrm{d}g \\ &\quad\lesssim \left|L_\gamma(\alpha)^c\right| + \frac{1}{\alpha^2} \int_{L_\beta(\alpha)} \mathcal{M}_{P,\beta}(f)(g)^2\, \mathrm{d}g.\end{aligned}$$

A standard integration with respect to α then implies that

$$\left\|\mathcal{S}_{P,\gamma}(f)\right\|_{\mathrm{L}^1(G)} \lesssim \left\|\mathcal{M}_{P,\beta}(f)\right\|_{\mathrm{L}^1(G)},$$

that is, the required estimate (9.10) holds.

We observe that

$$\begin{aligned}\int_{A_\beta(\alpha)} \mathcal{S}_{P,\gamma}(f)(g)^2 \,\mathrm{d}g &= \int_{A_\beta(\alpha)} \iint_{\Gamma^\gamma(g)} \left|\nabla\!\!\!/ (f * P_t)(h)\right|^2 \frac{t}{|B(o,t)|} \,\mathrm{d}t \,\mathrm{d}h \,\mathrm{d}g \\ &\lesssim \iint_{W_\gamma} \left|\nabla\!\!\!/ (f * P_t)(g)\right|^2 t \,\mathrm{d}t \,\mathrm{d}g \\ &\le \iint_{G\times\mathbb{R}_+} \left|\nabla\!\!\!/ (f * P_t)(g)\right|^2 |\eta(\mathrm{H}_t(g))|^2 t \,\mathrm{d}t \,\mathrm{d}g.\end{aligned}$$

From (9.12), it will therefore suffice to show that

$$\begin{aligned}\mathrm{I}_0 &:= \iint_{G\times\mathbb{R}_+} \left|\nabla\!\!\!/ (f * P_t)(g)\right|^2 |\eta(\mathrm{H}_t(g))|^2 t \,\mathrm{d}t \,\mathrm{d}g \\ &\lesssim \int_{L_\beta(\alpha)} \mathcal{M}_{P,\beta}(f)(g)^2 \,\mathrm{d}g + \alpha^2 \left|L_\gamma(\alpha)^c\right|.\end{aligned} \tag{9.13}$$

We note that $u : (g,t) \mapsto F * P_t(g)$ is harmonic on $G \times \mathbb{R}_+$ for all $F \in \mathsf{L}^1(G) + \mathsf{L}^\infty(G)$, in the sense that

$$\mathcal{L}\!\!\!/ \, u(g,t) = 0,$$

where $\mathcal{L}\!\!\!/ := \mathcal{L} - \partial_t^2$. Consequently,

$$\left|\nabla\!\!\!/ u(g,t)\right|^2 = -\frac{1}{2}\mathcal{L}\!\!\!/ \left(u^2(g,t)\right) \qquad \forall (g,t) \in G \times \mathbb{R}_+.$$

Further, by our remark on harmonicity, $\mathcal{L}\!\!\!/ \,\mathrm{H}_t = 0$, and so

$$\begin{aligned}\mathcal{L}\!\!\!/ \,\eta(\mathrm{H}_t(g)) &= \nabla\!\!\!/ \cdot (\eta'(\mathrm{H}_t(g))\nabla\!\!\!/ \mathrm{H}_t(g)) \\ &= \eta''(\mathrm{H}_t(g)) \left|\nabla\!\!\!/ \mathrm{H}_t(g))\right|^2.\end{aligned}$$

It follows that

$$\begin{aligned}\left|\nabla\!\!\!/ (f * P_t)(g)\eta(\mathrm{H}_t(g))\right|^2 &= -\frac{1}{2}\mathcal{L}\!\!\!/ \left(|f * P_t(g)\eta(\mathrm{H}_t(g))|^2\right) \\ &\quad - 4 f * P_t(g)\eta(\mathrm{H}_t(g))\nabla\!\!\!/ (f * P_t)(g) \cdot \nabla\!\!\!/ \eta(\mathrm{H}_t(g)) \\ &\quad - |f * P_t(g)|^2 \left|\nabla\!\!\!/ \eta(\mathrm{H}_t(g))\right|^2 \\ &\quad - |f * P_t(g)|^2 \eta(\mathrm{H}_t(g))\eta''(\mathrm{H}_t(g)) \left|\nabla\!\!\!/ \mathrm{H}_t(g))\right|^2.\end{aligned} \tag{9.14}$$

We estimate the second, third, and fourth terms on the right hand side of (9.14) as follows. First, by the arithmetic–geometric mean inequality and the chain rule,

$$\begin{aligned}
&\left|4f * P_t(g)\eta(\mathrm{H}_t(g))\nabla(f * P_t)(g)\cdot\nabla\eta(\mathrm{H}_t(g))\right| \\
&\quad\leq \frac{1}{2}\left|\nabla P_t * f(g)\right|^2 |\eta(\mathrm{H}_t(g))|^2 + 8\,|f * P_t(g)|^2 \left|\nabla\eta(\mathrm{H}_t(g))\right|^2 \\
&\quad\leq \frac{1}{2}\left|f * \nabla P_t(g)\right|^2 |\eta(\mathrm{H}_t(g))|^2 + 8\left\|\eta'\right\|_{\mathrm{L}^\infty(\mathbb{R})} |f * P_t(g)|^2 \left|\nabla\mathrm{H}_t(g)\right|^2
\end{aligned}$$

and we can move the first term on the right hand side of this inequality to the left hand side of (9.14). Next,

$$|f * P_t(g)|^2 \left|\nabla\eta(\mathrm{H}_t(g))\right|^2 \leq \left\|\eta'\right\|^2_{\mathrm{L}^\infty(\mathbb{R})} |f * P_t(g)|^2 \left|\nabla\mathrm{H}_t(g)\right|^2$$

and similarly,

$$\begin{aligned}
&|f * P_t(g)|^2 \,|\eta(\mathrm{H}_t(g))|\,\left|\eta''(\mathrm{H}_t(g))\right| \left|\nabla\mathrm{H}_t(g))\right|^2 \\
&\quad\leq \|\eta\|_{\mathrm{L}^\infty(\mathbb{R})} \left\|\eta''\right\|_{\mathrm{L}^\infty(\mathbb{R})} |f * P_t(g)|^2 \left|\nabla\mathrm{H}_t(g))\right|^2.
\end{aligned}$$

We conclude that

$$\begin{aligned}
&\left|\nabla(f * P_t)(g)\eta(\mathrm{H}_t(g))\right|^2 \\
&\quad\leq -\mathcal{L}\left(|f * P_t(g)\eta(\mathrm{H}_t(g))|^2\right) + C(\eta)\,|f * P_t(g)|^2 \left|\nabla\mathrm{H}_t(g))\right|^2 \qquad (9.15) \\
&\quad=: f_1(g,t) + f_2(g,t),
\end{aligned}$$

say. The proof of (9.13) is now straightforward.

Evidently, $\mathrm{I}_0 \leq \mathrm{I}_1 + \mathrm{I}_2$, where

$$\mathrm{I}_j = \left|\iint_{G\times\mathbb{R}_+} f_j(g,t)t\,\mathrm{d}t\,\mathrm{d}g\right|.$$

To treat the term I_1, we recall that $\mathcal{L} = \mathcal{L} - \partial_t^2$. Integration by parts and the decay of the Poisson integral $f * P_t$ at infinity imply that

$$\int_G \mathcal{L}\left(|f * P_t(g)\eta(\mathrm{H}_t(g))|^2\right)t\,\mathrm{d}g = 0$$

for all $t > 0$, and also that

$$\int_{\mathbb{R}_+} \partial_t^2\Big(|f * P_t(g)\eta(\mathrm{H}_t(g))|^2\Big)t\,\mathrm{d}t$$
$$= \Big[t\partial_t\Big(|f * P_t(g)\eta(\mathrm{H}_t(g))|^2\Big)\Big]_{t=0}^{t=\infty} - \int_{\mathbb{R}_+} \partial_t\Big(|f * P_t(g)\eta(\mathrm{H}_t(g))|^2\Big)\,\mathrm{d}t$$
$$= 2\Big[(f * P_t(g)\eta(\mathrm{H}_t(g)))\,t\partial_t\,(f * P_t(g)\eta(\mathrm{H}_t(g)))\Big]_{t=0}^{t=\infty}$$
$$- \Big[|f * P_t(g)\eta(\mathrm{H}_t(g))|^2\Big]_{t=0}^{t=\infty}$$
$$= \big|f(g)\chi_{L_\beta(\alpha)}(g)\big|^2;$$

many of the terms here when $t = 0$ vanish by our remarks before the enunciation of this proposition. Therefore

$$\mathrm{I}_1 = \int_{L_\beta(\alpha)} |f(g)|^2\,\mathrm{d}g,$$

which is the first term on the right hand side of (9.13).

Next, since $|f * P_t(g)| \le \mathcal{M}_{P,\beta}(g) \le \alpha$ when $(g,t) \in W_\beta$, and ∇P_t has mean 0,

$$\mathrm{I}_2 = \iint_{W_\beta} |f * P_t(g)|^2\,\big|\nabla \mathrm{H}_t(g)\big|^2\,t\,\mathrm{d}t\,\mathrm{d}g$$
$$\le \alpha^2 \iint_{W_\beta} \big|\nabla \mathrm{H}_t(g)\big|^2\,t\,\mathrm{d}t\,\mathrm{d}g$$
$$\le \alpha^2 \iint_{G\times\mathbb{R}_+} \big|\chi_{L_\beta(\alpha)} * t\nabla P_t(g))\big|^2\,\frac{\mathrm{d}t}{t}\,\mathrm{d}g$$
$$= \alpha^2 \iint_{G\times\mathbb{R}_+} \big|(1 - \chi_{L_\beta(\alpha)}) * t\nabla P_t(g))\big|^2\,\frac{\mathrm{d}t}{t}\,\mathrm{d}g$$
$$\eqsim \alpha^2\,\big|L_\beta(\alpha)^c\big|^2,$$

by Littlewood–Paley theory. This is the second term on the right hand side of (9.13), and the proposition is now proved. □

Remark 9.2 We summarise the first step of this proof as the application of harmonicity to estimate the desired square function as a sum of two terms in (9.15). The "main term" I_1 gives us the function f that we started with, while the "error term" I_2 gives us an expression that we can handle by using Littlewood–Paley arguments.

9.5.4 Part 3 of the Proof of Proposition 9.4

It remains to take a product group $\boldsymbol{G}$, prove the inclusion and inequality

$$\mathsf{H}^1_{\max,P,\beta}(\boldsymbol{G}) \subseteq \mathsf{H}^1_{\mathrm{sq},P,1}(\boldsymbol{G})$$

$$\left\| \mathcal{S}_{P,1} f \right\|_{\mathsf{L}^1(\boldsymbol{G})} \lesssim \left\| \mathcal{M}_{P,\beta} f \right\|_{\mathsf{L}^1(\boldsymbol{G})} \qquad \forall f \in \mathsf{H}^1_{\max,P,\beta}(\boldsymbol{G}).$$

Again we may and do assume that f is real-valued and smooth.

The initial definitions are the same as in the one-parameter case. Take $f \in \mathsf{L}^1(\boldsymbol{G})$ such that $\mathcal{M}_{P,\beta}(f) \in \mathsf{L}^1(\boldsymbol{G})$ and $\alpha > 0$. Define

$$L_\beta(\alpha) := \left\{ \boldsymbol{g} \in \boldsymbol{G} : \mathcal{M}_{P,\beta}(f)(\boldsymbol{g}) \le \alpha \right\},$$

$$A_\beta(\alpha) := \left\{ \boldsymbol{g} \in \boldsymbol{G} : \mathcal{M}_S(1 - \chi_{L_\beta(\alpha)})(\boldsymbol{g}) < \frac{1}{4} \right\},$$

where $\mathcal{M}_S$ denotes the strong maximal operator. By the same argument as in the one-parameter case,

$$A_\beta(\alpha) \subseteq L_\beta(\alpha) \qquad \text{and} \qquad \left| A_\beta(\alpha)^c \right| \lesssim \left| L_\beta(\alpha)^c \right|.$$

Define also

$$W_\beta := \bigcup_{\boldsymbol{g} \in A_\beta(\alpha)} \Gamma^\beta(\boldsymbol{g}) \qquad \text{and} \qquad \widetilde{W}_\beta := \bigcup_{\boldsymbol{h} \in L_\beta(\alpha)(f)} \Gamma^\beta(\boldsymbol{h}).$$

As in the one-parameter case, there exists $C_0 \in (0, 1)$ and $C_1 \in (0, C_0)$, such that

$$\chi_{L_1(\alpha)} * P_{\boldsymbol{t}}(\boldsymbol{g}) \ge C_0 \qquad \forall (\boldsymbol{g}, \boldsymbol{t}) \in W_\gamma$$

$$\chi_{L_\beta(\alpha)} * P_{\boldsymbol{t}}(\boldsymbol{g}) \le C_1 \qquad \forall (\boldsymbol{g}, \boldsymbol{t}) \in (\widetilde{W}_\beta)^c,$$

provided that β is large enough.

We let $\mathrm{H}_{\boldsymbol{t}} := \chi_{L_\beta(\alpha)} * P_{\boldsymbol{t}}$; here $t_1, t_2 \ge 0$. Take a smooth real-valued function η on $\mathbb{R}$ such that $\eta(s) = 1$ when $s \ge C_0$ and $\eta(s) = 0$ when $s \le C_1$. By definition,

$$t_1 \partial_{t_1} \mathrm{H}_{\boldsymbol{t}}(\boldsymbol{g}) = \chi_{L_\beta(\alpha)} * (Q_{t_1} \otimes P_{t_2})(\boldsymbol{g}),$$

and this is uniformly bounded for all $\boldsymbol{g} \in \boldsymbol{G}$ and $t_1, t_2 \ge 0$; further, by (9.4),

$$t_1 \partial_{t_1} \mathrm{H}_{\boldsymbol{t}}(\boldsymbol{g}) \to 0 \qquad \text{as } t_1 \to 0$$

for almost all $\boldsymbol{g} \in \boldsymbol{G}$.

Again, it will suffice to show that

$$\iint_{G\times T} \left|\nabla_1\nabla_2(f * P_t)(g)\right|^2 |\eta(\mathrm{H}_t(g))|^2\, t\, \mathrm{d}t\, \mathrm{d}g \\ \lesssim \int_{L_\beta(\alpha)} \mathcal{M}_{P,\beta}(f)(g)^2\, \mathrm{d}g + \alpha^2 \left|L_1(\alpha)^c\right|.$$

We do this by extending the computation for a single homogeneous group.

First, we fix the variables g_2 and t_2. By the one-parameter case,

$$\left\| \mathcal{S}^{[1]}_{P,\gamma}(f)(\cdot, g_2)\right\|_{\mathrm{L}^1(G_1)} \lesssim \left\| \mathcal{M}^{[1]}_{P,\beta}(f)(\cdot, g_2)\right\|_{\mathrm{L}^1(G_1)},$$

whence

$$\left\| \mathcal{S}^{[1]}_{P,\gamma}(f)\right\|_{\mathrm{L}^1(G)} \lesssim \left\| \mathcal{M}^{[1]}_{P,\beta}(f)\right\|_{\mathrm{L}^1(G)} \leq \left\| \mathcal{M}_{P,\beta}(f)\right\|_{\mathrm{L}^1(G)} \tag{9.16}$$

by integration over G_2 and the pointwise inequality $\mathcal{M}^{[1]}_{P,\beta}(f) \leq \mathcal{M}_{P,\beta}(f)$. A similar result holds for the Littlewood–Paley operator acting in the second variable only.

The function $(g, t) \mapsto f * P_t(g)$ is harmonic in the g_1 and t_1 variables, and in the g_2 and t_2 variables. This leads to a more complicated analogue of (9.15), with four terms, I_{11}, I_{21}, I_{12}, and I_{22}, where the subscript $i_1 i_2$ indicates a term like I_{i_1} in the first factor, and a term like I_{i_2} in the second factor.

There is one "main term" I_{11} with a double sub-Laplacian, namely,

$$\mathcal{L}_1\mathcal{L}_2\Big(|f * P_t(g)\eta(\mathrm{H}_t(g))|^2\Big).$$

When integrated, by iterating the argument used to treat I_1 in Sect. 9.5.3, I_{11} gives

$$\int_{L_1(\alpha)} |f(g)|^2\, \mathrm{d}g.$$

The "mixed terms" I_{21} and I_{12}, with a sub-Laplacian in one variable and a square function in the other, may be treated by (9.16) and its analogue with G_1 and G_2 interchanged. Finally, the "double error term" I_{22} may be treated using Littlewood–Paley theory, much as we treated I_2 before.

9.5.5 A Reproducing Formula

We shall use the discrete Calderón reproducing formula from [16, Theorem 2.9]. We first give a definition of the space $\mathsf{M}(G, r, g)$ of molecules of scale r near a point g

on a group G of homogeneous dimension Q, and then define the analogous space on a product group. In the more general setting of spaces of homogenous type, this space was introduced in [18]. Recall that $\varepsilon \in (0, 1]$ is a fixed parameter.

Definition 9.5 Fix $r > 0$ and $g \in G$. We say that a function f on G is in $\mathsf{M}(G, r, g)$ if there is a constant C such that

$$\begin{aligned} |f(h)| &\le C \frac{r^{\varepsilon}}{(r + \rho(g^{-1}h))^{Q+\varepsilon}} \\ \left|f(h) - f(h')\right| &\le C \frac{\rho(h^{-1}h')^{\varepsilon}}{(r + \rho(g^{-1}h) + \rho(g^{-1}h'))^{Q+\varepsilon}} \end{aligned} \tag{9.17}$$

for all $h, h' \in G$. If moreover f satisfies the cancellation condition

$$\int_G f(g)\, \mathrm{d}g = 0,$$

then we write $f \in \mathsf{M}_0(G, r, g)$. The norm $\|f\|_{\mathsf{M}(G,r,g)}$ is defined to be the least constant C such that the inequalities (9.17) both hold.

Clearly, the space $\mathsf{M}(G)$ of (9.1) is equal to $\mathsf{M}(G, 1, o)$, and $f \in \mathsf{M}(G)$ if and only if $f \in \mathsf{M}(G, r, g)$ for all $r > 0$ and all $g \in G$. Changing r or g changes the norms.

We now define the molecular space $\mathsf{M}(\boldsymbol{G}, \boldsymbol{r}, \boldsymbol{g})$ on the product group $\boldsymbol{G}$ as follows.

Definition 9.6 Fix $\boldsymbol{r} \in \boldsymbol{T}$ and $\boldsymbol{g} \in \boldsymbol{G}$. We say that $\psi : \boldsymbol{G} \to \mathbb{C}$ is in $\mathsf{M}(\boldsymbol{G}, \boldsymbol{r}, \boldsymbol{g})$ if $\psi(\cdot, h_2) \in \mathsf{M}(G_1, r_1, g_1)$ for all $h_2 \in G_2$ and $\psi(h_1, \cdot) \in \mathsf{M}(G_2, r_2, g_2)$ for all $h_1 \in G_1$, and

$$\begin{aligned} \|\psi(\cdot, h_2)\|_{\mathsf{M}(G_1,r_1,g_1)} &\le C \frac{r_2^{\varepsilon}}{(r_2 + \rho_2(g_2^{-1}h_2))^{Q_2+\varepsilon}} \\ \|\psi(h_1, \cdot)\|_{\mathsf{M}(G_2,r_2,g_2)} &\le C \frac{r_1^{\varepsilon}}{(r_1 + \rho_1(g_1^{-1}h_1))^{Q_1+\varepsilon}} \\ \left\|\psi(\cdot, h_2) - \psi(\cdot, h_2')\right\|_{\mathsf{M}(G_1,r_1,g_1)} &\le C \frac{\rho_2(h_2^{-1}h_2')^{\varepsilon}}{(r_2 + \rho_2(g_2^{-1}h_2) + \rho_2(g_2^{-1}h_2'))^{Q+\varepsilon}} \\ \left\|\psi(h_1, \cdot) - \psi(h_1', \cdot)\right\|_{\mathsf{M}(G_2,r_2,g_2)} &\le C \frac{\rho_1(h_1^{-1}h_1')^{\varepsilon}}{(r_1 + \rho_1(g_1^{-1}h_1) + \rho_1(g_1^{-1}h_1'))^{Q+\varepsilon}} \end{aligned} \tag{9.18}$$

for all $\boldsymbol{h}, \boldsymbol{h}' \in \boldsymbol{G}$. If moreover ψ satisfies the cancellation conditions

$$\int_{G_1} \psi(g_1, \cdot)\, \mathrm{d}g_1 = 0 \qquad \text{and} \qquad \int_{G_2} \psi(\cdot, g_2)\, \mathrm{d}g_2 = 0,$$

then we write $\psi \in \mathsf{M}_0(\boldsymbol{G}, \boldsymbol{r}, \boldsymbol{g})$. The norm $\|\psi\|_{\mathsf{M}(\boldsymbol{G},\boldsymbol{r},\boldsymbol{g})}$ is defined to be the least constant C such that the inequalities (9.18) above all hold.

Evidently, if $\psi_1 \in \mathsf{M}_0(G_1, r_1, g_1)$ and $\psi_2 \in \mathsf{M}_0(G_2, r_2, g_2)$, then $\psi_1 \otimes \psi_2 \in \mathsf{M}_0(\boldsymbol{G}, \boldsymbol{r}, \boldsymbol{g})$. It is easy to check that

$$\|f\|_{\mathsf{M}(\boldsymbol{G},\boldsymbol{r},\boldsymbol{g})} = \left\| r_1^{Q_1} r_2^{Q_2} \psi(\boldsymbol{g}\delta_{\boldsymbol{r}}(\cdot)) \right\|_{\mathsf{M}(\boldsymbol{G})}.$$

Hence the $\mathsf{L}^1(\boldsymbol{G})$ norms of elements of a bounded subset of $\mathsf{M}(\boldsymbol{G}, \boldsymbol{r}, \boldsymbol{g})$ are bounded.

We are now ready to state the version of the Calderón reproducing formula that we are going to use. Let $\sigma : \mathscr{P}(\boldsymbol{G}) \to \boldsymbol{G}$ be an arbitrary function such that $\sigma(R) \in \bar{R}$ for all $R \in \mathscr{P}(\boldsymbol{G})$ and let $\boldsymbol{\ell} : \mathscr{P}(\boldsymbol{G}) \to \boldsymbol{T}$ be the function such that $\ell_i(Q_1 \times Q_2) = \ell(Q_i)$, the "side-length" of Q_i. For $\varphi^{[1]} \in \mathsf{M}(G_1)$ and $\varphi^{[2]} \in \mathsf{M}(G_2)$, we take φ to be $\varphi^{[1]} \otimes \varphi^{[2]}$, and define φ_R to be the function $\boldsymbol{h} \mapsto [\bar{\varphi}]_{\boldsymbol{\ell}(R)}(\boldsymbol{h}^{-1}\sigma(R))$. Observe that

$$\int_{\boldsymbol{G}} f(\boldsymbol{h})\bar{\varphi}_R(\boldsymbol{h})\, \mathrm{d}\boldsymbol{h} = \int_{\boldsymbol{G}} f(\boldsymbol{h})\, [\varphi]_{\boldsymbol{\ell}(R)}\, (\boldsymbol{h}^{-1}\sigma(R))\, \mathrm{d}\boldsymbol{h} = f * [\varphi]_{\boldsymbol{\ell}(R)}(\sigma(R)). \tag{9.19}$$

The point of the following theorem is that the collection $\{|R|^{1/2}\varphi_R : R \in \mathscr{P}(\boldsymbol{G})\}$ is a well-behaved frame in $\mathsf{L}^2(\boldsymbol{G})$, with a well-behaved dual frame $\{|R|^{1/2}\tilde{\varphi}_R : R \in \mathscr{P}(\boldsymbol{G})\}$. By well-behaved, we mean that φ_R and $\tilde{\varphi}_R$ are concentrated near R, and certain molecular norms of φ_R and $\tilde{\varphi}_R$ are uniformly bounded in R *and* in σ.

Theorem 9.5 ([16, Theorem 2.9]) *Suppose that $\sigma : \mathscr{P}(\boldsymbol{G}) \to \boldsymbol{G}$ and $\varphi \in \mathsf{M}(\boldsymbol{G})$ are as discussed above. Then, after possible replacing φ by a normalised dilate of φ, there exist functions $\tilde{\varphi}_R$ in $\mathsf{M}_0(\boldsymbol{G})$, which may also depend on σ, such that*

$$\psi = \sum_{R\in\mathscr{P}(\boldsymbol{G})} |R|\, \langle \psi, \varphi_R\rangle\, \tilde{\varphi}_R$$

for every ψ in $\mathsf{M}_0(\boldsymbol{G})$. Further,

$$\left\|\varphi_R\right\|_{\mathsf{M}(\boldsymbol{G},\boldsymbol{\ell}(R),\sigma(R))} + \left\|\tilde{\varphi}_R\right\|_{\mathsf{M}_0(\boldsymbol{G},\boldsymbol{\ell}(R),\sigma(R))}$$

is uniformly bounded, irrespective of R and the choice of $\sigma(R)$.

As $\mathsf{L}^1(\boldsymbol{G})$ is a subspace of the dual space of $\mathsf{M}_0(\boldsymbol{G})$, by (9.19) and a duality argument,

$$f * \psi(\boldsymbol{g}) = \sum_{R\in\mathscr{P}(\boldsymbol{G})} \left(|R|\, f * [\varphi]_{\boldsymbol{\ell}(R)}(\sigma(R))\right) \tilde{\varphi}_R * \psi(\boldsymbol{g}) \qquad \forall \boldsymbol{g} \in \boldsymbol{G}$$

for all $\psi \in \mathsf{M}_0(\boldsymbol{G})$ and all $f \in \mathsf{L}^1(\boldsymbol{G})$.

9.5.6 Proof of Proposition 9.5

We are going to prove Proposition 9.5. Again, we need to extend what we know about homogeneous groups to product groups. We shall prove a stronger result concerning the grand maximal function, which we now introduce.

We define

$$\mathsf{F}(\boldsymbol{G}) := \left\{ \zeta^{[1]} \otimes \zeta^{[2]} : \left\| \zeta^{[1]} \right\|_{\mathsf{M}(G_1)} \leq 1, \left\| \zeta^{[2]} \right\|_{\mathsf{M}(G_2)} \leq 1 \right\},$$

and the grand maximal operator $\mathcal{G}$:

$$\mathcal{G}(f)(\boldsymbol{g}) := \sup\{|f * \zeta_{\boldsymbol{t}}(\boldsymbol{g})| : \zeta \in \mathsf{F}(\boldsymbol{G}), \boldsymbol{t} \in \boldsymbol{T}\} \qquad \forall \boldsymbol{g} \in \boldsymbol{G}$$

for all $f \in \mathsf{L}^1(\boldsymbol{G})$. We write $\mathcal{R}_{\boldsymbol{h}}$ for the operator of right translation by $\boldsymbol{h} \in \boldsymbol{G}$, that is, $\mathcal{R}_{\boldsymbol{h}}\zeta(\boldsymbol{g}) = \zeta(\boldsymbol{g}\boldsymbol{h})$ for all $\boldsymbol{g} \in \boldsymbol{G}$ and $\zeta \in \mathsf{F}(\boldsymbol{G})$. Since

$$\begin{aligned} \mathcal{M}_{\zeta,\beta}(f)(\boldsymbol{g}) &= \sup\{|f * \zeta_{\boldsymbol{t}}(\boldsymbol{g}')| : \boldsymbol{g}' \in R(\boldsymbol{g}, \beta\boldsymbol{t}), \boldsymbol{t} \in \boldsymbol{T}\} \\ &= \sup\{|f * (\mathcal{R}_{\boldsymbol{h}}\zeta)_{\boldsymbol{t}}(\boldsymbol{g})| : \boldsymbol{h} \in R(\boldsymbol{o}, \beta, \beta), \boldsymbol{t} \in \boldsymbol{T}\}, \end{aligned}$$

and $\mathcal{R}_{\boldsymbol{h}}\zeta$ is a uniformly bounded (β-dependent) multiple of a function in $\mathsf{F}(\boldsymbol{G})$ when $\boldsymbol{h} \in R(\boldsymbol{o}, \beta, \beta)$ we deduce that

$$\mathcal{M}_{\zeta,\beta} f(\boldsymbol{g}) \lesssim_\beta \mathcal{G}(f)(\boldsymbol{g}) \qquad \forall \boldsymbol{g} \in \boldsymbol{G}.$$

Take functions $\varphi^{[i]}$ on G_i such that $\|\varphi^{[i]}\|_{\mathsf{M}(G_i)} \leq 1$ and $\int_{G_i} \varphi^{[i]} \, \mathrm{d}g_i \neq 0$. From the discussion above, it will suffice to prove that there exists $\theta \in (0, 1)$ such that

$$\mathcal{M}_{\zeta,0} f(\boldsymbol{g}) \lesssim \left(\mathcal{M}_S(|\mathcal{M}_{\varphi,0}(f)(\boldsymbol{g})|^\theta) \right)^{1/\theta} \qquad \forall \boldsymbol{g} \in \boldsymbol{G} \tag{9.20}$$

for all $\beta \geq 0$, all $\zeta \in \mathsf{F}(\boldsymbol{G})$, and all $f \in \mathsf{L}^1(\boldsymbol{G})$, for then the $\mathsf{L}^{1/\theta}(\boldsymbol{G})$ boundedness of $\mathcal{M}_S$ shows that

$$\|\mathcal{G}(f)\|_{\mathsf{L}^1(\boldsymbol{G})} \lesssim \int_{\boldsymbol{G}} \left(\mathcal{M}_S(|\mathcal{M}_{\varphi,0}(f)|^\theta) \right)^{1/\theta} \mathrm{d}\boldsymbol{g} \lesssim \|\mathcal{M}_{\varphi,0}(f)\|_{\mathsf{L}^1(\boldsymbol{G})},$$

which implies the required result. We may assume that f is continuous, by mollification.

To prove (9.20), we make and confirm three claims, which together imply the result.

Claim 1 For θ less than but close to 1,

$$|f * \zeta_t| \lesssim_\theta \|\psi\|_{\mathsf{M}_0(\boldsymbol{G})} \big(\mathcal{M}_S\big(\big|\mathcal{M}_{\varphi,0}(f)\big|^\theta\big)\big)^{1/\theta} \tag{9.21}$$

for all $f \in \mathsf{L}^1(\boldsymbol{G})$, all $\boldsymbol{t} \in \boldsymbol{T}$ and all ζ of the form $\psi_1 \otimes \psi_2$, where $\psi_1 \in \mathsf{M}_0(G_1)$ and $\psi_2 \in \mathsf{M}_0(G_2)$. To prove this claim, we use Theorem 9.5, which tells us that

$$f * \psi_{\boldsymbol{t}}(\boldsymbol{g}) = \sum_{R \in \mathscr{P}(\boldsymbol{G})} |R| \, \langle f, \varphi_R \rangle \, \tilde{\varphi}_R * \psi_{\boldsymbol{t}}(\boldsymbol{g}). \tag{9.22}$$

On the one hand, once f is given, we may choose σ such that

$$|\langle f, \varphi_R \rangle| = \big|(f * [\varphi]_{\boldsymbol{\ell}(R)})(\sigma(R))\big| = \min\{\big|(f * [\varphi]_{\boldsymbol{\ell}(R)})(\boldsymbol{h})\big| : \boldsymbol{h} \in \bar{R}\},$$

and on the other, by the almost orthogonality estimate of [18, (4.4)], for all choices of σ and all choices of $\boldsymbol{h}$ in R,

$$\begin{aligned}
&|\tilde{\varphi}_R * \psi_{\boldsymbol{t}}(\boldsymbol{g})| \\
&\lesssim \mu_{\boldsymbol{j},\boldsymbol{t}}(\boldsymbol{h}^{-1}\boldsymbol{g}) \\
&\approx \frac{(\ell_1(R) \wedge t_1)^\varepsilon (\ell_1(R)t_1)^{-\varepsilon}}{((\ell_1(R) \wedge t_1)^{-1} + \rho_1(h_1^{-1}g_1))^{Q_1+\varepsilon}} \frac{(\ell_2(R) \wedge t_2)^\varepsilon (\ell_2(R)t_2)^{-\varepsilon}}{((\ell_2(R) \wedge t_2)^{-1} + \rho_1(h_2^{-1}g_2))^{Q_2+\varepsilon}}.
\end{aligned}$$

Here $\mu_{\boldsymbol{j},\boldsymbol{t}}$ is the least decreasing biradial majorant for all the functions $\tilde{\varphi}_R * \psi_{\boldsymbol{t}}(\boldsymbol{h}^{-1}\cdot)$ when $\boldsymbol{h} \in R \in \mathscr{P}^{\boldsymbol{j}}(\boldsymbol{G})$, and $a \wedge b$ denotes the minimum of a and b. Recall that the "sidelengths" of the cubes making the rectangle $R \in \mathscr{P}^{\boldsymbol{j}}(\boldsymbol{G})$ are κ^{j_1} and κ^{j_2}. Then, by also using (9.22) and (9.3), we see that

$$\begin{aligned}
|f * \psi_{\boldsymbol{t}}(\boldsymbol{g})|^\theta &\le \Bigg(\sum_{\boldsymbol{j} \in \mathbb{Z}^2} \sum_{R \in \mathscr{P}^{\boldsymbol{j}}(\boldsymbol{G})} |R| \min_{\boldsymbol{g} \in \bar{R}} \mathcal{M}_{\varphi,0}(f)(\boldsymbol{g}) \, |\tilde{\varphi}_R * \psi_{\boldsymbol{t}}(\boldsymbol{g})|\Bigg)^\theta \\
&\le \Bigg(\sum_{\boldsymbol{j} \in \mathbb{Z}^2} \sum_{R \in \mathscr{P}^{\boldsymbol{j}}(\boldsymbol{G})} |R|^\theta \Big(\min_{\boldsymbol{g} \in \bar{R}} \mathcal{M}_{\varphi,0}(f)(\boldsymbol{g})\Big)^\theta |\tilde{\varphi}_R * \psi_{\boldsymbol{t}}(\boldsymbol{g})|^\theta\Bigg) \\
&= \Bigg(\sum_{\boldsymbol{j} \in \mathbb{Z}^2} \sum_{R \in \mathscr{P}^{\boldsymbol{j}}(\boldsymbol{G})} |R|^{\theta-1} \int_R \Big(\min_{\boldsymbol{g} \in \bar{R}} \mathcal{M}_{\varphi,0}(f)(\boldsymbol{g})\Big)^\theta |\tilde{\varphi}_R * \psi_{\boldsymbol{t}}(\boldsymbol{g})|^\theta \, \mathrm{d}\boldsymbol{h}\Bigg) \qquad (9.23) \\
&\lesssim \Bigg(\sum_{\boldsymbol{j} \in \mathbb{Z}^2} (\kappa_1^{j_1 Q_1} \kappa_2^{j_2 Q_2})^{\theta-1} \int_{\boldsymbol{G}} \Big(\mathcal{M}_{\varphi,0}(f)(\boldsymbol{h})\Big)^\theta \mu_{\boldsymbol{j},\boldsymbol{t}}(\boldsymbol{h}^{-1}\boldsymbol{g})^\theta \, \mathrm{d}\boldsymbol{h}\Bigg) \\
&\le \Bigg(\sum_{\boldsymbol{j} \in \mathbb{Z}^2} (\kappa_1^{j_1 Q_1} \kappa_2^{j_2 Q_2})^{\theta-1} \left\|\mu_{\boldsymbol{j},\boldsymbol{t}}^\theta\right\|_{\mathsf{L}^1(\boldsymbol{G})} \mathcal{M}_S\left(\mathcal{M}_{\varphi,0}(f)\right)^\theta(\boldsymbol{g})\Bigg).
\end{aligned}$$

If $\max\{Q_1/(Q_1+\varepsilon), Q_2/(Q_2+\varepsilon)\} < \theta < 1$, then computation shows that

$$\sum_{\boldsymbol{j}\in\mathbb{Z}^2} (\kappa_1^{j_1 Q_1}\kappa_2^{j_2 Q_2})^{\theta-1} \left\| \mu_{\boldsymbol{j},\boldsymbol{t}}^{\theta} \right\|_{\mathsf{L}^1(\boldsymbol{G})} < \infty.$$

Thus the right-hand side of (9.23) is bounded by a multiple of

$$\left(\mathcal{M}_S\left(\left|\mathcal{M}_{\varphi,0}(f)\right|^{\theta}\right)\right)(\boldsymbol{g}^R),$$

which implies (9.21) and proves our claim.

Claim 2 For θ less than but close to 1,

$$|f * \zeta_{\boldsymbol{t}}(\boldsymbol{g})| \lesssim_\theta \|\zeta_1\|_{\mathsf{M}_0(G_1)} \left(\mathcal{M}_S\left(\left|\mathcal{M}_{\varphi,0}(f)\right|^{\theta}\right)(\boldsymbol{g})\right)^{1/\theta}$$

for all $f \in \mathsf{L}^1(\boldsymbol{G})$, all $\boldsymbol{t} \in \boldsymbol{T}$, where $\zeta = \psi_1 \otimes \varphi_2$; here $\psi_1 \in \mathsf{M}_0(G_1)$.

The proof of this claim involves use of a reproducing formula that involves the first variable only, namely,

$$f^{[1]} *_1 \psi_{t_1}^{[1]}(g_1) = \sum_{Q\in\mathcal{Q}(G_1)} |Q|\, f^{[1]} *_1 [\varphi^{[1]}]_{\ell(Q)}(\sigma(Q))\tilde{\varphi}_Q^{[1]} * \psi_{t_1}^{[1]}(g_1) \qquad \forall g_1 \in G_1$$

where $f^{[1]} \in \mathsf{L}^1(G_1)$. This implies that

$$f * \zeta_{\boldsymbol{t}}(\boldsymbol{g}) = \sum_{Q\in\mathcal{Q}(G_1)} |Q| f * ([\varphi^{[1]}]_{\ell(Q)} \otimes \varphi_{t_2}^{[2]})(\sigma(Q), g_2)\tilde{\varphi}_Q^{[1]} * \psi_{t_1}^{[1]}(g_1) \qquad \forall \boldsymbol{g} \in \boldsymbol{G},$$

when $f \in \mathsf{L}^1(\boldsymbol{G})$; a similar argument to that for Claim 1 may be used. We see that

$$\begin{aligned}
|\tilde{\varphi}_R * \zeta_{\boldsymbol{t}}(\boldsymbol{g})|^\theta &\le \left(\sum_{j_1\in\mathbb{Z}} \sum_{Q\in\mathcal{Q}^{j_1}(G_1)} |Q| \min_{g_1\in\bar{Q}} \mathcal{M}_{\varphi,0}(f)(g_1, g_2) \left|\tilde{\varphi}_Q^{[1]} * \psi_{t_1}^{[1]}(g_1)\right|\right)^\theta \\
&\le \left(\sum_{j_1\in\mathbb{Z}} (\kappa_1^{j_1 Q_1})^{\theta-1} \left\| (\mu_{j_1,t_1}^{[1]})^\theta \right\|_{\mathsf{L}^1(G_1)} (\mathcal{M}_1 \otimes \mathcal{I}_2)\left(|\mathcal{M}_{\varphi,0}(f)|^\theta\right)(\boldsymbol{g})\right) \\
&\lesssim \mathcal{M}_S\left(|\mathcal{M}_{\varphi,0}(f)|^\theta\right)(\boldsymbol{g}),
\end{aligned}$$

as claimed; here $\mathcal{I}_2$ denotes the identity operator acting on functions on G_2.

Claim 3 For θ less than but close to 1,

$$|f * \psi_{\boldsymbol{t}}(\boldsymbol{g})| \lesssim_\theta \|\psi_2\|_{\mathsf{M}_0(G_2)} \left(\mathcal{M}_S\left(\left|\mathcal{M}_{\varphi,0}(f)\right|^{\theta}\right)(\boldsymbol{g})\right)^{1/\theta}$$

for all $f \in \mathsf{L}^1(\boldsymbol{G})$, all $\boldsymbol{t} \in \boldsymbol{T}$, where now $\psi_1 = \varphi_1$ while $\psi_2 \in \mathsf{M}_0(G_2)$. The proof of this is a very minor modification of that of Claim 2.

To finish the proof, we must estimate $\mathcal{M}_{\zeta,0} f$, where $\zeta_1 \in \mathsf{M}(G_1)$ and $\zeta_2 \in \mathsf{M}(G_2)$. We write $\zeta_1 = c_1\varphi_1 + \psi_1$, where $\psi \in \mathsf{M}_0(G_1)$ and c_1 is chosen to make the integrals of both sides equal, and we decompose ζ_2 analogously. Then $\mathcal{M}_{\zeta,0} f(\boldsymbol{g})$ is dominated by a sum of four terms, each of which is bounded pointwise by $(\mathcal{M}_S(\mathcal{M}_{\varphi,0} f)^\theta)^{1/\theta}(\boldsymbol{g})$. This proves the desired inequality and hence Proposition 9.5.

9.6 Concluding Remarks

Many of our results can be proved in greater generality. For example, Proposition 9.5 should be true on much more general spaces of homogeneous type. Other results require the structure of stratified group that we have used here. These include Theorem 9.1 and Proposition 9.4. Indeed, the first relies on the Christ–Geller singular integral characterisation of the Hardy space on stratified groups, and the second on various properties of the Poisson kernel. It is an interesting challenge to extend either of these to a more substantial class of nilpotent Lie groups, let alone to general spaces of homogeneous type.

Acknowledgments The authors thank the unknown referees for their careful reading of and helpful comments about the paper.

J. Li and M. Cowling are supported by the Australian Research Council (ARC) through research grant DP220100285. L. Yan was supported by the NNSF of China, Grant No. 11871480, and by the Australian Research Council (ARC) through research grant DP190100970.

References

1. S.-Y.A. Chang, R. Fefferman, A continuous version of duality of H^1 with BMO on the bidisc. Ann. Math. **112**, 179–201 (1980)
2. P. Chen, X.T. Duong, J. Li, L.A. Ward, L. Yan, Product Hardy spaces associated to operators with heat kernel bounds on spaces of homogeneous type. Math. Z. **282**, 1033–1065 (2016)
3. P. Chen, M.G. Cowling, M. Lee, J. Li, A. Ottazzi, Flag Hardy space theory on Heisenberg groups and applications (2021). Available at `arXiv:2102.07371`
4. M. Christ, A $T(b)$ theorem with remarks on analytic capacity and the Cauchy integral. Colloq. Math. **61**, 601–628 (1990)
5. M. Christ, D. Geller, Singular integral characterisations of Hardy spaces on homogeneous groups. Duke. Math. J. **61**, 601–628 (1990)
6. R.R. Coifman, G. Weiss, *Analyse harmonique non-commutative sur certains espaces homogènes*. Lecture Notes in Math. 242 (Springer, Berlin, 1971)
7. R.R. Coifman, G. Weiss, Extensions of Hardy spaces and their use in analysis. Bull. Amer. Math. Soc. **83**, 569–645 (1977)

8. A. Cordoba, R. Fefferman, A geometric proof of the strong maximal theorem. Ann. Math. **102**, 95–100 (1975)
9. C. Fefferman, E.M. Stein, H^p spaces of several variables. Acta Math. **129**, 137–193 (1972)
10. R. Fefferman, Harmonic Analysis on product spaces. Ann. Math. **126**, 109–130 (1987)
11. R. Fefferman, E.M. Stein, Singular integrals on product spaces. Adv. Math. **45**, 117–143 (1982)
12. S.H. Ferguson, M.T. Lacey, A characterization of product BMO by commutators. Acta Math. **189**, 143–160 (2002)
13. G.B. Folland, E.M. Stein, *Hardy Spaces on Homogeneous Groups* (Princeton University Press, Princeton, 1982)
14. L. Grafakos, *Classical Fourier Analysis*, 3rd edn. Graduate Texts in Mathematics 249 (Springer, 2014)
15. R.F. Gundy, E.M. Stein, H^p theory for the poly-disc. Proc. Natl. Acad. Sci. U. S. A. **76**, 1026–1029 (1979)
16. Y. Han, J. Li, G. Lu, Multiparameter Hardy space theory on Carnot–Carathéodory spaces and product spaces of homogeneous type. Trans. Amer. Math. Soc. **365**, 319–360 (2013)
17. Y. Han, J. Li, M.C. Pereyra, L.A. Ward, Atomic decomposition of product Hardy spaces via wavelet bases on spaces of homogeneous type. New York J. Math. **27**, 1173–1239 (2021)
18. Y. Han, J. Li, L.A. Ward, Hardy space theory on spaces of homogeneous type via orthonormal wavelet bases. Appl. Comput. Harmonic Anal. **45**, 120–169 (2018)
19. W. Hebisch, A. Sikora, A smooth subadditive homogeneous norm on a homogeneous group. Stud. Math. **XCVI**, 231–236 (1990)
20. A. Hulanicki, A functional calculus for Rockland operators on nilpotent Lie groups. Stud. Math. **LXXVIII**, 253–266 (1984)
21. T. Hytönen, A. Kairema, Systems of pseudodyadic cubes in a doubling metric space. Colloq. Math. **126**, 1–33 (2012)
22. T. Hytönen, H. Martikainen, Non-homogeneous $T1$ theorem for bi-parameter singular integrals. Adv. Math. **261**, 220–273 (2014)
23. J.-L. Journé, Calderón–Zygmund operators on product spaces. Rev. Mat. Iberoam. **1**, 55–91 (1985)
24. M.T. Lacey, S. Petermichl, J. Pipher, B.D. Wick, Multiparameter Riesz commutators. Amer. J. Math. **131**, 731–769 (2009)
25. K.G. Merryfield, On the area integral, Carleson measures and H^p in the polydisc. Indiana Univ. Math. J. **34**, 663–685 (1985)
26. Y. Ou, A $T(b)$ theorem on product spaces. Trans. Amer. Math. Soc. **367**, 6159–6197 (2015)
27. Y. Ou, S. Petermichl, E. Strouse, Higher order Journé commutators and characterizations of multi-parameter BMO. Adv. Math. **291**, 24–58 (2016)
28. J. Pipher, Journé's covering lemma and its extension to higher dimensions. Duke Math. J. **53**, 683–690 (1986)
29. E. Sawyer, R. Wheeden, Weighted inequalities for fractional integrals on Euclidean and homogeneous spaces. Amer. J. Math. **114**, 813–874 (1992)
30. E.M. Stein, G. Weiss, On the theory of harmonic functions of several variables, I. The theory of H^p spaces. Acta Math. **103**, 25–62 (1960)
31. N.Th. Varopoulos, L. Saloff-Coste, T. Coulhon, *Analysis and Geometry on Groups*. Cambridge Tracts in Mathematics 100 (Cambridge University Press, Cambridge, 1992)

Chapter 10
A Sufficient Condition for Haar Multipliers in Triebel-Lizorkin Spaces

Gustavo Garrigós, Andreas Seeger, and Tino Ullrich

In memory of Guido Weiss.

Abstract We consider Haar multiplier operators T_m acting on Sobolev spaces, and more generally Triebel-Lizorkin spaces $F^s_{p,q}(\mathbb{R})$, for indices in which the Haar system is not unconditional. When m depends only on the Haar frequency, we give a sufficient condition for the boundedness of T_m in $F^s_{p,q}$, in terms of the variation norms $\|m\|_{V_u}$, which is optimal in u (up to endpoints) when $p, q > 1$.

Math Subject Class (2010) 46E35, 46B15, 42C40

10.1 Introduction

Consider the classical Haar system in $\mathbb{R}$,

$$\mathscr{H} = \{h_{j,\mu} \;:\; j \geq -1,\ \mu \in \mathbb{Z}\}, \tag{10.1}$$

where, if $h = \mathbb{1}_{[0,1/2)} - \mathbb{1}_{[1/2,1)}$, we let

$$h_{j,\mu}(x) = h(2^j x - \mu), \quad \text{for} \quad j = 0, 1, 2, \ldots, \quad \mu \in \mathbb{Z},$$

G. Garrigós (✉)
Departamento de Matemáticas, University of Murcia, Murcia, Spain
e-mail: gustavo.garrigos@um.es

A. Seeger
Department of Mathematics, University of Wisconsin-Madison, Madison, WI, USA
e-mail: seeger@math.wisc.edu

T. Ullrich
Fakultät für Mathematik, Technische Universität Chemnitz, Chemnitz, Germany
e-mail: tino.ullrich@mathematik.tu-chemnitz.de

E. Hernández et al. (eds.), *The Mathematical Heritage of Guido Weiss*,
Applied and Numerical Harmonic Analysis,
https://doi.org/10.1007/978-3-031-76793-7_10

while for $j = -1$ we let

$$h_{-1,\mu} = \mathbb{1}_{[\mu,\mu+1)}, \quad \mu \in \mathbb{Z}.$$

We shall refer to the elements of the family $\mathscr{H}_j = \{h_{j,\mu} : \mu \in \mathbb{Z}\}$ as *Haar functions of frequency* 2^j.

Let $F^s_{p,q}$ denote the usual Triebel-Lizorkin space in $\mathbb{R}$; see [16]. It is known from the work of Triebel [17, Theorem 2.9.ii] that $\mathscr{H}$ is an unconditional basis of $F^s_{p,q}(\mathbb{R})$ when s belongs to the range

$$\max\Big\{1/p - 1, 1/q - 1\Big\} < s < \min\Big\{1/p, 1/q, 1\Big\}. \tag{10.2}$$

That this range is actually optimal was shown by the last two authors in [12, 13]. More recently, we proved in [3] that $\mathscr{H}$ is a Schauder basis of $F^s_{p,q}(\mathbb{R})$ (with respect to natural enumerations) in the larger range

$$1/p - 1 < s < \min\Big\{1/p, 1\Big\}, \quad \text{(for all } 0 < q < \infty), \tag{10.3}$$

while at the endpoints (see [5]) the property holds if and only if

$$s = 1/p - 1 \quad \text{and} \quad 1/2 < p \leq 1, \tag{10.4}$$

also for all $0 < q < \infty$. These regions are depicted in Fig. 10.1.

We shall mainly be interested in values of the parameters outside the region of unconditionality. In that range, it becomes a natural question to find sufficient conditions on a sequence $\{m_{j,\mu}\}$ so that the mapping

$$f \longmapsto \sum_{j\geq 0}\sum_{\mu\in\mathbb{Z}} m_{j,\mu}\, 2^j \langle f, h_{j,\mu}\rangle h_{j,\mu},$$

defined say for $f \in \operatorname{span}\mathscr{H}$, extends as a bounded linear operator in the space $F^s_{p,q}$.

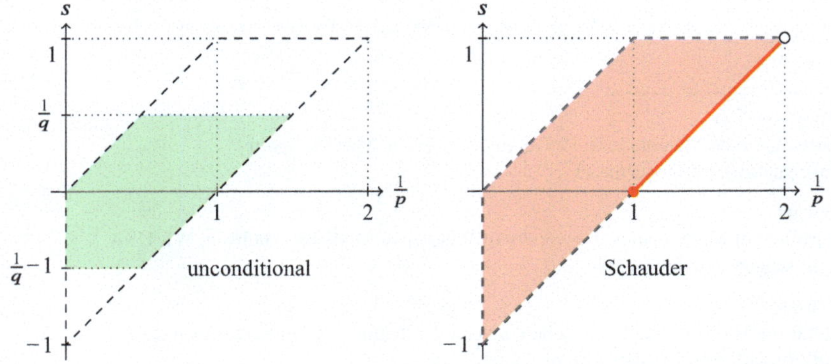

Fig. 10.1 Parameter domain for $\mathscr{H}$ to be an unconditional basis (left figure) or a Schauder basis (right figure) in $F^s_{p,q}(\mathbb{R})$

In this paper we regard this problem in the special case when the sequence is constant in each frequency level, namely, if $m = \{m(j)\}_{j\ge0}$, we consider the operators

$$T_m f = \sum_{j\ge0} m(j)\, \mathbb{D}_j f,$$

where $\mathbb{D}_j$ denotes the orthogonal projection onto the space generated by $\mathscr{H}_j$, that is

$$\mathbb{D}_j f = \sum_{\mu\in\mathbb{Z}} 2^j \langle f, h_{j,\mu}\rangle\, h_{j,\mu}, \quad j \ge 0.$$

It is well known that one can write

$$\mathbb{D}_j = \mathbb{E}_{j+1} - \mathbb{E}_j,$$

where $\mathbb{E}_j$ is the conditional expectation operator defined by

$$\mathbb{E}_j f(x) = \sum_{\mu\in\mathbb{Z}} \mathbb{1}_{I_{j,\mu}}(x)\, 2^j \int_{I_{j,\mu}} f(y)dy\,, \tag{10.5}$$

associated with the dyadic intervals $I_{j,\mu} = [\mu 2^{-j}, (\mu+1)2^{-j})$, $\mu \in \mathbb{Z}$.

The uniform boundedness of the operators $\mathbb{E}_N$ in $F^s_{p,q}$ (and $B^s_{p,q}$) has been thoroughly studied in the papers [3–5]. In particular, it is shown in those papers that $\mathscr{H}$ is a Schauder basis of $F^s_{p,q}$ (with respect to natural enumerations) if and only if

$$\sup_{N\ge0} \|\mathbb{E}_N\|_{F^s_{p,q}\to F^s_{p,q}} < \infty \quad \text{and} \quad \text{span}\,\mathscr{H} \text{ is dense in} F^s_{p,q},$$

and this in turn is equivalent to (s, p, q) belonging to the ranges in (10.3) and (10.4). In those cases, an elementary summation by parts argument and the σ-triangle inequality, with $\sigma = \min\{1, p, q\}$, imply that

$$\|T_m f\|_{F^s_{p,q}} \lesssim \|m\|_{\ell^\infty} + \Big[\sum_{j=1}^{\infty} |m(j) - m(j-1)|^\sigma\Big]^{\frac{1}{\sigma}}, \tag{10.6}$$

for all $f \in \text{span}\,\mathscr{H}$ with $\|f\|_{F^s_{p,q}} \le 1$.

We shall next formulate a stronger multiplier result which involves the Wiener space notion of sequences of bounded u-variation. We recall how these are defined. If $u \ge 1$, we let $\mathcal{V}_u(m)$ be the u-variation of the sequence $\{m(j)\}_{j\ge0}$, defined by

$$\mathcal{V}_u(m) = \sup\Big(\sum_{n=1}^{N} |m(j_n) - m(j_{n-1})|^u\Big)^{1/u}$$

with the supremum taken over all finite strings of numbers $\{j_0, \dots, j_N\}$ satisfying $j_{n-1} < j_n$ for $1 \le n \le N$, and $j_n \in \mathbb{N} \cup \{0\}$. Note that if $u = 1$ we simply have

$$\mathcal{V}_1(m) = \sum_{j=1}^{\infty} |m(j) - m(j-1)|.$$

We denote by V_u the space of all $m : \mathbb{N} \cup \{0\} \to \mathbb{C}$ for which

$$\|m\|_{V_u} := \|m\|_\infty + \mathcal{V}_u(m) < \infty.$$

In particular, if $1 \le u_1 \le u_2 < \infty$, it holds

$$V_1 \hookrightarrow V_{u_1} \hookrightarrow V_{u_2} \hookrightarrow \ell^\infty,$$

As an example, observe that $m_\alpha(n) = 1/(n+1)^\alpha$ belongs to V_1 for all $\alpha > 0$, while the alternate sequence $(-1)^n m_\alpha(n)$ belongs to V_u iff $\alpha > 1/u$.

We wish to find, in the region of exponents (s, p, q) where $\mathscr{H}$ is a conditional basis of $F^s_{p,q}$, the largest possible u for which $m \in V_u$ implies the boundedness of the operator T_m in $F^s_{p,q}$. The examples given in [12], based on multipliers taking the values 0 and 1 (suitable characteristic functions of finite sets of integers), show that for $1/u < s - 1/q$ there are $m \in V_u$ such that the corresponding operators T_m are unbounded on $F^s_{p,q}$; see also Sect. 10.3.3 below. Our main result in this note shows that, in the case $1 < p, q < \infty$, boundedness holds in the complementary range, except perhaps at the endpoint.

Theorem 10.1 *Let $1 < p < q < \infty$, $1/q \le s < 1/p$ and $u \ge 1$. Then*

$$\|T_m\|_{F^s_{p,q} \to F^s_{p,q}} + \|T_m\|_{F^{-s}_{p',q'} \to F^{-s}_{p',q'}} \le C\|m\|_{V_u}, \quad \textit{if} \quad \frac{1}{u} > s - \frac{1}{q}.$$

Remark 10.1 The appearance of the variation norms is inspired by a result of Coifman, Rubio de Francia and Semmes [2] on Fourier multipliers (which is based on the square function result of Rubio de Francia [8], see also [7, 14]). However the variation spaces come up in quite different ways in [2] where the variation norm is taken over dyadic intervals $[2^j, 2^{j+1})$, with a bound uniformly in j. This has no analogue in our situation as for each interval $[2^j, 2^{j+1})$ there is only one Haar frequency; instead our conditions involve the variation norms in the parameter j.

10.2 Subspaces of V_u

As in [2], in order to analyze functions in V_u it is convenient to consider certain subspaces R_u of V_u built on convex combinations of characteristic functions of unions of disjoint intervals. This is sketched in [2], but for the convenience of the reader we give a detailed exposition in the setting of variation spaces for functions on the integers.

For $1 \le u < \infty$, let r_u be the class of functions $g : \mathbb{N}_0 \to \mathbb{C}$ which are of the form

$$g = \sum_\nu a_\nu \chi_{I_\nu}, \quad \text{with} \quad (\sum_\nu |a_\nu|^u)^{1/u} \le 1,$$

where the I_ν are mutually disjoint intervals. Then R_u is the space of all sequences of the form

$$m = \sum_l c_l g_l, \quad \text{with} \quad g_l \in r_u \quad \text{and} \quad \sum |c_l| < \infty. \tag{10.7}$$

The norm $\|m\|_{R_u}$ is defined as the infimum of $\sum_l |c_l|$ over all representations as in (10.7). These definitions (for functions on the real line) can be found in [2]. The following result is a discrete analogue of [2, Lemme 2], whose proof is sketched for completeness.

Proposition 10.1 *For $\varepsilon > 0$, and $1 \le u < \infty$ we have*

$$R_u \subset V_u \subset R_{u+\varepsilon} \tag{10.8}$$

with continuous embedding.

Proof Consider first $g \in r_u$, with $g = \sum_\nu a_\nu \chi_{I_\nu}$. It is straightforward to see that $\mathcal{V}_u(g) \le 2\|a\|_{\ell^u}$, thus $R_u \subset V_u$.

For the second inclusion assume that $f \in V_u$, with

$$\mathcal{V}_u(f) = 1,$$

for some $u < \infty$. This implies that $\lim_{n\to\infty} f(n)$ exists and is finite.

Let $w(0) = 0$, and for $n \ge 1$, let $w(n)$ be the u-th power of the u-variation of f over $[0, n]$, that is

$$w(n) = \sup_{0 \le n_0 < n_1 < \cdots < n_N \le n} \sum_{i=1}^{N} |f(n_i) - f(n_{i-1})|^u.$$

Clearly w is positive, increasing and $\mathcal{V}_u(f) = \lim_{n\to\infty}[w(n)]^{1/u}$ so that w takes values in $[0, 1]$. There are two situations (i) $w(n) < 1$ for all $n \in \mathbb{N}$, and (ii) $w(n) = 1$ for $n \ge N_0$ (and some N_0). In what follows we assume (i) and omit the minor modification for (ii) (in the second case one works with finite sequences instead of infinite sequences).

As stated in [1, Theorem 2] and used in [2] one can express

$$f = \rho \circ w, \quad \text{where } \rho \in C^{1/u}[0, 1].$$

To verify this we may choose a strictly increasing sequence of nonnegative integers $\{j_n\}_{n=0}^{\infty}$ so that $j_0 = 0$ and

$$w(j_n) < w(j_{n+1}), \quad w(j_n) = w(k) \quad \text{for } j_n \le k < j_{n+1}.$$

This implies $f(k) = f(j_n)$ for $j_n \le k < j_{n+1}$. We now define a piecewise linear function $\rho(t)$, $0 \le t < 1$, as follows

$$\rho(t) = f(j_n) + \frac{f(j_{n+1}) - f(j_n)}{w(j_{n+1}) - w(j_n)}(t - w(j_n)), \quad w(j_n) \le t < w(j_{n+1}).$$

So $\rho(0) = f(0)$, and if we let $\rho(1) = \lim_{n\to\infty} f(n)$, then ρ is continuous in [0, 1]. Observe also that $\rho \circ w = f$, since for $j_n \le k < j_{n+1}$ we have

$$\rho(w(k)) = \rho(w(j_n)) = f(j_n) = f(k).$$

We now show the Hölder condition

$$|\rho(t) - \rho(t')| \le 3|t - t'|^{1/u}, \quad 0 \le t' < t \le 1. \tag{10.9}$$

By the definitions of ρ and w we have for $n' < n$

$$|\rho(w(j_n)) - \rho(w(j_{n'}))| = |f(j_n) - f(j_{n'})| \le (w(j_n) - w(j_{n'}))^{1/u}. \tag{10.10a}$$

For $w(j_n) \le t < w(j_{n+1})$

$$|\rho(t) - \rho(w(j_n))| = \left|\frac{f(j_{n+1}) - f(j_n)}{w(j_{n+1}) - w(j_n)}\right||t - w(j_n)| \tag{10.10b}$$

$$\le \left|w(j_{n+1}) - w(j_n)\right|^{-1+\frac{1}{u}} |t - w(j_n)| \le |t - w(j_n)|^{1/u},$$

since $u \ge 1$. Similarly

$$|\rho(w(j_{n+1})) - \rho(t)| = \left|\frac{f(j_{n+1}) - f(j_n)}{w(j_{n+1}) - w(j_n)}\right||w(j_{n+1}) - t| \tag{10.10c}$$

$$\le \left|w(j_{n+1}) - w(j_n)\right|^{-1+\frac{1}{u}} |w(j_{n+1}) - t| \le |w(j_{n+1}) - t|^{1/u}.$$

Combining the three cases we obtain (10.9) for all $t, t' \in [0, 1)$, and by continuity the result also holds true on [0, 1].

We now use the expansion of ρ in terms of the Haar system in [0, 1], that is $\{\mathbb{1}_{[0,1)}, h_{j,\mu}\}$ with $j \geq 0$ and $0 \leq \mu < 2^j$. Here

$$h_{j,\mu} = \mathbb{1}_{I^{\text{left}}_{j,\mu}} - \mathbb{1}_{I^{\text{right}}_{j,\mu}}$$

with $I^{\text{left}}_{j,\mu}$ and $I^{\text{right}}_{j,\mu}$ the left and right halves of $I_{j,\mu} = [2^{-j}\mu, 2^{-j}(\mu+1))$. Then

$$\rho(t) - \int_0^1 \rho(s)ds = \sum_{j=0}^{\infty} \rho_j(t)$$

where

$$\rho_j(t) = \sum_{\mu=0}^{2^j-1} 2^j \langle h_{j,\mu}, \rho \rangle h_{j,\mu}(t) = \rho_{j,1}(t) - \rho_{j,2}(t)$$

with

$$\rho_{j,1}(t) = \sum_{\mu=0}^{2^j-1} 2^j \langle h_{j,\mu}, \rho \rangle \mathbb{1}_{I^{\text{left}}_{j,\mu}}(t), \quad \text{and} \quad \rho_{j,2}(t) = \sum_{\mu=0}^{2^j-1} 2^j \langle h_{j,\mu}, \rho \rangle \mathbb{1}_{I^{\text{right}}_{j,\mu}}(t).$$

Now

$$|2^j \langle h_{j,\mu}, \rho \rangle| = 2^j \left| \int h_{j,\mu}(t) \big[\rho(t) - \rho\big(2^{-j}(\mu + \tfrac{1}{2})\big)\big] dt \right| \leq 3 \cdot 2^{-1/u} 2^{-j/u},$$

by (10.9). Thus, if $\varepsilon > 0$ we have

$$\Big(\sum_{\mu=0}^{2^j-1} |2^j \langle h_{j,\mu}, \rho \rangle|^{u+\varepsilon} \Big)^{\frac{1}{u+\varepsilon}} \leq 3 \cdot 2^{-\frac{1}{u}} 2^{-\frac{j}{u}} 2^{\frac{j}{u+\varepsilon}} =: c_{j,\varepsilon}.$$

Since w is increasing it is clear that the functions

$$n \mapsto \mathbb{1}_{I^{\text{left}}_{j,\mu}}\big(w(n)\big), \quad n \mapsto \mathbb{1}_{I^{\text{right}}_{j,\mu}}\big(w(n)\big)$$

are characteristic functions of intervals restricted to the integers. For fixed j these intervals are also mutually disjoint, so we see that

$$g_{j,\varepsilon} := \frac{1}{2c_{j,\varepsilon}} \big(\rho_j \circ w\big) \in r_{u+\varepsilon}.$$

Since $C_\varepsilon := 2\sum_{j\geq 0} |c_{j,\varepsilon}| < \infty$, it then follows that

$$f = \rho \circ w = \int_0^1 \rho + \sum_{j=0}^{\infty} 2c_{j,\varepsilon}\, g_{j,\varepsilon} \in R_{u+\varepsilon}, \quad \text{with } \|f\|_{R_{u+\varepsilon}} \leq 1 + C_\varepsilon.$$

Remark 10.2 Observe that the previous proof actually shows that, if $m \in V_u$ and $\varepsilon > 0$, then one can write $m = \sum_{j=0}^{\infty} c_j m_j$ with $m_j \in r_{u+\varepsilon}$ and $\sum_{j=1}^{\infty} |c_j|^\sigma < \infty$, for all $\sigma > 0$.

10.3 The Proof of Theorem 10.1

We shall actually prove a stronger result than Theorem 10.1, which provides optimal boundedness (up to endpoints) in a slightly larger region of indices. Given a fixed $q > 1$, we denote by $\mathbf{T}_q$ the open triangle in the plane $(1/p, s)$ with vertices $(1, 1)$, $(1/q, 1/q)$ and $(1 + 1/q, 1/q)$; see Fig. 10.2.

For this region we give the following result, which includes Theorem 10.1 as a special case.

Theorem 10.2 *Let* $1 < q < \infty$ *and* $(1/p, s) \in \mathbf{T}_q$. *Then, for all* $u \geq 1$ *such that* $1/u > s - 1/q$, *it holds*

$$\|T_m f\|_{F^s_{p,q} \to F^s_{p,q}} \leq c\, \|m\|_{V_u}, \quad m \in V_u. \tag{10.11}$$

Moreover, a necessary condition for (10.11) *to hold for all such* m *is that* $1/u \geq s - 1/q$.

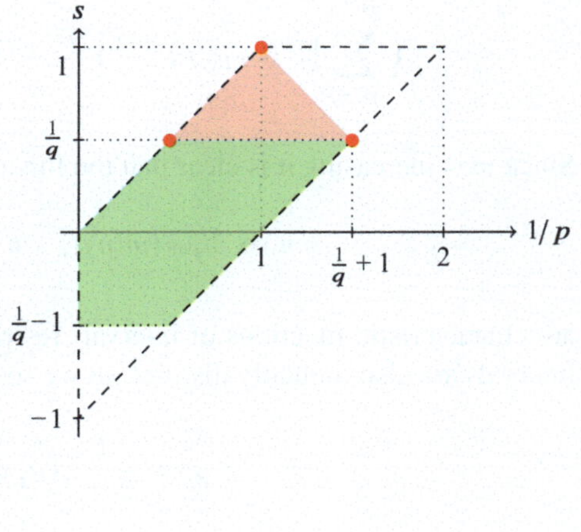

Fig. 10.2 In red, the region $\mathbf{T}_q$, for $q > 1$

In view of Proposition 10.1, we shall first consider sequences from the class r_u, that is multipliers m of the form

$$m[a, \mathcal{I}] = \sum_{\nu} a_\nu \mathbb{1}_{I_\nu}, \tag{10.12}$$

where $\mathcal{I} = \{I_\nu\}$ is a family of disjoint intervals and $a = \{a_\nu\}$ a sequence in ℓ^u. For these multipliers we have the following result.

Proposition 10.2 *Let $1 < q < \infty$ and $(1/p, s) \in \mathbf{T}_q$. Then, for all $u \geq 1$ such that $1/u > s - 1/q$, there exists $c = c(p, q, s, u) > 0$ such that*

$$\|T_{m[a,\mathcal{I}]} f\|_{F^s_{p,q}} \leq c\, \|a\|_{\ell^u} \|f\|_{F^s_{p,q}}, \quad \forall\, f \in F^s_{p,q}(\mathbb{R}),\ a \in \ell^u, \tag{10.13}$$

for every multiplier $m[a, \mathcal{I}]$ defined as in (10.12). *Moreover, a necessary condition for* (10.13) *to hold for all such $m[a, \mathcal{I}]$ is that $1/u \geq s - 1/q$.*

Remark 10.3 We emphasize that the constant c in (10.13) does not depend on the family of disjoint intervals $\mathcal{I} = \{I_\nu\}$.

In the next subsections we shall prove Proposition 10.2. For the sufficiency part we shall use complex interpolation applied to the bilinear operator

$$(a, f) \longmapsto \mathcal{T}[a, f] := T_{m[a,\mathcal{I}]} f.$$

For simplicity we shall remove the dependence on $\mathcal{I}$ in the subsequent notation, as it will be clear from the proofs that the involved constants do not depend on it.

10.3.1 Interpolation with Varying q and p = 1

We first prove an inequality for $p = 1$ and a fixed s near 1, namely

$$\|\mathcal{T}[a, f]\|_{F^s_{1,q}} \lesssim \|a\|_{\ell^u} \|f\|_{F^s_{1,q}}, \quad s < 1, \quad 1/u > 1 - 1/q. \tag{10.14}$$

Since the Haar system is an unconditional basis on $F^s_{1,1} = B^s_{1,1}$, $0 < s < 1$ (see Theorem 2.9 in [17]) we have

$$\|\mathcal{T}[a, f]\|_{F^s_{1,1}} \lesssim \|a\|_{\ell^\infty} \|f\|_{F^s_{1,1}}, \quad 0 < s < 1.$$

Next, the uniform boundedness of the operators $\mathbb{E}_N$ in F^s_{1,q_1} (see [3, Corollary 1.3]) and the trivial estimate in (10.6) (with $\sigma = 1$) imply that for any $q_1 \in (q, \infty)$

$$\|\mathcal{T}[a, f]\|_{F^s_{1,q_1}} \lesssim \|a\|_1 \|f\|_{F^s_{1,q_1}}, \quad 0 < s < 1.$$

By complex interpolation [6] we then obtain (10.14) for $1/u = (1-1/q)/(1-1/q_1)$, which after choosing q_1 large enough implies (10.14) whenever $1/u > 1 - 1/q$.

10.3.2 Interpolation with Fixed q

Let $q \in (1, \infty)$ be fixed, and let $(1/p, s) \in \mathbf{T}_q$. We shall prove (10.13) by interpolating sufficiently close to the upper vertex $(1, 1)$ and the lower segment $(1/p_1, 1/q)$ of $\mathbf{T}_q$ (Fig. 10.3).

Let $P = (1/p, s) \in \mathbf{T}_q$ and $\varepsilon_1 > \varepsilon_0 > 0$ be sufficiently small, to be chosen. Let $P_0 = (1, s_0)$ with $s_0 = 1 - \varepsilon_0$. Draw a line through P_0 and P, and let $P_1 = (1/p_1, s_1)$ be the intersection with the horizontal line $s_1 = 1/q - \varepsilon_1$. That is,

$$P = (1-\theta) \cdot P_0 + \theta \cdot P_1, \quad \text{with} \quad \theta = \frac{s_0 - s}{s_0 - s_1}.$$

Choosing $\varepsilon_0, \varepsilon_1$ sufficiently small we can guarantee that $s_1 < s < s_0$ (and hence $\theta \in (0, 1)$), and that P_1 lies in the green region. Next, take

$$\frac{1}{u_0} := 1 - \frac{1}{q} + \varepsilon_1 - \varepsilon_0 > 1 - \frac{1}{q}.$$

From the previous step and the unconditional basis property we have

$$\mathcal{T} : \ell^{u_0} \times F^{s_0}_{1,q} \to F^{s_0}_{1,q} \quad \text{and} \quad \mathcal{T} : \ell^{\infty} \times F^{s_1}_{p_1,q} \to F^{s_1}_{p_1,q}.$$

Using complex interpolation this yields

$$\|\mathcal{T}[a, f]\|_{F^s_{p,q}} \lesssim \|a\|_{\ell^u} \|f\|_{F^s_{p,q}}, \tag{10.15}$$

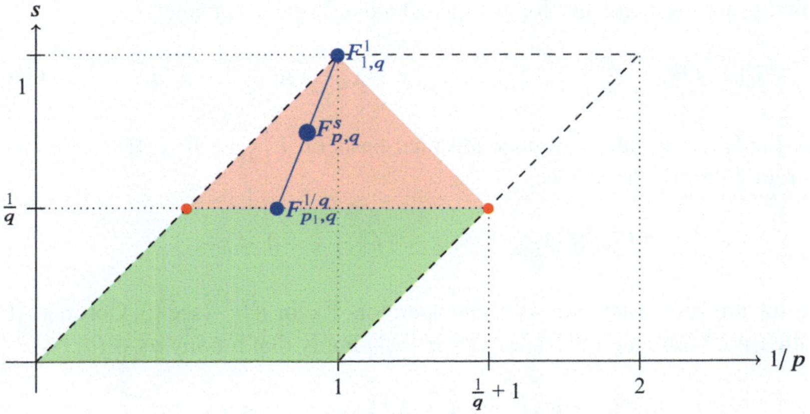

Fig. 10.3 Interpolation strategy for points $(1/p, s) \in \mathbf{T}_q$

with

$$\frac{1}{u} = \frac{1-\theta}{u_0} + \frac{\theta}{\infty} = \frac{s-s_1}{s_0-s_1} \cdot \frac{1}{u_0}.$$

Since the above definitions give $s_0 - s_1 = 1/u_0$, then we have

$$\frac{1}{u} = s - s_1 = s - \frac{1}{q} + \varepsilon_1.$$

Letting $\varepsilon_1 \searrow 0$ we deduce the validity of (10.15) whenever $\frac{1}{u} > s - \frac{1}{q}$. This completes the proof of the sufficient condition in Proposition 10.2.

10.3.3 Necessary Condition

Suppose first that $1 < p < q$ with $1/q < s < 1/p$. Then, the example constructed in [12, §5] gives a multiplier of the form $m = \mathbb{1}_E$, so that card $E = 2^N$ (with the elements in E being N-separated), and with the property that

$$\|T_m\|_{F^s_{p,q} \to F^s_{p,q}} \gtrsim 2^{N(s-\frac{1}{q})}. \tag{10.16}$$

Since we can write m in the form (10.12) (with $I_\nu = \{\nu\}$ and $a_\nu = 1$, for $\nu \in E$), then the validity of (10.13) will imply that

$$2^{N(s-\frac{1}{q})} \lesssim \|T_m\|_{F^s_{p,q} \to F^s_{p,q}} \lesssim \|a\|_{\ell^u} = 2^{N/u}.$$

Thus, we must necessarily have $1/u \geq s - 1/q$.

Arguing by interpolation as in [12, §7] one can show that (10.16) (with an ε loss) continues to hold for all $(1/p, s)$ with

$$\max\{1/q, 1/p - 1\} < s < \min\{1/p, 1\} \tag{10.17}$$

which is a larger region than $\mathbf{T}_q$; see Fig. 10.2.

To be more precise, let $P_1 = (1/p, s)$ belong to the open quadrilateral defined by (10.17), where we assume $p \leq 1$. We shall interpolate close to the points shown in Fig. 10.4.

Namely, given $\varepsilon > 0$, let $P_0 = (\frac{1}{q}, \frac{1}{q} - \varepsilon)$. Draw a segment from P_0 to P_1, and consider the convex combination of P_0 and P_1 with first coordinate $(1+\varepsilon)^{-1}$; i.e let $\theta \in (0, 1)$ and s_θ be such that

$$\left(\tfrac{1}{1+\varepsilon}, s_\theta\right) = (1-\theta)\left(\tfrac{1}{q}, \tfrac{1}{q} - \varepsilon\right) + \theta\left(\tfrac{1}{p}, s\right). \tag{10.18}$$

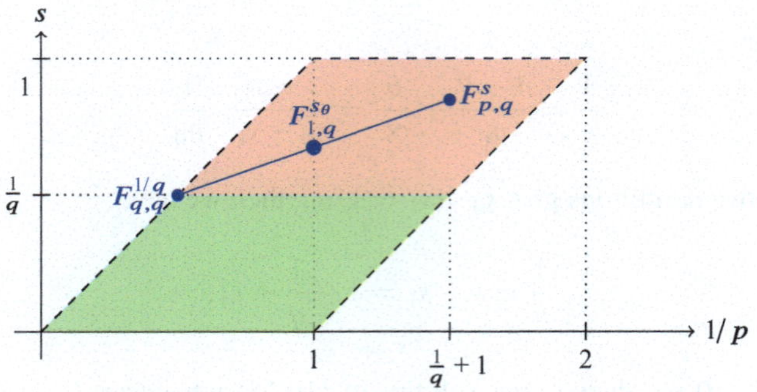

Fig. 10.4 Interpolation method for the necessary conditions

Then, by complex interpolation [6] we have

$$\|T_m\|_{F^{s_\theta}_{1+\varepsilon,q}} \lesssim \|T_m\|^{1-\theta}_{F^{\frac{1}{q}-\varepsilon}_{q,q}} \|T_m\|^{\theta}_{F^{s}_{p,q}}$$

By unconditionality, $\|T_m\|_{F^{\frac{1}{q}-\varepsilon}_{q,q}} \lesssim 1$, so we arrive at

$$\|T_m\|_{F^{s}_{p,q}} \gtrsim \|T_m\|^{1/\theta}_{F^{s_\theta}_{1+\varepsilon,q}} \gtrsim 2^{\frac{N}{\theta}(s_\theta-\frac{1}{q})},$$

the last bound due to (10.16). Now, solving for s_θ in (10.18) we see that

$$s_\theta - \frac{1}{q} = \big(s - \frac{1}{q}\big)\theta - (1-\theta)\varepsilon.$$

Thus,

$$\|T_m\|_{F^{s}_{p,q}} \gtrsim 2^{N\,[(s-\frac{1}{q})-\frac{1-\theta}{\theta}\,\varepsilon]}.$$

So, if (10.13) was true, arguing as above we would arrive at

$$\frac{1}{u} \geq (s - \frac{1}{q}) - \frac{1-\theta}{\theta}\,\varepsilon,$$

which letting $\varepsilon \searrow 0$ leads to $\frac{1}{u} \geq s - \frac{1}{q}$.

10.3.4 Conclusion of the Proof of Theorem 10.2

Let $q > 1$ and let $(1/p, s) \in \mathbf{T}_q$ be fixed. Let $1/u > s - 1/q$ and $m \in V_u$ with $u \geq 1$. Then, for some $u_1 > u$ we also have $1/u_1 > s - 1/q$. By Remark 10.2 we can write $m = \sum_{j=0}^{\infty} c_j m_j$ with $m_j \in r_{u_1}$ and $\sum_{j=0}^{\infty} |c_j|^\sigma \lesssim \|m\|_{V_u}^\sigma$, with $\sigma = \min\{1, p\}$. Then, using the σ-triangle inequality, we have

$$\|T_m f\|_{F^s_{p,q}}^\sigma \leq \sum_{j=0}^{\infty} |c_j|^\sigma \, \|T_{m_j} f\|_{F^s_{p,q}}^\sigma, \quad f \in F^s_{p,q}.$$

By Proposition 10.2, $\|T_{m_j} f\|_{F^s_{p,q}} \lesssim \|f\|_{F^s_{p,q}}$, for all $j \geq 0$, so we conclude that

$$\|T_m\|_{F^s_{p,q} \to F^s_{p,q}} \lesssim \|m\|_{V_u}.$$

Remark 10.4 When $p > 1$, the assertion

$$\|T_m\|_{F^{-s}_{p',q'} \to F^{-s}_{p',q'}} \lesssim \|m\|_{V_u}$$

stated in Theorem 10.1 follows from (10.11) by duality. So, when $q > 1$, the condition on V_u is optimal (up to endpoints) also in the lower triangle on the left of Fig. 10.1.

Remark 10.5 When $1/2 < p \leq 1$, we did not state any result for the right upper triangle in Fig. 10.2. It is also possible to obtain, by complex interpolation, a sufficient condition for multipliers of the form $m[a, \mathcal{I}]$ in terms of $\|a\|_{\ell^u}$, although in this range the value of u will no longer match the necessary condition from Sect. 10.3.3.

Remark 10.6 When $1/2 < q \leq 1$, one can also prove by interpolation, for multipliers of the form $m[a, \mathcal{I}]$, that

$$\|a\|_{\ell^u} < \infty, \qquad \frac{1}{u} > \frac{1}{q} - 1 - s,$$

is a sufficient condition in the open triangle with vertices $(0, -1)$, $(1/q, 1/q - 1)$ and $(1/q - 1, 1/q - 1)$; see Fig. 10.5. This matches the necessary condition from the examples in [12, §5] (except for the endpoint). In the remaining part of the figure, however, the sufficient condition obtained by interpolation will be weaker than this one.

Remark 10.7 It may be interesting to note that even for the special case of Sobolev spaces $H^s_p = F^s_{p,2}$, $1 < p < 2$, $1/2 \leq s < 1/p$ and $u < \frac{2}{2s-1}$ the use of Triebel-Lizorkin spaces $F^s_{p,q}$ with $q \neq 2$ is crucial. Such interpolation arguments were used for multiplier transformations in other contexts to establish endpoint results on

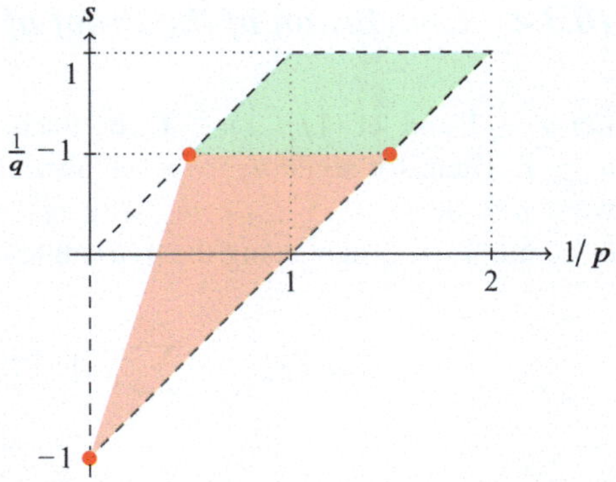

Fig. 10.5 Parameter domain for the cases $1/2 < q < 1$

Lorentz spaces $L_{p,2}$, see [9, 10] for basic versions, and [11, 15] for more advanced versions.

References

1. M. Bruneau, *Variation totale d'une fonction* (French). Lecture Notes in Mathematics, vol. 413 (Springer, Berlin, 1974)
2. R. Coifman, J.L. Rubio de Francia, S. Semmes, Multiplicateurs de Fourier de $L^p(\mathbb{R})$ et estimations quadratiques. C. R. Acad. Sci. Paris Sér. I Math. **306**(8), 351–354 (1988)
3. G. Garrigós, A. Seeger, T. Ullrich, The Haar system as a Schauder basis in spaces of Hardy-Sobolev type. J. Fourier Anal. Appl. **24**(5), 1319–1339 (2018)
4. G. Garrigós, A. Seeger, T. Ullrich, Basis properties of the Haar system in limiting Besov spaces, in *Geometric Aspects of Harmonic Analysis*, ed. by P. Ciatti, A. Martini. Springer Indam Series, vol. 45 (2021), pp 361–424
5. G. Garrigós, A. Seeger, T. Ullrich. The Haar system in Triebel-Lizorkin spaces: Endpoint results. J. Geom. Anal. **31**(9), 9045–9089 (2021)
6. N. Kalton, S. Mayboroda, M. Mitrea, Interpolation of Hardy-Sobolev-Besov-Triebel-Lizorkin spaces and applications to problems in partial differential equations, in *Interpolation Theory and Applications*. Contemp. Math., vol. 445 (American Mathematical Society, 2007), pp. 121–177
7. M.T. Lacey, *Issues Related to Rubio de Francia's Littlewood-Paley Inequality.* New York Jour. Math. Monographs, vol. 2 (State University of New York, University at Albany, Albany, NY, 2007), 36 pp.
8. J.L. Rubio de Francia, A Littlewood-Paley inequality for arbitrary intervals. Rev. Mat. Iberoamericana **1**(2), 1–14 (1985)
9. A. Seeger, A limit case of the Hörmander multiplier theorem. Monatsh. Math. **105**(2), 151–160 (1988)
10. A. Seeger, Estimates near L^1 for Fourier multipliers and maximal functions. Arch. Math. (Basel) **53**(2), 188–193 (1989)
11. A. Seeger, T. Tao, Sharp Lorentz space estimates for rough operators. Math. Ann. **320**(2), 381–415 (2001)
12. A. Seeger, T. Ullrich, Haar projection numbers and failure of unconditional convergence in Sobolev spaces. Math. Z. **285**, 91–119 (2017)
13. A. Seeger, T. Ullrich, Lower bounds for Haar projections: deterministic examples. Constr. Appr. **46**, 227–242 (2017)

14. P. Sjölin, A note on Littlewood-Paley decompositions with arbitrary intervals. J. Approx. Theory **48**(3), 328–334 (1986)
15. T. Tao, J. Wright, Endpoint multiplier theorems of Marcinkiewicz type. Rev. Mat. Iberoam. **17**(3), 521–558 (2001)
16. H. Triebel, *Theory of Function Spaces* (Birkhäuser Verlag, Basel, 1983)
17. H. Triebel, *Bases in Function Spaces, Sampling, Discrepancy, Numerical Integration*. EMS Tracts in Mathematics, vol. 11 (European Mathematical Society (EMS), Zürich, 2010)

Chapter 11
On Frames of Smooth, Compactly-Supported Wave Packets Adapted to Tilings of Frequency Space

Philip T. Gressman

Abstract We establish a broad notion of admissible tilings of frequency space which admit associated wave packet frames with elements which are smooth and compactly supported. The framework is designed to allow for tile geometries which are minimally constrained by the need to accommodate Schwartz tails on the Fourier side and goes beyond the usual scale of geometries ranging from Gabor to wavelet-type decompositions (Feichtinger and Fornasier, Ann Mat Pura Appl (4) 185(1), 105–131, 2006). The approach builds on techniques of Hernández et al. (J Geom Anal 12(4), 615–662, 2002) and Labate et al. (An approach to the study of wave packet systems. In Wavelets, frames and operator theory, pp. 215–235, 2004) as well as a classical result of Ingham (J Lond Math Soc 9(1), 29–32, 1934) characterizing the best-possible Fourier decay for functions of compact support.

11.1 Introduction

The article [6] of the author and E. Urheim establishes stable norm decay inequalities for multilinear oscillatory integral functionals through the use of certain continuous frame decompositions of $L^2(\mathbb{R})$ which behave like Gabor systems at low frequencies and so-called "wave atoms" at higher frequencies. Here the adjective "continuous" refers to decompositions which involve integrals over all translations of some countable family of window functions, where "discrete" would instead refer to summation over discrete lattices of translations. An issue of central concern there was the need for uniform control of frame constants independent of the frequency at

The author was partially supported by NSF grant DMS-2054602. This work is dedicated to the memory of Guido Weiss and his generous and engaging way of sharing mathematics with students.

P. T. Gressman (✉)
University of Pennsylvania, Department of Mathematics, Philadelphia, PA, USA
e-mail: gressman@math.upenn.edu

E. Hernández et al. (eds.), *The Mathematical Heritage of Guido Weiss*,
Applied and Numerical Harmonic Analysis,
https://doi.org/10.1007/978-3-031-76793-7_11

which the transition occurs. At the time, it was noted that the existence of analogous discrete frames with window functions that are smooth and compactly supported would likely have simplified the analysis and would also open interesting pathways to studying degenerate multilinear oscillatory integrals of a similar form.

The construction of continuous frames whose wave packets have compact *Fourier* support is essentially trivial and can be accomplished under extremely mild assumptions on the arrangement of frequency tiles (see, for example, the notion of bounded admissible partitions of unity as described in [5]). Unfortunately, the switch to compact *physical space* support introduces Schwartz tails on the frequency side which can, in extreme cases, have nontrivial effects. Passing from continuous to discrete frames introduces even more difficulties because the composition of analysis and synthesis operators for the frame is not translation-invariant and so a complete understanding of its properties does not follow from elementary Fourier analysis. The goal of the present article is to approach these difficulties directly for the purpose of establishing the existence of discrete frames of smooth, compactly-supported wave packets adapted to decompositions of frequency space which are as general as possible. A secondary but still important goal is to demonstrate existence of such discrete frames in a way that yields frame constants (and even choices of wave packets) which are uniform over very broad classes of decompositions so that, as was done for multilinear oscillatory integrals, it is possible to analyze functions or operators using parameter-dependent decompositions of frequency space while still proving inequalities that are independent of parameter.

Let $|\cdot|$ denote the ℓ^∞ norm on $\mathbb{R}^d$, i.e., $|(x_1,\ldots,x_d)| := \max\{|x_1|,\ldots,|x_d|\}$. Given any matrix $D \in \mathrm{GL}_d(\mathbb{R})$, we define a *tile* T to be a pair $(D,\xi) \in \mathrm{GL}_d(\mathbb{R}) \times \mathbb{R}^d$ and define its tile set $T_\square \subset \mathbb{R}^d$ to equal

$$T_\square := \left\{ \eta \in \mathbb{R}^d \;\middle|\; |D^{-1}(\eta - \xi)| \leq \frac{1}{2} \right\}.$$

Modulo only the finite symmetry group generated by permutation of coordinates and coordinate-wise reflection, the tile set $T_\square$ determines T (see Proposition 11.3), so it is only a small exaggeration to informally regard tiles themselves as subsets of $\mathbb{R}^d$. In particular, the notation $\mathbb{1}_T$ will always refer to the indicator function on $\mathbb{R}^d$ the tile set $T_\square$. The symbol $\mathcal{T}$ will denote the set of all tiles and $\dot{\mathcal{T}}$ indicates the set of all tiles centered at the origin, i.e., tiles of the form $(D,0)$. Given a tile T and a real number $t > 0$, the tile tT is defined to be the tile obtained by dilating T isotropically by a factor of t in each direction while preserving its center, i.e., if $T = (D,\xi)$, then $tT := (tD,\xi)$. Similarly, $\dot{T}$ is the translation of T which is centered at the origin, i.e., $\dot{T} = (D,0)$. Given any function φ on $\mathbb{R}^d$ and any tile $T = (D,\xi)$, we make the definitions

$$\varphi_{T,k}(x) := e^{2\pi i \xi\cdot x}\sqrt{|\det D|}\,\varphi(k + D^t x) \qquad \forall x, k \in \mathbb{R}^d, \tag{11.1}$$

$$\varphi^T(\xi') := \varphi(D^{-1}(\xi' - \xi)) \qquad \forall \xi' \in \mathbb{R}^d. \tag{11.2}$$

Here D^t is the usual transpose. Dilations of the first type $\varphi_{T,k}$ will primarily be applied to functions in physical space; dilations of the form φ^T are generally reserved for functions in frequency space. Here and throughout this work, the Fourier transform will be normalized so that

$$\widehat{f}(\xi) := \int e^{-2\pi i x\cdot\xi} f(x)dx$$

for any $f \in L^1(\mathbb{R}^d)$, and in particular, the notation $\widehat{\eta}^T$ is to be understood as $(\widehat{\eta})^T$.

The first theorem establishes that a discrete frame with smooth, compactly-supported window ψ adapted to an essentially arbitrary family of tiles $\mathcal{T}_*$ must exist whenever $\mathcal{T}_*$ is in some sense oversampling and admits a corresponding *continuous* frame adapted to $\mathcal{T}_*$ with compactly supported window g. The precise statement is as follows.

Theorem 11.1 (Discrete Frames from Continuous Frames) *Let $\mathcal{T}_*$ be any countable set of tiles in $\mathbb{R}^d$. Suppose that there exists a compactly supported $g \in L^2(\mathbb{R}^d)$, $A, B > 0$ and $\theta \in (0, 1)$ such that*

$$A \leq \sum_{T\in\mathcal{T}_*} |\widehat{g}^T(\xi)|^2 \mathbb{1}_{(1-\theta)T}(\xi) \leq \sum_{T\in\mathcal{T}_*} |\widehat{g}^T(\xi)|^2 \leq B \tag{11.3}$$

for almost every $\xi \in \mathbb{R}^d$. Then for all sufficiently small $\epsilon > 0$ (depending only on A, B, θ, and d), there exists a compactly supported $\psi \in C^\infty(\mathbb{R}^d)$ such that

$$(1-\epsilon)A||f||^2 \leq \sum_{\substack{T\in\mathcal{T}_*\\ k\in\mathbb{Z}^d}} |\langle f, \psi_{T,k}\rangle|^2 \leq (1+\epsilon)B||f||^2 \tag{11.4}$$

for all $f \in L^2(\mathbb{R}^d)$. This ψ may be taken to depend only on g, A, B, ϵ, θ, and d. In particular, it can be defined independently of the specific tiling $\mathcal{T}_*$.

To understand the significance of the hypotheses, it is helpful to recall the consequences of (11.4). Proposition 4.1 of Hernández et al. [9] establishes that any ψ satisfying (11.4) must also satisfy

$$\sum_{T\in\mathcal{T}_*} |\widehat{\psi}^T(\xi)|^2 \leq (1+\epsilon)B \text{ for a.e. } \xi \in \mathbb{R}^d. \tag{11.5}$$

An argument similar to the one establishing the upper bound (11.5) (testing the inequality on functions f whose Fourier support is concentrated on vanishingly small balls) also establishes that

$$\sum_{T\in\mathcal{T}_*} |\widehat{\psi}^T(\xi)|^2 \geq (1-\epsilon)A \text{ for a.e. } \xi \in \mathbb{R}^d.$$

Thus, modulo perturbations of the constants, the inequality (11.4) implies that (11.3) holds up to but not including the indicator function $\mathbb{1}_{(1-\theta)T}$. But Theorem 11.1 cannot hold without some additional restriction of this sort; even when $\theta = 0$, for example, the Balian-Low Theorem demonstrates that the analogue of Theorem 11.1 is false (see, for example, Theorem 3.1.d in [1] and the commentary immediately following that theorem; for any continuous ψ of compact support, its Zak transform is continuous and therefore always has a zero).

Theorem 11.1 can be used, for example, as a means to establish the existence of wavelet frames. Here the tiles would have the form $(2^j, 0)$ (for the one dimensional case) with j ranging over all integers. The tile sets have infinite overlap at the origin, but as far as the hypotheses (11.3) are concerned, this infinite overlap is easily managed by simply taking a function g with integral zero so that its Fourier transform vanishes at the origin. In fact, it is not especially difficult to see that *any* nontrivial choice of g which is compactly supported, at least C^1, and has integral zero will satisfy (11.3) for some nonzero values of A and B. Similar but more exotic tilings of frequency space adapted to the geometry of curves in the plane (for example, the usual decomposition of the neighborhood of a circle into boxes of size $2^{-j} \times 2^{-j/2}$ at distance roughly 2^{-j} to the circle) are possible as well provided that the Fourier transform of g vanishes on a suitably chosen parabola. This may be of some interest in the development of axiomatic approaches to decoupling theory (see Section 7.3 of Guth's survey [8] for some interesting observations suggesting the value that such an endeavor might have).

In some sense, the question as to whether a smooth, compactly supported window function exists for a particular tiling of frequency space comes down to a technical question about the effect of Schwartz tails, much like question of the effect of the indicator function $\mathbb{1}_{(1-\theta)T}$ in (11.3). This means in practice that some non-local criterion on tile geometry is necessary (non-local here meaning that one must compare tiles that are more than simply roughly adjacent or overlapping). A common approach when formulating such a criterion is to constrain the geometry of tiles to settings which are effectively spaces of homogeneous type and then to assume that the size of each tile is controlled by some slowly-growing function of its distance to the origin; see, for example, [2, 4, 12, 13]. Perhaps the most interesting example of tilings of this sort which is neither Gabor- nor wavelet-like is the case of so-called "wave atoms," in which frequency space is decomposed into isotropic balls of radius $\sim R^{1/2}$ at distance $\sim R$ to the origin for $R \geq 1$ (as is done for the high frequency decomposition in [6]).

What we seek to do here is to develop a notion of admissible tilings which is, in some sense, as broad as possible. This should mean, e.g., that it includes tiling geometries based on truly multiparameter tile geometries. While it must involve comparisons of tiles at some distance, the criterion used for comparison should also ideally be relatively easy to check (i.e., for any tile T_1, it should hopefully be necessary to compare its geometry to only finitely many other tiles T_2 even if there is no uniform bound on the number of comparisons as T_1 varies) and should not assume greater regularity than is necessary. These are the goals which motivate the second main theorem, Theorem 2, below.

To state the theorem, an auxiliary definition is in order. If $T_i := (D_i, \xi_i)$ are tiles for $i = 1, 2$, let us define

$$d(T_1, T_2) := \ln_2 \max\{|D_1^{-1} D_2|, |D_2^{-1} D_1|\},$$

where $|\cdot|$ here denotes the operator norm $\ell^\infty \to \ell^\infty$ and $\ln_2$ is the standard logarithm base 2. For obvious reasons, d is not literally a metric on tiles because it is independent of centers. However, it does become a genuine metric on the set of origin-centered tile sets (see Lemma 11.2). The choice of base 2 for logarithms is purely for convenience, as it means that tiles which differ in size by a factor of 2 have distance 1. Using this notion of distance, we now state the definition of *admissible tilings*, which are, roughly speaking, a decomposition of $\mathbb{R}^d$ into tile sets which are approximately disjoint and satisfy certain regularity properties guaranteeing that the shapes and sizes of tiles vary slowly at a local level. To make this definition, fix some constant $C > 1$, some $\theta \in (0, 1)$ and some nondecreasing function $K(t)$ on $[1, \infty)$ such that

$$K(1) \geq 1 \text{ and } \int_1^\infty \frac{\ln K(t)}{t^2} dt < \infty. \tag{11.6}$$

A collection of tiles $\mathcal{T}_* \subset \mathcal{T}$ will be called an *admissible tiling* of $\mathbb{R}^d$ with respect to (C, θ, K) when the following criteria hold:

1. (Covering) Almost every $\xi \in \mathbb{R}^d$ belongs to the tile set $(1 - \theta)T_\square$ for some $T \in \mathcal{T}_*$ and does not belong to $T_\square$ for more than C distinct tiles $T \in \mathcal{T}_*$.
2. (Regularity) There exists a function $\rho : \mathcal{T}_* \to [1, \infty)$ such that
   (a) If $T_1, T_2 \in \mathcal{T}_*$ and the tiles sets of T_2 and $\rho(T_1)T_1$ intersect, then $d(\dot{T}_1, \dot{T}_2) \leq C$. In other words, the proportionally enlarged tile set of $\rho(T_1)T_1$ only intersects tile sets $(T_2)_\square$ having similar shape and size as measured by the tile geometry metric.
   (b) For each $t \geq 1$, the set

   $$\left\{\dot{T}_\square \mid \rho(T) \leq t \text{ for some } T \in \mathcal{T}_*\right\},$$

   i.e., the set of origin-centered tile sets with the same shape as some $T \in \mathcal{T}_*$ having $\rho(T) \leq t$, can be covered by no more than $K(t)$ metric balls of radius C. Roughly speaking, this means the diversity of shapes and sizes of tiles grows subexponentially as a function of ρ.

While the definition of admissible tiling depends on the choice of C, θ, and $K(t)$, it should be noted that increasing C or $K(t)$ and decreasing θ preserves the property of admissibility without the need to change the function ρ.

The main theorem for admissible tilings is as follows.

Theorem 11.2 (Frames for Admissible Tilings) *For given $C > 1$, $\theta \in (0, 1)$, dimension $d \geq 1$, nondecreasing function $K : [1, \infty) \to [1, \infty)$ satisfying* (11.6), *and all $\epsilon > 0$ sufficiently small (depending on C, θ, d, and K) there exists a real function η on $\mathbb{R}^d$ which is C^∞ and compact support with the following properties:*

1. *This $\eta(x)$ is even as a function of each of each coordinate x_i and invariant under permutations of the x_i. In particular, for any tile T, this means that the set of translated functions $\{\eta_{T,k}\}_{k\in\mathbb{Z}^d}$ depends only on the tile set $T_\square$.*
2. *Suppose $\mathcal{T}_*$ is an admissible tiling of $\mathbb{R}^d$ with respect to (C, θ, K). Then*

$$(1-\epsilon)||f||^2 \le \sum_{\substack{T\in\mathcal{T}_*\\ k\in\mathbb{Z}^d}} \left|\langle f, \eta_{T,k}\rangle\right|^2 \le (C+\epsilon)||f||^2 \tag{11.7}$$

for all $f \in L^2(\mathbb{R}^d)$.

For a nontrivial example of a setting in which this theorem applies, one can take boxes in the Mikhlin-Hörmander tiling of $\mathbb{R}^d$, i.e.,

$$\mathcal{D} := \left\{ \prod_{i=1}^d [s_i 2^{j_1}, s_i 2^{j_i+1}] \mid j_1, \ldots, j_d \in \mathbb{Z},\ s_1, \ldots, s_d \in \{-1, 1\} \right\},$$

and generate an admissible tiling $\mathcal{T}_*$ by subdividing the tile $\prod_{i=1}^d [s_i 2^{j_1}, s_i 2^{j_i+1}]$ into approximately $\left[\ln_2\left(2+\max_{i=1,\ldots,d}|j_i|\right)\right]^{1+\delta}$ equal parts in each direction (for a total of approximately $\left[\ln_2\left(2+\max_{i=1,\ldots,d}|j_i|\right)\right]^{d(1+\delta)}$ congruent subtiles), dilating each box by a factor of $1+\delta_1$ for some small positive δ, and defining $\rho(T) := \left[\ln_2\left(2+\max_{i=1,\ldots,d}|j_i|\right)\right]^{1+\delta_2}$ for each tile $T \in \mathcal{T}_*$ which originally came from $\prod_{i=1}^d [s_i 2^{j_1}, s_i 2^{j_i+1}]$ (where again δ_2 is a small positive number). This logarithmic subdivision is in some sense necessitated by the fact that the single window function η promised by Theorem 2 generates a frame for *any* admissible tiling $\mathcal{T}_*$ with respect to (C, θ, K) and yields uniform frame bounds over all such $\mathcal{T}_*$. The extra smallness is necessary to compensate for the inability to formulate any sort of vanishing moment conditions which properly apply to all possible $\mathcal{T}_*$. The fact that only logarithmic-type subdivision is necessary is a consequence of a precise characterization of the maximal Schwartz tail decay which can be exhibited by smooth functions of compact support; without this additional analysis, exponentially finer subdivision is all that can be tolerated.

A second sample application of Theorem 2 which directly resolves the frame construction issue raised in [6] is described in the following proposition.

Proposition 11.1 *Fix any positive increasing function r on $[1, \infty)$ such that*

$$\liminf_{t\to\infty} \frac{\ln r(t)}{\ln\ln\ln t} > 1.$$

For all sufficiently small $c > 0$ (with threshold depending only on r and the dimension d) the following is true. Suppose that $\mathbb{R}^d$ is tiled by a family $\mathcal{B}$ of *nonoverlapping closed, isotropic boxes B each with side length $\ell(B) \geq 1$. If $\mathcal{B}$ has the property that any two adjacent boxes $B_1, B_2 \in \mathcal{B}$ (i.e., having common boundary points) also satisfy the comparability condition*

$$\max\{\ell(B_1), \ell(B_2)\} \leq \min\{\ell(B_1), \ell(B_2)\} \left[1 + \frac{1}{r(\max\{\ell(B_1), \ell(B_2)\})} \right], \tag{11.8}$$

then any collection $\mathcal{T}_*$ of tiles T whose tile sets are $(1+c)B$ for each $B \in \mathcal{B}$ must be an admissible tiling for some $(C, \theta, K(t), \rho)$ which depend only on r, d, and c.

The tilings $\mathcal{B}$ associated with the geometry in [6] all have the property that the boxes $B \in \mathcal{B}$ have integer side lengths and adjacent boxes have side lengths differing by at most 1, so taking $r(t) := \max\{1, t-1\}$ and applying the proposition and Theorem 2 provides for the existence of a single window function generating a frame for any possible geometry of this sort with constants that are independent of the particular tiling $\mathcal{B}$.

Proof of Proposition 11.1 First observe that one may, without loss of generality, replace r by any smaller function $\tilde{r}$ provided that it also satisfies the same hypotheses that r itself does. We will show that it is always possible to choose such an $\tilde{r}$ so that $\tilde{r}(tu) \leq u\tilde{r}(t)$ for all $t, u \geq 1$. To find this $\tilde{r}$, extend r to the domain $(0, \infty)$ by fixing $r(t) = r(1)$ for all $t \in (0, 1)$ and let $\tilde{r}(t) := \inf_{s \in (0,1]} s^{-1} r(st)$. Clearly $\tilde{r}(t) \leq r(t)$ for each $t \in (0, \infty)$ and $\tilde{r}$ is increasing because r is increasing. The function $\tilde{r}$ may be seen to satisfy $\tilde{r}(tu) \leq u\tilde{r}(t)$ for all $t > 0$ by simply comparing the infimum over $s \in (0, 1/u]$ of $s^{-1} r(stu)$ to the infimum over $s \in (0, 1]$ of $s^{-1} r(st)$. Because $r(t) \geq (\ln \ln t)^{1+\delta}$ for some $\delta > 0$ and all t greater than some t_0, it follows that when $t > t_0$,

$$\begin{aligned} \tilde{r}(t) &\geq \min \left\{ \inf_{t_0 t^{-1} < s \leq 1} s^{-1} (\ln \ln(st))^{1+\delta}, \inf_{s \in (0, t_0 t^{-1}]} s^{-1} r(1) \right\} \\ &= \min \left\{ \inf_{t_0 t^{-1} < s \leq 1} s^{-1} (\ln \ln(st))^{1+\delta}, t_0^{-1} r(1) t \right\}. \end{aligned}$$

Let $u_0 > 0$ be such that $u \ln u > (1+\delta)$ for all $u > u_0$. Differentiating the function $e^x (\ln(-x + \ln t))^{1+\delta}$ with respect to x shows that it is an increasing function on $x < -u_0 + \ln t$, which means that $s^{-1} (\ln \ln(st))^{1+\delta} \geq (\ln \ln t)^{1+\delta}$ provided $e^{u_0} t^{-1} < s \leq 1$. Assuming without loss of generality that $t_0 \geq e^{u_0}$,

$$\tilde{r}(t) \geq \min\{(\ln \ln t)^{1+\delta}, t_0^{-1} r(1) t\} = (\ln \ln t)^{1+\delta}$$

for all t sufficiently large. One may therefore assume that $r(tu) \leq ur(t)$ for all $t, u \geq 1$. Because r is increasing, $r(tu) \leq \max\{1, u\} r(t)$ for all $t \in [1, \infty)$ and $u \in (0, \infty)$.

Now let $B_1, \dots, B_{k+1}$ be any sequence of boxes in $\mathcal{B}$ such that B_j and B_{j+1} are adjacent or equal for each $j = 1, \dots, k$. For each $j = 1, \dots, k+1$, let

$$R_j := \max \left\{ \frac{\ell(B_j)}{\ell(B_1)}, \frac{\ell(B_1)}{\ell(B_j)} \right\} \geq 1.$$

It follows from the condition (11.8) and monotonicity of r that

$$\frac{R_{j+1}}{R_j} = \max \left\{ \frac{\ell(B_{j+1})}{\ell(B_1) R_j}, \frac{\ell(B_1)}{R_j \ell(B_{j+1})} \right\} \leq \max \left\{ \frac{\ell(B_{j+1})}{\ell(B_j)}, \frac{\ell(B_j)}{\ell(B_{j+1})} \right\}$$
$$\leq 1 + \frac{1}{r(\ell(B_{j+1}))}$$

for each $j = 1, \dots, k$. The inequality $r(ut) \leq \max\{1, u\} r(t)$ implies that

$$r(\ell(B_{j+1})) \geq \frac{r(\ell(B_1))}{\max\{1, \ell(B_1)/\ell(B_{j+1})\}} \geq \frac{r(\ell(B_1))}{R_{j+1}};$$

it follows that

$$\frac{R_{j+1}}{R_j} \leq 1 + \frac{R_{j+1}}{r(\ell(B_1))} \Rightarrow \frac{1}{R_{j+1}} \geq \frac{1}{R_j} - \frac{1}{r(\ell(B_1))}$$

for each $j = 1, \dots, k$. Since $R_1 = 1$, it follows that $R_{j+1} \leq (1 - j/r(\ell(B_1)))^{-1}$ for all j provided that provided that $k < r(\ell(B_1))$. In particular, if $k \leq r(\ell(B_1))/2$, then every box $B_1, \dots, B_{k+1}$ will have a side length which differs from $\ell(B_1)$ by at most a factor of 2. Because any two adjacent boxes in $\mathcal{B}$ have a ratio of sides bounded between $r(1)/(1 + r(1))$ and $(1 + r(1))/r(1)$, it follows that for any k_0, there is a constant C' depending only on k_0 and $r(1)$ such that any chain with $k \leq r(\ell(B_1))/2 + k_0$ have $(C')^{-1} \leq \ell(B_j)/\ell(B_1) \leq C'$ for all $j \leq k+1$. For definiteness, fix $k_0 := 2d + 1$.

The next objective is to show that when x is some point in $B_1 \in \mathcal{B}$ and $y \in \mathbb{R}^d$ satisfies $|x - y| \leq (2C')^{-1} \lceil r(\ell(B_1))/(2d) + 1 \rceil \ell(B_1)$, then $(C')^{-1} \leq \ell(B_1)/\ell(B') \leq C'$ for any box $B' \in \mathcal{B}$ containing y. The strategy is to build a finite sequence $x_1, \dots, x_{k+1}$ of points and a chain boxes $B_1, \dots, B_{k+1}$ of the sort just studied such that $x_1 := x$, $x_{k+1} := y$, and $x_j \in B_j$ for each $j = 1, \dots, k+1$. Here $k := d \lceil r(\ell(B_1))/(2d) + 1 \rceil \leq r(\ell(B_1))/2 + 2d$ is simply fixed depending on B_1. The points x_j will be selected to have the property that each x_j and x_{j+1} differ in at most one coordinate direction in such a way that $|x_j - x_{j+1}| \leq (2C')^{-1} \ell(B_1)$. Because $k + 1 \leq r(\ell(B_1))/2 + k_0$, once one shows that the points x_j belong to boxes B_j which are equal or adjacent for neighboring values of j, it will follow that y is contained in a box $B_{k+1} \in \mathcal{B}$ whose side length differs from $\ell(B_1)$ by at most a factor of C'. It will also follow that any box B_{k+2} adjacent to B_{k+1} has the same bounds on its size, which would mean that in fact that any box $B' \in \mathcal{B}$ containing y (if there happens to be more than one) must have the same bounds on its size.

To construct the sequence, let t_1 be the first coordinate of $y - x$, and let $\delta_1 := t_1 d/k$. By the size constraint on $y-x$, it must be the case that $|\delta_1| \leq (2C')^{-1}\ell(B_1)$. Because all boxes adjacent to B_1 have side length at least $(C')^{-1}\ell(B_1)$ (since $k \geq 1$), the path $t \mapsto x + te_1$ (where e_1 is the first standard coordinate vector) for $t \in [0, \delta_1]$ is too short to pass outside B_1, enter an adjacent box, and pass entirely through it, and exit it. Thus the point $x + \delta_1 e_1$ must reside in either B_1 or some box B_2 adjacent to B_1. Continuing on the same path $x + te_1$ up to $t_1 = \delta_1(k/d)$ in increments of size δ_1, we have constructed a sequence $x_1, \ldots, x_{1+k/d}$ such that each point x_j is contained in a box B_j so that B_j and B_{j+1} are either equal or adjacent for each j. The point $x_{1+k/d}$ agrees with y in the first coordinate direction and with x in each remaining direction. Repeating this process starting at the new point $x_{k/d+1}$ and moving in the second direction using k/d steps and so on gives the promised points and boxes and shows that every $B' \in \mathcal{B}$ which intersects $(1 + (C')^{-1}\lceil r(\ell(B_1))/(2d) + 1\rceil)B_1$ has $(C')^{-1} \leq \ell(B')/\ell(B_1) \leq C'$.

Now let $c \in (0, (C')^{-1}]$. For each $B \in \mathcal{B}$, let T_B be a tile whose tile set equals the dilation $(1 + c)B$, let $\mathcal{T}_*$ be the collection of all such T_B, let $\rho(T_B) := 1 + \frac{c}{2d(1+c)} r(\ell(B))$, and let $\theta := 1 - \frac{1}{1+c}$. Clearly every $\xi \in \mathbb{R}^d$ belongs to $(1 - \theta)T$ for some $T \in \mathcal{T}_*$. Suppose that the tile set of $\rho(T)T$ intersects the tile set of a tile $T' \in \mathcal{T}_*$. Fix $B'' \in \mathcal{B}$ to be any box in $\mathcal{B}$ containing a point in the intersection $(\rho(T)T)_\square \cap T'_\square$. The tile set $T'_\square$ is itself a $(1+c)$ dilation of some box $B' \in \mathcal{B}$ and $T'_\square$ intersects B'', so the side lengths of B' and B'' must be uniformly comparable, with ratio between $(C')^{-1}$ and C' (because $(1+c) \leq (1+(C')^{-1}\lceil r(\ell(B'))/(2d) + 1\rceil)$). Likewise $(\rho(T)T)_\square$ is the dilation of some box $B \in \mathcal{B}$ by a factor of $\rho(T)(1+c) = c + 1 + cr(\ell(B))/(2d) \leq 1 + (C')^{-1}\lceil r(\ell(B))/(2d) + 1\rceil$ and $(\rho(T)T)_\square$ intersects B'', so the side lengths of B and B'' are also uniformly comparable. This implies that the ratio of side lengths of $T_\square$ and $T'_\square$ must be between $(C')^{-2}$ and C'^2 and hence the distance $d(T, T')$ is uniformly bounded above for all such T and T'. This establishes the first regularity condition within the definition of admissible tilings. Moreover, when combined with Proposition 11.4 in Sect. 11.5, the fact that the boxes in $\mathcal{B}$ are nonoverlapping implies a uniform upper bound on the number of distinct tiles T whose tile sets can contain any fixed $\xi \in \mathbb{R}^d$ (since containing the same point ξ implies by the argument just given that the side lengths of the corresponding boxes $B \in \mathcal{B}$ are uniformly comparable).

Now all that remains to be established is the second regularity condition concerning the growth in size of the family of tiles with $\rho(T) \leq t$. The set of those tiles T with $\rho(T) \leq t$ are exactly all tiles whose side lengths ℓ satisfy $1 + \frac{c}{2d(1+c)} r(\ell) \leq t$. By assumption on r, there is some $\delta > 0$ such that $r(t) \geq (\ln \ln t)^{1+\delta}$ for all t larger than some t_0. Thus if ℓ is larger than the threshold t_0, it must be the case that

$$\ell \leq e^{e^{\left(\frac{2d(1+c)}{c}(t-1)\right)^{1/(1+\delta)}}}.$$

Since every isotropic tile at the origin of side length ℓ is at a metric distance at most 1 from a tile with side length equal to 2^j for some j, it follows that the maximum number $K(t)$ of unit metric balls required to cover all $\dot{T}_\square$ with $\rho(T) \leq t$ can be taken to depend logarithmically on the upper bound for ℓ just established. In particular, one can take

$$K(t) := C'' e^{\left(\frac{2d(1+c)}{c}(t-1)\right)^{1/(1+\delta)}}$$

for an appropriately-chosen constant C'' depending on r, c and d to give a K satisfying the remaining regularity condition as well as (11.6). □

The structure of the rest of this paper is as follows. Section 11.2 establishes an intermediate result concerning the relationship between discrete and continuous frames *a la* results of Hernández et al. [9] and Labate et al. [11]. Section 11.3 establishes the existence of certain smooth functions of compact support with optimal Fourier decay properties. Section 11.4 contains the proof of Theorem 11.1 using the results of the previous two sections. Section 11.5 establishes a number of important properties of the metric on tiles, and finally Sect. 11.6 is devoted to the proof of Theorem 2, which is formulated as an application of Theorem 11.1.

11.2 Discrete Frames as Approximations of Continuous Frames

The principal idea which underlies the proofs in this paper is that under appropriate conditions, one may regard discrete frames as approximations of continuous frames. This is an idea which goes back in various forms to work of Daubechies [3]. The approach followed here builds most directly on work of Hernández et al. [9] and Labate et al. [11]. The specific hypothesis (11.10) below also has clear connections to the uncertainty principle of Gröchenig and Malinnikova [7] and in particular indicates that one should expect Theorem 11.3 to produce only oversampled frames when the g_j are sufficiently regular (meaning that the frames will not in general be Riesz bases).

Theorem 11.3 (Comparing Discrete and Continuous Frames) *For each $d \geq 1$, there exists a positive constant c_d such that the following holds. Suppose $\{g_j\}_{j=1}^{\infty}$ is a sequence of functions in $L^2(\mathbb{R}^d)$ and (D_j, ξ_j) is a sequence of tiles in $\mathbb{R}^d$. If there exist positive constants $0 < A \leq B$ and $\epsilon \in [0, 1)$ such that the inequalities*

$$A \leq \sum_{j=1}^{\infty} |\widehat{g}_j(\xi)|^2 \leq B, \tag{11.9}$$

$$\sum_{j=1}^{\infty} |\widehat{g}_j(\xi)|^2 |D_j^{-1}(\xi - \xi_j)|^{d+1} \mathbb{1}_{|D_j^{-1}(\xi-\xi_j)|\geq \frac{1}{2}} \leq c_d \epsilon^2 A \tag{11.10}$$

hold for a.e. $\xi \in \mathbb{R}^d$*, then every* $f \in L^2(\mathbb{R}^d)$ *satisfies the inequalities*

$$(1-\epsilon)A||f||^2 \leq \sum_{j=1}^{\infty} \sum_{k\in\mathbb{Z}^d} \frac{\left|\left\langle f, \tau_{D_j^I k} g_j \right\rangle\right|^2}{|\det D_j|} \leq (1+\epsilon)B||f||^2.$$

Here $D_j^I := (D_j^t)^{-1}$ *for each* j *and* $\tau_p g_j(x) := g_j(x-p)$*.*

The essential step in proving Theorem 11.3 is the following lemma.

Lemma 11.1 *In any dimension* $d \geq 1$*, there is a constant* C_d *such that for any tile* $T = (D, \xi)$*, any* $\delta \in (0,1)$*, and any* $f, g \in L^2(\mathbb{R}^d)$*,*

$$\left| \sum_{k\in\mathbb{Z}^d} \frac{|\langle f, \tau_{D^I k} g\rangle|^2}{|\det D|} - \int_{\mathbb{R}^d} |\widehat{f}(\xi')|^2 |\widehat{g}(\xi')|^2 d\xi' \right|$$
$$\leq \delta \int_{\mathbb{R}^d} |\widehat{f}(\xi')|^2 |\widehat{g}(\xi')|^2 d\xi' + \frac{C_d}{\delta} \int_{\mathbb{R}^d} |\widehat{f}(\xi')|^2 |\widehat{g}(\xi')|^2 \omega^T(\xi') d\xi' \tag{11.11}$$

where $\omega(\xi') := \mathbb{1}_{|\xi'|\geq\frac{1}{2}} |\xi'|^{d+1}$*.*

Before proving this lemma, let us first see how the lemma very directly implies Theorem 11.3. Let $c_d := (4C_d)^{-1}$ with C_d being the constant from Lemma 11.1. The lemma can be applied for each j and the results summed to conclude that

$$\left[(1-\delta)A - \frac{\epsilon^2 A}{4\delta}\right] ||f||^2 \leq \sum_{j=1}^{\infty} \sum_{k\in\mathbb{Z}^d} \frac{\left|\left\langle f, \tau_{D_j^I k} g_j \right\rangle\right|^2}{|\det D_j|} \leq \left[(1+\delta)B + \frac{\epsilon^2 A}{4\delta}\right] ||f||^2$$

for any $\delta \in (0,1)$ and any $f \in L^2(\mathbb{R}^d)$ because the hypotheses (11.9) and (11.10) of Theorem 11.3 imply that

$$A||f||^2 \leq \int_{\mathbb{R}^d} |\widehat{f}(\xi')|^2 \sum_{j=1}^{\infty} |\widehat{g}_j(\xi')|^2 d\xi' \leq B||f||^2$$

and

$$\int_{\mathbb{R}^d} |\widehat{f}(\xi)|^2 \sum_{j=1}^{\infty} |\widehat{g}_j(\xi')|^2 |D_j^{-1}(\xi'-\xi_j)|^{d+1} \mathbb{1}_{|D_j^{-1}(\xi'-\xi_j)|\geq\frac{1}{2}} d\xi' \leq c_d \epsilon^2 A ||f||^2.$$

The theorem follows by fixing $\delta := \epsilon/2$, which is allowed because $\epsilon < 1$ (here one also uses the fact that $A \leq B$ to simplify the upper frame bound).

Proof of Lemma 11.1 It suffices to assume that the right-hand side of (11.11) is finite. For any $f, g \in L^2(\mathbb{R}^d)$, Plancherel's identity dictates that

$$\langle f, \tau_{D^I k} g\rangle = \int_{\mathbb{R}^d} f(x)\overline{g(x - D^I k)}dx = \int \widehat{f}(\xi')\overline{e^{-2\pi i \xi' \cdot D^I k}\widehat{g}(\xi')}d\xi',$$

and by writing $\mathbb{R}^d$ as a disjoint union of sets $D(\ell + [0, 1)^d)$ for $\ell \in \mathbb{Z}^d$ and applying the Dominated Convergence Theorem in the usual way, it follows that

$$\sum_{\ell \in \mathbb{Z}^d} \widehat{f}(\xi' - D\ell)\overline{\widehat{g}(\xi' - D\ell)} \tag{11.12}$$

is an integrable function on $D\mathbb{T}^d$ which converges absolutely for a.e. ξ' and that

$$\langle f, \tau_{D^I k} g\rangle = \int_{D\mathbb{T}^d} e^{2\pi i (D^{-1}\xi')\cdot k} \sum_{\ell \in \mathbb{Z}^d} \widehat{f}(\xi' - D\ell)\overline{\widehat{g}(\xi' - D\ell)}d\xi'$$

for each $k \in \mathbb{Z}^d$. Now by Parseval's identity,

$$\sum_{k \in \mathbb{Z}^d} \frac{|\langle f, \tau_{D^I k} g\rangle|^2}{|\det D|} = \int_{D\mathbb{T}^d} \left| \sum_{k \in \mathbb{Z}^d} \widehat{f}(\xi' - Dk)\overline{\widehat{g}(\xi' - Dk)} \right|^2 d\xi'. \tag{11.13}$$

Note in particular that Parseval's identity implies equality (11.13) when *either* side of this identity is known to be finite, so it also will trivially hold when either side is infinite. Since we may assume that f and g are chosen in such a way that the right-hand side of (11.11) is finite, it follows that

$$\sum_{k \in \mathbb{Z}^d} \frac{|\langle f, \tau_{D^I k} g\rangle|^2}{|\det D|} - \int_{\mathbb{R}^d} |\widehat{f}(\xi')|^2 |\widehat{g}(\xi')|^2 d\xi' =$$

$$\int_{D\mathbb{T}^d} \left[\left| \sum_{k \in \mathbb{Z}^d} \widehat{f}(\xi' - Dk)\overline{\widehat{g}(\xi' - Dk)} \right|^2 - \sum_{k \in \mathbb{Z}^d} \left| \widehat{f}(\xi' - Dk)\overline{\widehat{g}(\xi' - Dk)} \right|^2 \right] d\xi' \tag{11.14}$$

by a similar integral periodization computation to that above (note that the quantities subtracted from both sides of (11.14) are known to be finite and equal because the right-hand side of (11.11) is assumed to be finite, so (11.14) continues to make sense whether or not the quantities (11.13) are known to be finite). Because f and g both belong to L^2 and (11.12) consequently converges absolutely for a.e. ξ', one can write

$$\left| \sum_{k \in \mathbb{Z}^d} \widehat{f}(\xi' - Dk)\overline{\widehat{g}(\xi' - Dk)} \right|^2$$
$$= \sum_{k \in \mathbb{Z}^d} \sum_{\ell \in \mathbb{Z}^d} \widehat{f}(\xi' - Dk)\overline{\widehat{g}(\xi' - Dk)}\,\overline{\widehat{f}(\xi' - D\ell)}\widehat{g}(\xi' - D\ell)$$

with absolute convergence as a sum over $(k, \ell) \in \mathbb{Z}^d \times \mathbb{Z}^d$. The second term inside the integrand on the right-hand side of (11.14) is exactly equal to the sum of diagonal terms $k = \ell$, and so by the triangle inequality and (11.14), the magnitude of left-hand side of (11.11) is bounded above by

$$\int_{D\mathbb{T}^d} \sum_{k \in \mathbb{Z}^d} \sum_{\ell \in \mathbb{Z}^d \setminus \{k\}} \left| \widehat{f}(\xi' - Dk)\overline{\widehat{g}(\xi' - Dk)}\,\overline{\widehat{f}(\xi' - D\ell)}\widehat{g}(\xi' - D\ell) \right| d\xi'. \tag{11.15}$$

Now suppose that w is a strictly positive measurable function on $\mathbb{R}^d$. The precise form of w will be specified momentarily, but in any case, one may write

$$\left| \widehat{f}(\xi' - Dk)\overline{\widehat{g}(\xi' - Dk)}\,\overline{\widehat{f}(\xi' - D\ell)}\widehat{g}(\xi' - D\ell) \right|$$
$$\leq \frac{1}{2} \frac{|\widehat{f}(\xi' - Dk)\widehat{g}(\xi' - Dk)|^2 w(D^{-1}(\xi' - Dk - \xi))}{w(D^{-1}(\xi' - \xi) - \ell)}$$
$$+ \frac{1}{2} \frac{|\widehat{f}(\xi' - D\ell)\widehat{g}(\xi' - D\ell)|^2 w(D^{-1}(\xi' - D\ell - \xi))}{w(D^{-1}(\xi' - \xi) - k)}$$

for almost every ξ' by the standard inequality between arithmetic and geometric means. Applying this to the integrand of (11.15) and recognizing the symmetry between k and ℓ gives that the left-hand side of (11.11) is bounded above by

$$\int_{D\mathbb{T}^d} \sum_{k \in \mathbb{Z}^d} \sum_{\ell \in \mathbb{Z}^d \setminus \{k\}} \frac{|\widehat{f}(\xi' - Dk)\widehat{g}(\xi' - Dk)|^2 w(D^{-1}(\xi' - Dk - \xi))}{w(D^{-1}(\xi' - \xi) - \ell)} d\xi'$$
$$= \int_{\mathbb{R}^d} |\widehat{f}(\xi')|^2 |\widehat{g}(\xi')|^2 w(D^{-1}(\xi' - \xi)) \sum_{\ell' \in \mathbb{Z}^d \setminus \{0\}} \frac{1}{w(D^{-1}(\xi' - \xi) - \ell')} d\xi', \tag{11.16}$$

where the equality in (11.16) is a consequence of writing $\ell = \ell' + k$ and unperiodizing the integral. Now take w to have the form

$$w(\xi) = a\mathbb{1}_{|\xi| < \frac{1}{2}} + |\xi|^{d+1}\mathbb{1}_{|\xi| \geq \frac{1}{2}}$$

for a positive value of a which will depend on δ. The key computation required to complete the proof is an upper bound for the sum

$$\sum_{\ell' \in \mathbb{Z}^d \setminus \{0\}} \frac{1}{w(\xi' - \ell')}$$

in terms of $\xi' \in \mathbb{R}^d$. Now

$$\frac{1}{w(\xi)} = a^{-1} \mathbb{1}_{|\xi| < \frac{1}{2}} + |\xi|^{-d-1} \mathbb{1}_{|\xi| \geq \frac{1}{2}},$$

so

$$\begin{aligned} \sum_{\ell' \in \mathbb{Z}^d \setminus \{0\}} \frac{1}{w(\xi' - \ell')} &= \sum_{\ell' \in \mathbb{Z}^d \setminus \{0\}} \left[a^{-1} \mathbb{1}_{|\xi' - \ell'| < \frac{1}{2}} + |\xi' - \ell'|^{-d-1} \mathbb{1}_{|\xi' - \ell'| \geq \frac{1}{2}} \right] \\ &\leq a^{-1} \mathbb{1}_{|\xi'| > \frac{1}{2}} + \sum_{\ell' \in \mathbb{Z}^d} \frac{\mathbb{1}_{|\xi' - \ell'| \geq \frac{1}{2}}}{|\xi' - \ell'|^{d+1}} \end{aligned}$$

because $\mathbb{1}_{|\xi' - \ell'| < \frac{1}{2}}$ vanishes for all $\ell' \neq 0$ when $|\xi'| \leq \frac{1}{2}$ and otherwise is equal to 1 for at most one ℓ' for each fixed ξ'. Now

$$\begin{aligned} \sum_{\ell' \in \mathbb{Z}^d} \frac{\mathbb{1}_{|\xi' - \ell'| \geq \frac{1}{2}}}{|\xi' - \ell'|^{d+1}} &= \sum_{\ell' \in \mathbb{Z}^d} \mathbb{1}_{|\xi' - \ell'| \geq \frac{1}{2}} \int_{\frac{1}{2}}^{\infty} \mathbb{1}_{|\xi' - \ell'| < s} \frac{d+1}{s^{d+2}} ds \\ &= (d+1) \int_{\frac{1}{2}}^{\infty} \# \left\{ \ell' \in \mathbb{Z}^d \ \middle| \ \frac{1}{2} \leq |\xi' - \ell'| < s \right\} \frac{ds}{s^{d+2}}. \end{aligned}$$

Recalling that $|\cdot|$ denotes the ℓ^∞ norm on $\mathbb{R}^d$, the number of lattice points ℓ' inside the box $|\xi' - \ell'| < s$ never exceeds $(2s+1)^d$, which is itself less than $(4s)^d$ because $s > \frac{1}{2}$. Upon applying this inequality to the integrand immediately above, the sum in question can never exceed $2^{2d+1}(d+1)$ and therefore

$$\sum_{\ell' \in \mathbb{Z}^d \setminus \{0\}} \frac{1}{w(\xi' - \ell')} \leq a^{-1} \mathbb{1}_{|\xi'| > \frac{1}{2}} + 2^{2d+1}(d+1)$$

for every $\xi' \in \mathbb{R}^d$. Multiplying both sides by $w(\xi')$ gives that

$$\begin{aligned} w(\xi') \sum_{\ell' \in \mathbb{Z}^d \setminus \{0\}} \frac{1}{w(\xi' - \ell')} \\ &\leq 2^{2d+1}(d+1)a \mathbb{1}_{|\xi'|<\frac{1}{2}} + (2^{2d+1}(d+1) + a^{-1})|\xi|^{d+1} \mathbb{1}_{|\xi'|\geq \frac{1}{2}} \\ &\leq 2^{2d+1}(d+1)a + (2^{2d+1}(d+1) + a^{-1})|\xi'|^{d+1} \mathbb{1}_{|\xi'|\geq \frac{1}{2}}. \end{aligned}$$

For any $\delta \in (0, 1]$, fix $a := \delta 2^{-2d-1}/(d+1)$. In this case, $2^{2d+1}(d+1) \leq a^{-1}$, so

$$w(\xi') \sum_{\ell' \in \mathbb{Z}^d \setminus \{0\}} \frac{1}{w(\xi' - \ell')} \leq \delta + 2^{2d+2}(d+1)\delta^{-1}|\xi'|^{d+1} \mathbb{1}_{|\xi'|\geq \frac{1}{2}}. \tag{11.17}$$

Applying the inequality (11.17) with ξ' replaced by $D^{-1}(\xi' - \xi)$ to the integrand inside the right-hand side of (11.16) completes the lemma when one takes $C_d := 2^{2d+2}(d+1)$. □

11.3 Smooth Compactly-Supported Functions with Subexponential Fourier Decay

Before dealing directly with the proof of Theorem 11.1, we first record in this section a number of basic results concerning the Fourier decay of functions which are C^∞ and compactly supported. While it is widely known that such functions φ must have Fourier transforms $\widehat{\varphi}(\xi)$ which decay faster than $C_N|\xi|^{-N}$ for any positive N, it is perhaps much less widely known that there exists a sharp characterization, due to Ingham [10], of the possible Fourier decay rates exhibited by such functions φ. Proposition 11.2, included here for completeness and its intrinsic interest, records one direction of Ingham's characterization following essentially the same argument that appears there. For the converse direction, Ingham shows that when the integral in (11.18) is infinite, any function φ satisfying (11.19) must be quasi-analytic and therefore not compactly supported.

Proposition 11.2 (Rapid Fourier Decay of Compactly-Supported Functions) *Let $W : \mathbb{R} \to [1, \infty)$ be an even function nondecreasing on $[0, \infty)$ such that $W(t) = 1$ for all $t \in [0, t_*]$ for some positive t_*, $t^{-N}W(t) \to \infty$ as $t \to \infty$ for all $N > 0$, and*

$$C_W := \frac{1}{t_*} + \int_0^\infty \frac{\ln W(t)}{t^2} dt < \infty. \tag{11.18}$$

There exists an even, radial decreasing C^∞ function φ on $\mathbb{R}$ supported on the interval $[-\frac{\sqrt{e}}{\pi}C_W, \frac{\sqrt{e}}{\pi}C_W]$ such that φ and $\widehat{\varphi}$ are both real and nonnegative, $\int \varphi = 1$, and

$$|\widehat{\varphi}(\xi)| \leq \frac{1}{W(\xi)} \text{ for all } \xi \in \mathbb{R}. \tag{11.19}$$

Proof For each nonnegative integer j, let t_j be the infimum of all $t \in [t_*, \infty)$ such that $\ln W(t) \geq j$. By definition of t_j, if $t_j < t < t_{j+1}$ for some $j \geq 0$, then $j \leq \ln W(t) < j + 1$. It follows that for each positive N,

$$\sum_{j=0}^{N} j\left(\frac{1}{t_j} - \frac{1}{t_{j+1}}\right) \leq \sum_{j=0}^{N} \int_{t_j}^{t_{j+1}} \frac{\ln W(t)}{t^2} dt \leq \int_0^{t_{N+1}} \frac{\ln W(t)}{t^2} dt$$

and

$$\frac{N+1}{t_{N+1}} \leq \int_{t_{N+1}}^{\infty} \frac{\ln W(t)}{t^2} dt.$$

Summing the two inequalities, observing that $t_0 = t_*$, and letting $N \to \infty$ gives

$$\sum_{j=0}^{\infty} \frac{1}{t_j} \leq \frac{1}{t_*} + \int_0^{\infty} \frac{\ln W(t)}{t^2} dt = C_W.$$

For some $c > 0$ to be specified shortly, let $\eta_j(x) := c^{-1} t_j \mathbb{1}_{[-ct_j^{-1}/2, ct_j^{-1}/2]}(x)$ for each $j \geq 0$. Let φ_M equal the repeated convolution $\eta_0 * \eta_0 * \eta_1 * \eta_1 * \cdots * \eta_M * \eta_M$, and note that each φ_M is an even, nonnegative function which is radially decreasing, has integral 1, and is supported on the interval

$$\left[-\sum_{j=0}^{M} \frac{c}{t_j}, \sum_{j=1}^{M} \frac{c}{t_j}\right] \subset [-cC_W, cC_W].$$

The Fourier transform of φ_M is a product:

$$\widehat{\varphi_M}(\xi) = \prod_{j=0}^{M} \left[\frac{\sin ct_j^{-1}\pi\xi}{ct_j^{-1}\pi\xi}\right]^2. \tag{11.20}$$

Since $|\sin t|/|t| \leq 1$, these products are nonnegative and nonincreasing as a function of M for each fixed ξ, so there is pointwise convergence of $\widehat{\varphi_M}(\xi)$ as $M \to \infty$ for each ξ. Because $(\sin \pi\xi)^2/(\pi\xi)^2 \in L^1(\mathbb{R})$, the partial products also converge in $L^1(\mathbb{R})$ by dominated convergence. Then by Hausdorff-Young, the functions φ_M

necessarily converge uniformly as $M \to \infty$. Let φ be the uniform limit of the functions φ_M. Uniform convergence implies that φ is even, nonnegative, radially-decreasing, and supported on the interval $[-cC_W, cC_W]$. Moreover, the integral of φ must equal 1 because uniform convergence of these compactly-supported functions implies convergence in $L^1(\mathbb{R})$. Finally,

$$\int_{\mathbb{R}} e^{-2\pi i x\xi}\varphi(x)dx = \lim_{M\to\infty}\int_{\mathbb{R}} e^{-2\pi i x\xi}\varphi_M(x)dx = \prod_{j=0}^{\infty}\left|\frac{\sin ct_j^{-1}\pi\xi}{ct_j^{-1}\pi\xi}\right|^2$$

for each $\xi \in \mathbb{R}$ by virtue of convergence of the sequence $\{\varphi_M\}$ in L^1 as well. Now fix $c := e^{1/2}\pi^{-1}$. For any given $\xi \in \mathbb{R}$, if $|\xi| \geq t_{j_0}$, then one must have

$$\left|\frac{\sin ct_{j_0}^{-1}\pi\xi}{ct_{j_0}^{-1}\pi\xi}\right|^2 \leq e^{-1}.$$

Taking a product over $j \leq j_0$ gives that

$$|\widehat{\varphi}(\xi)| = \prod_{j=0}^{\infty}\left|\frac{\sin ct_j^{-1}\pi\xi}{ct_j^{-1}\pi\xi}\right|^2 \leq \prod_{j=0}^{j_0}\left|\frac{\sin ct_j^{-1}\pi\xi}{ct_j^{-1}\pi\xi}\right|^2 \leq \prod_{j=0}^{j_0} e^{-1} = e^{-j_0-1}$$

for all $|\xi| \geq t_{j_0}$. For each fixed $\xi \in \mathbb{R}\setminus[-t_*, t_*]$, if j_0 is the maximal index such that $|\xi| \geq t_{j_0}$, then $|\xi| < t_{j_0+1}$, meaning that $\ln W(\xi) < j_0 + 1$ (since the definition of t_{j_0+1} and the inequality $j_0 + 1 \leq \ln W(|\xi|)$ imply $t_{j_0+1} \leq |\xi|$). Therefore $|\xi| \geq t_*$ implies that $|\widehat{\varphi}(\xi)| \leq e^{-j_0-1} \leq (W(\xi))^{-1}$. The same inequality holds when $|\xi| \leq t_*$ as well simply because the L^1-norm of φ is one, so $|\widehat{\varphi}(\xi)| \leq 1 = 1/W(\xi)$. Finally, note that φ is C^∞ because its Fourier transform decays faster than $|\xi|^{-N}$ for any $N > 0$. □

The utility of Proposition 11.2 comes primarily through the following corollary, which establishes that one can always find nice C^∞ functions of compact support η whose Fourier transform is close to the indicator function $\mathbb{1}_{[-(1-\theta)/2,(1-\theta)/2]}$ with subexponential Schwartz tails.

Corollary 11.1 (Compactly-Supported Approximations of Tile Set Multipliers) *Let W be as in the previous proposition. There are constants C and C' depending only on W such that for any $\epsilon, \theta \in (0, 1)$ and any dimension $d \geq 1$, there is a real, C^∞ function η supported in $[-C\theta^{-1}\epsilon^{-1}, C\theta^{-1}\epsilon^{-1}]^d$ such that $\widehat{\eta}$ is real, nonnegative, bounded above by 1, even in each coordinate, invariant under permutation of coordinates, and satisfies*

$$\left(1 - \frac{C'}{W(\epsilon^{-1})}\right)^d \leq \widehat{\eta}(\xi) \text{ for all } |\xi| \leq \frac{1-\theta}{2} \tag{11.21}$$

and

$$\widehat{\eta}(\xi) \leq \frac{C'}{W(\epsilon^{-1}\xi)} \text{ for all } |\xi| \geq \frac{1}{2}. \tag{11.22}$$

Proof If $d > 1$, it suffices to fix $\eta(\xi)$ as the product $\eta(\xi_1)\cdots\eta(\xi_d)$ for the η constructed in the $d = 1$ case. It therefore suffices to assume that $d = 1$.

For φ as given by Proposition 11.2, let $\kappa := ||\varphi||_{L^2}^{-2}$ and consider the function

$$\widehat{\eta}(\xi) := \int_{|\xi'| \leq \frac{1}{2} - \frac{\theta}{6}} \left(\frac{\epsilon\theta}{3}\right)^{-1} \kappa |\widehat{\varphi}(3\theta^{-1}\epsilon^{-1}(\xi - \xi'))|^2 d\xi'.$$

This $\widehat{\eta}$ is exactly a convolution with a dilated version of $\kappa|\widehat{\varphi}|^2$ and the indicator function of the interval $|\xi'| \leq \frac{1}{2} - \frac{\theta}{6}$. As such, η will necessarily be a pointwise product of the inverse Fourier transform of $\mathbb{1}_{|\xi'| \leq 1/2 - \theta/6}$, which is a sinc function, and a dilated version of φ convolved with itself, which will be a C^∞ function supported in the interval $[-6\theta^{-1}\epsilon^{-1}\frac{\sqrt{e}}{\pi}C_W, 6\theta^{-1}\epsilon^{-1}\frac{\sqrt{e}}{\pi}C_W]$. Thus, η is real, even, C^∞, and supported on this same interval. The Fourier transform $\widehat{\eta}$ is everywhere nonnegative and bounded above by 1 by virtue of the choice of κ and the fact that the indicator function $\mathbb{1}_{|\xi'| \leq 1/2 - \theta/6}$ is simply bounded above by 1.

Now when $|\xi| \geq 1/2$ and $|\xi'| \leq \frac{1}{2} - \frac{\theta}{6}$, it follows that

$$|\xi - \xi'| \geq |\xi| - |\xi'| \geq |\xi| - \left(\frac{1}{2} - \frac{\theta}{6}\right) \geq |\xi| - |\xi| \left(1 - \frac{\theta}{3}\right) \geq \frac{|\xi|\theta}{3}.$$

A change of variables gives that

$$\begin{aligned}\widehat{\eta}(\xi) &\leq \int_{|\xi - \xi'| \geq \frac{\theta|\xi|}{3}} \left(\frac{\epsilon\theta}{3}\right)^{-1} \kappa \left|\widehat{\varphi}(3\theta^{-1}\epsilon^{-1}(\xi - \xi'))\right|^2 d\xi' \\ &\leq \int_{|\xi'| \geq \epsilon^{-1}|\xi|} \kappa \left|\widehat{\varphi}(\xi')\right|^2 d\xi' \leq \frac{\int \kappa\widehat{\varphi}(\xi')d\xi}{W(\epsilon^{-1}\xi)} = \frac{C'}{W(\epsilon^{-1}\xi)}\end{aligned}$$

for some C' depending only on W. Similarly, since

$$1 - \widehat{\eta}(\xi) = \int_{|\xi'| > \frac{1}{2} - \frac{\theta}{6}} \left(\frac{\epsilon\theta}{3}\right)^{-1} \kappa |\widehat{\varphi}(3\theta^{-1}\epsilon^{-1}(\xi - \xi'))|^2 d\xi' \tag{11.23}$$

(because the integral of $\left(\frac{\epsilon\theta}{3}\right)^{-1} \kappa|\widehat{\varphi}(3\theta^{-1}\epsilon^{-1}(\xi - \xi'))|^2$ with respect to ξ' is always 1), it follows by the same reasoning that when $|\xi| < \frac{1}{2}(1 - \theta)$ and $|\xi'| > \frac{1}{2} - \frac{\theta}{6}$, $|\xi' - \xi| \geq \frac{\theta}{3}$ on the support of the integral on the right-hand side of (11.23) and consequently

$$1 - \widehat{\eta}(\xi) \leq \int_{|\xi'| \geq \epsilon^{-1}} \kappa \left|\widehat{\varphi}_0(\xi')\right|^2 d\xi' \leq \frac{C'}{W(\epsilon^{-1})}.$$

This completes the proof of the corollary. □

11.4 Proof of Theorem 11.1

Proof of Theorem 11.1 The proof of Theorem 11.1 follows rather immediately from Theorem 11.3 and the constructions of the previous section. Recall the definition (11.2) of dilations $\widehat{\psi}^T$, and let ψ be the convolution of g with η as given by Corollary 11.1 for some W and ϵ to be chosen later (with θ taken to coincide with its value from the statement of Theorem 11.1). With this identification, $\widehat{\psi}^T(\xi) = \widehat{g}^T(\xi)\widehat{\eta}^T(\xi)$ for each $\xi \in \mathbb{R}^d$ and each $T \in \mathcal{T}_*$. Moreover ψ will necessarily be C^∞ and compactly supported because it is the convolution of the compactly supported functions g and η and η is smooth. First observe that

$$\sum_{T \in \mathcal{T}_*} |\widehat{\psi}^T(\xi)|^2 = \sum_{T \in \mathcal{T}_*} |\widehat{g}^T(\xi)|^2 |\widehat{\eta}^T(\xi)|^2 \leq \sum_{j=1}^{\infty} |\widehat{g}^T(\xi)|^2 \leq B$$

for almost every $\xi \in \mathbb{R}^d$ by virtue of the hypothesis (11.3). Likewise by (11.21),

$$\sum_{T \in \mathcal{T}_*} |\widehat{\psi}^T(\xi)|^2 = \sum_{T \in \mathcal{T}_*} |\widehat{g}^T(\xi)|^2 |\widehat{\eta}^T(\xi)|^2$$
$$\geq \left(1 - \frac{C'}{W(\epsilon^{-1})}\right)^{2d} \sum_{T \in \mathcal{T}_*} |\widehat{g}(\xi)|^2 \mathbb{1}_{(1-\theta)T}(\xi) \geq \left(1 - \frac{C'}{W(\epsilon^{-1})}\right)^{2d} A$$

for almost every $\xi \in \mathbb{R}^d$. Thus

$$\left(1 - \frac{C'}{W(\epsilon^{-1})}\right)^{2d} A \leq \sum_{T \in \mathcal{T}_*} |\widehat{\psi}^T(\xi)|^2 \leq B \tag{11.24}$$

for almost every $\xi \in \mathbb{R}^d$. This is exactly the assumption (11.9) of Theorem 11.3 with the $\widehat{g}_j$'s taken to be equal to $\widehat{\psi}^T$ for each $T \in \mathcal{T}_*$. Since $W(\epsilon^{-1})$ tends to infinity as $\epsilon \to 0^+$, the constant on the left-hand side of (11.24) can be made as close to A as desired for suitably small ϵ depending only on W, θ, d, and A.

Next consider the analogue of (11.10). If T is the tile (D, ξ), then

$$\sup_{\xi'} |D^{-1}(\xi' - \xi)|^{d+1} \mathbb{1}_{|D^{-1}(\xi'-\xi)| \geq \frac{1}{2}} |\widehat{\eta}^T(\xi')|^2 = \sup_{|\xi'| \geq \frac{1}{2}} |\xi'|^{d+1} |\widehat{\eta}(\xi')|^2$$

$$\leq \sup_{t \geq \frac{1}{2}} \frac{(C')^2 t^{d+1}}{(W(\epsilon^{-1}t))^2} = \epsilon^{d+1} \sup_{t \geq \frac{1}{2}} \frac{(C')^2 (\epsilon^{-1}t)^{d+1}}{(W(\epsilon^{-1}t))^2} \leq \epsilon^{d+1} \sup_{t \geq 0} \frac{(C')^2 t^{d+1}}{(W(t))^2},$$

which means that when $\omega(\xi') := |\xi'|^{d+1} \mathbb{1}_{|\xi'| \geq \frac{1}{2}}$,

$$\sum_{T \in \mathcal{T}_*} |\widehat{\psi}^T(\xi')|^2 \omega^T(\xi) \leq \left(\sum_{T \in \mathcal{T}_*} |\widehat{g}^T(\xi')|^2 \right) \epsilon^{d+1} \sup_{t \geq 0} \frac{(C')^2 t^{d+1}}{(W(t))^2}$$

$$\leq B(C')^2 \epsilon^{d+1} \sup_{t \geq 0} \frac{t^{d+1}}{(W(t))^2}.$$

In particular, for any fixed W, this right-hand side can be made as small as desired by choosing a sufficiently small ϵ (depending only on A, B and d, and θ), if, for example, $W(t) := \max\{1, e^{\sqrt{|t|}-1}\}$. Thus, Theorem 11.1 necessarily follows from Theorem 11.3. □

11.5 Geometry of Tile Space and Admissible Tilings

This section lays the groundwork for the proof of Theorem 2 by establishing some basic facts about tiles in $\mathbb{R}^d$. The key result of this section is Proposition 11.5, which establishes that, under the assumption that a set of tiles $\mathcal{T}_*$ is an admissible tiling, almost every point $\xi \in \mathbb{R}^d$ has the property that the number of tiles $T \in \mathcal{T}_*$ such that the tile set $tT_\square$ contains ξ grows at most subexponentially as a function of t. This is the key observation which is necessary to establish that there exists a *continuous* frame with C^∞ window of compact support which is adapted to $\mathcal{T}_*$.

Proposition 11.3 (Tile Sets and Symmetries) *Suppose that $D_1, D_2 \in \mathrm{GL}_d(\mathbb{R})$. The following are equivalent.*

1. *$|D_1^{-1} D_2| = |D_2^{-1} D_1| = 1$ (where, as before, $|\cdot|$ is the ℓ^∞ operator norm).*
2. *$D_2 = D_1 P$ for some matrix P for which there exists a permutation σ of $\{1, \ldots, d\}$ such that $|P_{ij}| = 1$ when $i = \sigma(j)$ and $P_{ij} = 0$ otherwise.*
3. *D_1 and D_2 generate identical tile sets at the origin, i.e., $(T_1)_\square = (T_2)_\square$ when $T_i := (D_i, 0)$ for $i = 1, 2$.*

Proof [1 ⇒ 2] Let $P := D_1^{-1} D_2$. By hypothesis, $|P| \leq 1$. Moreover, $P^{-1} = D_2^{-1} D_1$, so $|P^{-1}| \leq 1$ as well. However, since $I = PP^{-1}$ and $1 \leq |P||P^{-1}|$, it follows that $|P| = |P^{-1}| = 1$. But as $x = P^{-1} P x$ for each $x \in \mathbb{R}^d$, $|x| \leq$

$|P^{-1}||Px| \leq |Px|$, and consequently $|x| = |Px|$ for each x. Because the norm $|\cdot|$ is the ℓ^∞ norm, $|P| = \max_i \sum_j |P_{ij}|$. For each j, let e_j be the j-th standard basis vector. Since $1 = |e_j| = |Pe_j|$, it follows that there must be some index $\sigma(j) \in \{1, \ldots, d\}$ for which $|P_{\sigma(j)j}| = 1$. Since the norm of P is no more than 1, this implies that $P_{\sigma(j)j'} = 0$ when $j' \neq j$. This forces $\sigma(j) \neq \sigma(j')$ when $j \neq j'$, and consequently σ must be a permutation of $\{1, \ldots, d\}$.

[2 $\Rightarrow$ 3] The identity $D_2 = D_1 P$ implies that $D_2^{-1} = P^{-1} D_1^{-1}$. But P^{-1} must also clearly have the same form as P, namely some permutation of its rows converts it to a diagonal matrix with diagonal entries ± 1 only. As a consequence, $|P^{-1}x| = |x|$ for all $x \in \mathbb{R}^d$, and thus $|D_2^{-1}\eta| = |D_1^{-1}\eta|$ for all $\eta \in \mathbb{R}^d$, which in particular forces the tiles at the origin generated by D_1 and D_2 to be identical.

[3 $\Rightarrow$ 1] Invertibility of D_1 implies that a vector η has $|D_1^{-1}\eta| \leq \frac{1}{2}$ (which is exactly the definition of the tile set of the tile $(D_1, 0)$) if and only if $\eta = D_1\xi$ for some ξ with $|\xi| \leq \frac{1}{2}$, and equality of tile sets implies that every ξ with $|\xi| \leq \frac{1}{2}$ admits some ξ' also with $|\xi'| \leq \frac{1}{2}$ such that $D_1\xi = D_2\xi'$. This implies $\xi' = D_2^{-1}D_1\xi$, and consequently $|D_2^{-1}D_1\xi| \leq \frac{1}{2}$ for all ξ with $|\xi| \leq \frac{1}{2}$. This gives directly by the definition of the operator norm that $|D_2^{-1}D_1| \leq 1$. By symmetry, $|D_1^{-1}D_2| \leq 1$ as well. But since $I = D_2^{-1}D_1D_1^{-1}D_2$, this means that $1 \leq |D_2^{-1}D_1||D_1^{-1}D_2|$, giving that $1 = |D_1^{-1}D_2| = |D_2^{-1}D_1|$. □

Lemma 11.2 (Metric on Tile Sets) *For any $D_1, D_2 \in \mathrm{GL}_d(\mathbb{R})$, let*

$$d(D_1, D_2) := \ln_2 \max\{|D_1^{-1}D_2|, |D_2^{-1}D_1|\}.$$

Then $d(D_1, D_2)$ is nonnegative, symmetric, and satisfies the triangle inequality. Moreover, every $x \in \mathbb{R}^d$ satisfies

$$2^{-d(D_1,D_2)}|D_1^{-1}x| \leq |D_2^{-1}x| \leq 2^{d(D_1,D_2)}|D_1^{-1}x|. \tag{11.25}$$

Given any tiles $T_1, T_2 \in \dot{\mathcal{T}}$ with $T_i := (D_i, 0)$ for $i = 1, 2$, let

$$d(T_1, T_2) := d(D_1, D_2). \tag{11.26}$$

If T_1' and T_2' are any tiles in $\dot{\mathcal{T}}$ such that $(T_i)_\square = (T_i')_\square$ for $i = 1, 2$, then $d(T_1, T_2) = d(T_1', T_2')$ in particular d defines a metric on the space of tile sets at the origin.

Proof First we show the properties of d as defined on pairs of invertible matrices (D_1, D_2). Nonnegativity of d holds because $I = D_1^{-1}D_2D_2^{-1}D_1$ implies $1 \leq |D_1^{-1}D_2||D_2^{-1}D_1|$. Since the product is at least one, the maximum must be at least one as well, so $d(D_1, D_2) \geq 0$. In a similarly immediate way, taking logarithms of the inequalities $|D_1^{-1}D_3| \leq |D_1^{-1}D_2||D_2^{-1}D_3|$ and $|D_3^{-1}D_1| \leq |D_3^{-1}D_2||D_2^{-1}D_1|$

implies the triangle inequality. Finally, for any $D_1, D_2 \in \mathrm{GL}_d(\mathbb{R})$, it must be the case that

$$|D_2^{-1}x| = |D_2^{-1}D_1D_1^{-1}x| \leq |D_2^{-1}D_1||D_1^{-1}x| \leq 2^{d(D_1,D_2)}|D_1^{-1}x|$$

and

$$|D_1^{-1}x| = |D_1^{-1}D_2D_2^{-1}x| \leq |D_1^{-1}D_2||D_2^{-1}x| \leq 2^{d(D_1,D_2)}|D_2^{-1}x|$$

which together imply (11.25).

Now suppose $T_1, T_2 \in \dot{\mathcal{T}}$ are tiles at the origin generated by matrices D_1 and D_2, respectively. If D_1' and D_2' are any other matrices generating tiles T_1' and T_2' with the property that $(T_i)_\square = (T_i')_\square$ for $i = 1, 2$, then by Proposition 11.3, $d(D_1, D_1') = 0 = d(D_2, D_2')$, and by the triangle inequality it follows that $d(D_1, D_2) \leq d(D_1, D_1') + d(D_1', D_2') + d(D_2', D_2) = d(D_1', D_2')$. By symmetry $d(D_1, D_2) = d(D_1', D_2')$ and consequently $d((T_1)_\square, (T_2)_\square)$ is well-defined simply by fixing it to equal $d(D_1, D_2)$ for any choice of D_1 and D_2 generating the given tile sets $(T_1)_\square, (T_2)_\square$. Symmetry and the triangle inequality for d when regarded as a function of pairs of tile sets follow immediately from the corresponding inequalities for matrices, and $d((T_1)_\square, (T_2)_\square) = 0$ if and only if $\max\{|D_1^{-1}D_2|, |D_2^{-1}D_1|\} \leq 1$ for suitable representative D_1 and D_2, which as seen before implies that $|D_1^{-1}D_2| = |D_2^{-1}D_1| = 1$, meaning that $(T_1)_\square = (T_2)_\square$ (also by Proposition 11.3). □

We come now to the main task of this section, which is to count the overlap multiplicity of dilated tiles tT for $T \in \mathcal{T}_*$. When tiles are the similar shapes, there is an automatic algebraic bound on the overlap which comes directly from the fact that tiles have bounded multiplicity when not dilated. This is made precise in Proposition 11.4. Proposition 11.5 then extends the counting result to the full generality of admissible tilings.

Proposition 11.4 (Overlap of Dilated Tiles of Similar Shape) *Suppose that $\mathcal{T}_*$ is an admissible tiling of $\mathbb{R}^d$ and that $\mathcal{T}_*' \subset \mathcal{T}_*$ is a subset for which there exists some metric ball B_r of radius r inside the space of origin-centered tile sets such that $\dot{T}_\square$ belongs to B_r for all $T \in \mathcal{T}_*'$. Then for any $t \geq 1$ and any $\xi \in \mathbb{R}^d$, the subset of $\mathcal{T}_*'$ consisting of tiles T with the property that $\xi \in (tT)_\square$ is finite and has cardinality not exceeding $C(2^{2r+1}t)^d$.*

Proof Fix ξ and $T_1, \ldots, T_N \in \mathcal{T}_*'$ such that $\xi \in (tT_i)_\square$ for each i. Let each T_i have the form (D_i, ξ_i) for appropriate choices of D_i and ξ_i and let $T_0 := (D_0, \xi)$ be such that $(\dot{T}_0)_\square$ is the center of B_r (which may not belong to $\mathcal{T}_*$). By definition of the metric, $d(D_i, D_0) < r$ for each $i = 1, \ldots, N$.

The proof is essentially geometric. Suppose that η is any point in the tile set $(tT_i)_\square$. This implies $t^{-1}|D_i^{-1}(\eta - \xi_i)| \leq 1/2$. Because ξ also belongs to $(tT_i)_\square$, the triangle inequality implies that $t^{-1}|D_i^{-1}(\eta - \xi)| \leq 1$. By (11.25), this means that $t^{-1}|D_0^{-1}(\eta - \xi)| \leq 2^{d(D_0,D_i)} < 2^r$. In particular, this means that every tile set $(tT_i)_\square$ is contained in the tile set $(2^{r+1}tT_0)_\square$, which happens to have volume less than $(2^{r+1}t)^d|\det D_0|$. Since $t \geq 1$, the union $\bigcup_{i=1}^N (T_i)_\square$ is also contained this single large set tile centered at ξ. This means that

$$\int C^{-1}\sum_{i=1}^N \mathbb{1}_{T_i}(\xi')d\xi' \leq \int \mathbb{1}_{\bigcup_{i=1}^N (T_i)_\square}(\xi')d\xi' \leq \int \mathbb{1}_{2^{r+1}tT_0}(\xi')d\xi' \tag{11.27}$$

because the sum of indicator functions $\mathbb{1}_{T_i}(\xi')$ is assumed to be less than or equal to C for almost every ξ' and because the tile sets $(T_i)_\square$ are all contained in the tile set of $2^{r+1}tT_0$. But now the inequality $|D_i^{-1}x| \leq 2^r|D_0^{-1}x|$ for each x implies that the tile set $(\dot{T}_i)_\square$ must contain a parallelepiped with the same volume as $(2^{-r}T_0)_\square$ but centered at ξ_i. As a consequence, $|(T_i)_\square| \geq 2^{-rd}|\det D_0|$, where $|\cdot|$ on the left-hand side indicates Lebesgue measure. Therefore

$$NC^{-1}2^{-rd}|\det D_0| \leq \left|\bigcup_{i=1}^N (T_i)_\square\right| \leq (2^{r+1}t)^d|\det D_0|,$$

and hence it follows that $N \leq C\left(2^{2r+1}t\right)^d$ as asserted. □

Proposition 11.5 (Overlap of Dilated Tiles, Any Shape) *Suppose $\mathcal{T}_*$ is an admissible tiling of $\mathbb{R}^d$. There exists a constant $C' < \infty$ depending only on C and d such that the following holds. For any real $t \geq 1$ and $\xi \in \mathbb{R}^d$, let $O_\xi(t)$ be the set of tiles $T \in \mathcal{T}_*$ such that $\xi \in (tT)_\square$. Then for almost every $\xi \in \mathbb{R}^d$, $O_\xi(t)$ has cardinality not exceeding $C't^dK(t)$ for each $t \geq 1$.*

Proof Because $\mathcal{T}_*$ is admissible, almost every point ξ necessarily belongs to $(T_\xi)_\square$ for some $T_\xi \in \mathcal{T}_*$ (though note that ξ is not necessarily the *center* of the tile set but merely belongs to it). Divide the set $O_\xi(t)$ into two parts $O_\xi^{\mathrm{I}}(t)$ and $O_\xi^{\mathrm{II}}(t)$: the subset of those T for which $(\rho(T)T)_\square \cap (T_\xi)_\square$ is nonempty and those T for which $(\rho(T)T)_\square \cap (T_\xi)_\square$ is empty, respectively. For $T \in O_\xi^{\mathrm{I}}(t)$, the regularity conditions on $\mathcal{T}_*$ imply that $d(\dot{T}, \dot{T}_\xi) \leq C$. By Proposition 11.4, the cardinality of $O_\xi^{\mathrm{I}}(t)$ at most $C(2^{2C+1}t)^d$.

For $T \in O_\xi^{\mathrm{II}}(t)$, $(\rho(T)T)_\square \cap (T_\xi)_\square$ is empty, meaning in particular that it cannot contain ξ. Since $(tT)_\square \cap (T_\xi)_\square$ is *not* empty (it contains ξ), it follows that $t > \rho(T)$ for all $T \in O_\xi^{\mathrm{II}}(t)$. By admissibility of the tiling, this means that there exist finitely many tiles $T_1, \ldots, T_K$ centered at the origin with $K \leq K(t)$ such that every $T \in O_\xi^{\mathrm{II}}(t)$ has $d(\dot{T}, T_i) < C$ for some $i \in \{1, \ldots, K\}$. Applying Proposition 11.4 to the subset of $O_\xi^{\mathrm{II}}(t)$ whose translated tiles are close to T_i and summing over i gives that $\#O_\xi^{\mathrm{II}}(t) \leq C(2^{2C+1}t)^dK(t)$. Summing the estimates for $\#O_\xi^{\mathrm{I}}(t)$ and $\#O_\xi^{\mathrm{II}}(t)$ and using the fact that $K(t) \geq 1$ gives the proposition with $C' := 2C2^{d(2C+1)}$. □

11.6 Proof of Theorem 2

Proof of Theorem 2 Theorem 2 will be established as a consequence of Theorem 11.1 by taking $g = \eta$ for a suitable choice of η given by Corollary 11.1. Just as in the proof of Theorem 11.1, one has that

$$\sum_{T \in \mathcal{T}_*} |\widehat{\eta}^T(\xi')|^2 \mathbb{1}_{(1-\theta)T}(\xi') \geq \left(1 - \frac{C'}{W(\epsilon^{-1})}\right)^{2d} \sum_{T \in \mathcal{T}_*} \mathbb{1}_{(1-\theta)T}(\xi'), \tag{11.28}$$

for every $\xi' \in \mathbb{R}^d$, and the sum on the right-hand side is always at least one almost everywhere because $\mathcal{T}_*$ is admissible. As for the upper frame bound, it is useful to divide the sum into terms T for which $\xi' \in T_\square$ and those for which $\xi' \notin T_\square$. Then

$$\sum_{T \in \mathcal{T}_*} |\widehat{\eta}^T(\xi')|^2 \leq \sum_{T \in \mathcal{T}_*} \mathbb{1}_T(\xi') + \sum_{\xi' \notin T_\square} |\widehat{\eta}^T(\xi')|^2. \tag{11.29}$$

Admissibility of $\mathcal{T}_*$ means that the first sum on the right-hand side is at most C for almost every ξ'. As for the second sum, suppose that $T = (D, \xi)$ is a tile such that $\xi' \notin T_\square$. This means that $|D^{-1}(\xi' - \xi)| > \frac{1}{2}$. In this regime, the "tail" inequality (11.22) holds for η and consequently

$$|\widehat{\eta}^T(\xi')|^2 \leq \frac{(C')^2}{(W(\epsilon^{-1}|D^{-1}(\xi' - \xi)|))^2}.$$

This subsequently implies that

$$\begin{aligned} |\widehat{\eta}^T(\xi')|^2 &\leq \int_{2|D^{-1}(\xi'-\xi)|}^{4|D^{-1}(\xi'-\xi)|} \frac{(C')^2}{(W(\epsilon^{-1}|D^{-1}(\xi' - \xi)|))^2} dt \\ &\leq \int_1^\infty \frac{(C')^2 \mathbb{1}_{tT}(\xi')}{(W(\epsilon^{-1}t/4))^2} dt \end{aligned}$$

since $\mathbb{1}_{tT}(\xi')$ is one when $t \geq 2|D^{-1}(\xi' - \xi)| \geq 1$, the interval from $2|D^{-1}(\xi' - \xi)|$ to $4|D^{-1}(\xi' - \xi)|$ has length at least one, and since $W(\epsilon^{-1}t/4) \leq W(\epsilon^{-1}|D^{-1}(\xi' - \xi)|)$ when $t \leq 4|D^{-1}(\xi' - \xi)|$. Inserting this inequality into the right-hand side of (11.29) and interchanging summation and integration gives that almost every $\xi' \in \mathbb{R}^d$ satisfies

$$\begin{aligned} \sum_{T \in \mathcal{T}_*} |\widehat{\eta}^T(\xi')|^2 &\leq C + \int_1^\infty \frac{(C')^2 \sum_{T \in \mathcal{T}_*} \mathbb{1}_{tT}(\xi')}{(W(\epsilon^{-1}t/4\epsilon))^2} dt \\ &= C + \int_1^\infty \frac{(C')^2 \# O_{\xi'}(t)}{(W(\epsilon^{-1}t/4))^2} dt. \end{aligned} \tag{11.30}$$

Let $K(t)$ be the function from the definition of the admissible tiling $\mathcal{T}_*$ and fix

$$W(t) := \begin{cases} 1 & |t| \leq 1, \\ e^{\sqrt{|t|}-1} K(|t|) & |t| \geq 1. \end{cases}$$

Clearly this $W(t)$ is even and radially nondecreasing, has $W(t) = 1$ on $[-1, 1]$, grows faster than any power of t (because K is bounded below) and has

$$\int_0^\infty \frac{\ln W(t)}{t^2} dt = \int_1^\infty \frac{\sqrt{t} - 1 + \ln K(t)}{t^2} dt = 1 + \int_1^\infty \frac{\ln K(t)}{t^2} dt < \infty.$$

For this W and any $\epsilon \in (0, \frac{1}{4})$, it follows that

$$\int_1^\infty \frac{(C')^2 \# O_\xi(t)}{W(\epsilon^{-1}t/4)} dt \leq (C')^2 \int_1^\infty \frac{C' t^d}{e^{\sqrt{t}-1}} dt,$$

by virtue of the upper bound for $\# O_\xi(t)$ given by Proposition 11.5. Applying this inequality to (11.30) (noting that $(W(\epsilon^{-1}t/4))^{-2} \leq (W(\epsilon^{-1}/4))^{-1}(W(\epsilon^{-1}t/4))^{-1}$ on the domain of the integral) gives that

$$\sum_{T \in \mathcal{T}_*} |\widehat{\eta}^T(\xi)|^2 \leq C + C'' e^{-\frac{1}{2\sqrt{\epsilon}}+1}$$

for some constant C'' depending only on C, $K(t)$, and d. Combining this with (11.28) gives that

$$\left(1 - \frac{C'}{W(\epsilon^{-1})}\right)^{2d} \leq \sum_{T \in \mathcal{T}_*} |\widehat{\eta}^T(\xi')|^2 \mathbb{1}_{(1-\theta)T}(\xi') \leq \sum_{T \in \mathcal{T}_*} |\widehat{\eta}^T(\xi')|^2 \leq C + C'' e^{-\frac{1}{2\sqrt{\epsilon}}+1}$$

for almost every $\xi' \in \mathbb{R}^d$. For a suitable choice of ϵ (depending on C, K, θ, and d), the quantity on the left can be taken to be arbitrarily close to 1 and the quantity on the right can be taken as close as desired to C. Applying Theorem 11.1 for its own suitably small ϵ gives (11.4), which in this case corresponds to the frame condition (11.7) promised by Theorem 2. □

References

1. J.J. Benedetto, C. Heil, D.F. Walnut, Differentiation and the Balian-Low theorem. J. Fourier Anal. Appl. **1**(4), 355–402 (1995). MR1350699
2. D. Bytchenkoff, F. Voigtlaender, Design and properties of wave packet smoothness spaces. J. Math. Pures Appl. (9) **133**, 185–262 (2020). MR4044680

3. I. Daubechies, The wavelet transform, time-frequency localization and signal analysis. IEEE Trans. Inform. Theory **36**(5), 961–1005 (1990). MR1066587
4. L. Demanet, L. Ying, Wave atoms and sparsity of oscillatory patterns. Appl. Comput. Harmon. Anal. **23**(3), 368–387 (2007). MR2362408
5. H.G. Feichtinger, M. Fornasier, Flexible Gabor-wavelet atomic decompositions for L^2-Sobolev spaces. Ann. Mat. Pura Appl. (4) **185**(1), 105–131 (2006). MR2179584
6. P.T. Gressman, E. Urheim, Multilinear oscillatory integral operators and geometric stability. J. Geom. Anal. **31**(9), 8710–8734 (2021). MR4302195
7. K. Gröchenig, E. Malinnikova, Phase space localization of Riesz bases for $L^2(\mathbb{R}^d)$. Rev. Mat. Iberoam. **29**(1), 115–134 (2013). MR3010124
8. L. Guth, Decoupling estimates in Fourier analysis, arXiv:2207.00652 (2022)
9. E. Hernández, D. Labate, G. Weiss, A unified characterization of reproducing systems generated by a finite family. II. J. Geom. Anal. **12**(4), 615–662 (2002). MR1916862
10. A.E. Ingham, A note on Fourier transforms. J. Lond. Math. Soc. **9**(1), 29–32 (1934). MR1574706
11. D. Labate, G. Weiss, E. Wilson, An approach to the study of wave packet systems, in *Wavelets, Frames and Operator Theory* (2004), pp. 215–235. MR2066831
12. M. Nielsen, Frames for decomposition spaces generated by a single function. Collect. Math. **65**(2), 183–201 (2014). MR3189276
13. M. Nielsen, K.N. Rasmussen, Compactly supported frames for decomposition spaces. J. Fourier Anal. Appl. **18**(1), 87–117 (2012). MR2885560

Chapter 12
Singular Integral Operators and Hölder Spaces in Dunkl Setting

Yongsheng Han

Abstract We show the boundedness for the Dunkl-Calderón-Zygmund singular integral operators on the Hölder spaces. As consequences, the boundedness of the Dunkl-Riesz transforms on Hölder spaces is obtained.

12.1 Introduction

It is well known that the geometric considerations enter in a decisive way in harmonic analysis. Our purpose of this article is devoted to the study of the boundedness of singular integral operators on the Hölder spaces in the Dunkl setting which is associated with finite reflection groups on the Euclidean space. The Euclidean space $\mathbb{R}^N$, together with the Euclidean metric $\|x - y\|$ and the Dunkl measure ω become a space of homogeneous type in the sense of Coifman and Weiss. However, in the Dunkl setting, there are the Dunkl transform, translation and convolution operators conducted by finite reflection groups. Therefore, the Dunkl setting does not fall into the scope of spaces of homogeneous type. Recently, in [24], the Dunkl-Calderón-Zygmund singular integral operators whose kernels involving both the Euclidean metric and the so called Dunkl "metric" deduced by the finite reflection groups are introduced. This new class of singular integrals covers the well-known Dunkl-Riesz transforms, the Hörmander multipliers [1, 2, 10] and generalizes the classical Calderón-Zygmund singular integrals on spaces of homogeneous type. Moreover, the $T1$ theorem, Littlewood-Paley theory, Dunkl-Hardy spaces and the boundedness of this new class of singular integrals on the Dunkl-Hardy spaces are established. In this article, we show the boundedness of the Dunkl-Calderón-Zygmund singular integral operators on the Hölder spaces. To reach our goal, it is well known that the Littlewood-Paley theory provides a uniform treatment of function spaces and a powerful tool for proving the boundedness of

Y. Han (✉)
Department of Mathematics, Auburn University, Auburn, AL, USA
e-mail: hanyong@auburn.edu

E. Hernández et al. (eds.), *The Mathematical Heritage of Guido Weiss*,
Applied and Numerical Harmonic Analysis,
https://doi.org/10.1007/978-3-031-76793-7_12

Calderón-Zygmund singular integral operators on these function spaces. The key point is that before developing the classical Littlewood-Paley theory, one should establish the Calderón reproducing formula. To see this, let ψ be a Schwartz function satisfying the following conditions:

(i) $\operatorname{supp}\widehat{\psi} \subset \{\xi \in \mathbb{R}^N : 1/2 \leqslant |\xi| \leqslant 2\}$;
(ii) $|\widehat{\psi}(\xi)| \geqslant C > 0$ for $3/5 \leqslant |\xi| \leqslant 5/3$.

Calderón in [4] provided the following formula:

$$f = \sum_{k=-\infty}^{\infty} \phi_k * \psi_k * f, \tag{12.1}$$

where ϕ satisfies the properties similar to ψ, $\psi_k(x) = 2^{kN}\psi(2^k x)$ and the series converges in $L^2(\mathbb{R}^N)$. We point out that the Fourier transform is the main tool to show the above formula. Based on this formula, the square function of f is defined by:

$$S(f)(x) = \left\{ \sum_{k=-\infty}^{\infty} |\psi_k * f(x)|^2 \right\}^{1/2}.$$

The Littlewood-Paley $L^p, 1 < p < \infty$, theory then is given by the following estimates:

$$\|S(f)\|_p \sim \|f\|_p.$$

It is well known that the classical Hardy space $H^p(\mathbb{R}^N)$ can be characterized by such a square function. However, in applications, for example, to get the atomic decomposition and the dual space of $H^p(\mathbb{R}^N)$, the discrete square function is a powerful version. To this end, one needs the following discrete Calderón reproducing formula [11]:

$$f(x) = \sum_{k=-\infty}^{\infty} \sum_{Q} |Q| \phi_Q(x - x_Q) \psi_Q * f(x_Q), \tag{12.2}$$

where ϕ and ψ are same as in the formula (12.1), Q are all dyadic cubes with the side length 2^{-k}, $\phi_Q = \phi_k, \psi_Q = \psi_k$, x_Q are all dyadic points at the left lower corner of Q and the series converges in $L^p(\mathbb{R}^N), 1 < p < \infty$, $S_\infty\left(\mathbb{R}^N\right)$, Schwartz functions with all moment cancellation conditions, and $S'\left(\mathbb{R}^N\right)/\mathcal{P}\left(\mathbb{R}^N\right)$, the space of Schwartz distributions modulo the space of all polynomials. Again, the proof of this discrete Calderón reproducing formula depends on the compact support of the

Fourier transform. This discrete Calderón reproducing formula leads the following discrete Littlewood-Paley square function (see [12–14]):

$$S_d(f)(x) = \left\{ \sum_{k=-\infty}^{\infty} \sum_{Q} |\psi_Q * f(x_Q)|^2 \chi_Q(x) \right\}^{1/2}. \tag{12.3}$$

To generalize the classical Calderón-Zygmund singular integral operator theory to more general setting, Coifman and Weiss introduced spaces of homogeneous type, see [5, 6]. It is well known that the classical harmonic analysis rely on the Fourier transform. Needless to say, the Fourier transform does not exist on spaces of homogeneous type. However, David et al. in [7] developed the Littlewood-Paley theory for spaces of homogeneous type. To be precise, spaces they considered are (X, ρ, μ) with quasi-metric ρ satisfying some regularity properties and the measure μ satisfying the condition $\mu(B(x,r)) \sim r^n, n > 0$. Applying Coifman's approximation to the identity, that is, $\{S_k\}_{k=-\infty}^{\infty}$ is an approximation to the identity on $L^2(X, \mu)$ and together using Coifman's decomposition of the identity on L^2, namely

$$\begin{aligned} I = \sum_{k=-\infty}^{\infty} D_k &= \left(\sum_{k=-\infty}^{\infty} D_k \right) \left(\sum_{j=-\infty}^{\infty} D_j \right) \\ &= \sum_{|k-j| \leqslant M} D_k D_j + \sum_{|k-j| > M} D_k D_j = T_M + R_M, \end{aligned} \tag{12.4}$$

where $D_k = S_{k+1} - S_k$, David et al. proved that R_M is bounded on $L^p(X)$, $1 < p < \infty$, with the operator norm less than 1 if M is large enough and thus, T_M^{-1}, the inverse of T_M, is bounded on $L^p(X)$, $1 < p < \infty$. Denote $D_k^M = \sum_{|j| \leqslant M} D_{k+j}$ for $k \in \mathbb{Z}$, they obtained the following Calderón-type reproducing formulae:

$$f = \sum_{k=-\infty}^{\infty} T_M^{-1} D_k^M D_k(f) = \sum_{k=-\infty}^{\infty} D_k D_k^M T_M^{-1}(f) \tag{12.5}$$

where the series converge in $L^p(X)$, $1 < p < \infty$. Using this formula, they were able to obtain the Littlewood-Paley theory for the space $L^p(X)$: There exists a constant $C > 0$ such that for all $f \in L^p(X)$, $1 < p < \infty$,

$$C^{-1} \|f\|_{L^p(X)} \leqslant \left\| \{ \sum_{k=-\infty}^{\infty} |D_k(f)|^2 \}^{\frac{1}{2}} \right\|_{L^p(X)} \leqslant C \|f\|_{L^p(X)}.$$

Applying the above Littlewood-Paley estimates, they showed the remarkable $T1$ theorem on such spaces of homogeneous type. However, the presence of the operator

T_M^{-1} prevents one from developing Littlewood-Paley characterizations for other spaces, such as the Hardy spaces, by simply applying the Calderón-type formula in (12.5). To obtain the Calderón reproducing formula similar to one given in (12.1), in [17], the space of test functions, $\mathcal{M}(X)$, a suitable analogue of $S_\infty\left(\mathbb{R}^d\right)$, was introduced. The key idea is to show that T_M^{-1} is a Calderón-Zygmund operator whose kernel has some additional second order smoothness and is bounded on $\mathcal{M}(X)$. Since then $T_M^{-1}D_k^M$ satisfy nice estimates sufficiently similar to those satisfied by D_k itself. Precisely, the Calderón reproducing formula provided in [17] is given by the following: There exist families of operators $\{\widetilde{D}_k\}_{k=-\infty}^{\infty}$ and $\{\bar{D}_k\}_{k=-\infty}^{\infty}$ such that

$$f = \sum_{k=-\infty}^{\infty} \widetilde{D}_k D_k(f) = \sum_{k=-\infty}^{\infty} D_k \bar{D}_k(f), \tag{12.6}$$

where the series converge in the norms of the space $L^p(X)$, $1 < p < \infty$, the space $\mathcal{M}(X)$ and the dual space $(\mathcal{M}(X))'$, respectively. In [16], see also [20], applying Coifman's decomposition of the identity together with the boundedness of Calderón-Zygmund operators on the test function space, the following discrete Calderón reproducing formula was established:

$$\begin{aligned} f &= \sum_{k=-\infty}^{\infty} \sum_{Q} \mu(Q) \widetilde{\widetilde{D}}_k(x, x_Q) D_k(f)(x_Q) \\ &= \sum_{k=-\infty}^{\infty} \sum_{Q} \mu(Q) D_k(x, x_Q) \bar{\bar{D}}_k(f)(x_Q), \end{aligned}$$

where Q are dyadic cubes in the sense of Christ, x_Q are any fixed point in Q, and the series converge in the norms of the space $L^p(X)$, $1 < p < \infty$, the space $\mathcal{M}(X)$ and the dual space $(\mathcal{M}(X))'$, respectively. Using the above formulae, the Hardy space $H^p(X)$ theory was established, which includes the Littlewood-Paley characterization of $H^p(X)$, atomic decomposition, dual space of $H^p(X)$, and the $T1$ theorem of the boundedness of singular integrals on these spaces. See [15–17].

In [19, 20], motivated by the work of Nagel and Stein on the several complex variables in [23], they considered spaces of homogeneous type (X, ρ, μ), where the quasi-metric ρ satisfies some regularity properties and the measure μ satisfies the doubling and the reverse doubling conditions, that is, there are constants $\kappa > 0$ and $c \in (0, 1]$ such that

$$c\lambda^\kappa \mu(B(x, r)) \leqslant \mu(B(x, \lambda r)) \tag{12.7}$$

for all $x \in X$, $0 < r < \sup_{x,y \in X} \rho(x, y)/2$ and $1 \leqslant \lambda < \sup_{x,y \in X} \rho(x, y)/2r$.

Applying Coifman's approximation and decomposition of the identity together with some modified test functions, they provided the discrete Calderón reproducing formulae as in (12.7) and established the Hardy space theory in this setting. We would like to point out that the reverse doubling condition of μ and the Calderón-Zygmund operator theory play a crucial role for the boundedness of T_M^{-1} on the space of test functions. See [19, 20] for the Littlewood-Paley characterization of the Hardy space and its applications in this setting.

Notice that the regularity of the quasi-metric ρ is key fact for constructing Coifman's approximation to the identity and the reverse doubling condition is crucial to get the boundedness of T_M^{-1} on the space of test functions. To develop the Littlewood-Paley theory on space of homogeneous (X, ρ, μ), where the quasi-metric ρ has no any regularity and the measure μ satisfies the doubling condition only, a new approach is required.

Adapting the developed randomized dyadic structure on space of homogeneous type (X, ρ, μ) where ρ is the quasi-metric without any regularity and the measure μ satisfies the doubling condition only, in [3] Auscher and Hytönen builded a remarkable orthonormal basis of Hölder-continuous wavelet with exponential decay. Using this wavelet basis they provided a universal Calderón reproducing formula to study and develop function spaces theory and singular integrals. More precisely, they discussed $L^p, 1 < p < \infty$ spaces, BMO and gave a proof of the $T1$ theorem in this general setting. See more details in [3]. Applying Auscher-Hytönen's orthonormal basis, the Hardy space and the product Hardy space were developed in [21, 22].

In this article we consider the Dunkl setting. In [9] Dunkl introduced a similar transform, the Dunkl transform, which enjoys properties similar to the classical Fourier transform. This transform is defined by

$$\hat{f}(x) = c_h \int_{\mathbb{R}^N} E(x, -iy) f(y) h_\kappa^2(y) dy,$$

where the usual character $e^{-i\langle x,y\rangle}$ is replaced by $E(x, -iy) = V_\kappa(e^{-i\langle \cdot,y\rangle})(x)$ for some positive linear operator V_κ and the weight functions h_κ are invariant under a finite reflection group G on $\mathbb{R}^N$. Particularly, the Dunkl transform satisfies the Plancheral identity, namely, $\|\hat{f}\|_2 = \|f\|_2$ and if the parameter $\kappa = 0$, then $h_\kappa(x) = 1$ and $V_\kappa = id$, thus the Dunkl transform becomes the classical Fourier transform. In [25] the translation operator related to Dunkl transform then is defined on the Dunkl transform side by

$$\widehat{\tau_y f}(x) = E(y, -ix)\hat{f}(x)$$

for all $x, y \in \mathbb{R}^N$. When the function f is in the Schwartz class, the above equality holds pointwise. As an operator on $L^2(\mathbb{R}^N, h_\kappa^2)$, τ_y is bounded. However, it is not at all clear whether the translation operator can be defined for L^p functions with

$p \neq 2$. For $f, g \in L^2(\mathbb{R}^N, h_\kappa^2)$, their convolution can be defined in terms of the translation operator by

$$f *_\kappa g(x) = \int_{\mathbb{R}^N} f(y)\tau_x g^\vee(y) h_\kappa^2(y)dy,$$

where $g^\vee(y) = g(-y)$. Observe that there are two metrics in the Dunkl setting: the Euclidean metric $\|x - y\| = \{\sum_{j=1}^{N} |x_j - y_j|^2\}^{\frac{1}{2}}$ and the so-called Dunkl "metric" $d(x, y) = \min_{\sigma \in G} \|x - \sigma(y)\|$, the distance between two G-orbits $O(x)$ and $O(y)$. Obviously, $d(x, y) \leqslant \|x - y\|, d(x, y) = d(y, x)$ and $d(x, y) \leqslant d(x, z) + d(z, y)$ for all $x, y, z \in \mathbb{R}^N$. Moreover, $\omega(B(x, r)) \sim \omega(B(y, r))$ when $\|x - y\| \sim r$ and $\omega(B(x, r)) \leqslant \omega(B_d(x, r)) \leqslant |G|\omega(B(x, r))$, where $B(x, r) =: \{y \in \mathbb{R}^N : \|x - y\| < r\}$ and $B_d(x, r) =: \{y \in \mathbb{R}^N : d(x, y) < r\}$. The Dunkl measure is defined by $d\omega(x) = \prod_{\alpha \in R} |\langle \alpha, x\rangle|^{\kappa(\alpha)} dx$, where the multiplicity function κ defined on R (invariant under G) is fixed ≥ 0, see [2, 8]. Moreover, $d\omega(x)$ satisfies the doubling and reverse doubling properties, that is, there is a constant $C > 0$ such that for all $x \in \mathbb{R}^N, r > 0, \lambda \geqslant 1$ and $\mathbf{N} = N + \sum_{\alpha \in R} \kappa(\alpha)$,

$$C^{-1}\lambda^N \omega(B(x, r)) \leqslant \omega(B(x, \lambda r)) \leqslant C\lambda^{\mathbf{N}} \omega(B(x, r)).$$

It turns out that the Dunkl setting, $(\mathbb{R}^N, \|\cdot\|, \omega)$, is space of homogeneous type in the sense of Coifman and Wiess with the measure ω satisfying the doubling and the reverse doubling properties. Let $\dot{C}^\eta(\mathbb{R}^N)$ be the Hölder space of continuous functions f with

$$\|f\|_{\dot{C}^\eta} := \sup_{x \neq y} \frac{|f(x) - f(y)|}{\|x - y\|^\eta} < \infty.$$

We denote $\dot{C}_0^\eta(\mathbb{R}^N)$ by the Hölder space $\dot{C}^\eta(\mathbb{R}^N)$ with compact supports. The classical Calderón-Zygmund singular integral operator in $(\mathbb{R}^N, \|\cdot\|, \omega)$ is defined by the following

Definition 12.1 An operator $T : \dot{C}_0^\eta(\mathbb{R}^N) \to (\dot{C}_0^\eta(\mathbb{R}^N))'$ with $\eta > 0$, is said to be a classical Calderón-Zygmund singular integral operator if $K(x, y)$, the kernel of T, satisfies the following estimates: for some $0 < \varepsilon \leqslant 1$,

(i) $|K(x, y)| \leqslant \frac{C}{\omega(B(x, \|x-y\|))}$ for $x \neq y$;

(ii) $|K(x, y) - K(x', y)| \leqslant \left(\frac{\|x-x'\|}{\|x-y\|}\right)^\varepsilon \frac{C}{\omega(B(x, \|x-y\|))}$ for $\|x - x'\| \leqslant \frac{1}{2}\|x - y\|$;

(iii) $|K(x, y) - K(x, y')| \leqslant \left(\frac{\|y-y'\|}{\|x-y\|}\right)^\varepsilon \frac{C}{\omega(B(x, \|x-y\|))}$ for $\|y - y'\| \leqslant \frac{1}{2}\|x - y\|$.

Moreover, $\langle T(f), g\rangle = \int_{\mathbb{R}^N}\int_{\mathbb{R}^N} K(x,y)f(x)g(y)d\omega(x)d\omega(y)$ for $\mathrm{supp} f \cap \mathrm{supp} g = \emptyset$. T is said to be a classical Calderón-Zygmund operator if T is bounded on $L^2(\mathbb{R}^N)$.

Motivated by the Dunkl-Riesz transforms, in [24] they introduced the following Dunkl-Calderón-Zygmund singular integral operators.

Definition 12.2 An operator $T : \dot{C}_0^\eta(\mathbb{R}^N) \to (\dot{C}_0^\eta(\mathbb{R}^N))'$ with $\eta > 0$, is said to be a Dunkl-Calderón-Zygmund singular integral operator if $K(x, y)$, the kernel of T, satisfies the following estimates: for some $0 < \varepsilon \leqslant 1$,

$$|K(x,y)| \leqslant C\frac{1}{\omega(B(x,d(x,y)))}\Big(\frac{d(x,y)}{\|x-y\|}\Big)^{\varepsilon} \quad \text{for all} \quad x \neq y; \tag{12.8}$$

$$|K(x,y) - K(x,y')| \leqslant C\Big(\frac{\|y-y'\|}{\|x-y\|}\Big)^{\varepsilon}\frac{1}{\omega(B(x,d(x,y)))} \quad \text{for} \quad \|y-y'\| \leqslant d(x,y)/2; \tag{12.9}$$

$$|K(x',y) - K(x,y)| \leqslant C\Big(\frac{\|x-x'\|}{\|x-y\|}\Big)^{\varepsilon}\frac{1}{\omega(B(x,d(x,y)))} \quad \text{for} \quad \|x-x'\| \leqslant d(x,y)/2. \tag{12.10}$$

Moreover, $\langle T(f), g\rangle = \int_{\mathbb{R}^N}\int_{\mathbb{R}^N} K(x,y)f(x)g(y)d\omega(x)d\omega(y)$ for $\mathrm{supp} f \cap \mathrm{supp} g = \emptyset$. T is said to be a Dunkl-Calderón-Zygmund operator if T is bounded on $L^2(\mathbb{R}^N)$.

We remark that the size and regularity conditions of the Dunkl-Calderón-Zygmund singular integral operator are much weaker than the classical Calderón-Zygmund singular integral operators given in space of homogeneous type in the sense of Coifman and Weiss. Indeed, by the reverse doubling condition on the measure ω, $\omega(B(x,\|x-y\|)) = \omega(B(x, \frac{\|x-y\|}{d(x,y)}\cdot d(x,y)) \geqslant C\big(\frac{\|x-y\|}{d(x,y)}\big)^N \omega(B(x,d(x,y)))$. Thus, $\frac{1}{\omega(B(x,\|x-y\|))} \lesssim \Big(\frac{d(x,y)}{\|x-y\|}\Big)^N \frac{1}{\omega(B(x,d(x,y)))} \lesssim \Big(\frac{d(x,y)}{\|x-y\|}\Big)^{\varepsilon} \frac{1}{\omega(B(x,d(x,y)))}$ and if $\|x-x'\| \leqslant \frac{1}{2}d(x,y) \leqslant \frac{1}{2}\|x-y\|$, then, $\Big(\frac{\|x-x'\|}{\|x-y\|}\Big)^{\varepsilon}\frac{1}{\omega(B(x,\|x-y\|))} \leqslant C\Big(\frac{\|x-x'\|}{d(x,y)}\Big)^{\varepsilon}\frac{1}{\omega(B(x,d(x,y)))}\Big(\frac{d(x,y)}{\|x-y\|}\Big)^{\epsilon} \leqslant C\Big(\frac{\|x-x'\|}{\|x-y\|}\Big)^{\varepsilon} \frac{1}{\omega(B(x,d(x,y)))}$. Further, $K(x,y)$ is locally integrable in $\{\mathbb{R}^N \times \mathbb{R}^N : x \neq y\}$. Indeed, for any fixed $x \in \mathbb{R}^N$ and $0 < \delta < R < \infty$, by the doubling properties of the measure ω,

$$\begin{aligned}\int_{\delta<\|x-y\|<R} |K(x,y)|d\omega(y) &\leqslant C\frac{1}{\delta^{\varepsilon}}\int_{d(x,y)<R}\frac{d(x,y)^{\varepsilon}}{\omega(B(x,d(x,y))}d\omega(y)\\ &\leqslant C\Big(\frac{R}{\delta}\Big)^{\varepsilon} < \infty.\end{aligned}$$

Before state our results, we need to extend the definition of the Dunkl-Calderón-Zygmund operators to functions in $\dot{C}_b^\eta(\mathbb{R}^N)$, the bounded Hölder functions. The idea for doing this is to define $T(f)$ for $f \in \dot{C}_b^\eta(\mathbb{R}^N)$ as a distribution on $\dot{C}_{0,0}^\eta(\mathbb{R}^N) = \{g \in \dot{C}_0^\eta : \int_{\mathbb{R}^N} g(x)d\omega(x) = 0\}$. To this end, given $g \in \dot{C}_{0,0}^\eta(\mathbb{R}^N)$ with the support contained in the ball $B(x_0, R)$ for some $x_0 \in \mathbb{R}^N$ and $R > 0$. Let $\xi(x) = 1$ for $x \in B_d(x_0, 2R)$ and $\xi(x) = 0$ for $x \in \big(B_d(x_0, 4R)\big)^c$. Write $f(x) = \xi(x)f(x) + (1 - \xi(x))f(x)$ and formally, $\langle Tf, g\rangle = \langle T(f\xi), g\rangle + \langle T(1 - \xi)f, g\rangle$. The first term $\langle T(f\xi), g\rangle$ is well defined. By the cancellation condition of g and the fact that if $x \in \mathrm{supp} g$ and $y \notin B_d(x_0, 2R)$, we can write

$$\langle T(1 - \xi)f, g\rangle = \int_{\mathbb{R}^N}\int_{\mathbb{R}^N}[K(x, y) - K(x_0, y)](1 - \xi(y))f(y)g(x)d\omega(y)d\omega(x).$$

Observe that when $x \in B(x_0, R)$ and $y \notin B_d(x_0, 2R)$, $\|x - x_0\| \leqslant R \leqslant \frac{1}{2}d(x_0, y)$. Thus, $\int_{B(x_0,R)}\int_{(B_d(x_0,2R))^c} |K(x, y) - K(x_0, y)||g(x)|d\omega(y)d\omega(x) \leqslant C\|g\|_1$ and hence, $\langle Tf, g\rangle$ is well defined. Therefore, $T(f)$ is a distribution on $\left(\dot{C}_{0,0}^\eta(\mathbb{R}^N)\right)'$. Now $T(1) = 0$ means that for any $f \in \dot{C}_{0,0}^\eta(\mathbb{R}^N)$, $\langle T1, f\rangle = 0$. This is equivalent to $\int\limits_{\mathbb{R}^N} T^*(f)(x)dx = 0$ for any $f \in \dot{C}_{0,0}^\eta(\mathbb{R}^N)$. The definition of $T^*(1) = 0$ is defined similarly.

The *weak boundedness property* (WBP) in the Dunkl setting is defined by following

Definition 12.3 The Dunkl-Calderón-Zygmund singular integral operator T with the distribution kernel $K(x, y)$ is said to have the weak boundedness property if there exist $\eta > 0$ and $C < \infty$ such that

$$|\langle K, f\rangle| \leqslant C \max\{\omega(B(x_0, r)), \omega(B(y_0, r))\}$$

for all $f \in \dot{C}_0^\eta(\mathbb{R}^N \times \mathbb{R}^N)$ with $\mathrm{supp}(f) \subseteq B(x_0, r) \times B(y_0, r), x_0, y_0 \in \mathbb{R}^N$, $\|f\|_\infty \leqslant 1$, $\|f(\cdot, y)\|_{\dot{C}^\eta} \leqslant r^{-\eta}$ for all $y \in \mathbb{R}^N$ and $\|f(x, \cdot)\|_{\dot{C}^\eta} \leqslant r^{-\eta}$ for all $x \in \mathbb{R}^N$.

Now we can state the main result in this article.

Theorem 12.1 *Suppose that T is a Dunkl-Calderón-Zygmund singular integral operator with the kernel $K(x, y)$ satisfying the size condition* (12.8) *and the smoothness condition* (12.10) *only, and $T \in WBP$. Then T is a bounded operator on the Hölder space $\dot{C}^\alpha$ with $0 < \alpha < \varepsilon$ where ε is the exponent of the regularity of the kernel of T, if and only if $T(1)(x) = 0$. Moreover, there is a constant C such that*

$$\|T(f)\|_{\dot{C}^\alpha} \leqslant C\|f\|_{\dot{C}^\alpha}.$$

Before ending this section, we would like to quote what Meyer remarked in his preface to [8]. *"One is amazed by the dramatic changes that occurred in analysis during the twentieth century. In the 1930s complex methods and Fourier series played a seminal role. After many improvements, mostly achieved by the Calderón–Zygmund school, the action takes place today on spaces of homogeneous type. No group structure is available, the Fourier transform is missing, but a version of harmonic analysis is still present. Indeed the geometry is conducting the analysis."*

12.2 The Proof of Theorem 12.1

We begin with describing the Hölder's space $\dot{C}^\alpha(\mathbb{R}^N)$. It is easy to see that the norm of $\dot{C}^\alpha(\mathbb{R}^N)$ is not a norm, because the constant functions have norm zero. Therefore, we define $\dot{C}^\alpha(\mathbb{R}^N), 0 < \alpha < 1$, by the quotient space, modulo the constants. To show Theorem 12.1, we will establish the Littlewood-Paley characterization for $\dot{C}^\alpha(\mathbb{R}^N)$. To this end, we recall the Coifman approximation to the identity on $(\mathbb{R}^N, \|\cdot\|, \omega)$ with the Euclidean metric $\|x-y\|$ and the Dunkl measure ω. Denote $V_r(x) = \omega(B(x,r))$ for $r > 0, x \in \mathbb{R}^N$ and $V(x,y) = \omega(B(x, \|x-y\|))$ for $x, y \in \mathbb{R}^N$. Let $\theta : \mathbb{R} \mapsto [0,1]$ be a smooth function which is 1 for $\|x\| \leqslant 1$ and vanishes for $\|x\| \geqslant 2$. Set $T_k(f)(x) = \int_{\mathbb{R}^N} \theta(2^k\|x-y\|) f(y) d\omega(y), \quad k \in \mathbb{Z}$. Then $T_k(1)(x) \leqslant \int_{\|x-y\| \leqslant 2^{1-k}} d\omega(y) \leqslant C\omega(B(x, 2^{-k}))$ and while, $T_k(1)(x) \geqslant \int_{\|x-y\|<2^{-k}} d\omega(y) = \omega(B(x, 2^{-k}))$. This implies $T_k(1)(x) \sim \omega(B(x, 2^{-k})) = V_k(x)$. Since $V_k(x) \sim V_k(y)$ whenever $\|x-y\| \leqslant 2^{5-k}$, thus, $T_k\left(\frac{1}{T_k(1)}\right)(x) \sim \int_{\mathbb{R}^N} \theta(2^k\|x-y\|) \frac{1}{V_k(y)} d\omega(y) \sim \frac{1}{V_k(x)} T_k(1)(x) \sim 1$. Let M_k be the operator of multiplication by $M_k(x) := \frac{1}{T_k(1)(x)}$ and W_k be the operator of multiplication by $W_k(x) := \left[T_k\left(\frac{1}{T_k(1)}\right)(x)\right]^{-1}$. Coifman's approximation to the identity is constructed by $S_k = M_k T_k W_k T_k M_k$, where $S_k(x,y)$, the kernel of S_k, is given by

$$S_k(x, y) = \int_{\mathbb{R}^n} M_k(x)\theta(2^k\|x-z\|) W_k(z) \theta(2^k\|z-y\|) M_k(y) d\omega(z).$$

In [20], it was proved that kernels $S_k(x,y)$ defined on $\mathbb{R}^N \times \mathbb{R}^N$ satisfy the following properties.

(i) $S_k(x,y) = S_k(y,x)$;

(ii) $S_k(x,y) = 0$ if $\|x-y\| > 2^{(4-k)}$ and $|S_k(x,y)| \leqslant \dfrac{C}{V_k(x) + V_k(y)}$;

(iii) $|S_k(x,y) - S_k(x',y)| \leqslant C \dfrac{2^k\|x-x'\|}{V_k(x)+V_k(y)}$ for $\|x-x'\| \leqslant 2^{(8-k)}$;

(iv) $|S_k(x,y) - S_k(x,y')| \leqslant C \dfrac{2^k\|y-y'\|}{V_k(x)+V_k(y)}$ for $\|y-y'\| \leqslant 2^{(8-k)}$;

(v) $\big|[S_k(x,y)-S_k(x',y)]-[S_k(x,y')-S_k(x',y')]\big| \leqslant C\dfrac{2^k\|x-x'\|2^k\|y-y'\|}{V_k(x)+V_k(y)}$

for $\|x-x'\| \leqslant 2^{(8-k)}$ and $\|y-y'\| \leqslant 2^{(8-k)}$;

(vi) $\displaystyle\int_{\mathbb{R}^N} S_k(x,y)d\omega(x)=1 \qquad \text{for all } y\in\mathbb{R}^N;$

(vii) $\displaystyle\int_{\mathbb{R}^N} S_k(x,y)d\omega(y)=1 \qquad \text{for all } x\in\mathbb{R}^N.$

Coifman's decomposition of the identity on $L^2(\mathbb{R}^N,\omega)$ is given as follows. Let $D_k =: S_k - S_{k-1}$. The identity operator I on $L^2(\mathbb{R}^N,\omega)$ can be written as

$$I=\sum_{k=-\infty}^{\infty} D_k=\sum_{j=-\infty}^{\infty}\sum_{k=-\infty}^{\infty} D_kD_j=T_M^1+R_M^1,$$

where $T_M^1=\sum_{k\in\mathbb{Z}} D_kD_k^M$ with $D_k^M=\sum_{\{|j|\leqslant M\}} D_{k+j}$ and $R_M^1=\sum_{\{|k-j|>M\}} D_kD_j$.

Further, we decompose

$$\begin{aligned}
T_M^1(f)(x) &= \sum_{k\in\mathbb{Z}}\int D_k(x,y)D_k^M(f)(y)\omega(y)\\
&=\sum_{k\in\mathbb{Z}}\sum_{Q\in Q^{k+M}} \omega(Q)D_k(x,x_Q)D_k^M(f)(x_Q)\\
&\quad+\sum_{k\in\mathbb{Z}}\sum_{Q\in Q^{k+M}}\int_Q [D_k(x,y)D_k^M(f)(y)-D_k(x,x_Q)D_k^M(f)(x_Q)]d\omega(y)\\
&= T_M(f)(x)+R_M^2(f)(x).
\end{aligned}$$

where Q^{k+M} is the collection of all dyadic cubes in $\mathbb{R}^N$ such that for each $Q\in Q^{k+M}$ the side length of Q is 2^{-k-M} and x_Q is any fixed point in Q. Then Coifman's decomposition to the identity on $L^2(\mathbb{R}^N,\omega)$ can be written by $I = T_M+R_M^1+R_M^2$. To describe how Coifman's decomposition to the identity deduces the discrete Calderón reproducing formula, we recall the following definition of the test functions in space of homogeneous type $(\mathbb{R}^N, \|\cdot\|, \omega)$:

Definition 12.4 A function $f(x)$ defined on $\mathbb{R}^N$ is said to be a test function if there exits a constant C such that for $0<\beta\leqslant 1, \gamma>0, r>0$ and $x_0\in\mathbb{R}^N$,

(i) $|f(x)| \leqslant \dfrac{C}{V(x,r+\|x-x_0\|)}\Big(\dfrac{r}{r+\|x-x_0\|}\Big)^{\gamma};$

(ii) $|f(x)-f(x')| \leqslant C\Big(\dfrac{\|x-x'\|}{r+\|x-x_0\|}\Big)^{\beta}\dfrac{1}{V(x,r+\|x-x_0\|)}\Big(\dfrac{r}{r+\|x-x_0\|}\Big)^{\gamma},$

for $\quad \|x-x'\|\leqslant \dfrac{1}{2}(r+\|x-x_0\|);$

(iii) $\int_{\mathbb{R}^N} f(x)d\omega(x) = 0.$

We denote such a test function by $f \in \mathcal{M}(\beta, \gamma, r, x_0)$ and $\|f\|_{\mathcal{M}(\beta,\gamma,r,x_0)}$, the norm in $\mathcal{M}(\beta, \gamma, r, x_0)$, is defined by the smallest C satisfying the above conditions (i) and (ii).

The following result is crucial.

Theorem 12.2 *Suppose that T is a classical Calderón-Zygmund operator defined in Definition 12.4 with $T(1) = T^*(1) = 0, T \in WBP$ and $K(x, y)$, the kernel of T, satisfies the additional second order smoothness condition:*

$$|[K(x, y) - K(x', y)] - [K(x, y') - K(x', y')]| \leqslant \Big(\frac{\|x - x'\|\|y - y'\|}{\|x - y\|^2}\Big)^{\varepsilon} \frac{C}{\omega(B(x, \|x - y\|))}$$

for $\|x - x'\| \leqslant \frac{1}{2}\|x - y\|$ and $\|y - y'\| \leqslant \frac{1}{2}\|x - y\|$. Then T is bounded on the test function space $\mathcal{M}(\beta, \gamma, r, x_0)$ for fixed $r > 0, x_0 \in \mathbb{R}^N$, and $0 < \beta, \gamma < \varepsilon$. Moreover, there is a constant C such that

$$\|T(f)\|_{\mathcal{M}(\beta,\gamma,r,x_0)} \leqslant C\|f\|_{\mathcal{M}(\beta,\gamma,r,x_0)}.$$

We remark that the Theorem 12.2 was first given in [15] for the Calderón-Zygmund operators with the para-accretive functions in $\mathbb{R}^N$. See also [8, 14, 17, 18, 20] for such a theorem on spaces of homogeneous type.

Applying the Theorem 12.2 together with Coifman's decomposition for the identity operator, more precisely,

$$\|R^1_M(f) + R^2_M(f)\|_{\mathcal{M}(\beta,\gamma,r,x_0)} \leqslant C_M\|f\|_{\mathcal{M}(\beta,\gamma,r,x_0)}$$

and $C_M < 1$ for any fixed large M, then the discrete Calderón reproducing formula in space of homogeneous type $(\mathbb{R}^N, \|\cdot\|, \omega)$ is given by the following

Theorem 12.3 *Let $\{S_k\}_{k\in\mathbb{Z}}$ be a Coifman's approximations to the identity and set $D_k := S_k - S_{k-1}$. Then there exists a family of operators $\{\widetilde{D}_k\}_{k\in\mathbb{Z}}$ such that for any fixed $x_Q \in Q$ with $k \in \mathbb{Z}$ and Q are dyadic cubes with the side length $2^{(-M-k)}$ with a fixed large M,*

$$f(x) = \sum_{k=-\infty}^{\infty} \sum_{Q\in Q^k} \omega(Q)\widetilde{D}_k(x, x_Q)D_k(f)(x_Q),$$

where the series converge in $L^p(\omega)$, $1 < p < \infty$, $\mathcal{M}(\beta,\gamma,r,x_0)$, *and in* $(\mathcal{M}(\beta,\gamma,r,x_0))'$, *the dual of in* $\mathcal{M}(\beta,\gamma,r,x_0)$, *and moreover, the kernels of the operators* $\widetilde{D}_k$ *satisfy the following conditions: for any* $0 < \varepsilon < 1$,

(i) $|\widetilde{D}_k(x,y)| \leqslant C\frac{1}{V_k(x)+V_k(y)+V(x,y)}\Big(\frac{2^{-k}}{2^{-k}+\|x-y\|}\Big)^{\varepsilon}$;

(ii) $|\widetilde{D}_k(x,y)-\widetilde{D}_k(x',y)|$
$\leqslant C\Big(\frac{\|x-x'\|}{2^{-k}+\|x-x'\|}\Big)^{\varepsilon}\frac{1}{V_k(x)+V_k(y)+V(x,y)}\Big(\frac{2^{-k}}{2^{-k}+\|x-y\|}\Big)^{\varepsilon}$,
for $\|x-x'\| \leqslant (2^{-k}+\|x-y\|)/2$;

(iii) $\int_{\mathbb{R}^N}\widetilde{D}_k(x,y)d\omega(x)=0$ *for all* $y\in\mathbb{R}^N$;

(iv) $\int_{\mathbb{R}^N}\widetilde{D}_k(x,y)d\omega(y)=0$ *for all* $x\in\mathbb{R}^N$.

Similarly, there exists a family of linear operators $\{\widetilde{\widetilde{D}}_k\}_{k\in\mathbb{Z}}$ *such that for any fixed* $x_Q\in Q$,

$$f(x)=\sum_{k=-\infty}^{\infty}\sum_{Q\in Q^k}\omega(Q)D_k(x,x_Q)\widetilde{\widetilde{D}}_k(f)(x_Q),$$

where the kernels of the operators $\widetilde{\widetilde{D}}_k$ *satisfy the above conditions (i), (iii), (iv) and (ii) with x and y interchanged. See [20].*

We now introduce the Besov spaces as follows.

Definition 12.5 Suppose $f\in(\mathcal{M}(\beta,\gamma,r,x_0))'$ and $0<\alpha<1$. The Dunkl-Besov space $\dot{B}^{\alpha,\infty}_{\infty}$ is defined to be the collection of f such that

$$||f||_{\dot{B}^{\alpha,\infty}_{\infty}}=\sup_{\substack{k\in\mathbb{Z}\\ x\in\mathbb{R}^N}}2^{\alpha k}|D_kf(x)|<\infty$$

and the Dunkl-Besov space $\dot{B}^{-\alpha,1}_{1}$ to be the collection of f such that

$$||f||_{\dot{B}^{-\alpha,1}_{1}}=\sum_{k=-\infty}^{\infty}\sum_{Q\in Q^k}2^{-\alpha k}\omega(Q)|\widetilde{D}_k(f)(x_Q)|<\infty,$$

where D_k and $\widetilde{D}_k$ are given as in the Theorem 12.3.

We would like to point out that these definitions are well defined, that is, they are independent of the choices of the approximations S_k (and D_k) constructed by Coifman's approximation to the identity. Moreover, the dual space of $\dot{B}^{\alpha,\infty}_{\infty}$ is $\dot{B}^{-\alpha,1}_{1}$ for $0<\alpha<1$. See [20]. The following result gives the characterization of the Hölder space.

Proposition 12.1 *$f \in \dot{C}^\alpha(\mathbb{R}^N)$, $0 < \alpha < 1$, if and only if $f \in \dot{B}_\infty^{\alpha,\infty}$. Moreover, $\|f\|_{\dot{C}^\alpha} \sim \|f\|_{\dot{B}_\infty^{\alpha,\infty}}$.*

Suppose that $f \in \dot{C}^\alpha(\mathbb{R}^N)$, $0 < \alpha < 1$, and $g \in \mathcal{M}(\beta, \gamma, r, x_0)$ with $0 < \alpha < \gamma \leqslant 1$. Then by the cancellation and size conditions on g,

$$\Big|\int_{\mathbb{R}^N} f(x)g(x)d\omega(x)\Big| = \Big|\int_{\mathbb{R}^N} [f(x) - f(x_0)]g(x)d\omega(x)\Big| \leqslant C\|f\|_{\dot{C}^\alpha}\|g\|_{\mathcal{M}(\beta,\gamma,r,x_0)}.$$

Particularly, $|D_k(f)(x)| = \Big|\int_{\mathbb{R}^N} D_k(x,y)[f(y) - f(x)]d\omega(y)\Big| \leqslant C2^{-\alpha k}\|f\|_{\dot{C}^\alpha}$ and hence, $f \in \dot{B}_\infty^{\alpha,\infty}$. Moreover,

$$\|f\|_{\dot{B}_\infty^{\alpha,\infty}} \leqslant C\|f\|_{\dot{C}^\alpha}.$$

We would like to point out that if $f \in \dot{C}^\alpha(\mathbb{R}^N)$, $0 < \alpha < 1$, we can define $\langle f, g\rangle$, as a distribution for all $g \in \mathcal{M}(\beta, \gamma, r, x_0)$, by the bilinear form

$$\langle f, g\rangle = \sum_{k=-\infty}^{\infty} \sum_{Q\in Q^k} \omega(Q)\widetilde{D}_k(g)(x_Q)D_k(f)(x_Q).$$

Indeed, if $g \in \mathcal{M}(\beta, \gamma, r, x_0)$, then $g \in \dot{B}_1^{-\alpha,1}(\mathbb{R}^N, \omega)$ and

$$\sum_{k=-\infty}^{\infty} \sum_{Q\in Q^k} \omega(Q)|\widetilde{D}_k(g)(x_Q)D_k(f)(x_Q)| \leqslant C\|f\|_{\dot{B}_\infty^{\alpha,\infty}}\|g\|_{\dot{B}_1^{-\alpha,1}}.$$

To see that $f \in \dot{B}_\infty^{\alpha,\infty}$ implies $f \in \dot{C}^\alpha$, let $f \in \dot{B}_\infty^{\alpha,\infty}$ and $g \in \mathcal{M}(\beta, \gamma, r, x_0)$ with $0 < \alpha < \gamma$. Applying the Calderón reproducing formula given in the Theorem 12.3, we write

$$\langle f, g\rangle = \langle \sum_{k=-\infty}^{0} \sum_{Q\in Q^k} \omega(Q)\widetilde{D}_k(\cdot, x_Q)D_k(f)(x_Q), g\rangle$$

$$+ \langle\sum_{k=1}^{\infty} \sum_{Q\in Q^k} \omega(Q)\widetilde{D}_k(\cdot, x_Q)D_k(f)(x_Q), g\rangle =: I + II.$$

By the concellation of g, we can write

$$I = \langle \sum_{k=-\infty}^{0} \sum_{Q\in Q^k} \omega(Q)[\widetilde{D}_k(x, x_Q) - \widetilde{D}_k(x_0, x_Q)]D_k(f)(x_Q), g(x)\rangle.$$

Applying the smoothness condition on $\widetilde{D}_k$ yields that for $0 < \alpha < \varepsilon$,

$$\begin{aligned}&|\widetilde{D}_k(x, x_Q) - \widetilde{D}_k(x_0, x_Q)||D_k(f)(x_Q)|\\ &\leqslant C\|x - x_0\|^{\varepsilon} 2^{-k(\varepsilon-\alpha)}\Big\{\frac{1}{V_k(x) + V_k(x_Q) + V(x, x_Q)} \times \Big(\frac{2^{-k}}{2^{-k} + \|x - x_Q\|}\Big)^{\varepsilon}\\ &+ \frac{1}{V_k(x_0) + V_k(x_Q) + V(x_0, x_Q)}\Big(\frac{2^{-k}}{2^{-k} + \|x_0 - x_Q\|}\Big)^{\varepsilon}\Big\}.\end{aligned}$$

And thus, the series $\sum\limits_{k=-\infty}^{0}\sum\limits_{Q\in Q^k} \omega(Q)[\widetilde{D}_k(x, x_Q) - \widetilde{D}_k(x_0, x_Q)]D_k(f)(x_Q)$ converges unformly for $\|x - x_0\| \leqslant M$ with any fixed $M > 0$. Therefore, this series converges to a continuous function on $\mathbb{R}^N$. While for term II, the size estimate, $|D_k(f)(x_Q)| \leqslant C2^{-\alpha k}$, implies that $\sum\limits_{k=1}^{\infty}\sum\limits_{Q\in Q^k} \omega(Q)\widetilde{D}_k(x, x_Q)D_k(f)(x_Q)$ converges uniformly on $\mathbb{R}^N$ and thus, it is a continuous function for all x. Therefore, f is a continuous function. To see $f \in \dot{C}^{\alpha}(\mathbb{R}^N)$, given $x, x' \in \mathbb{R}^N$ and for some k_0 such that $2^{-k_0} \leqslant \|x - x'\| < 2^{1-k_0}$, split $f(x) - f(x')$ by $A = \sum\limits_{k\leqslant k_0}\sum\limits_{Q\in Q^k} \omega(Q)[\widetilde{D}_k(x, x_Q) - \widetilde{D}_k(x', x_Q)]D_k(f)(x_Q)$ and $B = \sum\limits_{k>k_0}\sum\limits_{Q\in Q^k} \omega(Q)[\widetilde{D}_k(x, x_Q) - \widetilde{D}_k(x', x_Q)]D_k(f)(x_Q)$. For the first series, applying the smoothness condition of $\widetilde{D}_k$ and the estimate of $D_k(f)(x_Q)$ yields $|A| \leqslant C\|f\|_{\dot{B}^{\alpha,\infty}_{\infty}} \sum\limits_{k\leqslant k_0} 2^{(\varepsilon-\alpha)k}\|x - x'\|^{\varepsilon} \leqslant C\|x - x'\|^{\alpha}\|f\|_{\dot{B}^{\alpha,\infty}_{\infty}}$. The estimates of $D_k(f)$ and the size condition of $\widetilde{D}_k$ give $|B| \leqslant C\sum\limits_{k>k_0} 2^{-\alpha k}\|f\|_{\dot{B}^{\alpha,\infty}_{\infty}} \leqslant C\|x - x'\|^{\alpha}\|f\|_{\dot{B}^{\alpha,\infty}_{\infty}}$. Moreover, $\|f\|_{\dot{C}^{\alpha}} \leqslant C\|f\|_{\dot{B}^{\alpha,\infty}_{\infty}}$.

We now show the Theorem 12.1. It's proof follows from following estimate: there exists a constant C such that if $f \in \dot{C}^{\alpha}(\mathbb{R}^N)$ with $0 < \alpha < \varepsilon$, where ε is the exponent of the regularity of the kernel of T, then

$$|\langle Tf(x), D_k(x, x_Q)\rangle| \leqslant C2^{-k\alpha}\|f\|_{C^{\alpha}}, \tag{12.11}$$

where $D_k(x, x_Q)$ are the functions in the Theorem 12.3 for $k \in \mathbb{Z}$ and $x_Q \in \mathbb{R}^N$.

To show the estimate in (12.11), suppose that $f(x) \in \dot{C}^{\alpha}(\mathbb{R}^N)$ with $0 < \alpha < \varepsilon \leqslant 1$. Set $1 = \xi(y) + \eta(y)$, where $\xi(y) = \theta\Big(\frac{d(y,x_Q)}{2^{5-k}}\Big)$ with $\theta \in C_0^{\infty}(\mathbb{R}), \theta(x) = 1$ for $\|x\| \leqslant 1$ and $\theta(x) = 0$ for $\|x\| \geqslant 2$. Applying the condition $T(1) = 0$ and the fact that $D_k(x, x_Q) \in \dot{C}^1_{0,0}(\mathbb{R}^N)$ for any fixed $k \in \mathbb{Z}$ and $x_Q \in \mathbb{R}^N$, we can write

$$\begin{aligned}\langle Tf(x), D_k(x, x_Q)\rangle &= \langle K(x, y), [f(y) - f(x_Q)]D_k(x, x_Q)\rangle\\ &= \langle K(x, y), (\xi(y) + \eta(y))[f(y) - f(x_Q)]D_k(x, x_Q)\rangle\\ &= \langle K(x, y), \xi(y)(f(y) - f(x))D_k(x, x_Q)\rangle\end{aligned}$$

$$
\begin{aligned}
&+ \langle K(x,y), \xi(y)[f(x) - f(x_Q)]D_k(x, x_Q)\rangle \\
&+ \langle K(x,y), \eta(y)[f(y) - f(x_Q)]D_k(x, x_Q)\rangle \\
&=: I_1 + I_2 + I_3.
\end{aligned}
$$

We first estimate the last term I_3. By the cancellation condition on $D_k(x, x_Q)$,

$$
\begin{aligned}
&\langle K(x,y), \eta(y)[f(y) - f(x_Q)]D_k(x, x_Q)\rangle \\
&= \langle K(x,y) - K(x_Q, y), \eta(y)[f(y) - f(x_Q)]D_k(x, x_Q)\rangle.
\end{aligned}
$$

Observing that by the support of $D_k(x, x_Q)$, $\|x - x_Q\| \leqslant 5 \cdot 2^{-k}$, we have $d(x, x_Q) \leqslant 5 \cdot 2^{-k}$. Since the support of $\eta(y)$ is contained in $\{y \mid d(y, x_Q) \geqslant 20 \cdot 2^{-k}\}$, so $d(y, x_Q) \geqslant \frac{1}{2}\|x - x_Q\|$ and hence, applying the smoothness condition on $K(x, y)$ and $f \in \dot{C}^\alpha, 0 < \alpha < \varepsilon$,

$$
\begin{aligned}
|I_3| \lesssim & \int_{\substack{d(y,x_Q) \geqslant 20 2^{-k} \\ d(x,x_Q) \leqslant 5 2^{-k}}} \frac{1}{\omega(B(x, d(y, x_Q)))} \Big(\frac{d(x, x_Q)}{\|y - x_Q\|}\Big)^\varepsilon \|y - x_Q\|^\alpha |D_k(x, x_Q)| \\
& d\omega(y) d\omega(x) \|f\|_{C^\alpha} \lesssim 2^{-k\alpha}\|f\|_{C^\alpha}.
\end{aligned}
$$

To estimate the term I_2, we claim that if x is in the support of $D_k(x, x_Q)$, then

$$
|T\xi(x)| \lesssim 1. \tag{12.12}
$$

To show (12.12), let $\psi \in C^\eta_{0,0}(\mathbb{R}^N)$ with $\mathrm{supp}(\psi) \subseteq B(x_Q, 5 \cdot 2^{-k})$. Applying the facts $\xi + \eta = 1$and $T(1) = 0$ yields

$$
\begin{aligned}
|\langle T\xi, \psi\rangle| &= |-\langle T\eta, \psi\rangle| \\
&= \Big| \int_{\mathbb{R}^N}\int_{\mathbb{R}^N} [K(x, y) - K(x_Q, y)]\eta(y)\psi(x) d\omega(y) d\omega(x) \Big|.
\end{aligned}
$$

Observe that the supports of $\eta(y)$ and $\psi(x)$ imply that $\|x - x_Q\| \leqslant \frac{1}{2} d(y, x_Q)$ and thus, the smoothness condition of K yields

$$
\begin{aligned}
&|\langle T\xi, \psi\rangle| \\
&\lesssim \iint_{d(y,x_Q) > 15 \cdot 2^{-k} \geqslant 2\|x - x_Q\|} \frac{1}{\omega(B(x, d(y, x_Q)))} \Big(\frac{d(x, x_Q)}{\|y - x_Q\|}\Big)^\varepsilon d\omega(y) |\psi(x)| d\omega(x) \\
&\lesssim \int_{\mathbb{R}^N} |\psi(x)| d\omega(x).
\end{aligned}
$$

This implies that $T\xi(x) = \sigma + \gamma(x)$ for $x \in B(x_Q, 5\cdot 2^{-k})$ with σ is a constant depending on ξ and $\|\gamma(x)\|_\infty \leqslant C_0$ for some constant C_0 independent of ξ. To estimate σ, choose $\varphi \in C_0^\eta(\mathbb{R}^N)$ with supp$\varphi \subseteq B(x_Q, 5\cdot 2^{-k})$, $0 \leqslant \varphi \leqslant 1$, $\|\varphi\|_\eta \leqslant r^{-\eta}$ and $\int_{\mathbb{R}^N}\varphi(x)d\omega(x) = C_1\omega(B(x_Q, 5\cdot 2^{-k}))$, for some constant C_1 independent of k. We then use $T \in WBP$ to get

$$\left|C_1\omega(B(x_Q,5\cdot 2^{-k}))\sigma + \int_{\mathbb{R}^N}\varphi(x)\gamma(x)d\omega(x)\right| = |\langle T\xi,\varphi\rangle| \leqslant C\omega(B(x_Q,5\cdot 2^{-k})),$$

which implies $|\sigma| \leqslant C_0 + \frac{C}{C_1}$ and hence, the proof of (12.12) is complete.

The estimate of the term I_2 then follows from (12.12). Indeed,

$$|I_2| \lesssim \|[f(x) - f(x_Q)]D_k(x, x_Q)\|_1 \lesssim 2^{-k\alpha}\|f\|_{C^\alpha}.$$

For the first term I_1, let $\lambda_\delta(x, y) = \theta(\frac{\|x-y\|}{\delta})$, $\delta > 0$. Then

$$\begin{aligned}
&\langle K(x, y), \xi(y)(f(y) - f(x))D_k(x, x_Q)\rangle\\
=&\langle K(x, y), (1 - \lambda_\delta(x, y))\xi(y)[f(y) - f(x)]D_k(x, x_Q)\rangle\\
&+ \langle K(x, y), \lambda_\delta(x, y)\xi(y)[f(y) - f(x)]D_k(x, x_Q)\rangle =: I_{1,\delta} + I_{2,\delta}.
\end{aligned} \tag{12.13}$$

Since K is locally integrable on $\Omega = \{(x, y) \in \mathbb{R}^N \times \mathbb{R}^N : x \neq y\}$. By the size condition on $K(x, y)$ and $f \in \dot{C}^\alpha$ together with the fact that if $\xi(y) \neq 0$, $\psi(x) \neq 0$ and $1 - \lambda_\delta(x, y) \neq 0$, then $\delta \leqslant \|x - y\|$ and $d(x, y) \leqslant 20\cdot 2^{-k}$. Thus, the first term on the right side of (12.13) satisfies

$$\begin{aligned}
&|I_{1,\delta}|\\
&= \left|\iint_{\substack{\|x-y\|\geqslant\delta\\ d(x,y)\leqslant 20\cdot 2^{-k}}} K(x, y)(1 - \lambda_\delta(x, y))\xi(y)[f(y) - f(x)]\right.\\
&\qquad\left.\times D_k(x, x_Q)(x)d\omega(y)d\omega(x)\right|\\
&\lesssim \iint_{d(x,y)\leqslant 20\cdot 2^{-k}} \frac{1}{\omega(B(x, d(x, y)))}\\
&\qquad\times\left(\frac{d(x, y)}{\|x - y\|}\right)^\alpha \|x - y\|^\alpha |D_k(x, x_Q)|d\omega(y)d\omega(x)\|f\|_{C^\alpha}\\
&\lesssim 2^{-k\alpha}\|f\|_{C^\alpha}.
\end{aligned}$$

It suffices to show that $\lim_{\delta\to 0} I_{2,\delta} = 0$, that is, for any fixed $k \in \mathbb{Z}$ and $x_Q \in \mathbb{R}^N$,

$$\lim_{\delta\to 0}\langle K(x, y), \lambda_\delta(x, y)\xi(y)[f(y) - f(x)]D_k(x, x_Q)\rangle = 0. \tag{12.14}$$

To this end, let $\{y_j\}_{j\in\mathbb{Z}} \in \mathbb{R}^N$ be the maximal collection of points satisfying

$$\frac{1}{2}\delta < \inf_{j\neq k} \|y_j - y_k\| \leqslant \delta. \tag{12.15}$$

Observing that $\{y_j\}_{j\in\mathbb{Z}}$ is a maximal collection, for each $x \in \mathbb{R}^N$ there exists a point y_j such that $\|x - y_j\| \leqslant \delta$. Let $\eta_j(y) = \theta(\frac{\|y-y_j\|}{\delta})$ and $\bar{\eta}_j(y) = [\sum_{i=1}^{\infty} \eta_i(y)]^{-1}\eta_j(y)$. To see that $\bar{\eta}_j$ is well defined, it suffices to show that for any $y \in \mathbb{R}^N$, there are only finitely many η_j with $\eta_j(y) \neq 0$. This follows from the following fact: $\eta_j(y) \neq 0$ if and only if $\|y - y_j\| \leqslant 2\delta$ and hence this implies that $B(y_j, \delta) \subseteq B(y, 4\delta)$. Inequality (12.15) shows $B(y_j, \frac{\delta}{4}) \cap B(y_k, \frac{\delta}{4}) = \emptyset$ for $j \neq k$ and hence there are at most C_0 points $y_j \in \mathbb{R}^N$ such that $B(y_j, \frac{\delta}{4}) \subseteq B(y, 4\delta)$. Now let $\Gamma = \{j : \bar{\eta}_j(y)\chi_0(y) \neq 0\}$. Then $\#\Gamma \leqslant Cr^N/\delta^N$ since $\mathrm{supp}(\chi_0) \subseteq \bigcup_{\sigma\in G} B(\sigma(x_0), 2r)$ and $\mathrm{supp}(\bar{\eta}_j) \subseteq B(y_j, 2\delta)$. We write

$$\begin{aligned}&\langle K(x, y), \lambda_\delta(x, y)\xi(y)[f(y) - f(x)]D_k(x, x_Q)\rangle\\ &= \sum_{j\in\Gamma} \langle K(x, y), \lambda_\delta(x, y)\bar{\eta}_j(y)\xi(y)[f(y) - f(x)]D_k(x, x_Q)\rangle.\end{aligned}$$

Note that we may choose $0 < \delta < 52^{-k}$ for any fixed $k \in \mathbb{Z}$ and hence, $\mathrm{supp}(\lambda_\delta(x, y)\bar{\eta}_j(y)\xi(y)[f(y) - f(x)]D_k(x, x_Q) \subseteq B(y_j, 4\delta) \times B(y_j, 2\delta)$ and

$$\|\lambda_\delta(x, y)\bar{\eta}_j(y)\xi(y)[f(y) - f(x)]D_k(x, x_Q)\|_\infty \leqslant C\delta^\eta,$$

where C is a constant depending only on θ, ϕ, ψ, x_Q, and k but not on δ and j.

We claim that

$$\|\lambda_\delta(\cdot, y)\bar{\eta}_j(y)\xi(y)[f(y) - f(\cdot)]D_k(\cdot, x_Q)\|_\eta \lesssim 1, \tag{12.16}$$

and

$$\|\lambda_\delta(x, \cdot)\bar{\eta}_j(\cdot)\xi(\cdot)[f(\cdot) - f(x)]D_k(x, x_Q)\|_\eta \lesssim 1. \tag{12.17}$$

Assuming (12.16) and (12.17) for the moment, since $T \in WBP$, we have

$$
\begin{aligned}
&|\langle K(x,y), \lambda_\delta(x,y)\xi(y)[f(y)-f(x)]D_k(x,x_Q)\rangle| \\
&\leqslant \sum_{j\in\Gamma} |\langle K(x,y), \lambda_\delta(x,y)\bar{\eta}_j(y)\xi(y)[f(y)-f(x)]D_k(x,x_Q)\rangle| \\
&\leqslant \sum_{j\in\Gamma} \omega(B(y_j,4\delta))\delta^\eta \lesssim \frac{2^{-kN}}{\delta^N} \sup_{j\in\Gamma} \omega(B(y_j,1))\delta^N\delta^\eta \lesssim \sup_{j\in\Gamma} \omega(B(y_j,1))2^{-kN}\delta^\eta,
\end{aligned}
$$

and hence, (12.14) holds.

It remains to show (12.16) and (12.17). We only show (12.16) since the proof for (12.17) is similar. To check (12.16), it suffices to show that for given $x_1, x_2 \in \mathbb{R}^N$ with $\|x_1 - x_2\| \leqslant \delta$,

$$
\begin{aligned}
&|\bar{\eta}_j(y)\chi_0(y)||\lambda_\delta(x_1,y)[\phi(y)-\phi(x_1)]\psi(x_1) - \lambda_\delta(x_2,y)[\phi(y)-\phi(x_2)]\psi(x_2)| \\
&\lesssim \|x_1 - x_2\|^\eta,
\end{aligned}
$$

since if $\|x_1 - x_2\| \geqslant \delta$, then the expansion on the left above is clearly bounded by

$$
\begin{aligned}
&|\bar{\eta}_j(y)\chi_0(y)|\{|\lambda_\delta(x_1,y)[\phi(y)-\phi(x_1)]\psi(x_1)| + |\lambda_\delta(x_2,y)[\phi(y)-\phi(x_2)]\psi(x_2)|\} \\
&\lesssim \delta^\eta \leqslant \|x_1 - x_2\|^\eta.
\end{aligned}
$$

By the construction of $\bar{\eta}_j$, it follows that $|\bar{\eta}_j(y)\chi_0(y)| \lesssim 1$ for all $y \in \mathbb{R}^N$. Thus

$$
\begin{aligned}
&\Big|\bar{\eta}_j(y)\xi(y)\Big|\Big|\lambda_\delta(x_1,y)[f(y)-f(x_1)]D_k(x_1,x_Q) - \lambda_\delta(x_2,y)[f(y)-f(x_2)] \\
&\quad D_k(x_2,x_Q)\Big| \\
&\lesssim \Big|\lambda_\delta(x_1,y) - \lambda_\delta(x_2,y)|[\phi(y) - f(x_1)]D_k(x_1,x_Q)\Big| \\
&+ \Big|\lambda_\delta(x_2,y)[f(x_1)-f(x_2)]D_k(x_1,x_Q)\Big| \\
&+ \Big|\lambda_\delta(x_2,y)[f(y)-f(x_2)][D_k(x_1,x_Q) - D_k(x_2,x_Q)]\Big| \\
&=: I_1 + I_2 + I_3.
\end{aligned}
$$

Since $\|x_1 - x_2\| \leqslant \delta$, if $\|x_1 - y\| > 4\delta$, then $\lambda_\delta(x_1,y) = \lambda_\delta(x_2,y) = 0$, so $I_1 = 0$. We may assume that $\|x_1 - y\| \leqslant 4\delta$, $\delta < 1$ and hence,

$$
I_1 \lesssim \left|\frac{\|x_1 - y\|}{\delta} - \frac{\|x_2 - y\|}{\delta}\right| \|x_1 - y\|^\eta \lesssim \delta^{\eta-1}\|x_1 - x_2\| \lesssim \|x_1 - x_2\|^\eta.
$$

It is easy to see that $I_2 + I_3 \lesssim \|x_1 - x_2\|^\eta$. The proofs of (12.16) and (12.17) are complete and we obtain the estimate in (12.11).

To finish the proof of the Theorem 12.1, observe that if $f \in \dot{C}^\alpha$ and $g \in \mathcal{M}(\beta, \gamma, r, x_0)$, as mentioned before, $\langle f, g\rangle$ can be defined by

$$\langle f, g\rangle = \sum_{k=-\infty}^{\infty} \sum_{Q \in Q^k} \omega(Q)\widetilde{D}_k(g)(x_Q)D_k(f)(x_Q).$$

This leads to define

$$\langle Tf, g\rangle = \sum_{k=-\infty}^{\infty} \sum_{Q \in Q^k} \omega(Q)\widetilde{D}_k(g)(x_Q)D_k(Tf)(x_Q).$$

To see this is well defined, by the estimate in (12.11), $|D_k(Tf)(x_Q)| \leqslant C2^{-\alpha k}\|f\|_{C^\alpha}$ and hence, $|\langle Tf, g\rangle| \leqslant C\|f\|_{\dot{C}^\alpha}\|g\|_{\dot{B}_1^{-\alpha,1}}$. Thus, $\langle Tf, g\rangle$ is well defined. Again, by the estimate in (12.11) and the characterization given in the Proposition 12.1, $Tf \in \dot{C}^\alpha$. Moreover, $\|Tf\|_{\dot{C}^\alpha} \leqslant C\|f\|_{\dot{C}^\alpha}$. The proof of the Theorem 12.1 is complete.

It is known that the Dunkl-Riesz transforms R_j, $1 \leqslant j \leqslant N$, are bounded on $L^2(\mathbb{R}^N, \omega)$ and $H^1(\mathbb{R}^N, \omega)$, respectively, see [1] and [2]. Moreover, they are the Dunkl-Calderón-Zygmund singular integral operators, see [24]. Therefore, R_j, $1 \leqslant j \leqslant N$, satisfy the conditions in the Theorem 12.1 and hence, as a direct consequence of the Theorem 12.1, we obtain the following

Theorem 12.4 *The Dunkl-Riesz transforms R_j, $1 \leqslant j \leqslant N$, are bounded on the Hölder space $\dot{C}^\alpha(\mathbb{R}^N)$, $0 < \alpha < 1$. Moreover, these exists a constant C such that for $1 \leqslant j \leqslant N$,*

$$\|R_j(f)\|_{\dot{C}^\alpha} \leqslant C\|f\|_{\dot{C}^\alpha}.$$

The following $T(1)$ theorem for the Dunkl-Calderón-Zygmund singular integral operators was provided in [24].

Theorem 12.5 *Suppose that T is a Dunkl-Calderón-Zygmund singular integral operator. Then T extends to a bounded operator on $L^2(\mathbb{R}^N, \omega)$ if and only if (a) $T(1)(x) \in BMO_d(\mathbb{R}^N, \omega)$; (b) $T^*(1)(x) \in BMO_d(\mathbb{R}^N, \omega)$; (c) $T \in WBP$.*

Acknowledgments The author would like to thank the referees for helpful comments which have improved the presentation of the paper.

References

1. B. Amri, M. Sifi, Riesz transforms for Dunkl transform. Ann. Math. Blaise Pascal **19**, 247–262 (2012)
2. J.-Ph. Anker, J. Dziubański, A. Hejna, Harmonic functions, conjugate harmonic functions and the Hardy H^1 in rational Dunkel setting. J. Fourier Anal. Appl. **25**(5), 2356–2418 (2019)
3. P. Auscher, T. Hytönen, Orthonormal bases of regular wavelets in spaces of homogeneous type. Appl. Comput. Harmon. Anal. **34**, 266–296 (2013)
4. A.P. Calderón, Intermediate spaces and interpolation, the complex method. Studia Math. **24**, 113–190 (1964)
5. R.R. Coifman, G. Weiss, *Analyse Harmonique Non-commutative sur Certains Espaces Homogènes*, Lecture Notes in Math., vol. 242 (Springer, Berlin, 1971)
6. R.R. Coifman, G. Weiss, Extensions of Hardy spaces and their use in analysis. Bull. Amer. Math. Soc. **83**, 569–645 (1977)
7. G. David, J.L. Journé, S. Semmes, Opérateurs de Calderón-Zygmund, fonctions paraaccrétives et interpolation. Rev. Mat. Iberoam. **1**, 1–56 (1985)
8. D. Deng, Y. Han, *Harmonic Analysis on Spaces of Homogeneous Type*, Lecture Notes in Math., vol. 1966 (Springer, Berlin, 2009), with a preface by Yves Meyer
9. C.F. Dunkl, Differential-difference operators associated to reflection groups. Trans. Amer. Math. **311**(1), 167–183 (1989)
10. J. Dziubański, A. Hejna, Hörmander multiplier theorem for the Dunkl transform. J. Funct. Anal. **277**, 2133–2159 (2019)
11. M. Frazier, B. Jawerth, Decomposition of Besov spaces. Indiana Univ. Math. J. **34**(4), 777–799 (1985)
12. M. Frazier, B. Jawerth, A discrete transform and decomposition of distribution spaces. J. Funct. Anal. **93**, 34–170 (1990)
13. M. Frazier, B. Jawerth, G. Weiss, *Littlewood-Paley Theory and the Study of Function Spaces*, CBMS Regional Conference Series, vol. 79 (A.M.S., Providence, RI, 1991)
14. J.E. Gilbert, Yo. Han, J.A. Hogan, J.D. lakey, D. Weiland, G. Weiss, Smooth molecular decompositions of functions and singular integral operators. Mem. Amer. Math. Soc. **156**(742), viii+74 pp. (2002)
15. Y. Han, Calderón-type reproducing formula and the Tb theorem. Rev. Mat. Iberoam. **10**, 51–91 (1994)
16. Y. Han, Plancherel-Pólya type inequality on spaces of homogeneous type and its applications. Proc. Amer. Math. Soc. **126**, 3315–3327 (1998)
17. Y. Han, E.T. Sawyer, Littlewood-Paley theory on spaces of homogeneous type and classical function spaces. Mem. Amer. Math. Soc. **110**(530), 1–126 (1994)
18. Y. Han, G. Weiss, *Function Spaces on Spaces of Homogeneous Type. Essays on Fourier Analysis in Honor of Elias M. Stein (Princeton, NJ, 1991)*. Princeton Math. Ser., vol. 42 (Princeton Univ. Press, Princeton, NJ, 1995), pp. 211–224
19. Y. Han, D. Müller, D. Yang, Littlewood-Paley characterizations of Hardy spaces on spaces of homogeneous type. Math. Nachr. **279**(13–14), 1505-1537 (2006)
20. Y. Han, D. Müller, D. Yang, A theory of Besov and Triebel-Lizorkin spaces on metric measure spaces modeled on Carnot–Carathéodory spaces. Abstr. Appl. Anal. **2008**, Article ID 893409, 250 pp. (2008)
21. Ya. Han, Yo. Han, J. Li, Criterion of the boundedness of singular integrals on spaces of homogeneous type. J. Funct. Anal. **271**, 3423–3464 (2016)
22. Y. Han, J. Li, L. Ward, Product H^p, CMO^p, VMO and duality via orthonormal bases on spaces of homogeneous type. Appl. Comput. Harmon. Anal. **45**(1), 120–169 (2018)

23. A. Nagel, E.M. Stein, The $\overline{\partial}_b$-complex on decoupled boundaries in $\mathbb{C}^n$. Ann. Math. **164**, 649–713 (2006)
24. C. Tan, Y. Han, Y. Han, M.-Y. Lee, J. Li, Criterion of the L2 boundedness in Dunkl setting and applications, Ann. Sc. Norm. Super. Pisa. CI. Sci. (5), https://doi.org/10.2422/2036-2145.202208_006
25. S. Thangvelu, Y. Xu, Convolution operator and maximal function for the Dunkl transform. J. Anal. Math. **97**, 25–55 (2005)

Chapter 13
An Homage to Guido Weiss and His Leadership of the Saint Louis Team: Commutators of Singular Integrals and Sobolev Inequalities

Cong Hoang, Kabe Moen, and Carlos Pérez

Abstract We extend some classical Sobolev type inequalities for linear and non-linear commutators.

13.1 Introduction

It is with great pleasure that we describe some of the results concerning commutators in [6], which has been one of the main mathematical contributions of Guido L. Weiss as part of the celebrated "Saint Louis team" during the 7th decade of the twentieth century, in connection with some new Sobolev type estimates in the spirit of our recent work [13]. This will be outlined in Sect. 13.3. Similarly, in Sect. 13.4 we will outline Sobolev type estimates as in [13] for certain nonlinear commutators introduced in another relevant paper [33]. Another very influential work of Guido is [36] which can be seen as model case of the work of Muckenhoupt and Wheeden [25] central in the applications of our results in [13].

We will begin with a perspective from our recent work [13] where the classical Gagliardo-Nirenberg-Sobolev (GNS) inequalities have been generalized and extended in several directions.

C. Hoang (✉)
Department of Mathematics, Florida Agricultural and Mechanical University, Tallahassee, FL, USA
e-mail: cong.hoang@famu.edu

K. Moen
Department of Mathematics, University of Alabama, Tuscaloosa, AL, USA
e-mail: kabe.moen@ua.edu

C. Pérez
Department of Mathematics, University of the Basque Country, Ikerbasque and BCAM, Bilbao, Spain
e-mail: cperez@bcamath.org

E. Hernández et al. (eds.), *The Mathematical Heritage of Guido Weiss*,
Applied and Numerical Harmonic Analysis,
https://doi.org/10.1007/978-3-031-76793-7_13

13.2 Recent Extensions of the Classical GNS Inequalities

13.2.1 Historical Remarks

The classical Sobolev inequalities from the 1930s of the last century which became the GNS inequalities in the 1960s of the same century have been improved in many directions. The purpose of this paper is to survey some recent improvements and extensions obtained by the authors in [13]. We begin by outlining the history of the GNS estimates in its different incarnations related to the context of L^p spaces and in $\mathbb{R}^n$ for $n \geq 2$ (see the recent treatment [16] Section 3.2.).

13.2.1.1 The Origins

Our starting point is commonly referred to as the GNS inequalities which are ubiquitous in harmonic analysis and P.D.E. Hereafter, we will use the notation of $p^* = \frac{np}{n-p}$ for the Sobolev exponent, often expressed in the following way

$$\frac{1}{p} - \frac{1}{p^*} = \frac{1}{n}.$$

Theorem A (The Classical GNS) *Let* $1 \leq p < n$. *Then*

$$\|f\|_{L^{p^*}(\mathbb{R}^n)} \leq c\,\|\nabla f\|_{L^p(\mathbb{R}^n)}, \quad f \in C_c^\infty(\mathbb{R}^n). \tag{13.1}$$

One common approach to proving (13.1), typically only when $1 < p < n$, is via the use of the well-known pointwise estimate (see [16, 34, 35])

$$|f(x)| \leq c\, I_1(|\nabla f|)(x) \tag{13.2}$$

combined with the boundedness (a well-known classical result)

$$I_1 : L^p(\mathbb{R}^n) \to L^{p^*}(\mathbb{R}^n) \tag{13.3}$$

for $p \in (1, n)$. I_1 is the Riesz potential operator of order $\alpha = 1$ given by

$$I_\alpha f(x) = \int_{\mathbb{R}^n} \frac{f(y)}{|x-y|^{n-\alpha}}\, dy, \quad 0 < \alpha < n.$$

At the endpoint $p = 1$ and $1^* = \frac{n}{n-1} = n'$, I_1 only satisfies the weak endpoint boundedness

$$I_1 : L^1(\mathbb{R}^n) \to L^{n',\infty}(\mathbb{R}^n),$$

yet the strong inequality (13.1) still holds in this case (more on this below).

13.2.1.2 The 1970s, the Influence of Muckenhoupt and Wheeden

A relevant extension of (13.3) and hence of (13.1) in the case $p > 1$ was obtained by B. Muckenhoupt and R. Wheeden [25]. They showed in this paper that I_1 satisfies weighted bounds of the form

$$\|wI_1 f\|_{L^{p^*}(\mathbb{R}^n)} \leq c\, \|wf\|_{L^p(\mathbb{R}^n)} \tag{13.4}$$

if and only if $w \in A_{p,p^*}$ namely

$$[w]_{A_{p,p^*}} = \sup_Q \left(⨍_Q w^{p^*} \right) \left(⨍_Q w^{-p'} \right)^{\frac{p^*}{p'}} < \infty.$$

Again, at the endpoint $p = 1$ only the weak boundedness holds, and Muckenhoupt and Wheeden proved that

$$\|I_1 f\|_{L^{n',\infty}(w^{n'})} \leq c\, \|wf\|_{L^1(\mathbb{R}^n)} \tag{13.5}$$

holds if and only if $w \in A_{1,n'}$, i.e.

$$\left(⨍_Q w^{n'} \right) \leq c \inf_Q (w^{n'})$$

where the smallest constant c is denoted as $[w]_{A_{1,n'}}$.

Combining (13.2) and (13.4) yields the following weighted GNS inequality.

Theorem B (The Weighted GNS) *Let* $1 \leq p < n$ *and let* $w \in A_{p,p^*}$. *Then*

$$\|wf\|_{L^{p^*}(\mathbb{R}^n)} \leq c\, \|w\nabla f\|_{L^p(\mathbb{R}^n)}, \qquad f \in C_c^\infty(\mathbb{R}^n). \tag{13.6}$$

When $p = 1$ and $w \in A_{1,n'}$, the operator I_1 only satisfies a weak-type inequality, yet the strong bound holds for the weighted GNS as we will detail in the next section.

13.2.1.3 The Influence of Modern Weighted Theory in Harmonic Analysis

Besides its dependence on the dimension and p, the constant c in (13.6) also depends on the constant $[w]_{A_{p,p^*}}$, and this relationship is of interest.

The sharp weighted bound for I_1 was proven in [17]

$$\|wI_1 f\|_{L^{p^*}(\mathbb{R}^n)} \leq c_p\, [w]_{A_{p,p^*}}^{\frac{1}{n'}\max\{1, \frac{p'}{p^*}\}} \|wf\|_{L^p(\mathbb{R}^n)},\ 1 < p < n, \tag{13.7}$$

which leads to the following quantitative weighted GNS inequality,

$$\|wf\|_{L^{p^*}(\mathbb{R}^n)} \leq c_p\,[w]_{A_{p,p^*}}^{\frac{1}{n'}\max\{1,\frac{p'}{p^*}\}}\,\|w\nabla f\|_{L^p(\mathbb{R}^n)}, \quad 1 < p < n. \tag{13.8}$$

Inequality (13.8) can be improved and the sharp version of the GNS inequalities is contained in the next theorem which can also be found in [17].

Theorem C (The Precise Weighted GNS) *Let* $1 \leq p < n$ *and let* $w \in A_{p,p^*}$. *Then*

$$\|wf\|_{L^{p^*}(\mathbb{R}^n)} \leq c_p\,[w]_{A_{p,p^*}}^{\frac{1}{n'}}\,\|w\nabla f\|_{L^p(\mathbb{R}^n)}, \quad 1 \leq p < n. \tag{13.9}$$

Notice that inequality (13.9) improves (13.8) in the exponent and holds at the endpoint $p = 1$. In fact, it is the endpoint $p = 1$ that is crucial in proving (13.9). We now briefly explain how to obtain the sharp inequality (13.9) by using the theory of extrapolation of weights. First, using the pointwise inequality (13.2), we claim the weak type endpoint inequality for $p = 1$ and $w \in A_{1,n'}$,

$$\|f\|_{L^{n',\infty}(w^{n'})} \leq c\,[w]_{A_{1,n'}}^{\frac{1}{n'}}\,\|w\nabla f\|_{L^1(\mathbb{R}^n)}.$$

Given any measure μ, the quantity $\|g\|_{L^{n',\infty}(\mu)}$ is equivalent to a norm since $n' > 1$. Since I_1 is a convolution operator with the kernel $k(x) = |x|^{1-n}$, we may use the Marcinkiewicz integral inequality for normed spaces to obtain

$$\begin{aligned}\|I_1 f\|_{L^{n',\infty}(\mu)} = \|k * f\|_{L^{n',\infty}(\mu)} &\leq c \int_{\mathbb{R}^n} |f(y)|\,\big\||\cdot - y|^{1-n}\big\|_{L^{n',\infty}(\mu)}\,dy \\ &\leq c \int_{\mathbb{R}^n} |f(y)| M\mu(y)^{\frac{1}{n'}}\,dy\end{aligned}$$

where we have used the fact that $\big\||\cdot - y|^{1-n}\big\|_{L^{n',\infty}(\mu)} \approx M\mu(y)^{\frac{1}{n'}}$, with M being the usual Hardy-Littlewood maximal function. Thus I_1 satisfies the weak bound

$$\|I_1 f\|_{L^{n',\infty}(w^{n'})} \leq c \int_{\mathbb{R}^n} |f| M(w^{n'})^{\frac{1}{n'}} \leq c\,[w]_{A_{1,n'}}^{\frac{1}{n'}}\,\|wf\|_{L^1(\mathbb{R}^n)}. \tag{13.10}$$

Inequality (13.10) now implies

$$\|I_1 f\|_{L^{p^*,\infty}(w^{p^*})} \leq c_p\,[w]_{A_{p,p^*}}^{\frac{1}{n'}}\,\|wf\|_{L^p(\mathbb{R}^n)}, \quad 1 \leq p < n, \tag{13.11}$$

by the sharp extrapolation theorem from [17, Theorem 2.1] which is based on the off-diagonal extrapolation theorem by Harboure et al. [12]. Inequality (13.11) now implies the weak Sobolev inequality

$$\|f\|_{L^{p^*,\infty}(w^{p^*})} \le c\|I_1(|\nabla f|)\|_{L^{p^*,\infty}(w^{p^*})} \le c_p[w]_{A_{p,p^*}}^{\frac{1}{n'}}\|w\nabla f\|_{L^p(\mathbb{R}^n)}. \qquad (13.12)$$

The truncation method of Maz'ya [23] (see also the survey article [11]) applied to (13.12) now implies the strong inequality with the same dependence on $[w]_{A_{p,p^*}}$, namely, inequality (13.9).

The sharpness of (13.9) follows from the following example. Let

$$f_\delta(x) = \exp(-|x|^\delta) \quad \text{and} \quad w_\delta(x) = |x|^{\frac{\delta-n}{p^*}}$$

where $\delta \in (0, 1)$. Then,

$$\|w_\delta f_\delta\|_{L^{p^*}(\mathbb{R}^n)} \approx \delta^{-\frac{1}{p^*}}, [w_\delta]_{A_{p,p^*}} \approx \delta^{-1}, \text{ and } \|w_\delta \nabla f_\delta\|_{L^p(\mathbb{R}^n)} \approx \delta^{\frac{1}{p'}}$$

showing that the power $\frac{1}{n'}$ in (13.9) cannot be improved by taking $\delta \to 0^+$.

13.2.2 The Toy Motivation: The Maximal Function Case

We are interested in extending inequality (13.9) to an inequality with an operator on the left hand side. We begin with the Hardy-Littlewood maximal function defined by

$$Mf(x) = \sup_{Q \ni x} \fint_Q |f(y)|\, dy.$$

The classical A_p theorem of Muckenhoupt showed that

$$M : L^p(w) \to L^p(w)$$

if and only if $w \in A_p$, i.e.

$$[w]_{A_p} = \sup_Q \left(\fint_Q w\right)\left(\fint_Q w^{1-p'}\right)^{p-1} < \infty.$$

There is an intimate connection between the A_p and A_{p,p^*} classes which will be very useful in what follows, namely

$$w \in A_{p,p^*} \text{ if and only if } w^{p^*} \in A_{\frac{p^*}{n'}}$$

and

$$[w]_{A_{p,p^*}} = [w^{p^*}]_{A_{\frac{p^*}{n'}}}.$$

Hence, if $w \in A_{p,p^*}$, then the weight w^{p^*} belongs to A_{p^*} since $A_{\frac{p^*}{n'}} \subset A_{p^*}$.

By Muckenhoupt's theorem we obtain the following Sobolev inequality

$$\|wMf\|_{L^{p^*}(\mathbb{R}^n)} \le c\,\|w\nabla f\|_{L^p(\mathbb{R}^n)}, \tag{13.13}$$

which is an extension of (13.6) since $Mf \ge |f|$, a.e. by the Lebesgue differentiation theorem.

Consider the problem of finding the dependence on $[w]_{A_{p,p^*}}$ of the constant c in inequality (13.13). One approach is to use Buckley's sharp bound in [1] which yields

$$\|Mf\|_{L^p(w)} \le cp'\,[w]_{A_p}^{\frac{1}{p-1}}\,\|f\|_{L^p(w)}. \tag{13.14}$$

This inequality leads to the following calculation,

$$\begin{aligned}\|wMf\|_{L^{p^*}(\mathbb{R}^n)} &= \|Mf\|_{L^{p^*}(w^{p^*})} \le c_p\,[w^{p^*}]_{A_{p^*}}^{\frac{1}{p^*-1}}\,\|f\|_{L^{p^*}(w^{p^*})}\\ &\le c_p\,[w^{p^*}]_{A_{p^*}}^{\frac{1}{p^*-1}}\,[w]_{A_{p,p^*}}^{\frac{1}{n'}}\,\|w\nabla f\|_{L^p(\mathbb{R}^n)} \le c_p\,[w]_{A_{p,p^*}}^{\frac{1}{p^*-1}+\frac{1}{n'}}\,\|w\nabla f\|_{L^p(\mathbb{R}^n)},\end{aligned}$$

using (13.9). We may improve these estimates by using the fact that the weight w^{p^*} belongs to the better class $A_{\frac{p^*}{n'}}$. Here we will use the following observation by J. Duoandikoetxea [9] (see also [21]) who showed that if $w \in A_q$ for $1 \le q < p$, namely a stronger condition, then

$$\|Mf\|_{L^p(w)} \le c_{p,q}\,[w]_{A_q}^{\frac{1}{p}}\,\|f\|_{L^p(w)}.$$

This leads to

$$\|wMf\|_{L^{p^*}(\mathbb{R}^n)} \le c\,[w]_{A_{p,p^*}}^{\frac{1}{p^*}+\frac{1}{n'}}\,\|w\nabla f\|_{L^p(\mathbb{R}^n)}, \tag{13.15}$$

using again (13.9). Observe that (13.15) also holds for $p = 1$. We do not know if the exponent is sharp as in the case of (13.9).

A similar argument works for iterations of M, namely

$$\|wM^m f\|_{L^{p^*}(\mathbb{R}^n)} \le c_{m,p}\,[w]_{A_{p,p^*}}^{\frac{m}{p^*}+\frac{1}{n'}}\,\|w\nabla f\|_{L^p(\mathbb{R}^n)}, \quad 1 \le p < n, \tag{13.16}$$

where M^m denotes the m-th iteration of the Hardy-Littlewood maximal operator, namely $M^m = \underbrace{M \circ \cdots \circ M}_{(m\, times)}$.

Recently, the estimates above have been significantly improved in [13] by using a different approach. In this work, it is shown that the bound

$$\|wMf\|_{L^{p^*}(\mathbb{R}^n)} \leq c_p\, [w]_{A_{p,p^*}}^{\frac{1}{n'}\max\{1,\frac{p'}{p^*}\}} \|w\nabla f\|_{L^p(\mathbb{R}^n)}, \tag{13.17}$$

holds when $p \in (1, n)$ although the result is worse near $p = 1$ than (13.15). This result follows applying M to (13.2) using the fact that $I_1 f \in A_1$ with dimensional constant, independent of f, namely

$$Mf(x) \leq c\, I_1(|\nabla f|)(x), \quad f \in C_c^\infty(\mathbb{R}^n).$$

The same pointwise bound holds for the iterated maximal operator, namely

$$M^m f(x) \leq c_m\, I_1(|\nabla f|)(x), \quad f \in C_c^\infty(\mathbb{R}^n),$$

and hence

$$\|wM^m f\|_{L^{p^*}(\mathbb{R}^n)} \leq c_{m,p}\, [w]_{A_{p,p^*}}^{\frac{1}{n'}\max\{1,\frac{p'}{p^*}\}} \|w\nabla f\|_{L^p(\mathbb{R}^n)}, \quad 1 < p < n. \tag{13.18}$$

Observe that (13.18) is an extension of (13.9) since, again, $M^m f \geq |f|$, a.e. Furthermore, the exponent is better than (13.16) since it does not depend on m.

When $p = 1$, we cannot use the truncation method mentioned earlier to derive the sharp estimate (13.9) since M and M^m are non-local operators. Nevertheless, (13.18) is also true, but with a worse constant, depending on m, namely

$$\|wM^m f\|_{L^{n'}(\mathbb{R}^n)} \leq c_m\, [w]_{A_{1,n'}}^{\frac{m+1}{n'}} \|w\nabla f\|_{L^1(\mathbb{R}^n)}. \tag{13.19}$$

It is possible to further improve (13.16) and hence (13.9). Indeed, if we define

$$M_r f = \left[M(|f|^r)\right]^{\frac{1}{r}} \quad \text{for } r \in (0, \infty),$$

then it is possible to prove in the range $p \in (1, n)$ that

$$\|wM_{n'} f\|_{L^{p^*}(\mathbb{R}^n)} \leq c_p\, [w]_{A_{p,p^*}}^{\frac{1}{n'}\max\{1,\frac{p'}{p^*}\}} \|w\nabla f\|_{L^p(\mathbb{R}^n)}, \tag{13.20}$$

by showing the pointwise estimate (see [13])

$$M_{n'} f(x) \leq c\, I_1(|\nabla f|)(x), \tag{13.21}$$

and using (13.7). Again, (13.20) improves the estimate that is obtained by using Buckley's inequality (13.14), namely the less desirable bound,

$$\|wM_{n'}f\|_{L^{p^*}(\mathbb{R}^n)} \leq c_p\,[w]_{A_{p,p^*}}^{\frac{1}{p^*-n'}+\frac{1}{n'}}\,\|w\nabla f\|_{L^p(\mathbb{R}^n)}.$$

Moreover, we cannot use $A_{\frac{p^*}{n'}} \subset A_{p^*}$ because $A_{\frac{p^*}{n'}}$ is the precise class of weights for which $M_{n'}$ is bounded.

Observe that inequality (13.20) is better than inequality (13.16). This follows from the fact that $M^k f \leq c_{r,k}\,M_r f$ since $M_r f \in A_1$ for $r > 1$, with constant independent of f.

Inequality (13.20) holds in the range $1 < p < n$, but it is false when $p = 1$. In this case we have two different replacements. First, we claim that

$$\|M_{n'}f\|_{L^{n',\infty}(w^{n'})} \leq c\,[w]_{A_{1,n'}}^{\frac{2}{n'}}\,\|w\nabla f\|_{L^1(\mathbb{R}^n)},$$

using the classical A_1 weak type (1, 1) estimate of Fefferman-Stein,

$$\|Mf\|_{L^{1,\infty}(u)} \leq c\,\|f\|_{L^1(Mu)},$$

combined with (13.9) when $p = 1$.

Alternatively, consider the quantity $\|wM_{n'}f\|_{L^{n',\infty}(\mathbb{R}^n)}$ which is very different from $\|M_{n'}f\|_{L^{n',\infty}(w^{n'})}$. We claim that for a constant c depending on the dimension and $[w]_{A_{1,n'}}$ we have

$$\|wM_{n'}f\|_{L^{n',\infty}(\mathbb{R}^n)} \leq c\,\|w\nabla f\|_{L^1(\mathbb{R}^n)}.$$

The difficulty is that the weight is inside the distribution level of the definition of the norm making the problem non-standard. However, it is proved in [7] (improved in [20] and then in [32]) that the Hardy-Littlewood maximal operator satisfies a Sawyer-type estimate of the form

$$\left\|\frac{Mf}{v}\right\|_{L^{1,\infty}(uv)} \leq c_{[u]_{A_1},[uv]_{A_\infty}}\,\|f\|_{L^1(u)}, \tag{13.22}$$

where $u \in A_1$ and $uv \in A_\infty$.

Applying this with $v = w^{-n'}$ and $u = \frac{1}{v}$, since $w^{n'} \in A_1$ we have

$$\begin{aligned}
\|wM_{n'}f\|_{L^{n',\infty}(\mathbb{R}^n)} &= \|w^{n'}M(|f|^{n'})\|_{L^{1,\infty}(\mathbb{R}^n)}^{\frac{1}{n'}} \\
&\leq c_{[w]_{A_{1,n'}}}\,\|w^{n'}|f|^{n'}\|_{L^1(\mathbb{R}^n)}^{\frac{1}{n'}} \\
&= c_{[w]_{A_{1,n'}}}\,\|wf\|_{L^{n'}} \leq c_{[w]_{A_{1,n'}}}\,\|w\nabla f\|_{L^1(\mathbb{R}^n)}
\end{aligned}$$

for which we have to use the inequalities (13.9) and (13.22). The explicit dependence on the constant is not ideal and we do not state it here.

Interestingly enough, inequality (13.21) is false for the pointwise larger operator $M_{n'+\epsilon}$, when $\epsilon > 0$. It is possible to replace $M_{n'}$ by a slightly larger maximal operator defined in the scale of the Lorentz space $L^{n',1}$. Namely, it is shown in [13], that

$$M_{n',1} f(x) \leq c\, I_1(|\nabla f|)(x), \tag{13.23}$$

which yields an improvement of (13.20) and hence of (13.9) when $p \in (1, n)$. Here we are using the normalized definition of the

$$M_{s,q} f(x) = \sup_{x \in Q} \frac{1}{|Q|^{1/s}} \left\| f\, \chi_Q \right\|_{L^{s,q}(dx)},$$

$0 < s, q \leq \infty$, where

$$\|f\|_{L^{s,q}(\mu)} = \left[s \int_0^\infty \left(t\, \mu\{x \in \mathbb{R}^n : |f(x)| > t\}^{1/s} \right)^q \frac{dt}{t} \right]^{1/q} < \infty,$$

if $q < \infty$, and $\sup_{0<t<\infty} t\, \mu\{x \in \mathbb{R}^n : |f(x)| > t\}^{1/s} < \infty$, when $q = \infty$. The proof of inequality (13.23) is based on the following improvement of the classical Poincaré-Sobolev inequality in Lorentz spaces due to O'Neil [26] and Peetre [29] (see also [22]),

$$\|f - f_Q\|_{L^{p^*,p}(Q)} \leq c\, \ell(Q) \left(\fint_Q |\nabla f|^p \right)^{\frac{1}{p}},$$

for $1 \leq p < n$, where we are using the normalized Lorentz average. In particular, when $p = 1$, we have the following

$$\|f - f_Q\|_{L^{n',1}(Q)} \leq c\, \ell(Q) \fint_Q |\nabla f|$$

from which we can derive (13.23) easily.

A natural question would be to find the largest maximal function M_Φ for which (13.23) holds, where Φ is a Young function and M_Φ is defined through a Luxemburg type norm.

13.2.3 The Case of Singular Integrals

Motivated by maximal operator results discussed above, we proved similar results for more complicated singular integral operators in [13]. This will be the case of several classical smooth convolution singular integrals but we will be considering operators beyond Calderón-Zygmund operators like the so-called rough singular integral operators T_Ω. Surprisingly, we found Sobolev type estimate as (13.17) or (13.20) for these operators which are defined by the expression

$$T_\Omega f(x) = p.v. \int_{\mathbb{R}^n} \frac{\Omega(y')}{|y|^n} f(x-y)\, dy$$

where $y' = \frac{y}{|y|} \in \mathbb{S}^{n-1}$ which are well defined if $\Omega \in L^1(\mathbb{S}^{n-1})$ and has zero average on $\mathbb{S}^{n-1}$. We are not able to work with the assumption $\Omega \in L^1(\mathbb{S}^{n-1})$ to obtain our results, but rather $\Omega \in L^{n,\infty}(\mathbb{S}^{n-1})$ which properly contains the spaces $L^r(\mathbb{S}^{n-1})$ for $r \geq n$.

One of our main results from [13] is the following.

Theorem 13.1 ([13]) *Let T_Ω be a rough singular integral convolution operator as above with $\Omega \in L^{n,\infty}(\mathbb{S}^{n-1})$. Then*

$$|T_\Omega f(x)| \leq c\, \|\Omega\|_{L^{n,\infty}(\mathbb{S}^{n-1})}\, I_1(|\nabla f|)(x), \quad f \in C_c^\infty(\mathbb{R}^n). \tag{13.24}$$

Hence, if $p \in (1, n)$ and if we let $w \in A_{p,p^*}$, we have

$$\|wT_\Omega f\|_{L^{p^*}(\mathbb{R}^n)} \leq c_p\, \|\Omega\|_{L^{n,\infty}(\mathbb{S}^{n-1})}\, [w]_{A_{p,p^*}}^{\frac{1}{n'}\max\{1,\frac{p'}{p^*}\}}\, \|w\nabla f\|_{L^p(\mathbb{R}^n)} \tag{13.25}$$

and

$$\|T_\Omega f\|_{L^{n',\infty}(w^{n'})} \leq c\, \|\Omega\|_{L^{n,\infty}(\mathbb{S}^{n-1})}\, [w]_{A_{1,n'}}^{\frac{1}{n'}}\, \|w\nabla f\|_{L^1(\mathbb{R}^n)}.$$

Similar results hold for rough maximal type operators and more exotic maximal operators such as the spherical maximal operator.

We note here that all the singular integral operators for which we have proven (13.24) have a cancellation on the kernel. We do not know if (13.24) also holds for Calderón-Zygmund operators of non-convolution type. We do however have reasonable bounds for these operators. Recall that these are a continuous linear operator $T : C_0^\infty(\mathbb{R}^n) \to \mathcal{D}'(\mathbb{R}^n)$ which extend to bounded operators on $L^2(\mathbb{R}^n)$, and whose distributional kernel coincides away from the diagonal $x = y$ in $\mathbb{R}^n \times \mathbb{R}^n$, with a function K so that

$$Tf(x) = \int_{\mathbb{R}^n} K(x,y) f(y) dy, \quad f \in C_0^\infty(\mathbb{R}^n), x \notin \operatorname{supp}(f),$$

and satisfies the standard estimates, namely, the size estimate

$$|K(x, y)| \leq \frac{c}{|x-y|^n} \tag{13.26}$$

and the regularity condition: for some $\epsilon > 0$,

$$|K(x, y) - K(z, y)| + |K(y, x) - K(y, z)| \leq c \frac{|x-z|^\epsilon}{|x-y|^{n+\epsilon}}, \tag{13.27}$$

whenever $2|x-z| < |x-y|$.

By the well-known classical theory, we have

$$T : L^p(\mathbb{R}^n) \to L^p(\mathbb{R}^n) \tag{13.28}$$

when $1 < p < \infty$ and

$$T : L^1(\mathbb{R}^n) \to L^{1,\infty}(\mathbb{R}^n). \tag{13.29}$$

We will be using the following result due to Coifman-Fefferman [5],

$$\|Tf\|_{L^p(w)} \leq c_{p,T}\, [w]_{A_\infty} \|Mf\|_{L^p(w)}, \tag{13.30}$$

for $p \in (0, \infty)$ and $w \in A_\infty$. The original proof does not yield this precise bound since some further ideas are required. See details in [28]. A similar result holds for the weak norm.

Now, arguing as in Sect. 13.2.2, using the sharp weighted estimate for Calderón-Zygmund operators, since $w^{p^*} \in A_{p^*}$, we have

$$\begin{aligned} \|wTf\|_{L^{p^*}(\mathbb{R}^n)} &\leq c_{p,T}\, [w^{p^*}]_{A_{p^*}}^{\max\{1, \frac{1}{p^*-1}\}} \|wf\|_{L^{p^*}(\mathbb{R}^n)} \\ &\leq c_{p,T}\, [w]_{A_{p,p^*}}^{\max\{1, \frac{1}{p^*-1}\}+\frac{1}{n'}} [w]_{A_{p,p^*}}^{\frac{1}{n'}} \|w\nabla f\|_{L^p(\mathbb{R}^n)} \end{aligned}$$

in the case $p \in (1, n)$, which is certainly less optimal than (13.25). Moreover, we may use the fact that $A_{\frac{p^*}{n'}}$ is a proper subclass of A_{p^*} and the estimate of Duoandikoetxea [9] (see also [18])

$$\|Tf\|_{L^p(w)} \leq c_{p,q,T}\, [w]_{A_q} \|f\|_{L^p(w)}, \quad 1 \leq q < p,$$

to obtain when $p \in (1, n)$

$$\|wTf\|_{L^{p^*}(\mathbb{R}^n)} \leq c_{p,T}\, [w]_{A_{p,p^*}}^{1+\frac{1}{n'}} \|w\nabla f\|_{L^p(\mathbb{R}^n)}. \tag{13.31}$$

On the other hand, if we use (13.30) instead, we claim

$$\|wTf\|_{L^{p^*}(\mathbb{R}^n)} \leq c_{p,T}\,[w^{p^*}]_{A_\infty}\,[w]_{A_{p,p^*}}^{\frac{1}{n'}\max\{1,\frac{p'}{p^*}\}}\,\|w\nabla f\|_{L^p(\mathbb{R}^n)}. \tag{13.32}$$

Indeed,

$$\begin{aligned}\|wTf\|_{L^{p^*}(\mathbb{R}^n)} &= \|Tf\|_{L^{p^*}(w^{p^*})} \leq c_{p,T}\,[w^{p^*}]_{A_\infty}\,\|Mf\|_{L^{p^*}(w^{p^*})}\\ &\leq c_{p,T}\,[w^{p^*}]_{A_\infty}\,[w]_{A_{p,p^*}}^{\frac{1}{n'}\max\{1,\frac{p'}{p^*}\}}\,\|w\nabla f\|_{L^p(\mathbb{R}^n)}\end{aligned}$$

by (13.17), which is also a worse result than (13.25).

Finally, using the main result in [19], we also have

$$\begin{aligned}\|Tf\|_{L^{n',\infty}(w^{n'})} &\leq c_T\,[w^{n'}]_{A_1}\,\|f\|_{L^{n'}(w^{n'})}\\ &\leq c_T\,[w]_{A_{1,n'}}^{1+\frac{1}{n'}}\,\|w\nabla f\|_{L^1(\mathbb{R}^n)},\end{aligned}$$

which is a better result than (13.31).

13.3 The Linear Commutators

Likely motivated by the work of A. P. Calderón on commutators, R. Coifman, R. Rochberg and G. Weiss introduced in [6] the operator

$$T_b f(x) = \int_{\mathbb{R}^n} (b(x) - b(y))K(x,y)f(y)\,dy,$$

where K is a kernel satisfying the standard Calderón-Zygmund estimates (see Sect. 13.2.3) and where b is any locally integrable function. The operator is called commutator since

$$T_b = [b,T] = bT - T(b\,\cdot)$$

where T is the Calderón-Zygmund singular integral operator associated to K. The main result from [6] states that $[b,T]$ is a bounded operator on $L^p(\mathbb{R}^n)$ if $1 < p < \infty$ and b is a BMO function. They proved that

$$\|T_b\|_{L^p(\mathbb{R}^n)\to L^p(\mathbb{R}^n)} \leq c_{p,T}\,\|b\|_{BMO}.$$

Moreover, they provided a new and beautiful characterization of BMO in the case of T being the Hilbert transform or any Riesz transform R_j, i.e.

$$\|b\|_{BMO} \approx \|T_b\|_{L^p(\mathbb{R}^n)\to L^p(\mathbb{R}^n)}.$$

These commutators have proved to be of great interest in many situations, for instance, the theory of non divergence elliptic equations with discontinuous coefficients [2, 3, 8]. We refer to [14] for a beautiful account.

A natural generalization of the commutator $[b, T]$ is given by

$$T_b^m f(x) = \int_{\mathbb{R}^n} (b(x) - b(y))^m K(x, y) f(y)\, dy,$$

where $m \in \mathbb{N}$, and the case $m = 0$ recaptures the Calderón-Zygmund singular integral operator. Observing that for $m \geq 2$, the iterated commutator

$$T_b^m = \overbrace{[b, \cdots, [b, T]]}^{(m\, times)}$$

is also a bounded operator

$$T_b^m : L^p(\mathbb{R}^n) \to L^p(\mathbb{R}^n),$$

whenever $1 < p < \infty$ and b is a BMO function.

It was shown in [30] that these operators are not of weak type $(1, 1)$ and that there is an intimate relationship with the iterated maximal function M^m. We will need the following result: for any $0 < p < \infty$ and any $w \in A_\infty$, we have

$$\|T_b^m f\|_{L^p(w)} \leq c_{m,p,T}\, [w]_{A_\infty}\, \|b\|_{BMO}^m \|M^{m+1}(f)\|_{L^p(w)}. \tag{13.33}$$

This inequality is sharp since M^{m+1} can not be replaced by the smaller operator M^m. $[w]_{A_\infty}$ denotes the usual A_∞ constant,

$$[w]_{A_\infty} := \sup_Q \frac{1}{w(Q)} \int_Q M(w\chi_Q)\, dx.$$

On the other hand, it follows from [4] that

$$\begin{aligned}\|wT_b^m f\|_{L^{p^*}(\mathbb{R}^n)} &= \|T_b^m f\|_{L^{p^*}(w^{p^*})}\\ &\leq c_{m,p,T}\, [w^{p^*}]_{A_{p^*}}^{(m+1)\max\{1, \frac{1}{p^*-1}\}} \|b\|_{BMO}^m \|f\|_{L^{p^*}(w^{p^*})},\end{aligned}$$

which combined with (13.9) yields

$$\|wT_b^m f\|_{L^{p^*}(\mathbb{R}^n)} \leq c_{m,p,T}\,[w]_{A_{p,p^*}}^{(m+1)\max\{1,\frac{1}{p^*-1}\}}\|b\|_{BMO}^m [w]_{A_{p,p^*}}^{\frac{1}{n'}}\|w\nabla f\|_{L^p(\mathbb{R}^n)}.$$

The above constant is worse than the one derived using (13.33) ($[w^{p^*}]_{A_\infty}$ times $[w]_{A_{p,p^*}}^{\frac{1}{n'}}$) at least in the case p is close to 1 and n is large.

Reasoning as in the proof of (13.32), for $p \in (1, n)$ we have

$$\|wT_b^m f\|_{L^{p^*}(\mathbb{R}^n)} \leq c_{m,p,T}\,[w^{p^*}]_{A_\infty}^{m+1}\,[w]_{A_{p,p^*}}^{\frac{1}{n'}\max\{1,\frac{p'}{p^*}\}}\,\|w\nabla f\|_{L^p(\mathbb{R}^n)}.$$

The case $p = 1$ has a different nature, observe that the exponent above blows up when $p \to 1$. We end this section with the following result whose proof is based on the A_1 results for commutators by C. Ortiz in [27].

Theorem 13.2 *Let $w \in A_{1,n'}$. Then*

$$\|T_b^m f\|_{L^{n',\infty}(w^{n'})} \leq c_T\,\|b\|_{BMO}^m\,[w]_{A_{1,n'}}^{m+1+\frac{1}{n'}}\,\|w\nabla f\|_{L^1(\mathbb{R}^n)}.$$

13.4 The Nonlinear Commutators

The second kind of commutators that we are going to consider were introduced by R. Rochberg and G. Weiss in [33]. These are nonlinear operators defined for appropriate functions by

$$f \mapsto Nf = N_T(f) := T(f\,\log|f|) - Tf\,\log|Tf|.$$

We will assume that T is a Calderón-Zygmund singular integral operator in this section. It is of interest to study this operator due to its relationship with the Jacobian mapping and with nonlinear P.D.E. as shown in [10, 15] (see also [24]).

The main result from [33] is the following L^p bound

$$\|Nf\|_{L^p(\mathbb{R}^n)} \leq c_{p,T}\;\|f\|_{L^p(\mathbb{R}^n)}\,, \quad 1 < p < \infty.$$

We have the following improvement from [31] that uses a completely different method.

Theorem 13.3 *Suppose that $0 < p < \infty$ and $w \in A_\infty$. Then,*

$$\|Nf\|_{L^p(w)} \leq c_{p,T}[w]_{A_\infty}\,\log\left(e + [w]_{A_\infty}\right)\,\|M^2 f\|_{L^p(w)}.$$

We briefly sketch the proof below. The idea is to split the nonlinear commutator N into simpler pieces:

$$T(f\ \log|f|) - Tf\ \log|Tf| = T(f\ \log\frac{|f|}{Mf}) + [\log Mf, T]f - Tf\ \log\frac{|Tf|}{Mf}$$
$$:= N_1 f + N_2 f + N_3 f.$$

To estimate N_1, we write

$$N_1 f = T\left(Mf\ \frac{f}{Mf}\ \log(\frac{|f|}{Mf})\right),$$

and we apply (13.33) with $m = 0$. Since $w \in A_\infty$, we have

$$\int_{\mathbb{R}^n} |N_1 f|^p\, w \le c\,[w]_{A_\infty}^p \int_{\mathbb{R}^n} \left(M\left(Mf\ \frac{f}{Mf}\ \log(\frac{|f|}{Mf})\right)\right)^p w$$
$$\le c\,[w]_{A_\infty}^p \int_{\mathbb{R}^n} (M^2 f)^p\, w$$

by utilizing the fact that $|t\ \log t| \le \frac{1}{e}$ when $0 < t \le 1$.

For N_2, we notice that it is of commutator type fixing $b_f = \log Mf \in BMO$ with dimensional constant, namely independent of f and hence we can apply again (13.33) (see [31] for details)

$$\int_{\mathbb{R}^n} |N_2 f|^p\, w = \int_{\mathbb{R}^n} |[\log Mf, T]f|^p\, w$$
$$\le c\ \|\log Mf\|_{BMO}^{2p}\ [w]_{A_\infty}^p \int_{\mathbb{R}^n} (M^2 f)^p\, w$$
$$\le c\,[w]_{A_\infty}^p \int_{\mathbb{R}^n} (M^2 f)^p\, w.$$

For N_3, we split $\mathbb{R}^n$ in two disjoint sets A and B where

$$A = \{y \in \mathbb{R}^n : |Tf(y)| \le Mf(y)\}$$

and $B = \mathbb{R}^n \setminus A$. Writing

$$N_3 f = Mf\ \frac{|Tf|}{Mf}\ \log\frac{|Tf|}{Mf}$$

we have that $|N_3 f| \le Mf$ on A, and on B we use $\log t \le \frac{t^\epsilon}{\epsilon}$, $t > 1$, $\epsilon > 0$ to get

$$\|N_3 f\|_{L^p(w)} \le \|N_3 f \chi_A\|_{L^p(w)} + \|N_3 f \chi_B\|_{L^p(w)}$$

$$\le \|Mf\|_{L^p(w)} + \frac{1}{\epsilon} \, \||Tf|^{(\epsilon+1)} \, (Mf)^{-\epsilon}\|_{L^p(w)}$$

$$= \|Mf\|_{L^p(w)} + \frac{1}{\epsilon} \, \|Tf\|^{1+\epsilon}_{L^{p(1+\epsilon)}(w(Mf)^{-\epsilon p})}$$

To apply (13.33) again with $m = 0$, we must verify that $(Mf)^{-\epsilon\, p}\, w \in A_\infty$ for ϵ small enough and with a constant independent of f. Being more precise, it can be shown that

$$[wu^{-1}]_{A_\infty} \le [w]_{A_\infty} \, [u]_{A_1},$$

whenever ϵ is chosen to be $\epsilon p < 1$. Indeed, we have by the well-known Coifman-Rochberg theorem that if $0 < \delta < 1$, $u = (Mf)^{1-\delta} \in A_1$ with $[(Mf)^{1-\delta}]_{A_1} \le \frac{c_n}{\delta}$. Hence, combining estimates,

$$[w(Mf)^{-\epsilon\, p}]_{A_\infty} \le \frac{c_n}{1-\epsilon p} \, [w]_{A_\infty},$$

from which we have

$$\|N_3 f\|_{L^p(w)} \le \|Mf\|_{L^p(w)} + \frac{1}{\epsilon} \, [w(Mf)^{-\epsilon p}]^{1+\epsilon}_{A_\infty} c^{1+\epsilon}_{p,T} \|Mf\|^{1+\epsilon}_{L^{p(1+\epsilon)}(w(Mf)^{-\epsilon p})}$$

$$\le \|Mf\|_{L^p(w)} + \frac{1}{\epsilon} \, [w]^{1+\epsilon}_{A_\infty} \, [(Mf)^{\epsilon p}]^{1+\epsilon}_{A_1} \, c^{1+\epsilon}_{p,T} \|Mf\|_{L^p(w)}$$

$$\le \|Mf\|_{L^p(w)} + \frac{1}{\epsilon} \, [w]^{1+\epsilon}_{A_\infty} \, (\frac{c_n}{1-p\epsilon})^{1+\epsilon} \, c^{1+\epsilon}_{p,T} \|Mf\|_{L^p(w)}.$$

Optimizing this estimate with $\epsilon = \frac{1}{p\log(e^2+[w]_{A_\infty})}$ we get finally

$$\|N_3 f\|_{L^p(w)} \le [w]_{A_\infty} \log\left(e + [w]_{A_\infty}\right) \, c_{p,T} \, \|Mf\|_{L^p(w)}$$

and hence

$$\|Nf\|_{L^p(w)} \le c_{p,T} \, [w]_{A_\infty} \log\left(e + [w]_{A_\infty}\right) \, \|M^2 f\|_{L^p(w)} \tag{13.34}$$

This yields the proof of Theorem 13.3.

Using (13.14) we have as an immediate consequence the following corollary.

Corollary 13.1 *Let $1 < p < \infty$ and $w \in A_p$. Then,*

$$\|Nf\|_{L^p(w)} \leq c_{p,T}\,[w]_{A_\infty} \log\left(e + [w]_{A_\infty}\right) [w]_{A_p}^{\frac{2}{p-1}} \,\|f\|_{L^p(w)}.$$

Another result that can be found in [31] is the following

Theorem 13.4 *Suppose that $1 < p < \infty$. Let w be any weight. Then*

$$\int_{\mathbb{R}^n} |Nf(x)|^p\, w(x)\, dx \leq c \int_{\mathbb{R}^n} |f(x)|^p\, M^{[2p]+1} w(x)\, dx.$$

The method used in [30, 31] cannot be applied here since the operator is non-linear. The splitting above has to be used.

Finally, combining (13.34) and the Sobolev type inequality (13.18) derived in [13]

$$\|wNf\|_{L^{p^*}(\mathbb{R}^n)} \leq c_{p,T}\,[w^{p^*}]_{A_\infty} \log\left(e + [w^{p^*}]_{A_\infty}\right) [w]_{A_{p,p^*}}^{\frac{1}{n'}\max\{1,\frac{p'}{p^*}\}} \,\|w\nabla f\|_{L^p(\mathbb{R}^n)}.$$

Observe that, as in the linear case, the exponent blows up when $p \to 1$. We state now a conjecture motivated by Theorem 13.2 without such a blow up.

Conjecture 13.1 Let $w \in A_{1,n'}$. Then

$$\|Nf\|_{L^{n',\infty}(w^{n'})} \leq c\,[w]_{A_{1,n'}}^{2+\frac{1}{n'}} \,\|w\nabla f\|_{L^1(\mathbb{R}^n)}.$$

Acknowledgments K.M. is supported by Simons Collaboration Grant for Mathematicians, 160427.

C. P. is supported by grant PID2020-113156GB-I00, Spanish Government; by the Basque Government through grant IT1615-22 and the BERC programme 2022–2025 program and by BCAM Severo Ochoa accreditation CEX2021-001142-S, Spanish Government. He is also very grateful to the Mittag-Leffler Institute where part of this research was carried out.

References

1. S. Buckley, Estimates for operator norms on weighted spaces and reverse Jensen inequalities. Trans. Amer. Math. Soc. **340**, 253–272 (1993)
2. F. Chiarenza, M. Frasca, P. Longo, Interior $W^{2,p}$ estimates for non divergence elliptic equations with discontinuous coefficients. Richerche Mat. **40**, 149–168 (1991)
3. F. Chiarenza, M. Frasca, P. Longo, $W^{2,p}$-solvability of the Dirichlet problem for nondivergence elliptic equations with VMO coefficients. Trans. Amer. Math. Soc. **334**, 841–853 (1993)
4. D. Chung, C. Pereyra, C. Pérez, Sharp bounds for general commutators on weighted Lebesgue spaces. Trans. Amer. Math. Soc. **364**, 1163–1177 (2012)

5. R. Coifman, C. Fefferman, Weighted norm inequalities for maximal functions and singular integrals. Studia Math. **51**, 241–250 (1974)
6. R.R. Coifman, R. Rochberg, G. Weiss, Factorization theorems for Hardy spaces in several variables. Ann. Math. (2) **103**(3), 611–635 (1976)
7. D. Cruz-Uribe, J.M. Martell, C. Pérez, Weighted weak-type inequalities and a conjecture of Sawyer. Int. Math. Res. Not. IMRN **30**, 1849–1871 (2005)
8. G. Di Fazio, M.A. Ragusa, Interior estimates in Morrey spaces for strong solutions to nondivergence form equations with discontinuous coefficients. J. Funct. Anal. **112**, 241–256 (1994)
9. J. Duoandikoetxea, Extrapolation of weights revisited: new proofs and sharp bounds. J. Funct. Anal. **260**, 1886–1901 (2011)
10. L. Greco, T. Iwaniec New inequalities for the Jacobian. Ann. Inst. Henri Poincaré **11**, 17–35 (1994)
11. P. Hajłasz, Sobolev inequalities, truncation method, and John domains. Papers on Analysis, Rep. Univ. Jyväskylä Dep. Math. Stat. **83**, 109–126 (2001)
12. E. Harboure, R. Macías, C. Segovia, Extrapolation results for classes of weights. Amer. J. Math. **110**, 383–397 (1988)
13. C. Hoang, K. Moen, C. Pérez, Pointwise estimates for rough operators with applications to weighted Sobolev inequalities. J. d'Analyse, to appear
14. T. Iwaniec, Let the beauty of harmonic analysis be revealed through nonlinear PDEs; A work of art in three sketches. Rev. Mat. Iberoam. **28**(3), 681–722 (2012)
15. T. Iwaniec, C. Sbordone, Weak minima of variational integrals. J. Reine Angew Math. **454**, 143–161 (1994)
16. J. Kinnunen, J. Lehrbäck, A. Vähäkangas, *Maximal Function Methods for Sobolev Spaces*, Mathematical Surveys and Monographs, vol. 257 (2021)
17. M.T. Lacey, K. Moen, C. Pérez, R.H. Torres, Sharp weighted bounds for fractional integral operators. J. Funct. Anal. **259**(5), 1073–1097 (2010)
18. A. Lerner, K. Moen, Mixed A_p-A_∞ estimates with one supremum. Studia Math. **219**(3), 247–267 (2013)
19. A. Lerner, S. Ombrosi, C. Pérez, Sharp A_1 bounds for Calderón-Zygmund operators and the relationship with a problem of Muckenhoupt and Wheeden. Int. Math. Res. Not. IMRN **6**, Art. ID rnm161, 11 pp., 42B20 (2008)
20. K. Li, S. Ombrosi, C. Pérez, Proof of an extension of E. Sawyer's conjecture about weighted mixed weak-type estimates. Math. Ann. **374**, 907 (2018)
21. T. Luque, C. Pérez, E. Rela, Reverse Hölder property for strong weights and general measures. J. Geom. Anal. **27**(1), 162–182 (2017)
22. J. Malý, L. Pick, An elementary proof of sharp Sobolev embeddings. Proc. Amer. Math. Soc. **130**(2), 555–563 (2002)
23. V. Maz'ya, *Sobolev Spaces with Applications to Elliptic Partial Differential Equations*, Second, revised and augmented edition (Springer, Berlin, 2011)
24. M. Milman, *Extrapolation and Optimal Decompositions*, Lect. Notes Math. 1580 (Springer, 1995)
25. B. Muckenhoupt, R. Wheeden, Weighted norm inequalities for fractional integrals. Trans. Amer. Math. Soc. **192**, 261–274 (1974)
26. R. O'Neil, Convolution operators and L(p,q) spaces. Duke Math. J. **30**, 129–142 (1963)
27. C. Ortiz-Caraballo, Quadratic A_1 bounds for commutators of singular integrals with BMO functions. Indiana Univ. Math. J. **60**(6), 2107–2130 (2011)
28. C. Ortiz-Caraballo, C. Pérez, E. Rela, Improving bounds for singular operators via Sharp Reverse Hölder Inequality for A_∞, in *Operator Theory: Advances and Applications*, vol. 229 (Springer Basel, 2013), pp. 303–321
29. J. Peetre, Espaces d'interpolation et théorèm de Soboleff. Ann. Inst. Fourier **16**, 279–317 (1966)
30. C. Pérez, Endpoint estimates for commutators of singular integral operators. J. Funct. Anal. (1) **128**, 163–185 (1995)

31. C. Pérez, Sharp estimates for commutators of singular integrals via iterations of the Hardy–Littlewood maximal function. J. Fourier Anal. Appl. (6) **3**, 108–146 (1997)
32. C. Pérez, E. Roure, Sawyer-type inequalities for Lorentz spaces. Math. Ann. **383**(1–2), 493–528 (2022)
33. R. Rochberg, G. Weiss, Derivatives of analytic families of Banach spaces. Ann. Math. (2) **118**(2), 315–347 (1983)
34. L. Saloff-Coste, *Aspects of Sobolev-Type Inequalities*. London Mathematical Society Lecture Note Series (2001)
35. E. Stein, *Singular Integrals and Differentiability Properties of Functions*, Princeton Mathematical Series, No. 30 (Princeton University Press, Princeton, 1970)
36. E. Stein, G. Weiss, Fractional integrals on n-dimensional Euclidean space. J. Math. Mech. **7**, 503–514 (1958)

Chapter 14
Time-Frequency Analysis Meets Adversarial Learning

Ilya Kavalerov, Ruijie Zheng, Rama Chellappa, and Wojciech Czaja

Dedicated to the memory of Guido Weiss

Abstract We propose a computational model of the auditory cortex inspired by recent developments in time-frequency analysis, as a defense against adversarial attacks on audio signals. Our proposed cortical network emulates the first stages of audio processing in human brain, by appropriately adapting the concept of the Fourier scattering transformation. To test the model's efficacy we apply several white-box iterative optimization-based adversarial attacks to an implementation of Amazon Alexa's HW network, and a modified version of this network with an integrated cortical representation, and show that the cortical features help defend against universal adversarial examples. At the same level of distortion, the adversarial noises found for the cortical network are always less effective for universal audio attacks.

I. Kavalerov
Google, Menlo Park, CA, USA
e-mail: ilyak@google.com

R. Zheng
Department of Computer Science, University of Maryland, College Park, MD, USA

R. Chellappa
Departments of Electrical and Computer Engineering and Biomedical Engineering, Johns Hopkins University, Baltimore, MD, USA

W. Czaja (✉)
Norbert Wiener Center for Harmonic Analysis and Applications, Department of Mathematics, University of Maryland, College Park, MD, USA
e-mail: wojtek@math.umd.edu

E. Hernández et al. (eds.), *The Mathematical Heritage of Guido Weiss*,
Applied and Numerical Harmonic Analysis,
https://doi.org/10.1007/978-3-031-76793-7_14

14.1 Introduction and Problem Description

The question of how to best represent a signal (such as an audio recording or an image) for a specific task (e.g., denoising or classification) is one of fundamental problems at the intersection of harmonic analysis and signal processing. Encoding schemes, like Huffman coding, that were invented for transmitting generic bits of information ignore whether data is semantically meaningful when encoding it. For example, when representing an audio recording of speech data for the ultimate goal of transcription, background noises can be considered as nuisances and left out. On the other hand when representing the same audio for the task of geographic localization, these background noises may contain relevant information and should not be discarded. This motivates the idea that optimal representations of data must depend on the task of interest, and should extract information that is semantically meaningful for the task, and discard information that is irrelevant from the perspective of the same objective. It is thus an exciting challenge to consider this problem for the many applications of audio signal processing, where semantic understanding plays an especially important role, and to consider the unique characteristics that should be exploited in the aim of optimally representing speech data to solve such tasks.

Over the last decade the rapid advancement of Deep Learning techniques, which come with the promise of learning hierarchical representations of data with minimal user design required, have been at the forefront of classification and denoising for numerous data modalities, including audio. Many researchers have advanced the performance of deep learning techniques by building-in semantically or physically meaningful priors to deep learning methods. Audio classification, automatic speech recognition, and—more recently—Large Language Models, such as ChatGPT, are among the primary and well publicised successes of machine learning in audio processing. In this work we aim to contribute along the same direction of improving deep learning methods by modifying them in such a way that they can take advantage of the physics and semantics behind the data to process. Current popular systems are driven by empirical rules that do not fully exploit the underlying physics of the data, for example many speech and audio systems fix STFT preprocessing before employing their networks. To meet this opportunity to innovate we propose to alter known deep learning methods to be more dependent on the semantic and physical underpinnings of the data to create better performing and more robust algorithms for sound classification. Our approach takes inspiration from Harmonic Analysis to defend against audio adversarial attacks.

With the increasing reliance on neural networks and artificial intelligence in decision making across a wide range of fields, a new type of problems emerged known collectively as adversarial attacks. Adversarial machine learning is thus a new field of study which aims at increasing our understanding about the adversarial attacks and providing defenses against such attacks. As voice assistant systems such as Amazon Alexa, Google Assistant, Apple Siri, and Microsoft Cortana become more ubiquitous and integrated into modern life, so does the risk of antagonists taking control of devices that we depend on. Thus adversarial attacks on audio [18] are one way in which a voice assistant could be subverted to send an unwanted message, bank transfer, or unlock a home. All the mentioned voice assistants rely on wake-word detection to initiate their automatic speech recognition (ASR) systems that give control over their functions. The wake word detection system is both a voice assistant's first line of defense against adversarial attacks and its least complex feature. An attacker's first step to manipulate a voice assistant would be to issue a human-imperceptible wake-word, or conversely to prevent a user's wake-words from being registered. The latter is called a denial-of-service, and can have safety critical applications, for example when a user needs to call for help but cannot physically reach a device.

A number of recent works have demonstrated that adversarial attacks for automatic speech recognition are a real possibility. The Carlini-Wagner attack [9, 10] can produce an adversarial waveform that is 99.9% similar to the original, but produces any other intended sentence by the ASR. Adversarial examples have since developed and improved in several ways. An important attribute for practical adversarial audio is that it be causal. Often an adversarial example is a function of the complete original audio that it is meant to modify, so thus original audio from hundreds of milliseconds in the future must be accessed to play an adversarial sound in the present. One solution is *universal adversarial examples* [21, 38], examples that can behave adversarially regardless of what original audio they are superimposed on. Universal adversarial audio examples have been demonstrated with as little as one training sample for ASR [1]. Another important attribute for practical adversarial audio attacks is that they must be robust, that is the adversarial examples remain effective despite some level of distortions. This goal is not completely orthogonal to universal adversarial examples, since examples cannot be universal without also being invariant to translation in time, and when superimposed with some other audio. The ultimate test of robustness is over-the-air attacks, where as shown in Fig. 14.1 the adversarial example is introduced to the environment generating the

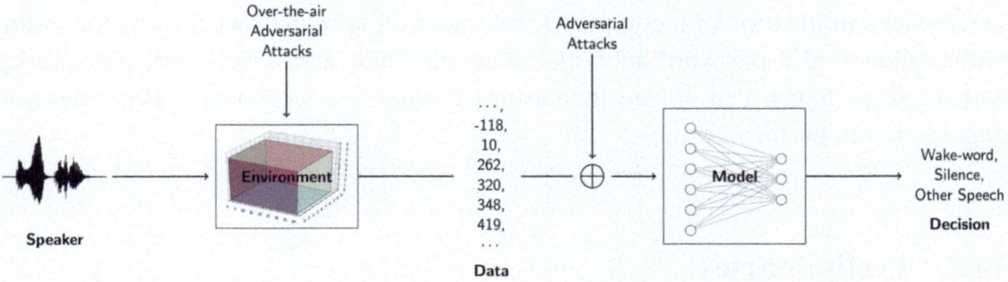

Fig. 14.1 The flow of information for a voice assistant wake-word detector and two attack vectors. The speaker generates audio for their intended command. The environment is an acoustic environment containing the speaker, other sources of audio like background and other speakers, distortions like reverberations, and the voice assistant's microphone and audio hardware including the ADC. The data (assumed to be 16 bit integer and 16 kHz) is then available to a model which ultimately makes the decision whether or not to "wake up" its general ASR capability

data for the model rather than to the data directly. Robust adversarial examples have been demonstrated to exist for simulated environment ASR systems [45], and more recently robust and universal examples have been generated for ASR systems, and even commercial wake word systems [27, 50].

Across many languages speech has a pattern of spectro-temporal modulations that differ from most environmental noise and distortions [14], and the brain has developed a hierarchical neural system that uniquely responds to these signals [28]. These specialized neuronal structures are in the auditory cortex, and for example are largely unresponsive to pure tones [28]. It has been demonstrated in animals that the auditory cortex encodes vocalization features while suppressing or being invariant to the noise features [23, 36], and invasive experiments in humans also have shown that auditory cortex neurons selectively suppress of acoustic features of noise in their neural responses [24, 36]. That is: the representation of speech in noise is often more separable in the cortical representation than at the level of spectrograms.

This characteristic of speech-noise separability found in the auditory cortex is exactly the same characteristic we would like to achieve in artificial intelligence-dependent voice assistants. For example, if a wake-word system had an intermediate representation that was tuned to speech the same way our brains are, then it would be necessary to perceptively modify speech to change that representation and ultimately the output of the wake-word model, making imperceptible adversarial examples impossible. Indeed the aforementioned biological findings have inspired a computational model of the mammalian auditory cortex to create such a representation [11, 47].

Our aim in this work is to incorporate the advantages of this cortical representation into a voice assistant and show the efficacy of this representation for defending against adversarial audio attacks. In this regard we turn our attention for inspiration to recently introduced scattering transformations [30]. In particular, the Fourier scattering transform [12, 13] is an amalgamation of time-frequency representations with neural network architectures, which has sufficient flexibility to be adapted for the purposes of defeating adversarial activity. Thus we propose and test a Fourier scattering method to defend against audio adversarial attacks. To achieve this goal we pursue a unification of a computational model of speech processing in the brain with commercial wake-word networks to create a new and novel cortical network, and we show that it can indeed increase resistance to adversarial noise without a degradation in performance.

14.2 Preliminaries

14.2.1 Time-Frequency Analysis

In 1946, Dennis Gabor proposed to study systems consisting of time and frequency shifts of a single function, for their usefulness in the analysis of information conveyed by communication channels [17]. The resulting theory led to a wealth of applications, ranging from frequency estimation and sampling theory, to uncertainty principles, and to auditory signal processing. More importantly, many new developments followed in the form of short-time Fourier transform, wavelet theory, Wigner distribution and many more. Guido Weiss with his collaborators provided a beautiful unifying characterization theory for a broad range of such resulting constructions. The key aspect of this unification was a consequence of the uniqueness property for almost periodic Fourier series that arise in the context of families formed by translations, see [20, 25, 26]. In addition to the cases covered by the general characterization theory, we also want to mention some individual examples of time-frequency transformations which were designed to specifically deal with auditory modeling and acoustic signal processing. In 1992, John Benedetto and Anthony Teolis proposed a Wavelet Auditory Model motivated by the structure of the mammalian auditory system and based on the theory of irregular sampling and frames for decoding of auditory patterns using cochlear filters [4–6]. A competing analytically tractable framework for acoustic signal processing in the early stages of the auditory system was proposed in [58]. It employed wavelet representations, multiresolution processing, and convex projections to reconstruct the input stimuli. A combined time-frequency and time-scale representation was proposed in [56], where the role of chroma was established both as a useful representation of acoustic signals and as a method in the formation of pitch estimate. Finally, auditory cortical representations of acoustic spectra have also been demonstrated to play significant role in extraction and representation of spectro-temporal information, see, e.g., [51].

It is this last aspect—the role of multiscale cortical representations with specially designed filters, that we want to take advantage of in our defense strategy against wake-word adversarial attacks. We describe this transformation in detail in the next section.

We close this brief historical overview of the role of Time Frequency Analysis in cortical auditory models with a remark that mathematical characterizations of neural systems reach far beyond representation theory. Petitot and Tondut pioneered geometric models in describing the functional architecture of the primary visual cortex [42, 43], laying foundations for a new field of neurogeometry in biological cybernetics. This approach was adapted to a bio-inspired geometric model for sound reconstruction by Boscain et al. [8]. On the other hand differential equations also played a significant role in cochlear modeling, see e.g., the work of Reimann [46] or Duits and Franken [15, 16], among many other approaches. In this paper however, we focus on the role of multiscale time-frequency representations, for their natural connection to neural networks.

14.2.2 Cortical Transform

The details of human hearing have been studied for centuries. In this section we highlight some of the consensus of the last few decades of modeling for the first stages of how the brain makes sense of speech, cf., [28]. In the early stage, acoustic signals are transformed into 2 dimensional spectrograms in the cochlea, which acts as a filter bank, and filtered by the hair cell and lateral inhibitory network [28, 32]. These lower auditory areas are tono-topically organized: each cochlear location is associated with a best frequency it responds to, with time as the other axis this representation can be interpreted as an auditory image [28]. As signals move out to the auditory cortex (A1), what was essentially 1 dimensional processing in the cochlea becomes 2 dimensional in A1 [47]. Invasive experiments have measured the responses of ferrets' A1 neurons and characterized them with corresponding spectro-temporal receptive fields (STRFs) [34]. An STRF of a neuron indicates the frequencies and time lags that correlate with an increased response of that neuron; these two dimensions are often referred to as scale (spectral) and rate (temporal).

A computational model for these receptive fields can be estimated from the measurements of these neurons responses [47, 52]. On the scale axis the transfer function of these neurons is well approximated by the second derivative of a Gaussian function [47]. Neurons are tuned to be responsive to a variety of rates and scales [34]. For scale features we parameterize the approximation of the neural response as:

$$R_{2,\psi}(y) = (\Omega_2 y/\psi)^2 \exp(1 - (\Omega_2 y/\psi)^2), \tag{14.1}$$

where $y = 1, \ldots, L$ and $\psi \in \Psi$—the selected scale range of frequency. The constant Ω_2 is the tuned cycles per octave (24) divided by $2L$ [47]. We use $R_{2,\psi}$ to define $r_{2,\psi}(f)$ via

$$r_{2,\psi}(y) = \mathcal{F}^{-1}\{R_{2,\psi}(f)\}, \tag{14.2}$$

where $\mathcal{F}$ represents the Fourier transform [47]. On the rate axis, the response is modeled by a function with a central excitatory band surrounded by inhibitory side bands, this is approximated with:

$$r_{1,\omega}(t) = (t/\Omega_1\omega)^2 \omega \exp(-\alpha t/\Omega\omega) \sin(2\pi t\, \Omega_1\omega), \tag{14.3}$$

where $t = 1, \ldots, L$, $\omega \in \mathcal{W}$, and temporal and spectral features, resp., L and $\mathcal{W}$, are appropriately selected, and $\alpha = 3.5$ was chosen empirically during the fitting stage and the constant Ω_1 is the sample rate of frames per second (125). These functions approximate neural responses collected from ferrets listening to human speech [47, 52]. Finally the 2D STRF filters are

$$f(\omega, \psi, \phi) = \begin{cases} r_{2,\psi} \otimes r_{1,\omega}, & \phi = 1 \\ r_{2,\psi} \otimes \overline{r_{1,\omega}}, & \phi = -1 \end{cases}, \tag{14.4}$$

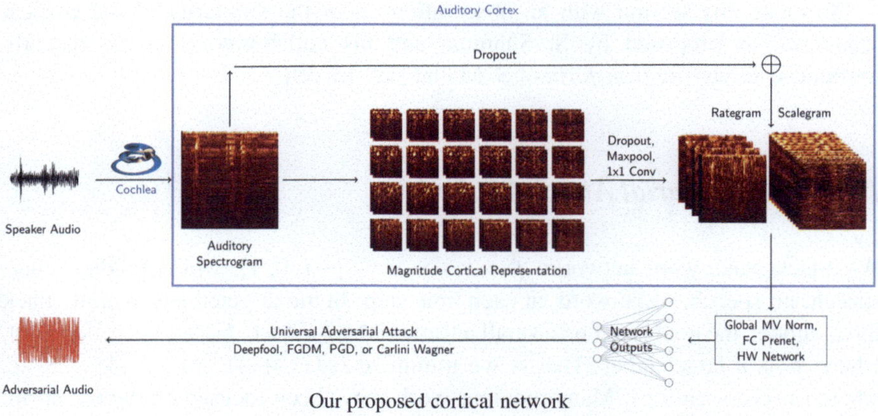

Fig. 14.2 Our proposed cortical network emulates the first two stages of audio processing in the brain (labeled in light blue). Integrating this features extraction directly into the network gives us two advantages over pre-processing: (1) The feature representation can be enhanced with learning (dropout with 1×1 convolutions) (2) We can backpropagate through these layers are find adversarial audio directly without any lossy transformations (such as Griffin-Lim)

with $\omega \in \mathcal{W}, \psi \in \Psi$. When phase is -1 the rate filter is conjugated. These filters with $\psi \in \{0.25, 0.5, 1, 2, 4, 8\}$ (cycles/octave) and $\omega \in \{4, 8, 16, 32\}$ (Hz) are displayed in Fig. 14.2.

Taking these biophysical observations, we can build an audio processing pipeline that transforms sound into a two dimensional auditory spectrogram, and then applies the STRFs in Eq. (14.4) to form a higher dimensional *cortical representation* from the magnitude response. This abstraction captures important physiological observations: selectivity to combined spectral and temporal features, and with proper choice of $\mathcal{W}$ the temporal dynamics of phase locking which decreases to less than 30 Hz in the cortex [11] (something only possible with large filters). This pipeline can be seen in Fig. 14.2.

The cortical transform of audio is a high dimensional tensor with dimensions: time $\times$ frequency $\times$ rate $\times$ scale. This can be reduced to two 3D tensors by reducing the coefficients using max pooling across the rate (*scalegram*) or scale (*rategram*) [32]. Such a pre-processing feature extraction pipeline has previously been applied successfully to speaker detection [33] and phoneme classification [3, 32]. The cortical representation captures the distinctive dynamics of phonemes, which are easily visually discernible in rategrams and scalegrams [32]. The magnitude cortical response carries enough information to reconstruct the spectrogram that generated it [11]. Because of its ability to capture speech specific features, inverting the magnitude cortical response has been applied to speech enhancement since it has high fidelity in the reconstruction of speech and low fidelity in the reconstruction of background noise [31].

We close this section with an observation about the similarity of the cortical transform, as proposed by S. Shamma and his collaborators, to the recently introduced *scattering transform* of S. Mallat [12, 13, 30].

14.2.3 Adversarial Attack Methods

We attack wake-word networks $f : x \mapsto y \in \{-1, 0, 1\}$, which predict other-speech, no speech, wake-word, at each time step. In the targeted adversarial attack threat model, the goal is to find a small adversarial noise δ s.t. $f(x+\delta) = y' \neq f(x)$, where y' is a target label. That is we minimize $l(f(x + \delta), y')$ s.t. $\|\delta\|_\infty \leq \epsilon$, where l is cross-entropy. Many previous works have been focused on the ℓ_∞ norm. However, in audio adversarial attacks, ℓ_∞ may not be the optimal choice since the perceptual difference between two different audio inputs cannot be well captured by the simple ℓ_∞ norm. So in order to design a capture this perceptual difference, Yao et al. and Szuerly and Kolter have proposed a perceptual loss $l_{\text{percept}}(x, \delta)$ based on psychoacoustics effects, which is added as an extra term into the attack objective [45, 54].

To solve this minimization problem, the Fast Gradient Sign Method (FGSM) finds the direction of the sign of the gradient and takes a corresponding step in this direction: $x = x - \epsilon\ \text{sign}(\nabla l(f(x), y'))$ [18]. More advanced adversarial example generation algorithms have adopted an iterative scheme to solve the optimization task [9, 29, 37, 38], including the projected gradient descent (PGD) attack, the DeepFool attack, as well as Carlini-Wagner (CW) attack. The DeepFool attack iteratively makes a linear approximation of the decision boundary and then takes a step normal to the tangent plane [37]. The CW attack solves a single optimization algorithm $l(f(x+\delta), y')+c\|\delta\|$, where the loss function l has been carefully chosen [9]. The PGD attack projects the perturbed input to the constrained ϵ-ball to enforce the ℓ_∞ constraint. In our paper, we consider these three attack algorithms as "strong" attacks and the goal is to show the efficacy of the cortical network to defend against these adversarial attacks.

14.2.4 Wake-Word Detection

Since we study defending against adversarial attacks on the wake-word systems of voice assistants, we choose a baseline network from the Amazon Alexa team. A variety of networks have been published by team members at Amazon and cybersecurity practitioners [22, 55], and the time-delayed bottleneck highway network (TDB-HW) is the latest in the literature [19, 41, 53]. We choose this network as a baseline because it is a larger and more powerful network than some others in the literature [7]. It consists of a feature extractor stage and a classifier stage. Before the feature extraction stage, auditory spectrogram (a log-mel filter bank spectrogram, abbreviated LFBE) features are fed into a two layer FC prenet with dropout [53]. The feature extraction stage of TDB-HW consists of four highway blocks. Then, a bottleneck with 20 left (200 ms) and 10 right (100 ms) contexts is applied. Finally, six highway blocks and one FC layer are applied to provide a classification. This is the network we refer to as a "baseline".

14.3 Proposed Architecture

For our cortical network, the input LFBE features are convolved with 48 size 32×32 complex STRF filters from Eq. (14.4). These 48 filters are shown in Fig. 14.2, and have parameters which mimic the response measured in auditory cortex neurons [47]. After applying an absolute value and dropout, max-pooling is applied along the rate and scale axes of the features to yield a scalegram and a rategram. These are concatenated, a 1×1 convolution is applied, and the original spectrogram with substantial dropout (0.9) is added to this 3D tensor. A small prenet is applied before the TDB-HW network is applied. The residual addition is motivated by the loss of 2D high frequency features by the STRFs, and a way to recover some of that information. The STRFs are not learned during training, so the total learnable parameters are similar between the two models.

14.4 Experimental Results and Discussion

14.4.1 Methods

We created several monaural 15 hour training datasets and 3 hour validation and test sets. We create our train/val/test splits from different speakers of wake-words from the Kaggle Alexa dataset and speakers in the Librispeech dataset [2, 40]. For additional background noise we use the DEMAND dataset kitchen, living room, laundry room, hallway, office, cafeteria, cafe environments. We extract words from our positive and negative datasets, pitch shift and time stretch these words, and mix them in a room, randomizing all these variables according to reasonable distributions [48, 49].

We evaluate the merit of the cortical network as a defense to adversarial examples on Universal FGSM, PGD, Deepfool, and Carlini-Wagner attacks [9, 18, 29, 37, 38]. We use the standard implementation of these in the IBM Adversarial Robustness Toolbox (ART) [39]. The FGSM, Deepfool, and Carlini-Wagner attacks were performed for 4000 iterations and evaluated every 400, PGD was performed for 250 examples for 100 iterations each, evaluated every 25 examples. For each trial, the minimum accuracy noise was recorded on a test set, and the average minimum accuracy of 12 trials over the three networks is show in Fig. 14.3.

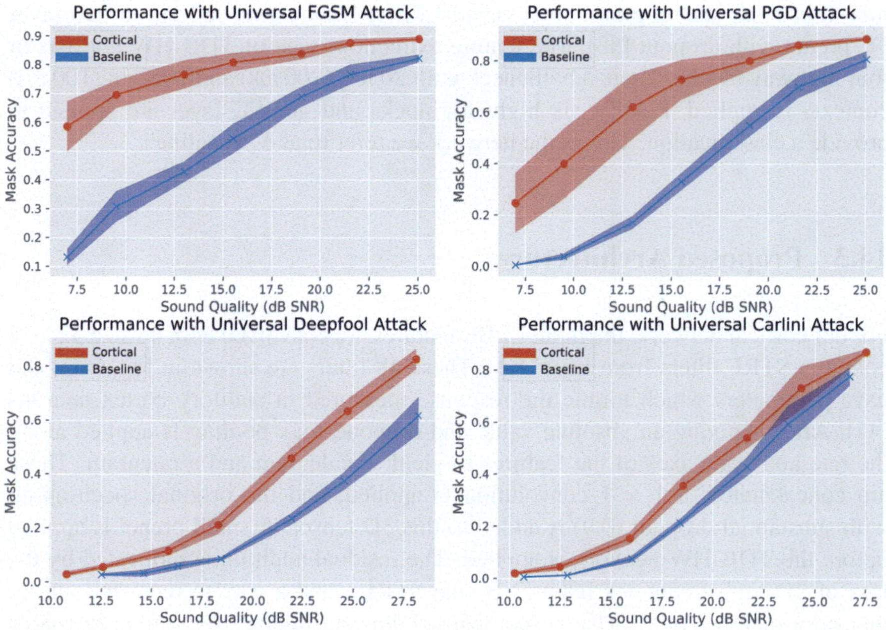

Fig. 14.3 Three baseline and three cortical networks are attacked in four trials each with various SOTA attacks from IBM's ART toolbox [39]. Each point in each figure corresponds to the hyperparameters for a single type of attack constrained to an ℓ_∞ ball. Each point is the mean of 12 trials, showing the mean test mask accuracy for the best (minimum accuracy) adversarial noise found during the attack trial on a different attack training set. 1 standard deviation is shaded

To prepare our dataset we developed an algorithm to extract separate words from speech, which we use to isolate single words from the Librispeech dataset for negative example words. We use this algorithm and fine tune it manually when necessary on the Alexa dataset to extract start/end times positive words (but keep surrounding audio including noise), and we use it unsupervised on Librispeech to extract negative words. Positive words are augmented randomly to have duration between 400 and 900 ms (drawn from a distribution fitted on the training speakers), and to have a base frequency drawn from $\mathcal{U}([80, 350] \cap [f_{\text{source}} - 60, f_{\text{source}} + 60])$ Hz using the high quality C++ sound library Rubberband [48]. Negative words have modified length multiplied by $\mathcal{U}([0.8, 1.2])$, and modified pitch multiplied by $\mathcal{U}([0.8, 1.4])$ using the same library. To form a single sound clip, words are randomly selected to be positive or negative, and arranged with silence ($\mathcal{U}(50, 150)$ ms) to form a multi channel clip. Background noise form DEMAND is added as an additional channel. To form the single channel audio the channels are normalized to -15 dBFS and simulated in a bedroom or living room with absorption $\mathcal{U}([0.4, 0.9])$ with random distances from the simulated microphone [49]. Finally the single channel is normalized -15 dBFS and quantized to 16-bits.

To extract words from audio in an unsupervised method on Librispeech we used the following algorithm. Original audio was filtered for speech base frequencies 85–255 Hz and normalized, then the energy of this signal was convolved with a 100 ms Hann window at 25 ms hop, giving an envelope for the signal's energy. Local min/max extrema were found, and segments were marked around triplets of low-high-low extrema points. Neighboring segments were joined if less than 350 ms, and segments longer than 1.2 s were discarded. Then words segments were sorted by the quietest silences preceding and following the word, and were extracted from the original unmodified audio.

14.4.2 Analysis

The baseline and cortical networks were both able to train to similar competitive rates as seen in the Detection Error Tradeoff (DET) curves in Fig. 14.4. These curves are specific to the dataset that we created, but for comparison, the company PicoVoice has created a dataset in a similar manner to ours and tested the CMU Sphinx Open Source Toolkit PocketSphinx, the open source KITT.AI, and their own commercial product PicoVoice [44]. They advertise a miss rate of 6% at 1 false alarm per 10 hours for their own product, 32% for KITT.AI, and 48% for PocketSphinx. On our dataset our miss rate is 7% at 1 false alarm per 10 hours for our baseline and 10% for the cortical network. However this is an extreme mode of operation, at 1 false alarm per 1 hour our miss rate is 5% for both models.

The FGSM and PGD attacks have the slowest accuracy decay rate as the strength of the noise grows on the cortical network. This is likely due to the more dispersed patterns of attacks that these two methods create. Because of the STRFs' selectivity for changing phoneme dynamics [32, 35], the cortical network should be more invariant to stationary noises as is the human auditory cortex [24], and is only slowly impacted by this type of noise. We see the noises generated by the Deepfool and Carlini-Wagner attacks are very textured in Fig. 14.6. For the same search in a given

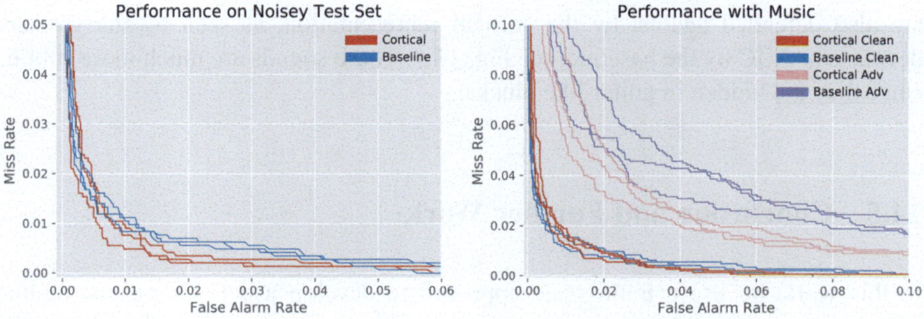

Fig. 14.4 False alarm and Miss rates of the networks that we compare. One line is plotted for each of three training trials for each of the two architectures. Left is the test set performance, Right is the performance with adversarial music [27]

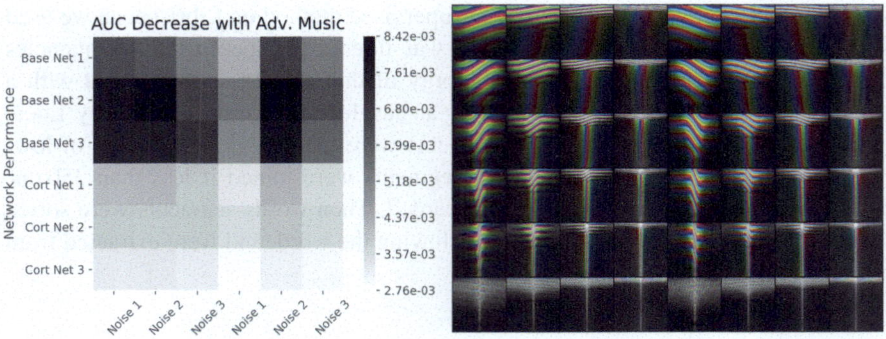

Fig. 14.5 The Left shows the AUC of the baseline networks decreases the most with adversarial music. The Right shows the 48 cortical filters used in the cortical network. Along the rows are the six scale parameters $\psi \in \{0.25, 0.5, 1, 2, 4, 8\}$, the left four columns are the rate parameters $\omega \in \{4, 8, 16, 32\}$ with phase $\phi = +1$ and the four right columns have phase $\phi = -1$ and the same scale $\times$ rate parameters. Colorization with hue = $(\angle z + \pi)/(2\pi) + 1/2$ and lightness = $1 - 1/(1 + |z|^{0.3})$

ℓ_∞ size ball, the cortical network requires a more elaborate noise than the baseline network off of the auditory spectrogram (and still performs better at the same SNR). The noise for the baseline is more stationary and has a higher impact on speaking frequencies. However, all the adversarial noises cannot be said to be imperceptible to the human ear, though speech is still clearly heard and understood by a human, the fricative like quality of the adversarial noises is perceptible even at 25dB SNR.

There are also qualitative differences between the adversarial attacks on baseline and cortical networks for the FGSM and PGD attacks in the mel spectrogram domain. There is a slightly higher dependence of modifying speaker frequencies in the baseline network over the cortical network. This is consistent with our hypothesis that the cortical representation of speech is more invariant to noise, at least for speaker frequencies (below 1 kHz). In the PGD attacks we see more horizontal lines in the baseline, and some diagonal lines in the cortical. This is consistent with our expectation that cortical features filter out stationary noises.

In the right hand side of Fig. 14.4 we create adversarial music attacks [27]. These are also defended against by the cortical representation, as seen by the greater decrease in AUC by the baseline, see Fig. 14.5. These sounds are much more subtle since they are hidden in guitar like plucks.

14.5 Conclusion and Further Work

In this work we use a multi-scale approach to develop a defense against audio adversarial attacks. We propose a unification of a computational model of speech processing in the brain with commercial wake-word networks to create a cortical network, and show that it can increase resistance to adversarial noise without a

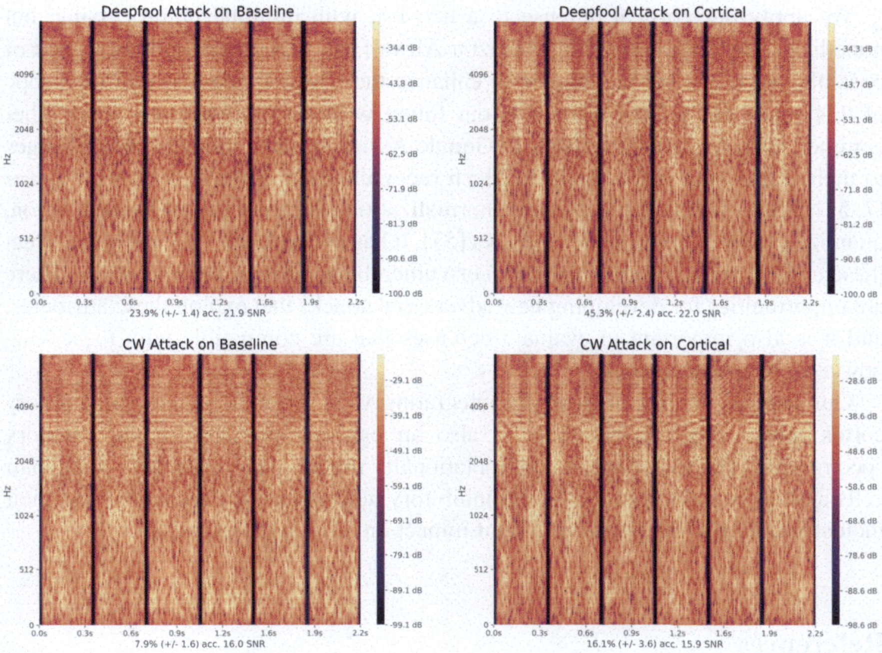

Fig. 14.6 Qualitative differences between the adversarial attacks on baseline and cortical networks for the Deepfool and Carlini-Wagner attacks in the mel spectrogram domain. Each subfigure concatenates the best noise from 6 trials, and reports the SNR and accuracy on these 6 noises on the x-axis. The cortical network attacks appear to pulsate more in time while the baseline network attacks are more stationary. The cortical attacks also exhibit more diagonal lines at resonant frequencies

degradation in performance. We integrated knowledge from the cortical representation of speech studied in Neuroscience to develop a defense against adversarial attacks on audio. We enhanced a neural networks used for wake-word detection by integrating this biological representation for sound processing, and found that the accuracy of the architecture did not degrade, while resistance to adversarial noise increased several times.

We apply several white-box iterative optimization-based adversarial attacks to an implementation of Amazon Alexa's HW network, and a modified version of this network with an integrated cortical representation, and show that the cortical features help defend against universal adversarial examples. At the same level of distortion, the adversarial noises found for the cortical network are always less effective for universal attacks. Further work could start by refining the auditory preprocessing, a more faithful auditory spectrogram would simulate the transduction of the hair cells and the reduction of the lateral inhibitory network. We also constrained our analysis to highly effective but not imperceptible adversarial noise. Further work could create more subtle noises that degrade performance gradually, enforcing imperceptibility with perceptual masking [45].

We apply our cortical defense to a network with minimal preprocessing, but nonetheless reportedly used by Amazon Alexa [19]. (Testing of a broad range of possible preprocessing techniques to enhance the network goes beyond the scope of this paper, but will be a topic of our future work on this subject.) Some other commercial voice assistants such as Google Assistant make miniaturizing changes to their wake-word detection and speech recognition algorithms to save on power [7, 57]. These include local dynamic normalization instead of global normalization, quantized weights, and smaller features [57]. It has not been investigated what effect these choices make on the robustness or vulnerability of their ASR systems. There are opportunities for developing new adversarial attacks that exploit these attributes, and it is also important to evaluate defenses that are compatible with these low-power architecture choices.

Our cortical defense focuses on integrating what is known about the auditory cortex into a defense, but there is also an opportunity to refine the auditory spectrogram. The techniques to computationally mimic the transduction of the hair cells and the reduction of the lateral inhibitory network are differentiable and their inclusion could be investigated in their impact on adversarial examples.

References

1. S. Abdoli, L.G. Hafemann, J. Rony, I.B. Ayed, P. Cardinal, A.L. Koerich, Universal adversarial audio perturbations. arXiv preprint arXiv:1908.03173, 2019
2. Alexa Dataset. https://kaggle.com/aanhari/alexa-dataset
3. J. Anden, V. Lostanlen, S. Mallat, Joint time-frequency scattering for audio classification, in *2015 IEEE 25th International Workshop on Machine Learning for Signal Processing (MLSP)* (2015), pp. 1–6
4. J.J. Benedetto, From a wavelet auditory model to definitions of the Fourier transform, in *Wavelets and Their Applications* (1994), pp. 1–17
5. J.J. Benedetto, A. Teolis, An auditory motivated time-scale signal representation, in *Proceedings of the IEEE-SP International Symposium on Time-Frequency and Time-Scale Analysis* (1992), pp. 49–52
6. J.J. Benedetto, A. Teolis: A wavelet auditory model and data compression. Appl. Comput. Harmonic Anal. **1**, 3–28 (1993)
7. T. Bluche, M. Primet, T. Gisselbrecht, Small-footprint open-vocabulary keyword spotting with quantized LSTM networks. arXiv preprint arXiv:2002.10851, 2020
8. U. Boscain, D. Prandi, L. Sacchelli, G. Turco, A bio-inspired geometric model for sound reconstruction. J. Math. Neurosci. **11**, 2 (2021)
9. N. Carlini, D. Wagner, Towards evaluating the robustness of neural networks, in *2017 IEEE Symposium on Security and Privacy (SP)* (2017), pp. 39–57
10. N. Carlini, D. Wagner, Audio adversarial examples: targeted attacks on speech-to-text, in *2018 IEEE Security and Privacy Workshops (SPW)* (2018), pp. 1–7
11. T. Chi, P. Ru, S.A. Shamma, Multiresolution spectrotemporal analysis of complex sounds. J. Acoust. Soc. Am. **118**(2), 887–906 (2005)
12. W. Czaja, W. Li, Analysis of time-frequency scattering transforms. Appl. Comput. Harmonic Anal. **47**, 149–171 (2019)
13. W. Czaja, W. Li, Rotationally invariant time–frequency scattering transforms. J. Fourier Anal. Appl. **26**, 1–23 (2020)

14. N. Ding, A.D. Patel, L. Chen, H. Butler, C. Luo, D. Poeppel, Temporal modulations in speech and music. Neurosci. Biobehav. Rev. **81**(Pt B), 181–187 (2017)
15. R. Duits, E. Franken, Left-invariant parabolic evolutions on SE(2) and contour enhancement via invertible orientation scores. Part I: Linear left-invariant diffusion equations on SE. Q. Appl. Math. **68**, 255–292 (2010)
16. R. Duits, E. Franken, Left-invariant parabolic evolutions on SE(2) and contour enhancement via invertible orientation scores. Part II: nonlinear left-invariant diffusions on invertible orientation scores. Q. Appl. Math. **68**, 293–331 (2010)
17. D. Gabor, Theory of communication. Part 1: The analysis of information. J. Inst. Electr. Eng. Part 3 Radio Commun. Eng. **93**, 429–441 (1946)
18. I.J. Goodfellow, J. Shlens, C. Szegedy, Explaining and harnessing adversarial examples. arXiv preprint arXiv:1412.6572, 2014
19. J. Guo, K. Kumatani, M. Sun, M. Wu, A. Raju, N. Ström, A. Mandal, Time-delayed bottleneck highway networks using a DFT feature for keyword spotting, in *2018 IEEE International Conference on Acoustics, Speech and Signal Processing (ICASSP)* (2018), pp. 5489–5493
20. E. Hernández, D. Labate, G. Weiss, A unified characterization of reproducing systems generated by a finite family, II. J. Geom. Anal. **12**, 615–662 (2002)
21. H. Hirano, K. Takemoto, Simple iterative method for generating targeted universal adversarial perturbations, in *Proceedings of 25th International Symposium on Artificial Life and Robotics (AROB 25th 2020)* (2019), pp. 426–430
22. jhautry/echo-dot. https://github.com/jhautry/echo-dot. Library Catalog: github.com
23. A.J.E. Kell, J.H. McDermott, Invariance to background noise as a signature of non-primary auditory cortex. Nat. Commun. **10**(1), 3958 (2019). Number: 1 Publisher: Nature Publishing Group
24. B. Khalighinejad, J.L. Herrero, A.D. Mehta, N. Mesgarani, Adaptation of the human auditory cortex to changing background noise. Nat. Commun. **10**(1), 2509 (2019). Number: 1 Publisher: Nature Publishing Group
25. D. Labate, G. Weiss, Continuous and discrete reproducing systems that arise from translations. theory and applications of composite wavelets, in *Four Short Courses on Harmonic Analysis* (Birkhäuser, Boston, 2010), pp. 85–130
26. D. Labate, G. Weiss, E. Wilson, An approach to the study of wave packet systems, in *Contemp. Math., Wavelets, Frames, Operator Theory*, vol. 345 (2004), pp. 215–235
27. J. Li, S. Qu, X. Li, J. Szurley, J.Z. Kolter, F. Metze, Adversarial music: Real world audio adversary against wake-word detection system, in *Advances in Neural Information Processing Systems*, vol. 32 (2019)
28. R. Lyon, *Human and Machine Hearing: Extracting Meaning from Sound* (Cambridge University Press, 2017)
29. A. Madry, A. Makelov, L. Schmidt, D. Tsipras, A. Vladu, Towards deep learning models resistant to adversarial attacks, in *International Conference on Learning Representations* (2018)
30. S. Mallat, Group invariant scattering. Commun. Pure Appl. Math. **65**, 1331–1398 (2012)
31. N. Mesgarani, S. Shamma, Denoising in the domain of spectrotemporal modulations. EURASIP J. Audio Speech Music Process. **2007**(1), 042357 (2007)
32. N. Mesgarani, S. Shamma, Speech processing with a cortical representation of audio, in *2011 IEEE International Conference on Acoustics, Speech and Signal Processing (ICASSP)* (2011), pp. 5872–5875
33. N. Mesgarani, M. Slaney, S. Shamma, Discrimination of speech from nonspeech based on multiscale spectro-temporal modulations. IEEE Trans. Audio Speech Lang. Process. **14**(3), 920–930 (2006). Conference Name: IEEE Transactions on Audio, Speech, and Language Processing
34. N. Mesgarani, S.V. David, J.B. Fritz, S.A. Shamma, Phoneme representation and classification in primary auditory cortex. J. Acoust. Soc. Am. **123**(2), 899–909 (2008). Publisher: Acoustical Society of America

35. N. Mesgarani, C. Cheung, K. Johnson, E.F. Chang, Phonetic feature encoding in human superior temporal gyrus. Science **343**(6174), 1006–1010 (2014)
36. N. Mesgarani, S.V. David, J.B. Fritz, S.A. Shamma, Mechanisms of noise robust representation of speech in primary auditory cortex. Proc. Natl. Acad. Sci. **111**(18), 6792–6797 (2014). Publisher: National Academy of Sciences Section: Biological Sciences
37. S.M. Moosavi-Dezfooli, A. Fawzi, P. Frossard, DeepFool: A simple and accurate method to fool deep neural networks, in *2016 IEEE Conference on Computer Vision and Pattern Recognition (CVPR)* (2016), pp. 2574–2582
38. S.M. Moosavi-Dezfooli, A. Fawzi, O. Fawzi, P. Frossard, Universal adversarial perturbations, in *2017 IEEE Conference on Computer Vision and Pattern Recognition (CVPR)* (IEEE, Honolulu, HI, 2017), pp. 86–94
39. M.I. Nicolae, M. Sinn, M.N. Tran, B. Buesser, A. Rawat, M. Wistuba, V. Zantedeschi, N. Baracaldo, B. Chen, H. Ludwig, I. Molloy, B. Edwards, Adversarial robustness toolbox v1.2.0. arXiv preprint arXiv:1807.01069, 2018
40. V. Panayotov, G. Chen, D. Povey, S. Khudanpur, Librispeech: an ASR corpus based on public domain audio books, in *Acoustics, Speech and Signal Processing (ICASSP), 2015 IEEE International Conference on* (IEEE, 2015), pp. 5206–5210
41. S. Panchapagesan, M. Sun, A. Khare, S. Matsoukas, A. Mandal, B. Hoffmeister, S. Vitaladevuni, Multi-task learning and weighted cross-entropy for dnn-based keyword spotting, in *Interspeech 2016* (2016), pp. 760–764
42. J. Petitot, The neurogeometry of pinwheels as a sub-Riemannian contact structure. J. Physiol. Paris **97**, 265–309 (2003)
43. J. Petitot, Y. Tondut, Vers une neurogéométrie. fibrations corticales, structures de contact et contours subjectifs modaux. Math. Sci. Hum. **145**, 5–101 (1999)
44. Picovoice/wake-word-benchmark (2020). https://github.com/Picovoice/wake-word-benchmark. Original-date: 2018-04-25T00:12:46Z
45. Y. Qin, N. Carlini, G. Cottrell, I. Goodfellow, C. Raffel, Imperceptible, robust, and targeted adversarial examples for automatic speech recognition, in *International Conference on Machine Learning*. PMLR (2019), pp. 5231–5240
46. H.M. Reimann, Signal processing in the cochlea: the structure equation. J. Math. Neurosci. **1**, 5 (2011)
47. P. Ru, Multiscale multirate spectro-temporal auditory model. University of Maryland College Park, USA (2001)
48. Rubber Band Audio Time Stretcher Library. https://breakfastquay.com/rubberband/
49. R. Scheibler, E. Bezzam, I. Dokmanić, Pyroomacoustics: A Python package for audio room simulations and array processing algorithms, in *2018 IEEE International Conference on Acoustics, Speech and Signal Processing (ICASSP)* (2018), pp. 351–355
50. L. Schonherr, T. Eisenhofer, S. Zeiler, T. Holz, D. Kolossa, Imperio: robust over-the-air adversarial examples for automatic speech recognition systems. Proceedings of the 36th Annual Computer Security Applications Conference (2020), pp. 843–855
51. S. Shamma, Auditory cortical representation of complex acoustic spectra as inferred from the ripple analysis method. Netw. Comput. Neural Syst. **7**, 439–476 (1996)
52. S. Shamma, H. Versnel, Ripple analysis in ferret primary auditory cortex III. prediction of unit responses to arbitrary spectral profiles. Tech. rep., University of Maryland, College Park, 1995
53. M. Sun, V. Nagaraja, B. Hoffmeister, S. Vitaladevuni, Model shrinking for embedded keyword spotting, in *2015 IEEE 14th International Conference on Machine Learning and Applications (ICMLA)* (2015), pp. 369–374
54. J. Szurley, J.Z. Kolter, Perceptual based adversarial audio attacks. arXiv preprint arXiv:1906.06355, 2019
55. testandroidtests. https://github.com/testandroidtests. Library Catalog: github.com
56. G.H. Wakefield, Time-pitch representations: acoustic signal processing and auditory representations, in *Proceedings of the IEEE-SP International Symposium on Time-Frequency and Time-Scale Analysis* (1998), p. 98TH8380

57. P. Warden, D. Situnayake, *TinyML: Machine Learning with TensorFlow Lite on Arduino and Ultra-Low-Power Microcontrollers* (O'Reilly Media, 2019)
58. X. Yang, K. Wang, S. Shamma, Auditory representations of acoustic signals. IEEE Trans. Inf. Theory **38**, 824–839 (1992)

Chapter 15
On Low-Rank Convex-Convex Quadratic Fractional Programming

Ilya Krishtal and Brendan Miller

Abstract We present an efficient algorithm for solving fractional programming problems whose objective functions are the ratio of a low-rank quadratic to a positive definite quadratic with convex constraints. The proposed algorithm for these convex-convex problems is based on the Shen-Yu Quadratic Transform (Shen and Yu, IEEE Trans Signal Process 66(10):2616–2630, 2018) which finds stationary points of concave-convex sum-of-ratios problems. We further use elements of the algorithm proposed by Shen and Yu and the classic Dinkelbach approach to ensure convergence. We show that our algorithm performs better than previous algorithms for low-rank problems.

15.1 Introduction

Methods of fractional programming encompass a large range of techniques to solve problems of the form

$$\begin{aligned} \max \quad & F(x) = \sum_{m=1}^{M} \frac{N_m(x)}{D_m(x)} \\ \text{s.t.} \quad & x \in X \subset \mathbb{R}^n \end{aligned} \tag{15.1}$$

where $N_m(x)$, $D_m(x) : \mathbb{R}^n \to \mathbb{R}$ are continuous functions, and X is a closed, convex set. A convention of the field, which we adopt throughout this paper, is that $N_m(x) \geq 0$ and $D_m(x) > 0$ for all $x \in X$. The problem is called *single-ratio* if $M = 1$ and a *sum-of-ratios* fractional programming problem if $M > 1$. Fractional programming problems arise in many different applications, such as finance and portfolio analysis,

I. Krishtal (✉) · B. Miller
Northern Illinois University, Dekalb, IL, USA
e-mail: ikrishtal@niu.edu; bmiller14@niu.edu

E. Hernández et al. (eds.), *The Mathematical Heritage of Guido Weiss*,
Applied and Numerical Harmonic Analysis,
https://doi.org/10.1007/978-3-031-76793-7_15

government contracting, and engineering (see [1, 2, 10, 11]). We were led to a single-ratio problem in the process of studying the problem of optimal sensor placement for dynamical sampling on graphs [8]. In particular, the relative error of reconstructing a signal on a graph from spatio-temporal samples in the presence of noise can be bounded above by a ratio of two quadratics depending on the eigenvalues of a certain frame operator.

Fractional programs have a long history. Below we outline a few milestones in the development of the theory.

In 1962, Charnes and Cooper showed in [3] that linear fractional programming problems, which are single-ratio fractional programming problems whose numerator, denominator, and constraints are linear, can be solved by simplex method.

In 1967, Dinkelbach [4] showed that any concave-convex single-ratio fractional programming problem can be solved efficiently by consecutively solving several concave maximization problems. The approach used by Dinkelbach was so useful that it has become standard and even seen many applications to fractional programs which are not concave-convex. In the latter case, however, the method involves solving a succession of non-concave maximization problems and becomes rather expensive.

In 1997, Lo and MacKinlay [11] proposed the problem of maximizing the ratio of two convex quadratic functions in the context of portfolio analysis. The following decade saw two papers published [7, 14] which presented algorithms for solving this problem exactly. Both follow the standard Dinkelbach approach, with different methods of handling the most expensive aspect of the problem: the need to solve multiple non-convex quadratic programming problems. To do so, the former paper by Gotoh and Konnoh implemented a branch-and-bound technique while the latter by Yamamoto and Konnoh iteratively approximated the quadratic by a piece-wise linear function, which is maximized by standard mixed integer linear programming techniques.

In recent years, there has also been much research on the sum-of-ratios problem. For example, [1, 2, 5] all produce branch-and-bound type algorithms for solving the concave-convex sum-of-ratios problem. Similarly, [10, 12] propose efficient algorithms for solving quadratically constrained quadratic sum-of-ratios type problems. Of particular interest to us is the Shen-Yu Quadratic Transform introduced in [13] which, like the Dinkelbach method, iteratively solves a concave programming problem to converge to a stationary point of the concave-convex sum-of-ratios problem.

In this paper, we propose an algorithm for solving the single-ratio convex-convex quadratic programming problem which can be effectively utilized when the numerator has low rank. We divide the feasible region into several subregions in which the Shen-Yu Quadratic Transform method can be applied successively until global convergence. Since the methods of Shen and Yu only guarantee convergence to a stationary point, we use the mixed integer linear programming techniques employed in [14] to check if a given stationary point is a global optimum. Although, as we will see empirically, this rather expensive procedure can often be omitted altogether (with a minimal chance of error). This yields a very efficient algorithm which, with high probability, converges to the global solution of a convex-convex quadratic programming problem by successively solving concave maximization problems. The only known algorithms for solving such problems involve successively maximizing a nonconvex quadratic programming problems.

The remainder of this paper is organized as follows. In Sect. 15.2, we state the problem we focus on and outline the classic Dinkelbach approach for ratio maximization. We then convert our problem into a sum-of-ratios problem and recall a recent Shen-Yu scheme for solving such problems. Section 15.3 is the centerpiece of the paper. In it, we present a natural way of subdividing the feasible region of our problem into a number of subregions thereby replacing a convex-convex sum-of-ratios problem with a finite number of much simpler concave-convex sum-of-ratios problems. This results in a region checking algorithm (Algorithm 15.3) which encompasses our approach to solving single-ratio quadratic convex-convex problems. Numerical experiments described in Sect. 15.4 illustrate the effectiveness of our approach in comparison with other algorithms. Finally, concluding remarks are presented in Sect. 15.5.

15.2 Basic Analysis

We study the problem of the form

$$\begin{aligned} \max \quad & F(x) = \frac{x^T Q x}{x^T P x} \\ \text{s.t.} \quad & Ax \leq b \end{aligned} \tag{15.2}$$

where Q is an $n \times n$ positive semidefinite matrix, P is an $n \times n$ positive definite matrix, $A \in \mathbb{R}^{T \times n}$, $b \in \mathbb{R}^T$. We will denote the feasible region of this problem by $X \subset \mathbb{R}^n$. Such optimization problems are nonconcave in general, and thus require expensive algorithms to solve. We recall the standard Dinkelbach method which utilizes the function

$$\pi(\lambda) := \max_{x \in X} x^T (Q - \lambda P) x, \quad \lambda > 0.$$

It is convenient to introduce the following notation:

$$x(\lambda) = \arg\max_{x \in X} x^T (Q - \lambda P) x, \quad \lambda > 0.$$

Theorem 15.1 *The function $\pi(\lambda)$ is convex and strictly decreasing in λ. Furthermore, $\pi(\lambda) = 0$ if and only if $x(\lambda)$ maximizes* (15.2) *in X.*

The algorithms of [7, 14] are root-finding algorithms which use a scheme developed by Ibaraki [9] to search for the root of π. These algorithms become expensive because computing $\pi(\lambda)$ is a nonconvex quadratic programming problem, whose difficulty is larger for smaller values of λ. We will utilize this Theorem of Dinkelbach in our algorithm, but only to check if a local maximum of (15.2) is indeed a global maximum. We summarize the Ibaraki scheme here as Algorithm 15.1 for reference, but omit the explanation of convergence.

input : Matrices Q and P, linear inequality constraints $Ax \leq b$, and a tolerance ε.

Find λ^u with $\pi(\lambda^u) < 0$ and λ^l with $\pi(\lambda^l) > 0$;

repeat

Compute $\overline{\lambda}$ as follows:

$$\overline{\lambda} = \begin{cases} -\frac{\pi(\lambda^u)}{\Delta\pi} + \lambda^u & \text{if } x(\lambda^u)^T Px(\lambda^u) + \Delta\pi \neq 0 \\ \frac{\pi(\lambda^u)}{x(\lambda^u)^T Px(\lambda^u)} + \lambda^u & \text{otherwise,} \end{cases}$$

where $\Delta\pi = (\pi(\lambda^u) - \pi(\lambda^l))/(\lambda^u - \lambda^l)$;

Compute $\pi(\overline{\lambda})$;

Update $\lambda^l = \overline{\lambda}$ if $\pi(\overline{\lambda}) > 0$, or $\lambda^u = \overline{\lambda}$ if $\pi(\overline{\lambda}) < 0$;

until $|\pi(\overline{\lambda})| < \varepsilon$;

Algorithm 15.1: Interpolated binary search (Ibaraki scheme)

The algorithms presented in [7, 14] both utilize Algorithm 15.1, but employ different methods of computing $\pi(\overline{\lambda})$. We will use contemporary software to solve these non-convex quadratic programming problems to ensure the most accurate solutions.

Suppose now that Q can be written as

$$Q = \sum_{m=1}^{M} q_m q_m^T,$$

which is a sum of rank-one matrices. Note that Q always admits such a decomposition as we can take $M = rank(Q)$ and $q_m = \sqrt{\lambda_m} v_m$ where v_m is the eigenvector of Q associated to the nonzero eigenvalue λ_m. Then we may rewrite the objective function in the following way:

$$\begin{aligned} F(x) &= \frac{\sum_{m=1}^{M} x^T (q_m q_m^T) x}{x^T Px} \\ &= \sum_{m=1}^{M} \frac{\langle q_m, x\rangle^2}{x^T Px}. \end{aligned} \tag{15.3}$$

This reformulation of the objective function converts a single-ratio fractional programming problem (15.2) into a convex-convex sum-of-ratios fractional programming problem which we will refer to as (15.3).

Next, we summarize a method for suboptimally solving sum-of-ratios fractional programming problems (15.1) that was introduced in [13].

Definition 15.1 (See [13]) Given the sum-of-ratios fractional programming problem *(15.1)*, the Shen-Yu Quadratic Transform of this problem is defined to be

$$g(x, y) = \sum_{m=1}^{M} \left(2y_m\sqrt{N_m(x)} - y_m^2 D_m(x)\right). \tag{15.4}$$

Theorem 15.2 ([13]) *The sum-of-ratios problem* (15.1) *and its quadratic transform* (15.4) *maximization problem are equivalent. More precisely, the maximum of* (15.4) *occurs at* (x^*, y^*) *where* x^* *maximizes* (15.1) *and* $y^* = (y_m^*)_{m=1}^M$ *satisfies*

$$y_m^* = \frac{\sqrt{N_m(x^*)}}{D_m(x^*)}.$$

The following Lemma is useful for deriving the Shen-Yu scheme.

Lemma 15.1 *Let* $g(x, y)$ *be as in* (15.4). *If* $x_0 \in X$, *then* $y_0 := \arg\max_{y\in\mathbb{R}^M} g(x_0, y)$ *can be found analytically and is given by* $y_0 = (y_m^0)_{m=1}^M$, *where*

$$y_m^0 = \frac{\sqrt{N_m(x_0)}}{D_m(x_0)}.$$

Proof It suffices to maximize each term in the sum in (15.4) individually. Clearly, the vertex of the quadratic $f(y_m) = 2y_m\sqrt{N_m(x_0)} - y_m^2 D(x_0)$ occurs at the point

$$y_m^0 = \frac{\sqrt{N_m(x_0)}}{D_m(x_0)}$$

as desired. □

We may now state and prove a theorem from which, when taken together with Theorem 15.2, an algorithm for suboptimally solving (15.1) is naturally derived.

Theorem 15.3 (Shen, Yu [13]) *Consider the sum-of-ratios problem* (15.1) *and suppose that* $x_0 \in X$. *Let* $y_0 = (y_m^0)_{m=1}^M$ *with* $y_m^0 = \sqrt{N_m(x_0)}/D_m(x_0)$. *If*

$$x^* = \arg\max_{x\in X} g(x, y_0) = \arg\max_{x\in X} \sum_{m=1}^{M} \left(2y_m^0\sqrt{N_m(x)} - (y_m^0)^2 D_m(x)\right),$$

then $F(x^*) \geq F(x_0)$.

Proof Let x_0, y_0, and x^* be as above and set

$$y_m^* = \frac{\sqrt{N_m(x^*)}}{D_m(x^*)}$$

and $y^* = (y_m^*)_{m=1}^M$. Then we have the following string of inequalities

$$\begin{aligned} F(x_0) &= g(x_0, y_0) \\ &\leq g(x^*, y_0) \\ &\leq g(x^*, y^*) \\ &= F(x^*) \end{aligned}$$

where the third line follows from Lemma 15.1, and the last by a direct computation. The proof is complete. □

The above Theorem guarantees that replacing x_0 by x^* and y_0 by y^* improves the value of the objective function with each iteration, and thus this method (with the scheme written explicitly in Algorithm 15.2) converges. It is clear that the algorithm converges to a local maximum, say (x^*, y^*), of the Quadratic Transform g. It follows easily that the value x^* is indeed a local maximum of the objective F as well.

input : Functions N_m and D_m, and a compact, convex set X.

Find an initial $x_0 \in X$;
repeat
  set $y_m^0 = \sqrt{N_m(x_0)}/D_m(x_0)$;
  solve $x^* = \arg\max_{x \in X} \sum_{m=1}^M \left(2y_m^0\sqrt{N_m(x)} - (y_m^0)^2 D_m(x)\right)$;
  update $x_0 = x^*$;
until convergence;

Algorithm 15.2: Shen-Yu iterative algorithm

Corollary 15.1 *Algorithm* 15.2 *converges to a stationary point of the sum-of-ratios fractional programming problem* (15.1).

Although Algorithm 15.2 is guaranteed to converge to a stationary point, its usefulness is limited to the difficulty of maximizing the Quadratic Transform over the variable x. In the case when each N_m is concave and each D_m is convex, this can be done by any method of concave programming. When the objective has the form (15.3), the Quadratic Transform becomes

$$g(x, y) = \sum_{m=1}^{M} 2y_m|\langle q_m, x\rangle| - y_m^2 x^T P x \tag{15.5}$$

which is not, in general, concave in x. However, in the event that the absolute values around each linear term in the sum can be dropped (i.e. each $\langle q_m, x\rangle$ is either nonnegative or nonpositive valued on X), then we may apply Algorithm 15.2 effectively.

15.3 Region Checking Algorithm

Henceforth we will assume the objective function F is as in (15.2) and, therefore, it can be rewritten in the sum-of-ratios form (15.3). The first observation to make is that combining Algorithm 15.2 with the Dinkelbach method yields an algorithm which converges to the global maximum of (15.3). Indeed, if x^* is the local optimum found by Algorithm 15.2, we may set $\lambda^* = F(x^*)$ and compute both $\pi(\lambda^*)$ and $x(\lambda^*)$. If $|\pi(\lambda^*)| > \varepsilon$ where ε is some tolerance, we again run Algorithm 15.2 with $x(\lambda^*)$ as the initial feasible point and repeat. This method is guaranteed to converge to the globally optimal solution since

$$x(\lambda^*)^T(Q - \lambda^* P)x(\lambda^*) > x^{*T}(Q - \lambda^* P)x^*$$

implies that

$$F(x(\lambda^*)) > F(x^*) \geq F(x_0)$$

where x_0 is the initial feasible point used in Algorithm 15.2.

Finding the value of $\pi(\lambda^*)$ and the vector $x(\lambda^*)$ can be achieved by the methods introduced in [7] and refined in [14], but maximizing the Quadratic Transform as in Algorithm 15.2 cannot. To circumvent this issue, we divide the feasible region into at most $2^{rank(Q)}$ subregions and perform this algorithm independently in each subregion. Suppose, as before, that

$$Q = \sum_{m=1}^{M} q_m q_m^T,$$

and for each $1 \leq m \leq M$ define

$$R_m^0 = \{x \in X \mid \langle q_m, x\rangle \leq 0\}, \quad R_m^1 = \{x \in X \mid \langle q_m, x\rangle \geq 0\}.$$

For each binary sequence $(n_m)_{m=1}^M = n \in \{0, 1\}^M$, we denote by R^n a (possibly empty) subregion of X given by

$$R^n = \bigcap_{m=1}^{M} R_m^{n_m}.$$

A straightforward observation then yields the following result.

Lemma 15.2 *The Quadratic Transform* (15.5) *of* (15.3) *is concave in the x variable over each nonempty subregion R^n for $n \in \{0, 1\}^M$.*

Proof Let $g(x, y)$ be as in (15.5) and fix a binary sequence $n \in \{0, 1\}^M$. Then, by definition of R^n, we see that for each $1 \leq m \leq M$ we have either $\langle q_m, x\rangle \geq 0$ or $\langle q_m, x\rangle \leq 0$ for all $x \in R^n$. This implies that $|\langle q_m, x\rangle|$ is linear over R^n, and hence $g(x, y)$ is a sum of concave functions, which is itself concave. □

We can now formulate our method for solving the convex-convex quadratic fractional programming problem as Algorithm 15.3.

input : Matrices $Q = \sum_{m=1}^{M} q_m q_m^T$, P, linear constraints $Ax \leq b$, and a tolerance ε.
output: Global Solution of (15.2)
for $n \in \{0, 1\}^M$ **do**
    if $R^n \neq \emptyset$ **then**
        Find an initial $x_n \in R^n$;
        repeat
            Find stationary point x^* of (15.2) via Algorithm 15.2 with initial point x_n;
            Set $\lambda = F(x^*)$;
            Compute $\pi(\lambda)$ and $x(\lambda)$;
            Update $x_n = x(\lambda)$;
        until $|\pi(\lambda)| < \varepsilon$;
    end
end
Determine max $F(x_n)$;

Algorithm 15.3: Region-checking algorithm for convex-convex quadratic fractional programming

Remark 15.1 The utility of Algorithm 15.3 is, in full generality, limited by the rank of Q or, more precisely, by the number of nonempty subregions (which is controlled by the rank of Q). It is also worth noting that different decompositions of Q may yield different numbers of nonempty subregions of the feasible region. We leave the question of how to find better decompositions of Q beyond the scope of this paper. We do, however, mention explicitly the case when Q is a totally nonnegative matrix and the optimization problem in question is

$$\begin{aligned} \max \quad & F(x) = \frac{x^T Q x}{x^T P x} \\ \text{s.t.} \quad & Ax = b \\ & 0 \leq x \leq \alpha. \end{aligned}$$

In this case, we write $Q = LDL^T$ for D a diagonal matrix and L a lower triangular matrix. This gives a decomposition of Q as

$$Q = \sum_{m=1}^{M} d_m \ell_m \ell_m^T,$$

where ℓ_m and d_m are, respectively, the mth column of L and mth diagonal entry of D. Since Q is totally nonnegative, the entries of each vector ℓ_m are nonnegative [6], in which case $\langle \ell_m, x \rangle \geq 0$ for all feasible x. Thus, there is only one subregion R^n of the feasible region which is nonempty.

It is quite possible that Algorithm 15.3 is considerably slower than the algorithms proposed in [7] and [14] given the potentially large number of times $\pi(\lambda)$ is computed. However, as we will show empirically in the next section, often the first local maximum found by Algorithm 15.3 in a given region is, in fact, the global maximum of the region. Thus, we will also compare the efficiency and accuracy of Algorithm 15.3 without computing $\pi(\lambda)$ and assuming each x^* found by Algorithm 15.2 is a global maximum of the region. This modification of Algorithm 15.3 is Algorithm 15.4 below.

input : Matrices $Q = \sum_{m=1}^{M} q_m q_m^T$, P, linear constraints $Ax \leq b$, and a tolerance ε.
output: Local Solution of (15.2)

for $n \in \{0, 1\}^M$ **do**
  if $R^n \neq \emptyset$ **then**
    Find an initial $x_n \in R^n$;
    Find stationary point x^* of (15.2) via Algorithm 15.2 with initial point x_n;
    Update $x_n = x^*$;
  end
end
Determine max $F(x_n)$;

Algorithm 15.4: Fast region-checking algorithm

Remark 15.2 Suppose $Q = qq^T$ is a rank-one matrix. We write the objective function as

$$F(x) = \frac{\langle q, x \rangle^2}{x^T P x}.$$

In this case, there are only two subregions of the feasible region: R^0 and R^1. Also, we may equivalently maximize the square-root of the objective, which is given by

$$\sqrt{F(x)} = \begin{cases} -\langle q, x \rangle / \sqrt{x^T P x} & x \in R^0 \\ \langle q, x \rangle / \sqrt{x^T P x} & x \in R^1. \end{cases}$$

Thus, $\sqrt{F(x)}$ is a concave-convex fractional programming problem in each subregion of the feasible region, and hence can be solved by two applications of Algorithm 15.1 where computing $\pi(\lambda)$ is a concave programming problem. This method is superior to Algorithm 15.4 as it was shown in [13] that Algorithm 15.2 is slower than the standard Dinkelbach method for standard single-ratio concave-convex fractional programs.

15.4 Numerical Experiments

In this section we conduct several numerical experiments on the following optimization problem:

$$\begin{aligned} \max \quad & F(x) = \frac{x^T Q x}{x^T P x} \\ \text{s.t.} \quad & Ax \leq b, \quad A \in \mathbb{R}^{T \times n}, \quad b \in \mathbb{R}^T \\ & \sum_{i=1}^{n} x_i = 1 \\ & 0 \leq x_i \leq 0.1, \quad i = 1, ..., n \end{aligned}$$

where A and b have random entries in the interval $[0, 1]$ so that the vector $(1/n, ..., 1/n)$ is feasible, and $Q = X^T X$ and $P = Y^T Y$ where X and Y are, respectively, an $M \times n$ and an $n \times n$ random matrix with entries in $[0, 1]$. Unless otherwise specified, we will always decompose Q as a sum of rank one matrices according to its eigendecomposition as noted in Sect. 15.2.

We first demonstrate the efficiency and accuracy of Algorithms 15.3 and 15.4 against Algorithm 15.1 for various combinations of (n, M, T). Next, we examine the average number of nonempty subregions in Algorithms 15.3 and 15.4. Finally, we demonstrate the accuracy of Algorithm 15.4 in a full-rank example (i.e. $r = n$, which has several local maxima) when there are a small number of subregions (see Remark 15.1).

All computation was done in MATLAB (on AMD A6-7400K Radeon R5 4.09 GHz processor), using Gurobi 9.5 interface to solve the nonconvex quadratic programming problems involved in Algorithms 15.1 and 15.3. The tolerance ε is always set as $\varepsilon = 10^{-3}$. All values in the forthcoming tables are averages of five tests.

15.4.1 Algorithm Comparison

We first give a demonstration of how differing combinations of (n, M, T) affect the computation time of Algorithm 15.3 in a single nonempty region of the feasible set.

Table 15.1 shows that the computation time for Algorithm 15.3 increases sharply with the number of inequality constraints due to the increasing complexity of solving the nonconvex quadratic subproblems. Likewise, the computation time increases with both the number of variables and the rank of Q, albeit not as sharply. Second, none of the problems solved in this experiment took more than two iterations of Algorithm 15.2 to converge. In fact, 209 of the 240 problems solved in this experiment converged in just one iteration of Algorithm 15.2. This suggests that the expensive procedure of computing $\pi(\lambda)$ can safely be dropped from the algorithm, if a low probability of missing the exact solution may be tolerated by the application.

In the next two experiments, we fix $T = 10$. We now compare Algorithm 15.3 with Algorithm 15.1. We show the results for Algorithm 15.3 both converging to the global solution and forcing only one iteration per region without computing $\pi(\lambda)$.

Table 15.2 shows the comparison of Algorithms 15.1, 15.3, and 15.4. There are several things to note about these results. First, the error incurred from performing only one iteration per region is negligible; it is always under one percent and quite often is under 0.1%. This implies that one need only perform Algorithm 15.2 once in each subregion of the feasible region, making irrelevant the need to compute $\pi(\lambda)$. Second, for a given value of M, the computation time needed to complete Algorithm 15.2 increases steadily as n increases, but does so at a much slower rate for Algorithm 15.4.

In fact, Table 15.3 shows that Algorithm 15.4 can be used efficiently for large scale problems when the number of subregions is less than 130. For comparison, the

Table 15.1 Efficiency of Algorithm 15.3 for one region

		CPU time (sec)				Iterations			
		$T = 1$	$T = 10$	$T = 30$	$T = 50$	$T = 1$	$T = 10$	$T = 30$	$T = 50$
$n = 25$	$M = 2$	0.4657	0.3689	0.7372	0.7063	1.2	1	1	1
	5	1.074	0.5517	1.183	0.7725	1	1.2	1	1
	7	1.037	2.339	1.473	1.337	1	1	1	1
	10	0.5639	1.253	9.06	1.849	1.2	1	1.4	1.2
50	2	2.022	2.512	2.403	5.308	1.2	1.2	1	1.2
	5	3.208	6.842	15.71	19.66	1	1	1.2	1
	7	60.85	5.055	6.284	31.49	1.4	1	1.2	1
	10	8.032	126.5	663.4	49.52	1.2	1	1.4	1
75	2	4.261	7.612	11.04	7.077	1	1	1.2	1
	5	5.299	32.95	58.67	79.35	1	1.2	1.4	1
	7	125.9	62.65	48.03	273.3	1.4	1.2	1.4	1
	10	23.8	20.38	93.34	904.7	1	1	1	1.4

Table 15.2 Comparison of algorithms in CPU time (sec) Error = $(F(x_{opt}) - F(x))/F(x_{opt})$

		CPU time (sec)			Alg 4 error
		Alg 1	Alg 3	Alg 4	Error (%)
$n = 10$	$M = 2$	0.01372	0.07733	0.09616	0
	3	0.006822	0.01969	0.1347	0
	4	0.005291	0.02188	0.2637	0
	5	0.01414	0.02462	0.5047	0
	7	0.01159	–	2.006	0
	10	0.01003	–	16.67	0
25	2	0.5673	0.9934	0.1193	0
	3	0.6599	1.266	0.1932	0.08
	4	1.068	3.975	0.3618	0
	5	1.105	26.85	0.7185	0.06
	7	0.7177	–	2.976	0.02
	10	0.8667	–	23.35	0.02
50	2	4.932	4.649	0.1331	0.04
	3	5.258	11.6	0.2104	0.05
	4	9.485	72.51	0.4216	0.02
	5	7.618	223.9	0.91	0.01
	7	5.573	–	3.18	0.03
	10	8.004	–	26.5	0
75	2	12.63	17.12	0.1229	0.01
	3	14.43	273.3	0.346	0
	4	30.47	211.2	0.531	0.2
	5	45.63	868.8	1.249	0.07
	7	34.33	–	4.407	0
	10	53.35	–	35.55	0.05

Table 15.3 CPU time (sec) for Algorithm 15.4

			Algorithm 15.4
$n = 250$	$M = 7$	$T = 10$	22.32
500	7	10	117.34
750	7	10	357.42
1000	7	10	743.24

authors in [14] state that the case when $n = 500$ is within reach via their algorithm only by employing an elaborate local search.

15.4.2 Accuracy of Algorithm 4

The numerical experiments above call into question the accuracy of Algorithm 15.4 in high-rank problems with a small number of nonempty regions. Table 15.1 shows

Table 15.4 Average number of subregions checked by Algorithm 15.4

		Number of subregions				
		$T = n/2$	$T = n$	$T = 3n/2$	$T = 2n$	$T = 5n/2$
$n = 30$	$M = 2$	2	2	2	1	1
	3	4	4	3.8	2	1
	5	16	16	12	3	1
	7	64	64	49.4	11	1
50	2	2	2	2	1.8	1
	3	4	4	4	2.8	1
	5	16	16	14.4	5	1
	7	64	64	64	19.4	1
100	2	2	2	2	1.4	1
	3	4	4	4	2.8	1
	5	16	16	16	6.6	1
	7	64	64	64	23.4	1
150	2	2	2	2	1.2	1
	3	4	4	4	2.8	1
	5	16	16	16	8	1
	7	64	64	64	13.6	1

that there is typically only one local maximum per subregion, and this may seem to be attributable to the large number of subregions relative to the number of local maxima. Thus, one might conclude that it may be disadvantageous to choose a decomposition of Q with fewer terms as this leads to fewer possible subregions. We aim to show that this is not the case.

The number of subregions is, of course, determined by the linear constraints. Table 15.4 shows that the number of subregions is usually $2^{rank(Q)-1}$, and this decreases only when the number of inequality constraints is much larger than the number of variables. This implies that the number of subregions checked by the algorithm increases exponentially with the rank of Q, and thus the accuracy of Algorithm 15.4 could be a product of the brute-force nature of checking each region.

For illustration we construct a full-rank example with a small number of subregions to check. For a given value of n, we generate n random, linearly independent vectors $(q_i)_{i=1}^n$ with values in the unit interval $[0, 1]$. We construct the matrix Q as

$$Q = \sum_{i=1}^{n} q_i q_i^T .$$

Note that since the q_i are constructed to be linearly independent, the matrix Q will be invertible. Using this decomposition of Q and the constraints as before, there will be only one non-empty subregion of the feasible region. This is because the entries

Table 15.5 CPU time (sec) and error in full rank example

		CPU time (sec)		Alg 4 error
		Alg 1	Alg 4	Error (%)
$n = 20$	$T = 1$	0.2189	0.01942	0
	10	0.6094	0.02111	0
	30	0.9902	0.02257	0.02936
	50	1.766	0.04667	0
35	1	1.423	0.02659	0.08328
	10	6.945	0.02921	0.01075
	30	7.726	0.03527	0.05105
	50	20.61	0.05936	0.02039
50	1	6.25	0.03352	0
	10	6.753	0.03664	0.02419
	30	25.31	0.04325	0.01203
	50	189	0.09199	0.01428

of each q_i are positive, and hence $\langle q_i, x\rangle \geq 0$ for all feasible x. All other matrices are constructed in the same manner as before.

It is shown in Table 15.5 that the accuracy of Algorithm 15.4 remains quite high using this decomposition of Q. So, the accuracy of Algorithm 15.4 should not be attributed primarily to the number of subregions. Therefore, since choosing a decomposition of Q which results in few nonempty subregions yields a faster algorithm, it is advantageous to choose one which yields the fewest number of nonempty subregions of the feasible region.

15.5 Conclusion

We have presented an efficient and accurate algorithm for globally maximizing low-rank convex-convex quadratic fractional programming problems. We have also demonstrated that this algorithm can be utilized in high-rank problems if the numerator admits a decomposition which divides the feasible region into a small number of subregions. Although Algorithm 15.4 is only guaranteed to converge to a local maximum of (15.2), we have shown heuristically that it almost always converges to the global solution.

To guarantee global convergence of Algorithm 15.4, one needs only to have a method to determine if there is a feasible $x \in X \cap R$ such that $F(x) > F(x^*)$ where x^* is the local maximum found in the subregion R by Algorithm 15.2. One way this may be done is by using a solver to maximize the nonconvex quadratic

$$G(x) = x^T(Q - F(x^*)P)x$$

with the added quadratic constraint that $G(x) > 0$, and artificially terminating the solver once a feasible x is found. If no such x can be found, then x^* is the global solution in the subregion.

Finally, we remark that Algorithm 15.4 is also applicable to the quadratic sum-of-ratios problems, i.e. in the case when the matrix P in the denominator of (15.3) is allowed to vary with m (in fact, the denominators need not be quadratic, they just need to be convex).

Acknowledgments Both authors of the paper were supported in part by the NSF grant DMS-2208031. The authors are grateful to the anonymous referees for their suggestions that helped improve the exposition. The paper is dedicated to the everlasting memory of Guido L. Weiss whose research, teaching, and friendship has inspired generations.

References

1. H.P. Benson, Global optimization algorithm for the nonlinear sum of ratios problem. J. Optim. Theory Appl. **112**(1), 1–29 (2002)
2. H.P. Benson, Using concave envelopes to globally solve the nonlinear sum of ratios problem. J. Global Optim. **22**(1/4), 343–364 (2002)
3. A. Charnes, W.W. Cooper, Programming with linear fractional functionals. Naval Res. Logist. Q. **10**(1), 273–274 (1962)
4. W. Dinkelbach, On nonlinear fractional programming. Manag. Sci. **13**(7), 492–498 (1967)
5. L. Gao, S.K. Mishra, J. Shi, An extension of branch-and-bound algorithm for solving sum-of-nonlinear-ratios problem. Optim. Lett. **6**(2), 221–230 (2010)
6. K.R. Goodearl, T.H. Lenagan, Lu decomposition of totally nonnegative matrices. Linear Algebra Appl. **436**(7), 2554–2566 (2012)
7. J.-Y. Gotoh, H. Konno, Maximization of the ratio of two convex quadratic functions over a polytope. Comput. Optim. Appl. **20**(1), 43–60 (2001)
8. L. Huang, D. Needell, S. Tang, Robust recovery of bandlimited graph signals via randomized dynamical sampling. arXiv:2109.14079 (2021)
9. T. Ibaraki, Parametric approaches to fractional programs. Math. Program. **26**(3), 345–362 (1983)
10. H. Jiao, S. Liu, An efficient algorithm for quadratic sum-of-ratios fractional programs problem. Numer. Funct. Anal. Optim. **38**(11), 1426–1445 (2017)
11. A. Lo, A.C. MacKinlay. Maximizing predictability in the stock and bond markets. Macroecon. Dynamics **1**, 102 (1997)
12. S.-J. Qu, K.-C. Zhang, J.-K. Zhao, An efficient algorithm for globally minimizing sum of quadratic ratios problem with nonconvex quadratic constraints. Appl. Math. Comput. **189**(2), 1624–1636 (2007)
13. K. Shen, W. Yu, Fractional programming for communication systems–part I: Power control and beamforming. IEEE Trans. Signal Process. **66**(10), 2616–2630 (2018)
14. R. Yamamoto, H. Konno, An efficient algorithm for solving convex–convex quadratic fractional programs. J. Optim. Theory Appl. **133**(2), 241–255 (2007)

Chapter 16
Crystalline Measures and Wave Front Sets

Yves Meyer

Guido, your were a guide, you were a friend.

Abstract Lars Hörmander's classical result on the product of two distributions paves the road to the construction of new crystalline measures on $\mathbb{R}^2$.

16.1 Definition of Crystalline Measures

Let us fix some notations which are needed to define crystalline measures. The euclidean norm of $a \in \mathbb{R}^n$ is denoted by $|a|$. The Dirac mass at $a \in \mathbb{R}^n$ is denoted by δ_a or $\delta_a(x)$. It is defined by $\langle \delta_a, f \rangle = f(a)$ for any continuous function f. A *purely atomic measure* (or an atomic measure) is a series $\mu = \sum_{\lambda \in \Lambda} c(\lambda)\delta_\lambda$ of Dirac measures δ_λ at $\lambda \in \mathbb{R}^n$ where the coefficients $c(\lambda)$ are real or complex numbers and $\sum_{\{\lambda \in \Lambda; |\lambda| \le R\}} |c(\lambda)|$ is finite for any $R \ge 0$. A subset $\Lambda \subset \mathbb{R}^n$ is *locally finite* if $\Lambda \cap B$ is finite for any bounded set B. Equivalently Λ is *locally finite* if it can be ordered as a sequence $\{\lambda_j,\ j = 1, 2, \ldots\}$ such that $|\lambda_j|$ tends to infinity with j. A measure μ is a *tempered distribution* if it has a polynomial growth at infinity in the sense given by Laurent Schwartz in [11]. For instance the measure $\sum_1^\infty 2^k \delta_k$ is not a tempered distribution while $\sum_1^\infty k^3 \delta_k$ and $\sum_1^\infty 2^k[\delta_{(k+2^{-k})} - \delta_k]$ are tempered distributions. The *Fourier transform* $\mathcal{F}(f) = \widehat{f}$ of a function $f \in L^1(\mathbb{R}^n)$ is defined by

$$\widehat{f}(y) = \int_{\mathbb{R}^n} \exp(-2\pi\, i x \cdot y) f(x)\, dx, \quad \forall y \in \mathbb{R}^n. \tag{16.1}$$

Y. Meyer (✉)
Académie des Sciences, Paris, France
e-mail: yves.meyer305@orange.fr

E. Hernández et al. (eds.), *The Mathematical Heritage of Guido Weiss*,
Applied and Numerical Harmonic Analysis,
https://doi.org/10.1007/978-3-031-76793-7_16

The *distributional Fourier transform* $\widehat{\mu}$ of a tempered distribution μ is defined by the following rule: for any test function ϕ belonging to the Schwartz class $\mathcal{S}(\mathbb{R}^n)$ one has $\langle \widehat{\mu}, \phi \rangle = \langle \mu, \widehat{\phi} \rangle$.

Definition 16.1 An atomic measure μ on $\mathbb{R}^n$ is a crystalline measure if the three following conditions are satisfied:

(a) μ is a tempered distribution,
(b) the support of μ is a locally finite set Λ,
(c) and the distributional Fourier transform $\widehat{\mu}$ of μ is also an atomic measure supported by a locally finite set F.

The simplest example of a crystalline measure is the Dirac comb $\nu = \sum_{k \in \mathbb{Z}} \delta_k$. Then $\widehat{\nu} = \nu$ and for any function ϕ in the Schwartz class we have $\sum_{k \in \mathbb{Z}} \phi(k) = \langle \nu, \phi \rangle = \langle \nu, \widehat{\phi} \rangle = \sum_{m \in \mathbb{Z}} \widehat{\phi}(m)$ which is the standard Poisson summation formula. More generally for any lattice $\Gamma \subset \mathbb{R}^n$ the atomic measure $\nu_\Gamma = \sum_{\gamma \in \Gamma} \delta_\gamma$ is a crystalline measure and we have $\widehat{\nu_\Gamma} = c_\Gamma \nu_{\Gamma^*}$ where Γ^* is the dual lattice of Γ and $1/c_\Gamma$ is the volume of a fundamental domain of Γ. Then for any $y \in \mathbb{R}^n$ the translated measure $\nu_\Gamma * \delta_y$ is also named a Dirac comb. We have for any function ϕ in the Schwartz class $\langle \widehat{\mu}, \phi \rangle = \langle \mu, \widehat{\phi} \rangle$ and if μ is a crystalline measure this is a new Poisson summation formula. The collection of all crystalline measures on $\mathbb{R}^n$ is a vector space. The product $P(x)d\mu(x)$ between a finite trigonometric sum $P(x) = \sum_{y \in F} a(y) \exp(2\pi i x \cdot y)$ and a crystalline measure $d\mu(x)$ is still a crystalline measure.

Definition 16.2 An atomic measure μ is a trivial crystalline measure if $\mu = \sum_1^N P_j \nu_j$ where each P_j, $1 \le j \le N$, is a finite trigonometric sum and each ν_j is a Dirac comb.

Crystalline measures were introduced simultaneously by Andrew Guinand in [2] and by Jean-Pierre Kahane and Szolem Mandelbrojt in [4]. Kahane and Mandelbrojt were studying meromorphic functions $\phi(s)$ in the complex plane enjoying the following three properties: (a) if $\Re s > s_0$ $\phi(s)$ is the sum of a general Dirichlet series $\sum_1^\infty c_j \lambda_j^{-s}$, where $0 < \lambda_1 < \lambda_2 < \cdots$ is an increasing sequence of positive real numbers, (b) either $\phi(s)$ is an entire function or $\phi(s)$ is a meromorphic function in the complex plane and the unique pole of $\phi(s)$ is $s = 1$, and (c) $\phi(s)$ satisfies the same functional equation as the Riemann zeta function. Kahane and Mandelbrojt proved the following theorem:

Theorem 16.1 *Let us assume that properties* (a), (b), *and* (c) *are satisfied by the meromorphic function* $\phi(s) = \sum_1^\infty c_j \lambda_j^{-s}$ *and that there exist a* $\beta > 0$ *and some intervals* J_k *whose lengths tend to infinity such that* $\lambda_{j+1} - \lambda_j \ge \beta > 0$ *for* $j \in J_k$. *Then there exists a constant* c *such that* $\mu = c\,\delta_0 + \sum_1^\infty c_j(\delta_{\lambda_j} + \delta_{-\lambda_j})$ *is a crystalline measure which satisfies* $\widehat{\mu} = \mu$.

From the results obtained by Nir Lev and Alexander Olevskii [6, 7], it can be conjectured that the only crystalline measures μ of Theorem 16.1 are generalized Dirac combs. Then the meromorphic function $\phi(s)$ of Theorem 16.1 reduces to the

Riemann zeta function after some trivial modifications. The converse implication is much easier. Let us assume that $\mu = \sum_{\lambda\in\Lambda} c(\lambda)\delta_\lambda$ is a crystalline measure and that there exists a $s_0 > 0$ such that $\sum_{\{\lambda\in\Lambda,\lambda\neq 0\}} |c(\lambda)||\lambda|^{-s_0}$ is finite. A similar condition is imposed to $\widehat{\mu}$. Then the corresponding zeta function is the sum of the Dirichlet series $\zeta(\mu, s) = \sum_{\{\lambda\in\Lambda,\lambda\neq 0\}} c(\lambda)|\lambda|^{-s}$, $s \in \mathbb{C}$, $\Re s > s_0$. This function $\zeta(\mu, s)$ is obviously analytic in the open half plane defined by $s \in \mathbb{C}$, $\Re s > s_0$. If $n = 1$ and if μ is the Dirac comb then $\zeta(\mu, s)$ is two times the Riemann zeta function. If μ is a n−dimensional Dirac comb then $\zeta(\mu, s)$ is the Epstein zeta function. Kahane and Mandelbrojt made the following observation:

Theorem 16.2 *Let $\mu = \sum_{\lambda\in\Lambda} c(\lambda)\delta_\lambda$ be a crystalline measure on $\mathbb{R}^n$ and let $\widehat{\mu} = \sum_{y\in F} a(y)\delta_y$ be the distributional Fourier transform μ. We consider $\xi(\mu, s) = \pi^{-s/2}\Gamma(s/2)\zeta(\mu, s)$. If the above mentioned properties are satisfied then $\xi(\mu, s) - \frac{2a(0)}{n-s} - \frac{2c(0)}{s} = E(\mu, s)$ is an entire function and we have $\xi(\mu, s) = \xi(\widehat{\mu}, n - s)$ identically on $\mathbb{C} \setminus \{0, n\}$.*

As mentioned by Kahane and Mandelbrojt Theorem 16.2 can be traced back to Riemann and Titchmarsh [4]. If $n = 1$ and if $\widehat{\mu} = \mu$ then $\zeta(\mu, s)$ satisfies the same functional equation as the Riemann zeta function. The first non trivial crystalline measure satisfying $\widehat{\mu} = \mu$ was proposed by Guinand in [2]. Guinand defined a sequence γ_k, $k = 0, 1, \ldots$, of rational numbers by

$$\sum_0^\infty \gamma_k q^k = \prod_1^\infty (1 - q^n)(1 + q^{2n})^{2/3}(1 + q^n)^{1/3} \tag{16.2}$$

where $|q| < 1$. We have $\gamma_0 = 1, \gamma_1 = -2/3, \gamma_2 = -4/9, \gamma_3 = -40/81, \gamma_4 = -160/243, \gamma_5 = 268/729, \ldots$ and $|\gamma_k| \leq Ck^{1/3}$. Let $\lambda_k = \sqrt{k + 1/9}, k = 0, 1, \ldots$. Then the Guinand's crystalline measure is $\mu_G = \sum_0^\infty \gamma_k(\delta_{\lambda_k} + \delta_{-\lambda_k})$.

Theorem 16.3 *The Guinand measure μ_G satisfies $\widehat{\mu}_G = \mu_G$.*

This was announced in [2] and proved in [9]. Nir Lev and Alexander Olevskii gave other examples of non trivial crystalline measures [8]. Pavel Kurasov and Peter Sarnak constructed a Delone set $\Lambda \subset \mathbb{R}$ which is not of the form $\Lambda = \cup_1^N (a_j\mathbb{Z}+b_j)$ and such that $\sigma_\Lambda = \sum_{\lambda\in\Lambda} \delta_\lambda$ is a crystalline measure [5]. Maryna Viazovska could bridge the gap between crystalline measures and sharp sphere packings [12]. The tensor product $\mu = \mu_1 \otimes \ldots \otimes \mu_n$ between n crystalline measures on the line is a crystalline measure on $\mathbb{R}^n$. Is it the unique way to construct crystalline measures on $\mathbb{R}^n$? A counter example is given in [10] for $n = 2$. But the crystalline measure μ which is constructed in [10] is periodic with respect to the first variable. It satisfies $\mu * \delta_y = \mu$ if $y = (1, 0)$. The anonymous referee of [10] is wondering if one could get rid of this limitation. An answer is given by our Theorem 16.6. We begin the discussion with $n = 2$ and treat the general case in the last section. A classical theorem by Lars Hörmander on the product of two distributions is the Ariadne Thread of this note. It is surprising that the construction of non trivial crystalline measures is much simpler in dimension two than in dimension one.

16.2 The Product of Two Distributions

The pointwise product $u\,v$ between two distributions $u, v \in \mathcal{D}'(\mathbb{R}^n)$ does not exist in general. However this product exists if the two distributions u and v are transverse in a sense given by two-microlocal analysis. This is Theorem 16.4. The definition of the pointwise product of two distributions is quite natural when this product exists. Let ϕ be a function in the Schwartz class $\mathcal{S}(\mathbb{R}^n)$ such that $\int \phi = 1$ and let ϕ_ϵ, $0 < \epsilon \leq 1$, be given by $\phi_\epsilon(x) = \epsilon^{-n}\phi(x/\epsilon)$.

Definition 16.3 Let u and v be two tempered distributions. Let ϕ be a function in the Schwartz class $\mathcal{S}(\mathbb{R}^n)$ such that $\int \phi = 1$. Let us assume that the pointwise product $(u * \phi_\epsilon)(v * \phi_\epsilon)$ tends to a limit in $\mathcal{S}'(\mathbb{R}^n)$ as ϵ tends to 0 and that this limit does not depend on the choice of ϕ. Then the pointwise product uv between u and v exists and is defined by $u\,v = \lim_{\epsilon \to 0}(u * \phi_\epsilon)(v * \phi_\epsilon)$.

Following Hörmander's seminal work [3] *two-microlocal analysis* is used now. The *singular support* $F_u \subset \mathbb{R}^n$ of a distribution $u \in \mathcal{D}'(\mathbb{R}^n)$ is a closed set which is defined by its complement $\mathbb{R}^n \setminus F_u$. This complement is the largest open set $\Omega \subset \mathbb{R}^n$ on which u coincides with a C^∞ function. The unit sphere of $\mathbb{R}^n$ which is centered at 0 is denoted by $\mathbb{S}^{n-1}$. The *wave front set* $WF(u)$ of u is a closed subset of $F_u \times \mathbb{S}^{n-1}$. Here again $WF(u)$ is defined by its complement in $F_u \times \mathbb{S}^{n-1}$. A pair (x_0, ξ_0), $x_0 \in \mathbb{R}^n$, $|\xi_0| = 1$, does not belong to $WF(u)$ if there exists (i) a compactly supported smooth function ϕ on $\mathbb{R}^n$ such that $\phi(x_0) = 1$ and (ii) a circular cone $C_\phi^\epsilon \subset \mathbb{R}^n$ defined by $C_\phi^\epsilon = \{\xi; \xi \cdot \xi_0 \geq (1-\epsilon)|\xi|\}$ for some $\epsilon \in (0, 1)$ such that the Fourier transform of the product ϕu once restricted to C_ϕ^ϵ is rapidly decreasing at infinity. It is easily checked that $WF(u)$ is closed. If u is real valued, $(x_0, \xi_0) \in WF(u)$ is equivalent to $(x_0, -\xi_0) \in WF(u)$. The wave front set $WF(u)$ of u is empty if and only if u is a C^∞ function. Many authors define the wave front set as a conic subset of $F_u \times (\mathbb{R}^n \setminus \{0\})$. A conic set $E \subset \mathbb{R}^n \times (\mathbb{R}^n \setminus \{0\})$ is defined by the following property: $(x, \xi) \in E$ and $\lambda > 0$ imply $(x, \lambda\xi) \in E$. These two viewpoints result in the same definition if a half line from 0 is identified with its intersection with $\mathbb{S}^{n-1}$.

Remark 16.1 Let $C \subset \mathbb{R}^n$ be a closed cone and let u be a tempered distribution whose Fourier transform $\widehat{u}$ is supported by C. Then the wave front set $WF(u)$ of a is contained in $\mathbb{R}^n \times (C \cap \mathbb{S}^{n-1})$.

Theorem 16.4 *Let u and v be two distributions on $\mathbb{R}^n$. Let us assume that for any (x, ξ) belonging to $WF(u)$ we have $(x, -\xi) \notin WF(v)$. Then the pointwise product $u\,v$ makes sense.*

Hörmander's sufficient condition prevents us from defining $|u|^2$ if u is a distribution which is not a C^∞ function. Indeed $|u|^2 = u\overline{u}$ and if $(x, \xi) \in WF(u)$ we have $(x, -\xi) \in WF(\overline{u})$. Furthermore Hörmander proved that the wave front set of $u\,v$ is contained in the union $E \cup F \cup G$ where

(*a*) E is the set of all $(x, \frac{\xi+\xi'}{|\xi+\xi'|})$ such that (*i*) $x \in F_u \cap F_v$, (*ii*) $(x, \xi) \in WF(u)$ and (*iii*) $(x, \xi') \in WF(v)$,
(*b*) F is the set of all (x, ξ') such that x belongs to the support of u and $(x, \xi') \in WF(v)$,
(*c*) G is the set of all (x', ξ) such that x' belongs to the support of v and $(x', \xi) \in WF(u)$.

Let us observe that $x \in F_u \cap F_v$, $(x, \xi) \in WF(u)$ and $(x, \xi') \in WF(v)$ imply $\xi + \xi' \neq 0$. This is Hörmander's sufficient condition. Therefore $\frac{\xi+\xi'}{|\xi+\xi'|}$ makes sense. If $F_u \cap F_v = \emptyset$ Hörmander's theorem is trivial: locally the product $u\,v$ reduces to the product between a distribution and a C^∞ function. Let us consider $u(x) = |x|$. Then $WF(u) = \{0\} \times \mathbb{S}^{n-1}$. However u^2 is C^∞. This example shows that Hörmander's sufficient condition is not necessary.

Corollary 16.1 *Let $C_1 \subset \mathbb{R}^n$ and $C_2 \subset \mathbb{R}^n$ be two closed cones such that $C_1 \cap (-C_2) = \{0\}$. Let u_1 and u_2 be two tempered distributions. If the Fourier transform of u_1 is supported by C_1 and the Fourier transform of u_2 is supported by C_2 then the pointwise product $u = u_1 u_2$ makes sense and the Fourier transform of u is supported by $C_1 + C_2$.*

This follows from Hörmander's theorem and Remark 16.1. Corollary 16.1 is a well known fact which can be proved directly. Corollary 16.1 applies to the measures σ_A and σ_B which are constructed below. The product $\mu = \sigma_A\,\sigma_B$ is the crystalline measure we are looking for.

Another instance of Hörmander's theorem is given by the following example. We assume that:

(i) $n = 2$, $\Gamma_1 \subset \mathbb{R}^2$ and $\Gamma_2 \subset \mathbb{R}^2$ are the graphs of two smooth functions $g_1 : \mathbb{R} \mapsto \mathbb{R}$ and $g_2 : \mathbb{R} \mapsto \mathbb{R}$,
(ii) $\Gamma_1 \cap \Gamma_2$ is reduced to a single point $x_0 = (a, b)$ and the values of the arclenghts on Γ_1 and Γ_2 at $x_0 = (a, b)$ are denoted by $s_1(a)$ and $s_2(a)$,
(iii) $g_2'(a) \neq g_1'(a)$ which implies that Γ_1 and Γ_2 are transverse at x_0,
(iv) u is a measure supported by Γ_1, it is given by $u = \omega_1\, ds_{\Gamma_1}$ where ds_{Γ_1} is the arc length measure on Γ_1, and the density ω_1 of u is a smooth function,
(v) v is defined similarly with respect to Γ_2.

The transversality between Γ_1 and Γ_2 at x_0 implies Hörmander's sufficient condition. Indeed the wave front set of u is contained in the set of pairs (x, ξ) such that $x \in \Gamma_1$ and ξ is normal to Γ_1 at x and the same property is valid for v. Therefore the product $u\,v$ makes sense. However Hörmander's theorem is not needed to define $u\,v$. As it is proved below a straightforward calculation yields the following result:

Lemma 16.1 *If u and v satisfy these five properties then the product $u\,v$ makes sense and is given by*

$$u\,v = c\,\delta_{x_0},\quad c = \omega_1(s_1(a))\omega_2(s_2(a))\frac{\sqrt{1+g_1'(a)^2}\sqrt{1+g_2'(a)^2}}{|g_2'(a)-g_1'(a)|}. \tag{16.3}$$

If Γ_1 and Γ_2 are orthogonal at x_0 and if $\omega_1 = \omega_2 = 1$ we have $g_1'(a)g_2'(a) = -1$ and (3) simply reduces to $u\,v = \delta_{x_0}$. Lemma 16.1 is the only result of this section which is used below. Hörmander's theorem gives the right perspective on Lemma 16.1 since the transversality assumption $g_2'(a) \neq g_1'(a)$ of Lemma 16.1 is exactly Hörmander's sufficient condition. Here is a sketch of the proof of Lemma 16.1. For easing the notations one assumes $\omega_1 = \omega_2 = 1$. For $\epsilon \in (0,1]$ one considers the function $u_\epsilon(x) = \epsilon^{-1}\sqrt{1+g_1'(x_1)^2}\chi_{u,\epsilon}(x)$ where $\chi_{u,\epsilon}$ is the indicator function of $\{x|\, g_1(x_1) \leq x_2 \leq g_1(x_1)+\epsilon\}$. The distributional limit of $u_\epsilon(x)$ as ϵ tends to 0 is ds_{Γ_1}. One defines $\chi_{v,\epsilon}(x)$ similarly. It now suffices to prove that $\lim_{\epsilon\to 0}\epsilon^{-2}\chi_{u,\epsilon}(x)\chi_{v,\epsilon}(x) = |g_2'(a)-g_1'(a)|^{-1}\delta_{x_0}$. Since g_1 and g_2 are smooth, it amounts to computing the area of the parallelogram which is defined by $\{x|\, g_1(a)+(x_1-a)g_1'(a) \leq x_2 \leq g_1(a)+(x_1-a)g_1'(a)+\epsilon\}$ and $\{x|\, g_2(a)+(x_1-a)g_2'(a) \leq x_2 \leq g_2(a)+(x_1-a)g_2'(a)+\epsilon\}$. This area is given by$|g_2'(a)-g_1'(a)|^{-1}\epsilon^2$ which ends the proof.

16.3 Sketch of the Construction

The core of the construction of the crystalline measure μ of Theorem 16.5 is a probability measure $\widetilde{\sigma}$ on the two dimensional torus. This measure $\widetilde{\sigma} = \widetilde{\sigma}_\epsilon$ depends on a small parameter ϵ and has three conflicting properties: (*a*) $\widetilde{\sigma}$ is a probability measure on the two dimensional torus $\mathbb{T}^2 = (\mathbb{R}/\mathbb{Z})^2$, (*b*) $\widetilde{\sigma}$ is supported by a smooth closed curve Γ of $\mathbb{T}^2$, and (*c*) the Fourier series of $\widetilde{\sigma}$ is

$$\widetilde{\sigma}(x) = \sum_{k\in\mathcal{B}_\epsilon} c(k)\exp(2\pi i k\cdot x) \tag{16.4}$$

where $\mathcal{B}_\epsilon$ is a cone with an arbitrarily small aperture as ϵ tends to 0. More precisely $\mathcal{B}_\epsilon = \{x : |x| \leq (1+\epsilon)|x\cdot e|\}$ for $e = (1/\sqrt{2}, 1/\sqrt{2})$ and $\epsilon > 0$. The construction of $\widetilde{\sigma} = \widetilde{\sigma}(\epsilon)$ can be achieved for any positive ϵ. This is proved at the end of Sect. 16.6. The existence of $\widetilde{\sigma}$ is obvious when Γ is defined by $x_1 + x_2 = 0$. In this example $\widetilde{\sigma}$ is simply the Haar measure on Γ and $\mathcal{B}$ is the first diagonal. Then the corresponding crystalline measure of Theorem 16.5 is a general Dirac comb. In our construction $\epsilon = \sqrt{2}-1$. Then $\mathcal{B}_\epsilon$ is the union between the first and the third quadrant, $\widetilde{\sigma}$ is a Ahern measure (Definition 16.6), and the index ϵ will be dropped. Actually $\widetilde{\sigma}$ depends on a second parameter $r \in [0,1)$. A family Γ_r, $0 \leq r < 1$, of smooth closed curves of $\mathbb{T}^2$ is constructed together with a family $\widetilde{\sigma}_r$, $0 \le r < 1$, of Ahern measures satisfying (b). The measure $\widetilde{\sigma}_r$, is supported by Γ_r. Moreover $\widetilde{\sigma}_0$ is the Haar measure on $\Gamma_0 = \{x_1 + x_2 = 0\}$. Once $\widetilde{\sigma}_r$ is constructed basic linear algebra yield the crystalline measure μ. Hörmander's theorem illuminates the construction. Here is the sketch of this construction. For easing the notations the subscript r is dropped. We denote by σ the measure on $\mathbb{R}^2$ which is the $\mathbb{Z}^2$ periodic version of $\widetilde{\sigma}$. Then σ is supported by the periodic version of Γ. The periodic version of Γ is a disjoint union of curves Γ_m, $m \in \mathbb{Z}$. We then consider two invertible linear operators $A : \mathbb{R}^2 \mapsto \mathbb{R}^2$ and $B : \mathbb{R}^2 \mapsto \mathbb{R}^2$. We assume $\det A = \det B = 1$. The image measures of σ by A and B are denoted by σ_A and σ_B. Then $\sigma_A = \sigma \circ A^{-1}$ is supported by the disjoint union of the curves $A(\Gamma_m)$, $m \in \mathbb{Z}$, and is periodic with respect to the lattice $\Lambda_A = A(\mathbb{Z}^2)$. Similarly $\sigma_B = \sigma \circ B^{-1}$ is supported by the disjoint union of the curves $B(\Gamma_m)$, $m \in \mathbb{Z}$, and is periodic with respect to the lattice $\Lambda_B = B(\mathbb{Z}^2)$. We fix A and B in such a way that the support $\bigcup_{m\in\mathbb{Z}} A(\Gamma_m)$ of σ_A be transverse to the support $\bigcup_{m\in\mathbb{Z}} B(\Gamma_m)$ of σ_B at every point belonging to the intersection between these two supports. It is easily checked that this intersection Λ is locally finite. We use (16.3) to define the pointwise product $\mu = \sigma_A \sigma_B$. Then $\mu = \sigma_A \sigma_B$ makes sense and is supported by Λ. This Λ depends on A, B, and r. In what follow A and B are fixed and the parameter $r > 0$ which enters in the construction of $\sigma = \sigma_r$ is small, depending on A and B. The Fourier series of σ is identical to the Fourier series of $\widetilde{\sigma}$. It is given by (16.4). The automorphisms A and B can be fixed in such a way that the wave front set condition of Hörmander's theorem is satisfied by σ_A and σ_B. It implies that $\widehat{\mu}$ is also supported by a locally finite set. Then the pointwise product $\mu = \sigma_A \sigma_B$ is a crystalline measure. For most choices of A, B, and r this μ is neither a tensor product between two crystalline measures on the line nor a periodic crystalline measure. This is proved in Sect. 16.6.

16.4 Construction of σ

We return to the definition of σ_r. Let $r \in [0, 1)$. Let $\phi_r : \mathbb{R} \mapsto \mathbb{R}$ be the odd and decreasing function of the real variable θ defined by $\phi_r(0) = 0$ and $\frac{d\phi_r}{d\theta} = -\frac{1-r^2}{1+r^2-2r\cos(2\pi\theta)}$. We have $\phi_r(\theta + 1) = \phi_r(\theta) - 1$ and $\exp(-2\pi i \phi_r) = \frac{\exp(2\pi i\theta)-r}{1-r\exp(2\pi i\theta)}$. Let us observe that $U(z) = \frac{z-r}{1-rz}$ is holomorphic in the open unit disc $\mathbb{D}$ and is unimodular on its boundary. It is a Blaschke factor. The function $\psi_r(\theta) = \phi_r(\theta)+\theta$ is periodic of period 1 and together with its derivative tends uniformly to 0 as r tends to 0. Indeed $\|\psi_r\|_\infty \le r/2$ and $\|\frac{d}{d\theta}\psi_r\|_\infty \le 2r/(1-r)$. A finite Blaschke product $U(z)$ could be used as well instead of a Blaschke factor and the continuous function ϕ_U would then be defined by $\exp(-2\pi i\phi_U(\theta)) = U(\exp(2\pi i\theta))$. We return to the construction of σ_r. Let $\Gamma_m \subset \mathbb{R}^2$, $m \in \mathbb{Z}$, be the curve in the plane defined by $x_2 = \phi_r(x_1) + m$ or equivalently by $x_2 = \psi_r(x_1) - x_1 + m$ and let $C = \bigcup_{m\in\mathbb{Z}} \Gamma_m$ be the union of these disjoint curves. The curve Γ_m is contained in the strip W_m defined by $|x_1 + x_2 - m| \le r/2$. Since $0 \le r < 1$ these strips are pairwise disjoint. It implies the following:

Lemma 16.2 *If $x \in \Gamma_m$, $y \in \Gamma_l$ and $l \neq m$ we have $|x - y| \ge \frac{1-r}{\sqrt{2}}$.*

Definition 16.4 The measure $\sigma_{(r,m)}$ on Γ_m is defined as the image of the Lebesgue measure dx_1 on the real axis by the map $x_1 \mapsto (x_1, \phi_r(x_1) + m)$.

For any compactly supported continuous function f on $\mathbb{R}^2$ we have

$$\langle \sigma_{(r,m)}, f \rangle = \int_{\mathbb{R}} f(x_1, \phi_r(x_1) + m) dx_1. \tag{16.5}$$

Then $\sigma_{(r,m)}(x_1 + 1, x_2 - 1) = \sigma_{r,m)}(x_1, x_2)$.

Definition 16.5 Let $\sigma = \sum_{-\infty}^{+\infty} \sigma_{(r,m)}$.

From now on the index r is dropped: we write σ instead of σ_r. The measure σ is $\mathbb{Z}^2$-periodic and is supported by C. The Fourier coefficients of σ are computed in Lemma 16.5. We have $\widehat{\sigma}(0) = 1$.

Lemma 16.3 *The wave front set of the measure σ is the set WF_σ consisting of all pairs $(x, \xi) \in \mathbb{R}^2 \times \mathbb{S}$ such that $x \in C$ and ξ is normal to C at x.*

This is a general fact which does not depend on the definition of the curves Γ_m. Let $S^+ \subset \mathbb{R}^2$ be the angular sector defined by $x_1 \ge 0, x_2 \ge 0$. Similarly S^- is defined by $x_1 \le 0, x_2 \le 0$ and $S = S^+ \cup S^-$. Then we have:

Lemma 16.4 *If (x, ξ) belongs to WF_σ then x belongs to C and ξ belongs to S.*

If (x, ξ) belongs to WF_σ then x belongs to C and ξ is normal to C at x. But the function ϕ_r is decreasing. Therefore a vector which is normal to C at x belongs to S. We can say more. Let S_r^+ be the sector defined by $0 \le \frac{1-r}{1+r} x_1 \le x_2 \le \frac{1+r}{1-r} x_1$. The sector $S_r^- = -S_r^+$ is defined by $\frac{1+r}{1-r} x_1 \le x_2 \le \frac{1-r}{1+r} x_1 \le 0$. Then S can be replaced by $S_r^+ \cup S_r^- = S_r \subset S$ in Lemma 16.4. Indeed a vector which is normal to C at x belongs to S_r. Here is another precise result. The measure σ can be viewed as a measure $\widetilde{\sigma}$ on the two dimensional torus $\mathbb{T}^2 = \mathbb{R}^2/\mathbb{Z}^2$. Let $\mathbb{D}^2 \subset \mathbb{C}^2$ be the bidisc defined by $|z_1| < 1, |z_2| < 1$. The distinguished boundary of $\mathbb{D}^2$ is $\mathbb{T}^2$. Here $\mathbb{T}^2$ is identified to $\{(z_1, z_2), |z_1| = |z_2| = 1\}$. We claim that $\widetilde{\sigma}$ is an Ahern measure [1].

Definition 16.6 If $n \ge 2$ a real valued measure on $\mathbb{T}^n$ is a Ahern measure if it is the real part of the trace on $\mathbb{T}^n$ of a holomorphic distribution on $\mathbb{D}^n$.

Let $\mathbf{B} = S \cap \mathbb{Z}^2$. Then we have:

Lemma 16.5 *The Fourier coefficients $\widehat{\widetilde{\sigma}}(k)$ of $\widetilde{\sigma}$ vanish if $k \notin \mathbf{B}$.*

Therefore $\widetilde{\sigma}$ is a Ahern measure. In other words for any $k \in \mathbb{Z}^2$ we have $\widehat{\widetilde{\sigma}}(k) = 0$ if $k_1 < 0$ and $k_2 > 0$ or if $k_1 > 0$ and $k_2 < 0$. Let us treat the first case. Let Q be the square defined by $0 \le x_1 < 1$, $-1 \le x_2 < 0$. Then the Fourier coefficients of $\widetilde{\sigma}$ are given by $\widehat{\widetilde{\sigma}}(k) = \int_Q \exp(-2\pi i x \cdot k)\, d\sigma$. But the restriction of σ to Q is identical to the restriction of $\sigma_{(r,0)}$ to Q. Indeed $\Gamma_m \cap Q = \emptyset$ if $m \neq 0$. Using (5) we obtain $\widehat{\widetilde{\sigma}}(k) = \int_0^1 \exp[-2\pi i(k_1 x_1 + k_2 \phi_r(x_1))]dx_1$. Since $\exp(-2\pi i\phi_r) = B$ is the boundary value of an analytic function the same is true for $\exp(-2\pi i k_2 \phi_r) = B^{k_2}$ for any $k_2 \geq 1$. Then the Fourier coefficients of B^{k_2} vanish on $(-\infty, -1]$. We have $\widehat{\widetilde{\sigma}}(k) = 0$ if $k_1 < 0$ and $k_2 > 0$. The second case where $k_1 > 0$ and $k_2 < 0$ is similar. The Fourier coefficients of σ do not vanish in $S \setminus S_r$ but have a rapid decay at infinity in any closed conical set contained in the interior of $S \setminus S_r$. Using the Taylor series expansion of $\exp(-2\pi i\phi_r)$ the Fourier coefficients of $\widetilde{\sigma}$ can be computed in S.

Lemma 16.6 *We have* $(\frac{z-r}{1-rz})^k = \sum_0^\infty P_{q,k}(r)z^q$ *where* $P_{q,k}(r)$ *is a polynomial of degree* $q+k$. *The coefficient of* r^{q+k} *in* $P_{q,k}(r)$ *is* $\frac{k(k+1)\cdots(k+q-1)}{q!}(-1)^k$.

This is easily checked. Indeed $(1-rz)^{-k} = \sum_0^\infty \frac{k(k+1)\cdots(k+q-1)}{q!} r^q z^q$ and it suffices to multiply this series by the polynomial $(z-r)^k$ to conclude.

Definition 16.7 The support of $\widehat{\widetilde{\sigma}}$ is denoted by E.

Lemma 16.7 *There exists a countable set* $R \subset [0, 1]$ *such that for any* $r \notin R$ *we have* $E = S$.

Indeed R is defined by $R = \cup_{q,k} R_{q,k}$ where $R_{q,k}$ is the set of roots $r \in [0, 1]$ of the polynomial $P_{q,k}(r)$. Lemma 16.7 implies $E + E = \mathbb{Z}^2$.

16.5 The Definition of σ_A and σ_B

The measure σ_A is defined as the image measure of σ by an invertible linear map A and similarly σ_B is the image measure of σ by B. The matrices of these linear maps are also denoted by A and B. Here is the definition of A and B. Let us consider $A = \begin{pmatrix} b_2 & -b_1 \\ -a_2 & a_1 \end{pmatrix}$ and $B = \begin{pmatrix} b_4 & -b_3 \\ -a_4 & a_3 \end{pmatrix}$ where

(i) $a_1 > 0, a_2 > 0, a_3 > 0, a_4 > 0$,
(ii) $b_1 > 0, b_2 > 0$,
(iii) $b_3 < 0, b_4 < 0$,
(iv) $\det A = a_1 b_2 - a_2 b_1 = 1$,
(v) $\det B = a_3 b_4 - a_4 b_3 = 1$.

It implies $(A^*)^{-1} = \begin{pmatrix} a_1 & a_2 \\ b_1 & b_2 \end{pmatrix}$ and $(B^*)^{-1} = \begin{pmatrix} a_3 & a_4 \\ b_3 & b_4 \end{pmatrix}$.

The distributional Fourier transform of σ_A is $\widehat{\sigma} \circ A^*$. The support of $\widehat{\sigma} \circ A^*$ is contained in $(A^*)^{-1}(S) = S_A$. This $S_A \subset S$ is a narrower sector defined by $0 < c \leq x_2/x_1 \leq C$ where $C > c > 0$. Similarly The distributional Fourier transform of σ_B is $\widehat{\sigma} \circ B^*$ and is supported by $(B^*)^{-1}(S) = S_B \subset S$. This sector S_B is defined by $0 < c' \leq -x_2/x_1 \leq C'$. Let us observe that $S_A \cap S_B = \{0\}$. Since S_A and S_B are symmetric with respect to 0 this is Hörmander's sufficient condition. For instance if $\lambda > 1$ and $a_1 = \lambda, a_2 = \lambda - 1, b_1 = \lambda + 1, b_2 = \lambda$, the angular sector S_A is delimited by the lines $x_2 = \frac{\lambda}{\lambda-1}x_1$ and $x_2 = \frac{\lambda+1}{\lambda}x_1$. The aperture of S_A is $O(\lambda^{-2})$ and tends to 0 as λ tends to ∞. Similarly if $\lambda' > 1$ and $a_3 = \lambda' + 1, a_4 = \lambda', b_3 = -\lambda', b_4 = -\lambda' + 1$, the angular sector S_B is delimited by the lines $x_2 = -\frac{\lambda'}{\lambda'+1}x_1$ and $x_2 = -\frac{\lambda'-1}{\lambda'}x_1$. The aperture of S_B is $O(\lambda'^{-2})$ and tends to 0 as λ' tends to ∞.

We can assume that

$$(A^*)^{-1}(\mathbb{Z}^2) \cap (B^*)^{-1}(\mathbb{Z}^2) = \{0\}. \tag{16.6}$$

Indeed this is equivalent to the following condition: the vectors (a_1, b_1), (a_2, b_2), (a_3, b_3) and (a_4, b_4) are $\mathbb{Q}-$linearly independent. Let us observe that (16.6) is equivalent to $A(\mathbb{Z}^2) \cap B(\mathbb{Z}^2) = \{0\}$. In our example it amounts to say that $1, \lambda, \lambda'$ are $\mathbb{Q}-$linearly independent. This example suffices to our construction.

Lemma 16.8 *The mapping $P : S_A \times S_B \mapsto \mathbb{R}^2$ which maps (x, y) to $x + y$ is a proper map.*

We argue by contradiction. If one can find two sequences $x_k \in S_A$, $y_k \in S_B$ such that $|x_k| \to \infty$ and $|x_k + y_k| \leq C$ then we have $||x_k| - |y_k|| \leq C$. It implies $|y_k|/|x_k| \to 1$ as $k \to \infty$. Extracting a subsequence we can assume $x_k/|x_k| \to x'$ and $y_k/|y_k| \to y'$. Finally $x' \in S_A$, $y' \in S_B$, $|x'| = |y'| = 1$, and $y' = -x'$ which contradicts $S_A \cap S_B = \{0\}$. Here again Hörmander's sufficient condition plays a seminal role.

Lemma 16.9 *For any $k \in \mathbb{Z}$ and for any $l \in \mathbb{Z}$ the curve $A(\Gamma_k)$ is transverse to the curve $B(\Gamma_l)$.*

More precisely we have:

Lemma 16.10 *There exists a $\beta > 0$ depending on A, B, and r such that the curve $A(\Gamma_k)$ is the graph of a decreasing function $g_{k,A}$ such that $g'_{k,A} \leq -\beta$ and each curve $B(\Gamma_k)$ is the graph of an increasing function $h_{k,B}$ such that $h'_{k,A} \geq \beta$.*

Indeed the curve $A(\Gamma_k)$ admits a parametric representation given by $x_1 = b_2 t - b_1\phi(t)$ and $x_2 = -a_2 t + a_1\phi(t)$, $t \in \mathbb{R}$. Then x_1 is an increasing function of t and this function defines a diffeomorphism of $\mathbb{R}$. Next x_2 is a decreasing function of t. Finally x_2 is a decreasing function of x_1. Moreover $g_{k,A} : \mathbb{R} \mapsto \mathbb{R}$ is a diffeomorphism. The second assertion of is similar. Therefore the intersection $A(\Gamma_k) \cap B(\Gamma_l)$ is reduced to a point denoted by $\gamma_{k,l}$.

Lemma 16.11 *The set $\Lambda = \{\gamma_{k,l};\, k \in \mathbb{Z}, l \in \mathbb{Z}\}$ is a Delone set.*

Before proving Lemma 16.11 let us observe that $\Lambda = \Lambda(A, B, r)$ depends on A, B, and r. We first prove that Λ is uniformly discrete and then that Λ is relatively dense. If $x, y \in \Lambda$ are distinct three cases can occur. Either x and y belong to the same curve $A(\Gamma_k)$, or to the same $B(\Gamma_l)$ or to distinct curves. In the first case

$x \in B(\Gamma_l)$, $y \in B(\Gamma_m)$ and $l \neq m$. Then $|x - y| \geq c_B > 0$ by Lemma 16.2. The two other cases are similar.

We now prove that Λ is relatively dense. Let L_m be the line defined by $x_1 + x_2 = m$. Let us write $\mathbf{e} = (1, 1)$, $\mathbf{f} = (A^*)^{-1}(\mathbf{e})$, $\mathbf{g} = (B^*)^{-1}(\mathbf{e})$ and let us denote by M the lattice generated by $\mathbf{f}$ and $\mathbf{g}$. Let M^* be the dual lattice of M. Then M^* is also the set of all intersections $\widetilde{\gamma_{k,l}} = A(L_k) \cap B(L_l)$, $k, l \in \mathbb{Z}$.

Lemma 16.12 *Let $r \in [0, 1)$. There exists a constant C depending only on the matrices A and B such that for any $k, l \in \mathbb{Z}$ we have*

$$|\gamma_{k,l} - \widetilde{\gamma_{k,l}}| \leq Cr. \tag{16.7}$$

Since M^* is a lattice Lemma 16.12 implies that Λ is relatively dense. Moreover Λ is a small pertubation of M^* if r is small. Let us prove Lemma 16.12. We already know that the curve Γ_m is contained in the strip W_m defined by $|x_1 + x_2 - m| \leq r/2$. Let us consider the intersection $K(k, l) = A(W_k) \cap B(W_l)$. The center of the parallelogram $K(k, l)$ is $\widetilde{\gamma_{k,l}}$. The diameter of this parallelogram $K(k, l)$ does not exceed Cr since the width of W_k does not exceed r. Therefore $K(k, l) \subset B(\widetilde{\gamma_{k,l}}, Cr)$ where $B(x, R)$ denotes the ball centered at x with radius R. Moreover $\gamma_{k,l} = A(\Gamma_k) \cap B(\Gamma_l)$ belongs to $K(k, l)$. It implies $|\gamma_{k,l} - \widetilde{\gamma_{k,l}}| \leq Cr$ and Lemma 16.12 is fully proved. One can observe that $K(k, l) = \widetilde{\gamma_{k,l}} + K(0, 0)$.

16.6 Transversality

The pushforward measure $\sigma_A = \sigma \circ A^{-1}$ (the image measure of σ by A) is carried by $A(C)$ and is periodic with respect to the lattice $\Lambda_A = A(\mathbb{Z}^2)$. Similarly the pushforward measure $\sigma_B = \sigma \circ B^{-1}$ is carried by $B(C)$ and is periodic with respect to the lattice $\Lambda_B = B(\mathbb{Z}^2)$.

Definition 16.8 The atomic measure μ is defined as the pointwise product between these two image measures σ_A and σ_B.

This product makes sense since the support of σ_A is transverse to the support of σ_B.

Theorem 16.5 *This product measure $\mu = \sigma_A \sigma_B$ is a crystalline measure.*

The existence of μ follows from the transversality between $A(\Gamma_k)$ and $B(\Gamma_l)$. The support of μ is the Delone set $\Lambda = \{\gamma_{k,l};\, k \in \mathbb{Z}, l \in \mathbb{Z}\}$ and $\mu = \sum c(k, l)\delta_{\gamma_{k,l}}$. If $\gamma_{k,l} = (\alpha_{k,l}, \beta_{k,l})$ we have

$$c(k, l) = (h'_{l,B}(\alpha_{k,l}) - g'_{k,A}(\alpha_{k,l}))^{-1} > 0 \tag{16.8}$$

where g' denotes the derivative of g and h' denotes the derivative of h. Therefore we have $\mu \geq 0$.

It remains to prove that the support of the Fourier transform of μ is locally finite. Let $\Lambda_{A^*} = (A^*)^{-1}(\mathbb{Z}^2)$ be the dual lattice of Λ_A. As above $S_A = (A^*)^{-1}(S)$. The definition of Λ_{B^*} is similar. The distributional Fourier transform of σ_A is $\widehat{\sigma} \circ A^*$. The following remark is obvious:

Lemma 16.13 *If E denotes the support of $\widehat{\sigma}$ we have $E \subset S$ and $(A^*)^{-1}(E)$ is the support of $\widehat{\sigma}_A$.*

Therefore the support of $\widehat{\sigma}_A$ is contained in $(A^*)^{-1}(\mathbf{B}) = S_A \cap \Lambda_{A^*}$. We have $\widehat{\sigma} = \sum_{k \in E} c(k)\delta_k$ which implies $\widehat{\sigma}_A = \sum_{k \in E} c(k)\delta_{(A^*)^{-1}(k)}$. Similarly the Fourier transform of σ_B is supported by $S_B \cap \Lambda_{B^*}$ and $\widehat{\sigma}_B = \sum_{k \in E} c(k)\delta_{(B^*)^{-1}(k)}$. Since μ is the product between σ_A and σ_B we have

$$\widehat{\mu} = \sum_{k \in E, l \in E} c(k)c(l)\delta_{(A^*)^{-1}(k)+(B^*)^{-1}(l)}. \tag{16.9}$$

By (16.6) we know that $k \in \mathbb{Z}^2$ and $l \in \mathbb{Z}^2$ are uniquely defined by the sum $(A^*)^{-1}(k) + (B^*)^{-1}(l)$. We can conclude:

Lemma 16.14 *The support of $\widehat{\mu}$ is $(A^*)^{-1}(E) + (B^*)^{-1}(E)$ which is contained in $S_A \cap \Lambda_{A^*} + S_B \cap \Lambda_{B^*}$.*

Lemma 16.8 implies that $S_A \cap \Lambda_{A^*} + S_B \cap \Lambda_{B^*}$ is locally finite which ends the proof.

Let us observe that one could use ϕ_r for defining the curves $A(\Gamma_k)$ and a different function ϕ_s for defining the curves $B(\Gamma_l)$. One could use as well the phases of two arbitrary Blaschke products. Then the spectrum of μ is always contained in the locally finite set $S_A \cap \Lambda_{A^*} + S_B \cap \Lambda_{B^*}$ which does not depend on the two inner functions which are used in the construction. The measure μ and its support Λ obviously depend on these inner functions.

Let us check that μ is not a periodic measure if $r \notin R$. The set R is defined by Lemma 16.7. More precisely we have;

Theorem 16.6 *If $r \notin R$ we cannot have $\mu(x - y) = \mu(x)$ for some $y \neq 0$.*

Theorem 16.6 is proved by contradiction. First $\mu(x-y) = \mu(x)$ implies $[\exp(2\pi i y \cdot x) - 1]\widehat{\mu}(x) = 0$. The support of $\widehat{\mu}$ is $(A^*)^{-1}(E)+(B^*)^{-1}(E)$ and we have $0 \in E$. Therefore $\exp(2\pi i y \cdot x) = 1$ on $(A^*)^{-1}(E)$ and on $(B^*)^{-1}(E)$. We know that $E + E = \mathbb{Z}^2$. Therefore $\exp(2\pi i y \cdot x) = 1$ on $(A^*)^{-1}(\mathbb{Z}^2)$ and on $(B^*)^{-1}(\mathbb{Z}^2)$. It implies $A^{-1}(y) \in \mathbb{Z}^2$ and $B^{-1}(y) \in \mathbb{Z}^2$. Finally $y \in A(\mathbb{Z}^2) \cap B(\mathbb{Z}^2) = \{0\}$.

Theorem 16.7 *If $r > 0$ is small enough (depending on A and B) μ is not given by a tensor product $\mu = \mu_1 \otimes \mu_2$ between two crystalline measures on the line.*

The proof relies on Lemma 16.15. If $x = (x_1, x_2) \in \mathbb{R}^2$ we write $x_1 = p_1(x), x_2 = p_2(x)$. The closed ball centered at 0 with radius $\alpha > 0$ is denoted by $B(\alpha)$.

Lemma 16.15 *Let $F \subset \mathbb{R}^2$ be a lattice. Let us assume that both $p_1 : F \mapsto \mathbb{R}$ and $p_2 : F \mapsto \mathbb{R}$ have a dense range. Then there exists a positive α such that for any Delone set $\Lambda \subset \mathbb{R}^2$ the following property holds: if $F \subset \Lambda + B(\alpha)$, and if $\Lambda \subset F + B(\alpha)$ we cannot have $\Lambda = U \times V$ where U and V are two closed sets of real numbers.*

Before proving Lemma 16.15 let us show that the crystalline measure μ is not a tensor product. Let $F = M^*$ (M^* is defined in the proof of Lemma 16.12). Using (16.6) and Lemma 16.12 we see that the hypotheses of Lemma 16.15 are satisfied. As above Λ is the support of the crystalline measure μ. If we had $\mu = \mu_1 \otimes \mu_2$ it would imply $\Lambda = U \times V$ where U is the support of μ_1 and V is the support of μ_2. If r is small enough Lemma 16.15 yields the required contradiction.

We prove Lemma 16.15 by contradiction. Let us assume that $\Lambda = U \times V$. We have $p_1(F) \subset p_1(\Lambda) + [-\alpha, \alpha]$. But $p_1(\Lambda) = U$ and $U + [-\alpha, \alpha]$ is closed. Since $p_1(F)$ is a dense set of real numbers, we have $U + [-\alpha, \alpha] = \mathbb{R}$. Similarly we obtain $V + [-\alpha, \alpha] = \mathbb{R}$. These two properties imply $U \times V + [-\alpha, \alpha]^2 = \mathbb{R}^2$. Finally $\mathbb{R}^2 = \Lambda + [-\alpha, \alpha]^2 \subset F + [-2\alpha, 2\alpha]^2 = \mathbb{R}^2$ which cannot hold if α is small enough. This critical value of α only depends on F.

This proof leaves open the following issue: Is μ given in a new coordinate system by a tensor product $\mu_1 \otimes \mu_2$ between two crystalline measures? This issue will be addressed in a forthcoming contribution.

If $N \in \mathbb{N}$ and if

$$A = \begin{pmatrix} N & -(N+1) \\ -(N-1) & N \end{pmatrix} \tag{16.10}$$

then A belongs to $SL(2, \mathbb{Z})$ and is an automorphism of $\mathbb{T}^2$. The pushforward measure $\widetilde{\sigma} \circ A^{-1}$ is still supported by a closed curve of $\mathbb{T}^2$ and its Fourier coefficients vanish outside the thin angular sector delimited by the two lines $x_2 = \frac{N+1}{N} x_1$ and $x_2 = \frac{N}{N-1} x_1$. The aperture of this angular sector is $O(N^{-2})$ which can be arbitrarily small as it was announced in Sect. 16.3.

16.7 Generalization

To construct new crystalline measures on $\mathbb{R}^n$ one adapts the preceding tools. The following lemma replaces Lemma 16.8.

Lemma 16.16 *Let $e_1, \ldots, e_n$ be a basis of $\mathbb{R}^n$. Let us assume that $|e_j| = 1$, $1 \leq j \leq n$. For $\epsilon \in (0, 1)$ let C_j^ϵ be the cone defined by $|x \cdot e_j| \geq (1 - \epsilon)|x|$. Then for ϵ small enough the map $\Phi : C_1^\epsilon \times \ldots \times C_n^\epsilon \mapsto \mathbb{R}^n$ defined by $\Phi(x_1, \ldots, x_n) = x_1 + \cdots + x_n$ is a proper map.*

One starts with n smooth inner functions $B_1, \ldots, B_n$ on the polydisc $\mathbb{D}^{n-1}$. Let $\phi_j : \mathbb{T}^{n-1} \mapsto \mathbb{T}$ be a smooth determination of the phase of B_j and $\Gamma_j \subset \mathbb{T}^n$ be the graph of $-\phi_j$. The measure $\widetilde{\sigma}_j$ is defined as the image of the Haar measure on $\mathbb{T}^{n-1}$ by the map $x' \mapsto x = (x', -\phi_j(x'))$. Then $\widetilde{\sigma}_j$ is an Ahern measure supported by Γ_j. Let $S \subset \mathbb{R}^n$ be defined by $x_1 \geq 0, \ldots, x_n \geq 0$. The Fourier coefficients $\widehat{\widetilde{\sigma}}_j(k)$ of $\widetilde{\sigma}_j$ vanish if $k \notin S \cup (-S)$. The measure $\widetilde{\sigma}_j$ can also be considered as a $\mathbb{Z}^n$ periodic measure σ_j on $\mathbb{R}^n$. For each $n \times n$ non singular matrix $A(j)$ we consider the corresponding pushforward measure $\sigma_j^{A(j)}$. One can tune these n matrices $A(j)$ so that the Fourier coefficients $\sigma_j^{A(j)}$ vanish outside C_j^ϵ. Then if ϵ is small enough the pointwise product $\mu = \sigma_1^{A(1)} \ldots \sigma_n^{A(n)}$ makes sense and is a crystalline measure. The details will be published in a forthcoming paper.

Acknowledgments Let me thank the referees for a careful reading of the manuscript. They suggested many improvements to this manuscript. This work was supported by a grant from the Simons Foundation (601950, YM).

References

1. P.R. Ahern, Inner functions in the polydisc and measures on the torus. Michigan Math. J. **20**, 33–37 (1973)
2. A.P. Guinand, Concordance and the harmonic analysis of sequences. Acta Math. **101**, 235–271 (1959)
3. L. Hörmander, *The Analysis of Linear Partial Differential Operators: I Distribution Theory and Fourier Analysis*, 2nd edn. (Springer, Berlin, 1990)
4. J.P. Kahane, S. Mandelbrojt, Sur l'équation fonctionnelle de Riemann et la formule sommatoire de Poisson. Ann. Sci. l'E.N.S. Tome **75**, 57–80 (1958)
5. P. Kurasov, P. Sarnak, Stable Polynomials and crystalline measures. https://arxiv.org/abs/2004.05678v1
6. N. Lev, A. Olevskii, Measures with uniformly discrete support and spectrum. C. R. Math. Acad. Sci. Paris **351**, 599–603 (2013)
7. N. Lev, A. Olevskii. Quasicrystals and Poisson's summation formula. Invent. Math. **200**(2), 585–606 (2015)
8. N. Lev, A. Olevskii, Quasicrystals with discrete support and spectrum. Rev. Mat. Iberoamer. **32**(4), 1341–1352 (2016)
9. Y. Meyer, Measures with locally finite support and spectrum. Rev. Mat. Iberoamer. **33**, 1025–1036 (2017)
10. Y. Meyer, Crystalline measures in two dimensions. Publ. Mat. **67**(1), 469–480 (2023)
11. L. Schwartz, *Théorie des Distributions* (Hermann, Paris, 1950)
12. M. Viazovska, Sharp sphere packings. Proc. Int. Cong. of Math. **2**, 473–484 (2018)

Chapter 17
Muckenhoupt Matrix Weights for General Bases

Morten Nielsen and Hrvoje Šikić

Dedicated to Guido L. Weiss

Abstract We study matrix weights defined on the multivariate torus $\mathbb{T}^d$. We formulate the theory of matrix weights in the context of general coverings and in connection with the corresponding averaging operators. On this level of generality we prove the diagonalization theorem of Bloom and a form of its converse for the full range of Lipschitz functions. We incorporate within our setting the studies of log-preserving matrix weights, the BMO distance theorem for matrix weights, and the almost diagonal matrix weights.

17.1 Introduction

In this paper we study matrix weights, built via quite general covering basis, on the multivariate torus $\mathbb{T}^d$. The significance of weights in the studies of the boundedness of the Hardy-Littlewood maximal operators was brought to light in the seminal paper by Muckenhoupt [14], where the notion of the A_p–weight was introduced. Among the early papers let us also mention Hunt, Muckenhoupt and Wheeden [10] and Coifman and Fefferman [4]. Here we emphasize the approach that was very much used by the members of the Analysis group at Washington University in St. Louis (see, in particular, Coifman et al. [5]), studying the boundedness of the commutator operator, based on the Fefferman duality theorem and the BMO spaces. It inspired the work of S. Bloom, who in his PhD dissertation and subsequent

M. Nielsen
Department of Mathematical Sciences, Aalborg University, Aalborg, Denmark
e-mail: mnielsen@math.aau.dk

H. Šikić (✉)
Department of Mathematics, Faculty of Science, University of Zagreb, Zagreb, Croatia
e-mail: hsikic@math.hr

E. Hernández et al. (eds.), *The Mathematical Heritage of Guido Weiss*,
Applied and Numerical Harmonic Analysis,
https://doi.org/10.1007/978-3-031-76793-7_17

publications (see [2] and [3]) introduced a theory of matrix-valued weights, inluding the important class of log-preserving unitary matrices. This pioneering work was ahead of its time and technically demanding, and it took almost two decades before the revival of the theory of matrix-valued A_p–weights and the introduction of new ideas; for us here the following papers are particularly relevant, Nazarov and Treĭl′ [15], Treĭl′ and Volberg [24], Volberg [25], and Goldberg [9]. In one of his main results, the diagonalization theorem, Bloom relied on the Hilbert transform, but Nazarov and Treĭl′ showed that for the basis of cubes they can connect the matrix A_p–condition to the uniform boundedness of the very natural class of averaging operators. We take this approach here, as well.

We approached the subject inspired by our work on the stability of Schauder bases in the context of shift invariant spaces (see [17]), where the distance to the BMO spaces naturally appears. It guided us toward the analysis of the Garnett-Jones distance theorem, where we developed the A_2–decomposition property for scalar weights, but for the very general class of coverings (see [18]). We were able to extend this theory to matrix weights, but only for cubes, in [19]. Among others, and in order to develop this theory, we needed to extend the Bloom method of the log-preserving matrices to the multivariate case. In [19] we succeded in doing so for functions in Lip_1. These developments present us with a natural question. Can we built our theory, based on general coverings, for matrix weights and so that we incorporate the Bloom method in a multivariate setting and for the full range of Lipschitz-type functions? The last of the three requirements (i.e, general covering, multivariate setting, Lipschitz range) was stated as an open question in [19]. In this paper we give positive answer to this question. The crucial steps involve the development of the averaging operators method for the general class of coverings, application of this method to the general form of the Bloom diagonalization theorem (with only a mild continuity condition imposed on the covering) and a form of its converse, and a new method to treat almost diagonal matrix weights that include some results interesting even in the case of scalar weights (see Proposition 17.4 in this paper).

In Sect. 17.2 we introduce our theory of matrix weights in the context of general coverings and prove that the A_p–condition is equivalent to the uniform boundedness of averaging operators on weighted L^p–spaces. In Sect. 17.3 we deal with the class of log-preserving continuous matrix weights in our general context and we prove the diagonalization theorem. In Sect. 17.4 we prove that the Bloom method of log-preserving matrices holds for general coverings, in a multivariate setting and for the full Lipschitz range. Our method of the study of the BMO distance extends almost verbatim to the general framework, as outlined in short Sect. 17.5. In the sixth, and final, section we give the treatment of almost diagonal matrix weights within our theory.

17.2 Muckenhoupt Matrix Weights

In this section we introduce weights and matrix weights on the torus $\mathbb{T}^d$. The domain $\mathbb{T}^d$ can be replaced with $\mathbb{R}^d$ for all the results in the present section, but it will be essential for the proofs of our main results, Theorems 17.2 and 17.3, that the domain is bounded with finite measure.

Let us first consider the notion of a basis for $\mathbb{T}^d$ and corresponding scalar Muckenhoupt classes, see also Jawerth [11] for further details.

Definition 17.1 A basis for $\mathbb{T}^d$, $d \in \mathbb{N}$, is any collection $\mathcal{B}$ of measurable, bounded subsets of $\mathbb{T}^d$ with non-empty interior. A structured basis for $\mathbb{T}^d$ is a basis such that there exists a sequence $E_j \in \mathcal{B}$, $j \in \mathbb{N}$, with $\cup_j \mathring{E}_j = \mathbb{T}^d$ and $E_j \subseteq E_{j+1}$, $j \in \mathbb{N}$.

One can easily verify that standard collections such as the Euclidean balls $\mathcal{E}$, cubes $\mathcal{Q}$, dyadic cubes $\mathcal{D}$, and the collection of rectangles $\mathcal{R}$ (aligned with the coordinate axes in $\mathbb{T}^d$) are all structured bases in the sense of Definition 17.1.

Let $\mathcal{B}$ be a structured basis for $\mathbb{T}^d$, $d \in \mathbb{N}$. A (measurable) scalar weight $w : \mathbb{T}^d \to (0, \infty)$ is said to satisfy the Muckenhoupt A_p-condition, $1 < p < \infty$, provided

$$[w]_{A_p(\mathcal{B})} := \sup_{E \in \mathcal{B}} \fint_E w(x)\,\mathrm{d}x \cdot \left[\fint_E w^{-\frac{1}{p-1}}(x)\,\mathrm{d}x \right]^{p-1} < \infty, \tag{17.1}$$

where for any measurable subset $E \subset \mathbb{T}^d$ of positive measure, we define

$$f_E := \frac{1}{|E|} \int_E f(x)\,\mathrm{d}x,$$

and we use the notation

$$\fint_E f(x)\,\mathrm{d}x := \frac{1}{|E|} \int_E f(x)\,\mathrm{d}x.$$

We denote the class of such Muckenhoupt weights by $A_p(\mathcal{B})$. Even though the scalar A_p-conditions are quite involved, they are still very much operational since quite large classes of, e.g., polynomial weights are known to satisfy the respective conditions, see e.g. [20].

Now consider a matrix-valued function $W : \mathbb{T}^d \to \mathbb{C}^{N \times N}$, which is measurable and strictly positive definite a.e. We will refer to such a function as a matrix weight.

The matrix A_p condition was introduced and studied in [15, 24, 25] and it is considerably more complicated than the scalar condition, and there are no known straightforward sufficient conditions on a matrix weight to ensure membership in the A_p class except in very special cases (e.g., for diagonal weights and for weights with strong pointwise bounds on its spectrum).

We define the following family of metrics:

$$\rho_t(x) = |W^{1/p}(t)x|, \qquad x \in \mathbb{C}^N, \quad t \in \mathbb{T}^d,$$

where $|\cdot|$ denotes Euclidean norm on $\mathbb{C}^N$, with the dual metric given by

$$\rho_t^*(x) := \sup_{y \neq 0} \frac{|\langle x, y \rangle|}{\rho_t(y)} = |W^{-1/p}(t)x|, \qquad x \in \mathbb{C}^N, \quad t \in \mathbb{T}^d.$$

Assuming suitable local integrability of ρ_t, we now calculate the average

$$\rho_{p,E}(x) := \left(\frac{1}{|E|} \int_E [\rho_t(x)]^p \, \mathrm{d}t \right)^{1/p},$$

and likewise for the dual metric

$$\rho_{p',E}^*(x) := \left(\frac{1}{|E|} \int_E [\rho_t^*(x)]^{p'} \, \mathrm{d}t \right)^{1/p'},$$

with p' being p's Hölder conjugate, i.e., $1 = 1/p + 1/p'$.

The A_p condition can then be stated as follows.

Definition 17.2 Let $\mathcal{B}$ be a structured basis for $\mathbb{T}^d$. For $1 < p < \infty$, we say that W is an $\mathbf{A}_p(\mathcal{B})$ matrix weight if $W \colon \mathbb{T}^d \to \mathbb{C}^{N \times N}$ is measurable and positive definite a.e. such that W and $W^{-p'/p}$ are locally integrable, with p' the conjugate exponent to p, provided

$$[W]_{\mathbf{A}_p(\mathcal{B})} := \sup_{E \in \mathcal{B}} \sup_{v \in \mathbb{C}^N \setminus \{0\}} \frac{\rho_{p',E}^*(v)}{(\rho_{p,E})^*(v)} < +\infty. \tag{17.2}$$

where the integrability assumptions ensure that both $\rho_{p,E}$ and $\rho_{p',E}^*$ are well-defined. The family of such matrix weights is denoted by $\mathbf{A}_p(\mathcal{B})$. In the special case $N = 1$ (1×1-matrices), one can verify that $\mathbf{A}_p(\mathcal{B}) = A_p(\mathcal{B})$.

S. Roudenko introduced an equivalent matrix A_p condition in [22] which is often more straightforward to verify. Let $1 < p < \infty$ and let p' be the dual exponent, $1/p + 1/p' = 1$. The matrix A_p condition holds if and only if $W \colon \mathbb{T}^d \to \mathbb{C}^{N \times N}$ is measurable and positive definite a.e. and satisfies

$$[W]_{\widetilde{\mathbf{A}_p(\mathcal{B})}} := \sup_{E \in \mathcal{B}} \int_E \left(\int_E \left\| W^{1/p}(x) W^{-1/p}(t) \right\|^{p'} \frac{\mathrm{d}t}{|E|} \right)^{p/p'} \frac{\mathrm{d}x}{|E|} < \infty. \tag{17.3}$$

The norm $\|\cdot\|$ appearing in the integral is any matrix norm on the $N \times N$ matrices. It can be verified that $[W]_{\widetilde{\mathbf{A}_p(\mathcal{B})}} \asymp [W]_{\mathbf{A}_p(\mathcal{B})}$, uniformly for $W \in \mathbf{A}_p(\mathcal{B})$.

Remark 17.1 The equivalence $[\cdot]_{\widetilde{\mathbf{A}_p(\mathcal{B})}} \asymp [\cdot]_{\mathbf{A}_p(\mathcal{B})}$ was proved by Roudenko in [22] in the case where $\mathcal{B}$ is the family of cubes in $\mathbb{R}^d$, but the reader can easily verify that Roudenko's proof applies verbatim for general bases for $\mathbb{T}^d$.

The following lemma, which may be of independent interest, will demonstrate the operational nature of (17.3) and the result will be used later in the proof of Theorem 17.3. We let $L^\infty_{N\times N}$ denote the family of $N \times N$-matrices defined on $\mathbb{T}^d$ with entries in $L^\infty(\mathbb{T}^d)$.

Lemma 17.1 *Suppose $V \in \mathbf{A}_p(\mathcal{B})$, $1 < p < \infty$, and $B = \exp(B')$ with $B' \in L^\infty_{N\times N}$ satisfying $[V, B] = 0$, then $W = VB \in \mathbf{A}_p(\mathcal{B})$.*

Proof Calling on (17.3), $V \in \mathbf{A}_p(\mathcal{B})$ is equivalent to the uniform estimate over $E \in \mathcal{B}$,

$$\int_E \left(\int_E \| V^{1/p}(x) V^{-1/p}(t) \|^{p'} \frac{\mathrm{d}t}{|E|} \right)^{p/p'} \frac{\mathrm{d}x}{|E|} \le C_V < \infty,$$

with p' the dual exponent to p, i.e., $1/p + 1/p' = 1$. Now, we notice, since V and B commute, $(BV)^{1/p} = V^{1/p} B^{1/p} = B^{1/p} V^{1/p}$. And, similarly, since V^{-1} and B^{-1} commute, $(BV)^{-1/p} = V^{-1/p} B^{-1/p}$. Hence,

$$\begin{aligned} \|(BV)^{1/p}(x)(BV)^{-1/p}(t)\| &= \|B^{1/p}(x)\big[V^{1/p}(x)V^{-1/p}(t)\big]B^{-1/p}(t)\| \\ &\le C^2 \|V^{1/p}(x)V^{-1/p}(t)\|, \end{aligned}$$

with $C := \sup_{u\in\mathbb{T}^d}\{\|B^{-1/p}(u)\|, \|B^{1/p}(u)\|\}$. We obtain,

$$\begin{aligned} &\int_E \left(\int_E \|(BV)^{1/p}(x)(BV)^{-1/p}(t)\|^{p'} \frac{\mathrm{d}t}{|E|} \right)^{p/p'} \frac{\mathrm{d}x}{|E|} \\ &\qquad \le C^{2p} \int_E \left(\int_E \|V^{1/p}(x)V^{-1/p}(t)\|^{p'} \frac{\mathrm{d}t}{|E|} \right)^{p/p'} \\ &\qquad \frac{\mathrm{d}x}{|E|} \\ &\qquad \le C^{2p} C_V < \infty, \end{aligned}$$

with the bound independent of $E \in \mathcal{B}$. We conclude, using (17.3), that $VB = BV \in \mathbf{A}_p(\mathcal{B})$. □

We will also need weighted vector-valued L^p-spaces, $1 < p < \infty$. For $W\colon \mathbb{T}^d \to \mathbb{C}^{N\times N}$ a matrix-valued function, which is measurable and positive definite a.e., let $L^p(W)$ denote the family of measurable functions $f\colon \mathbb{T}^d \to \mathbb{C}^N$ with

$$\|f\|_{L^p(W)} := \left(\int_{\mathbb{T}^d} |W^{1/p} f|^p \, \mathrm{d}x \right)^{1/p} < \infty.$$

In order to turn $L^p(W)$ into a Banach space, one has to factorize over $\{f\colon \mathbb{T}^d \to \mathbb{C}^N;\ \|f\|_{L^p(W)} = 0\}$. It can be verified that the dual space to $L^p(W)$ is $L^{p'}(W^{-p'/p})$, see [25] for further details.

Muckenhoupt (matrix) weights can often be characterised by studying various operators on associated weighted L^p-spaces, see [10, 24]. The Hilbert transform and Riesz transforms are often central to such characterizations, but we note that the structure of such operators turns out to be adapted to the case where we have a basis $\mathcal{B}$ of cubes. However, in this paper we aim to study more general bases $\mathcal{B}$, so we will mainly focus our attention to the study of certain averaging operators.

We let $\mathcal{B}$ denote a structured basis for $\mathbb{T}^d$. For $E \in \mathcal{B}$, and a locally integrable vector-valued function $f : \mathbb{T}^d \to \mathbb{C}^N$, we consider the vector-valued operator

$$A_E f(x) := \chi_E(x)\frac{1}{|E|}\int_E f(t)\,\mathrm{d}t,$$

where the integral is applied coordinate-wise.

Nazarov and Treĭl′ made the fundamental discovery in [15], for the basis of cubes, that uniform boundedness of A_E on $L^p(W)$ is closely related to the matrix A_p-condition. The following result extends the result of Nazarov and Treĭl′ to general bases. The proof given below is inspired by Goldberg [9, Proposition 2.1].

Theorem 17.1 *Let $W\colon \mathbb{T}^d \to \mathbb{C}^{N\times N}$ be a matrix weight, $1 < p < \infty$, and let $\mathcal{B}$ be a structured basis for $\mathbb{T}^d$. Then $W \in \mathbf{A}_p(\mathcal{B})$ if and only if the operators $\{A_E\}_{E\in\mathcal{B}}$ are uniformly bounded on $L^p(W)$. Moreover, the supremum of the operator norms of $\{A_E\}_{E\in\mathcal{B}}$ on $L^p(W)$ is equal to the A_p-constant $[W]_{\mathbf{A}_p(\mathcal{B})}$ given by* (17.2).

Proof Let $E \in \mathcal{B}$, and let $f \in L^p(W)$. The $L^p(W)$-norm of $A_E f$ is given by

$$\frac{1}{|E|}\left(\int_E \left|W^{1/p}(y)\int_E f(t)\,\mathrm{d}t\right|^p \mathrm{d}y\right)^{1/p},$$

which equals $|E|^{-1/p'}\rho_{p,B}\big(\int_E f\,\mathrm{d}t\big)$. Hence, using the duality of $L^p(W)$ and $L^{p'}(W^{-p'/p})$,

$$\begin{aligned}
\sup_{f:\|f\|_{L^p(W)}=1} \|A_E f\|_{L^p(W)} &= \sup_{f:\|f\|_{L^p(W)}=1}\ \sup_{v\in\mathbb{C}^N\setminus\{0\}} |E|^{-1/p'}\frac{\int_E\langle v, f(t)\rangle \mathrm{d}t}{(\rho_{p,E})^*(v)}\\
&= \sup_{v\in\mathbb{C}^N\setminus\{0\}} |E|^{-1/p'}\frac{\|\chi_E v\|_{L^{p'}(W^{-p'/p})}}{(\rho_{p,E})^*(v)}\\
&= \sup_{v\in\mathbb{C}^N\setminus\{0\}} \frac{\rho^*_{p',E}(v)}{(\rho_{p,E})^*(v)},
\end{aligned}$$

and it follows that

$$\sup_{E\in\mathcal{B}} \sup_{f:\|f\|_{L^p(W)}=1} \|A_E f\|_{L^p(W)} = [W]_{\mathbf{A}_p(\mathcal{B})}.$$

□

17.3 Log Preserving Continuous Matrix Weights

Consider a matrix weight $W : \mathbb{T}^d \to \mathbb{C}^{N\times N}$. Since W is measurable positive definite a.e., it is always possible to find corresponding measurable matrix functions $D : \mathbb{T}^d \to \mathbb{C}^{N\times N}$ and $U : \mathbb{T}^d \to \mathbb{C}^{N\times N}$, such that D is a diagonal matrix function and U is unitary a.e., satisfying

$$W(t) = U^*(t)D(t)U(t), \qquad a.e.\ t \in \mathbb{T}^d,$$

see, e.g., [21] for a proof of this fact. Below, we will derive certain properties of the weight W by imposing various (mild) assumptions on the functions U and D.

Let us first define the class of functions of bounded mean oscillation. Traditionally this space has been considered for means taken over Euclidean balls or cubes, but it may be considered for any structured basis, see [18] for further details.

Definition 17.3 Let $f \in L^1_{\text{loc}}(\mathbb{T}^d)$, and let $\mathcal{B}$ be a structured basis for $\mathbb{T}^d$. We say that $f \in BMO(\mathcal{B})$ provided that

$$\|f\|_{BMO(\mathcal{B})} := \sup_{E\in\mathcal{B}} \frac{1}{|E|}\int_E |f(x) - f_E|\,\mathrm{d}x < \infty. \tag{17.4}$$

We have the following associated definition of matrix BMO.

Definition 17.4 We say that a matrix weight $W : \mathbb{T}^d \to \mathbb{C}^{N\times N}$, $W = (w_{ij})$, is in $BMO(\mathcal{B})$ provided that each entry $w_{ij} : \mathbb{T}^d \to \mathbb{C}$, $i, j \in \{1, \dots, N\}$, is a scalar valued $BMO(\mathcal{B})$ function. The BMO-norm of such W is defined by

$$\max_{i,j} \|w_{ij}\|_{BMO(\mathcal{B})}.$$

In the scalar case, one can obtain Muckenhoupt weights by exponentiating BMO-function. However, Bloom [2] has constructed a symmetric $S \in BMO(Q)$ for which $\exp(tS) \notin \mathbf{A}_2(Q)$ for any $t > 0$. To avoid such ill-conditioned matrices, we follow Bloom [2] and define so-called log-preserving unitary matrices.

Definition 17.5 A unitary matrix function $U : \mathbb{T}^d \to \mathbb{C}^{N\times N}$ is called log-preserving, denoted $U \in LP$, if for any diagonal matrix $\Lambda \in BMO(\mathcal{B})$, there exists an $\alpha > 0$ depending only on the $BMO(\mathcal{B})$-norm of Λ, and the matrix U, such that

$$U^* \exp(\alpha \Lambda) U \in \mathbf{A}_2(\mathcal{B}),$$

with the corresponding A_2-constant depending only on the $BMO(\mathcal{B})$-norm of Λ, and the matrix U.

Examples of log-preserving can be constructed by calling on Corollary 17.1 presented below.

We now present a generalisation of Bloom's diagonalization result [2, Theorem 4.9] in the multivariate setting, where the generalisation is obtained by relying on averaging operators A_E in place of the Hilbert transform as considered by Bloom in [2]. This approach immediately opens for generalizations to other bases $\mathcal{B}$ besides the cubes due to the full generality of Theorem 17.1. We will need the following technical lemma.

Lemma 17.2 *Let $\mathcal{B}$ be a structure basis for $\mathbb{T}^d$, and let $W \in \mathbf{A}_2(\mathcal{B})$. Then there exists a constant $C := C([W]_{\mathbf{A}_2(\mathbb{T}^d)})$ such that for $f \in L^2(W)$ and $E \in \mathcal{B}$, we have*

$$A_E\langle W A_E f, A_E f\rangle \le C A_E\langle Wf, f\rangle = C\chi_E \frac{1}{|E|}\int_E \langle W(x)f(x), f(x)\rangle\, \mathrm{d}x.$$

Proof Let $E \in \mathcal{B}$. Notice that $A_E f = A_E(\chi_E f)$. We have

$$\begin{aligned}
A_E\langle W A_E f, A_E f\rangle &= A_E\langle W A_E(\chi_E f), A_E(\chi_E f)\rangle \\
&\le \frac{1}{|E|}\int_{\mathbb{T}^d} \langle W A_E(\chi_E f), A_E(\chi_E f)\rangle\, \mathrm{d}x \\
&\le \frac{C}{|E|}\int_{\mathbb{T}^d} \langle W(\chi_E f), \chi_E f\rangle\, \mathrm{d}x \\
&= C A_E\langle W(\chi_E f), \chi_E f\rangle \\
&\le C A_E\langle Wf, f\rangle,
\end{aligned}$$

where we used Theorem 17.1 for the second inequality. □

We now give a generalisation of Bloom's diagonalization result [2, Theorem 4.9] in the multivariate setting for more general bases $\mathcal{B}$. We will apply the diagonalization result in Sect. 17.5 to study various distance to L^∞ results for matrix weights.

Theorem 17.2 *Suppose $U \colon \mathbb{T}^d \to \mathbb{C}^{N\times N}$ is a log-preserving matrix, with the mapping being continuous. Let $\mathcal{B}$ be a structured basis for $\mathbb{T}^d$ having the following continuity property: for any $\epsilon > 0$ there exists $\delta > 0$ such that $E \in \mathcal{B}$ with $|E| < \delta$ implies $diam(E) < \epsilon$. Then there exists a constant $C := C(U)$ such that for matrix weights $W = U^* D U \in \mathbf{A}_2(\mathcal{B})$, with $D = diag\{\lambda_1, \lambda_2, \ldots, \lambda_N\}$, we have $[\lambda_i]_{A_2(\mathcal{B})} \le C[W]_{\mathbf{A}_2(\mathcal{B})}$, $i = 1, 2, \ldots, N$.*

Proof We have to obtain a uniform bound for the product

$$A_E(\lambda_i)A_E(\lambda_i^{-1}), \tag{17.5}$$

for $i = 1, 2, \dots, N$, as B varies over $\mathcal{B}$. Now, it is known, see e.g. [9, Corollary 2.3], that $\|W\| \asymp \text{trace}(D) \in A_2(\mathcal{B})$, since $W \in \mathbf{A}_2(\mathcal{B})$, and $\|W^{-1}\| \asymp \text{trace}(D^{-1}) \in A_2(\mathcal{B})$ since $W^{-1} \in \mathbf{A}_2(\mathcal{B})$. In particular, trace$(D)$, trace$(D^{-1}) \in L^1(\mathbb{T}^d)$, and since the diagonal elements in D and D^{-1} are all positive, we deduce that $\lambda_i, \lambda_i^{-1} \in L^1(\mathbb{T}^d)$. Hence, the product in Eq. (17.5) is automatically bounded for all E with large measure, say, with $|E| \geq \epsilon_0$ for some fixed $\epsilon_0 > 0$ that we may choose.

Fix an arbitrary small $E \in \mathcal{B}$, with $|E| < \epsilon_0$ and pick some $x_0 \in E$. We notice that for a fixed unitary matrix V, $[V^*WV]_{\mathbf{A}_2(\mathcal{B})} = [W]_{\mathbf{A}_2(\mathcal{B})}$, and V^*WV obviously preserves the spectrum of W. By replacing W by $U(x_0)U^*DUU^*(x_0)$, we may, without loss of generality, assume that $U(x_0) = Id$. Consider $\epsilon > 0$ (to be specified later). Let us write $U = (u_{ij})_{i,j=1}^N$. By the continuity, and consequently, by uniform continuity of U on $\mathbb{T}^d$, there exists $\delta > 0$, independent of x_0, such that

$$|u_{ij}(x) - \delta_{ij}| < \epsilon, \tag{17.6}$$

for $|x - x_0| < \delta$. We now adjust the value of $\epsilon_0 > 0$ to ensure diam$(E) < \delta$.

For notational convenience, we first focus on the λ_1-case. We use Lemma 17.2 with the specific function $f = \lambda_1^{-1}U^*e_1$, i.e., $f_r = \lambda_1^{-1}\bar{u}_{1r}$, $r = 1, 2, \dots, N$. We obtain

$$\langle Wf, f\rangle = \lambda_1^{-2}\langle U^*DUU^*e_1, U^*e_1\rangle = \lambda_1^{-1}.$$

Hence,

$$A_E\langle WA_Ef, A_Ef\rangle \leq CA_E\langle Wf, f\rangle = CA_E(\lambda_1^{-1}),$$

where $C := C([W]_{\mathbf{A}_2(\mathbb{T}^d)})$. Let $\tilde{D}$ be as D, but with 0 in place of λ_1, and consider $\tilde{W} := U^*\tilde{D}U$. Clearly, $\tilde{W} := (\tilde{w}_{ij})_{i,j=1}^N$ is non-negative definite as the corresponding eigenvalues are all non-negative, so $A_E\langle \tilde{W}A_Ef, A_Ef\rangle \geq 0$, and consequently,

$$A_E\langle WA_Ef, A_Ef\rangle - A_E\langle \tilde{W}A_Ef, A_Ef\rangle \leq CA_E(\lambda_1^{-1}). \tag{17.7}$$

We now expand the matrix product $W = U^*DU$ to obtain,

$$w_{ij} = \sum_{k=1}^{N} u_{kj}\bar{u}_{ki}\lambda_k = \lambda_1 u_{1j}\bar{u}_{1i} + \tilde{w}_{ij}.$$

By linearity of A_E, we can then calculate $A_E\langle WA_E f, A_E f\rangle$ as follows

$$A_E\langle WA_E f, A_E f\rangle = \sum_{r,s=1}^{N} A_E(w_{rs})A_E(\bar{f}_r)A_E(f_s)$$
$$= \sum_{r,s=1}^{N} A_E(\lambda_1 u_{1j}\bar{u}_{1i})A_E(\bar{f}_r)A_E(f_s) + A_E\langle \tilde{W}A_E f, A_E f\rangle.$$

Hence, by substituting this into Eq. (17.7), we obtain

$$\sum_{r,s=1}^{N} A_E(\lambda_1 u_{1r}\bar{u}_{1s})A_E(\bar{f}_r)A_E(f_s) \le CA_E(\lambda_1^{-1}). \tag{17.8}$$

Let us recall that $f_r = \lambda_1^{-1}\bar{u}_{1r}$. We estimate the individual terms on the left-hand side of Eq. (17.8) by considering the following cases.

Case a.: $r = s = 1$. Here, calling on the estimate (17.6) with $i = j = 1$,

$$A_E(\lambda_1|u_{11}|^2)|A_E(\lambda_1^{-1}u_{11})|^2 \ge (1-\epsilon)^4 A_E(\lambda_i)A_E(\lambda_1^{-1})^2.$$

Case b.: $r = 1$, $s \neq 1$. Now, again calling on the estimate (17.6),

$$|A_E(\lambda_1\bar{u}_{11}u_{1s})A_E(\lambda_1^{-1}u_{11})A_E(\lambda_1^{-1}\bar{u}_{1s})| \le \epsilon^2 A_E(\lambda_i)A(\lambda_i^{-1})^2.$$

Case c.: $r \neq 1$, $s = 1$. The estimate is identical to Case b.
Case d.: $r \neq 1$, $s \neq 1$. The terms in this case are all bound by $\epsilon^4 A_E(\lambda_1)A_E(\lambda_i^{-1})^2$.

By combining all the estimates of the terms in Eq. (17.8), we obtain

$$A_E(\lambda_1)A_E(\lambda_1^{-1})\Big[(1-\epsilon)^4 + 2(N-1)\epsilon^2 - (N-1)^2\epsilon^4\Big] \le C.$$

Now we adjust the value of $\epsilon > 0$ to ensure that $(1-\epsilon)^4 + 2(N-1)\epsilon^2 - (N-1)^2\epsilon^4 \ge \frac{1}{2}$, which implies the wanted estimate

$$A_E(\lambda_1)A_E(\lambda_1^{-1}) \le 2C,$$

with C independent of E for E with small measure. We conclude that $\lambda_1 \in A_2(\mathcal{B})$ with $[\lambda_i]_{A_2(\mathcal{B})} \le \max\{2C, \epsilon_0^{-2}\}$. The estimates for $\lambda_2, \ldots, \lambda_N$ are similar. □

17.4 Lipschitz Continuous Matrix Weights

In this section we will focus on result that can be seen as a converse to the diagonalisation theorem. However, we need to impose slightly stronger assumptions on the unitary matrix function U.

Recall that a function $f : \mathbb{T}^d \to \mathbb{C}$ is said to be in $\mathrm{Lip}_\gamma(\mathbb{T}^d)$, $0 < \gamma \leq 1$, provided there exists a Lipschitz constant $C_\gamma < \infty$ such that

$$|f(x) - f(y)| \leq C_\gamma |x - y|^\gamma, \qquad x, y \in \mathbb{T}^d.$$

We can now state the main result, Theorem 17.3. The univariate case with $p = 2$ was proven by Bloom [2]. The multivariate case with $\alpha = 1$, and basis $\mathcal{B}$ given by the cubes $\mathcal{Q}$, was considered earlier by the authors in [19]. The following result allows for anisotropic bases $\mathcal{B}$, a less restrictive Lipschitz condition, and the extension of the result to arbitrary p with $1 < p < \infty$.

A well-known useful observation is that for any linear operator T, $1 < p < \infty$, we have

$$\|Tf\|_{L^p(W)} = \|W^{1/p}Tf\|_{L^p(Id)} = \|W^{1/p}TW^{-1/p}g\|_{L^p(Id)},$$

with $g = W^{1/p}f$ satisfying $\|f\|_{L^p(W)} = \|g\|_{L^p(Id)}$. Hence, boundedness of T on $L^p(W)$ is equivalent to boundedness of the conjugate operator $W^{1/p}TW^{-1/p}$ on the unweighted space $L^p(\mathbb{C}^N) := L^p(Id)$.

Theorem 17.3 *Suppose $\mathcal{B}$ is a structured basis for $\mathbb{T}^d$. Let $\alpha \in (0, 1]$, $1 < p_0 < \infty$, and let $W = U^*DU$, with $D = diag(e^{\lambda_1}, \ldots, e^{\lambda_N})$, where $e^{\lambda_i} \in A_{p_0}(\mathcal{B})$. Suppose $U = (u_{ij})$ is a unitary matrix function with entries $u_{ij} \in Lip_\alpha(\mathbb{T}^d)$, $i, j \in \{1, 2, \ldots, N\}$, for which there exist $C > 0$, and $0 \leq \beta < \alpha$, such that*

$$diam(E)^{d+\beta} \leq C|E|, \qquad E \in \mathcal{B}. \tag{17.9}$$

Then there exists $\eta > 0$ such that $W^\eta = U^*D^\eta U \in \mathbf{A}_{p_0}(\mathcal{B})$, where the value of η and the A_{p_0} constant of W^η depend only on the Lipschitz constants of $\{u_{ij}\}_{ij}$, β and on the A_{p_0} constants of the scalar functions $\{e^{\lambda_i}\}_i$.

Proof We first assume that the elements of D have been normalized such that $\|e^{\lambda_i}\|_{L^1(\mathbb{T}^d)} = 1$ for $i = 1, 2 \ldots, N$. For $\tau > 0$, we have $W^\tau = U^*D^\tau U$, with

$$D^\tau = \mathrm{diag}(e^{\tau\lambda_1}, \ldots, e^{\tau\lambda_N}).$$

For use later, we notice that for any linear operator T on $L^p(W^\tau)$, with $f \in L^p(W^\tau)$,

$$\begin{aligned}
\|Tf\|^p_{L^p(W^\tau)} &= \int_{\mathbb{T}^d} \langle W^{2\tau/p}Tf, Tf\rangle^{p/2}\,\mathrm{d}x \\
&= \int_{\mathbb{T}^d} \left\langle \exp\left(\frac{2\tau}{p}D\right)UTf, UTf\right\rangle^{p/2} \mathrm{d}x \\
&= \int_{\mathbb{T}^d} \left(\sum_k |(UTf)_k|^2 \exp\left(\frac{2\tau\lambda_k}{p}\right)\right)^{p/2} \mathrm{d}x \\
&\geq \int_{\mathbb{T}^d} |(UTf(x))_k|^p \exp(\tau\lambda_k)\,\mathrm{d}x, \qquad k = 1, 2, \ldots, N. \qquad (17.10)
\end{aligned}$$

Now fix $E \in \mathcal{B}$ and let f be any smooth vector-valued function. Let us fix p with $\max\{2, p_0\} < p < \infty$. We notice that for any $0 < \tau \leq 1$ the scalar Muckenhoupt conditions $e^{\tau\lambda_k} \in A_p(\mathcal{B})$ for $k = 1, \ldots, N$ are satisfied, which can be verified directly using Hölder's inequality, using the assumption that $e^{\lambda_k} \in A_{p_0}(\mathcal{B})$. For use later, we notice that for a scalar-valued function g, we can use Theorem 17.1 in the scalar case to conclude that

$$\int_{\mathbb{T}^d} |A_E g|^p e^{\tau\lambda_r}\,\mathrm{d}x \leq K \int_{\mathbb{T}^d} |g|^p e^{\tau\lambda_r}\,\mathrm{d}x, \qquad r = 1, 2, \ldots, N, \qquad (17.11)$$

where $K^{1/p}$ is any upper bound on the A_p-constants of $e^{\tau\lambda_1}, \ldots, e^{\tau\lambda_N}$. Notice, in particular, that K does not depend on $E \in \mathcal{B}$.

We now write

$$U(x)A_E f(x) = U(x)\frac{\chi_E(x)}{|E|}\int_E f(t)\,\mathrm{d}t = \frac{\chi_E(x)}{|E|}\int_E U(x)U^*(t)(Uf)(t)\,\mathrm{d}t,$$

where the matrix $U(x)U^*(t)$ is continuous in t, and equals the identity at $t = x$. With δ_{ik} the Kronecker delta, and recalling that $|u_{ij}| \leq 1$ since U is unitary,

$$\begin{aligned}
|(U(x)U^*(t))_{ik} - \delta_{ik}| &= \left|\sum_r u_{ir}(x)\bar{u}_{kr}(x) - \sum_r u_{ir}(x)\bar{u}_{kr}(x)\right| \\
&\leq \sum_r |u_{ir}(x)||\bar{u}_{kr}(t) - \bar{u}_{kr}(x)| \\
&\leq \sum_r |\bar{u}_{kr}(t) - \bar{u}_{kr}(x)| \\
&\leq M|x - t|^\alpha,
\end{aligned}$$

with M any upper bound for the $\alpha-$Lipschitz constants of the collection $\{\bar{u}_{ij}\}_{ij}$. Consequently,

$$|(U(x)U^*(t))_{ik}| \leq \delta_{ik} + M|x-t|^{\alpha}. \tag{17.12}$$

We now employ the estimate (17.12) to deduce the following

$$\begin{aligned}
|(U(x)A_E f)(x))_r| &= \frac{\chi_E(x)}{|E|}\left|\int_E \sum_k (U(x)U^*(t))_{rk}(Uf)_k(t)\,\mathrm{d}t\right| \\
&\leq \frac{\chi_E(x)}{|E|}\int_E \sum_k (\delta_{rk} + M|x-t|^{\alpha})|(Uf)_k(t)|\,\mathrm{d}t \\
&\leq A_E(|(Uf)_r|)(x) + M\sum_k \frac{1}{|E|}\int_E |x-t|^{\alpha}|(Uf)_k(t)|\,\mathrm{d}t \\
&\leq A_E(|(Uf)_r|)(x) + CM\sum_k \int_E |x-t|^{\alpha-d-\beta}|(Uf)_k(t)|\,\mathrm{d}t \\
&\leq A_E(|(Uf)_r|)(x) + CM\sum_k \int_{\mathbb{T}^d} |x-t|^{\alpha-d-\beta}|(Uf)_k(t)|\,\mathrm{d}t,
\end{aligned} \tag{17.13}$$

where we used the assumption (17.9) ensuring that $|x-t|^{d+\beta} \leq \operatorname{diam}(E)^{d+\beta} \leq C|E|$ for $x, t \in B$ and some constant $C > 0$.

If $d = 1$, we may skip the following step and continue the estimate (17.13) directly as indicated below in (17.16) as $p > 2$. For $d \geq 2$, however, we need an additional estimate. We fix $q > 1$ such that $q(d+\beta-\alpha) < d$ and we redefine p by $1/p + 1/q = 1$, where still ensure that $p > \max\{2, p_0\}$. Pick $s > 1$ such that $sq(d+\beta-\alpha) < d$ and let $1/s + 1/s' = 1$. Then

$$\begin{aligned}
&\int_{\mathbb{T}^d} |x-t|^{q(\alpha-d-\beta)} \exp(-\tau q/p\lambda_k)\,\mathrm{d}t \\
&\leq \left(\int_{\mathbb{T}^d} |x-t|^{sq(\alpha-d-\beta)}\,\mathrm{d}t\right)^{1/s}\left(\int_{\mathbb{T}^d} \exp\left(-\frac{s'\tau q}{p}\lambda_k\right)\mathrm{d}t\right)^{1/s'}.
\end{aligned}$$

We now adjust the value of τ so $\tau s' = 1$, which ensures that $e^{\tau s'\lambda_k} \in A_p$, $k = 1, \ldots, N$, and implies that $\exp(-s'\tau q\lambda_k/p) \in L^1(\mathbb{T}^d)$, $k = 1, \ldots, N$, with norms that depend only on the $A_p(\mathcal{B})$-constants of $\{e^{\lambda_k}\}$. Consequently, with these choices,

$$\int_{\mathbb{T}^d} |x-t|^{q(\alpha-d-\beta)} \exp(-\tau q/p\lambda_k)\,\mathrm{d}t \leq L, \tag{17.14}$$

for some constant $L < \infty$. We now continue from the estimate (17.13),

$$
\begin{aligned}
&|(U(x)A_E f)(x))_r| \qquad (17.15)\\
&\le A_E(|(Uf)_r|)(x) + CM\sum_k \int_{\mathbb{T}^d} |x-t|^{\alpha-d-\beta}|(Uf)_k(t)|e^{\tau\lambda_k/p}e^{-\tau\lambda_k/p}\,\mathrm{d}t\\
&\le A_E(|(Uf)_r|)(x) + CM\sum_k \left(\int_{\mathbb{T}^d} |(Uf)_k(t)|^p e^{\tau\lambda_k}\,\mathrm{d}t\right)^{1/p}\\
&\qquad\qquad \times\left(\int_{\mathbb{T}^d} |x-t|^{q(\alpha-d-\beta)}e^{-\tau q\lambda_k/p}\,\mathrm{d}t\right)^{1/q}\\
&\le A_E(|(Uf)_r|)(x) + CML^{1/q}\sum_k \left(\int_{\mathbb{T}^d} |(Uf)_k(t)|^p e^{\tau\lambda_k}\,\mathrm{d}t\right)^{1/p}\\
&\le A_E(|(Uf)_r|)(x) + \tilde{C}\|f\|_{L^p(W)}, \qquad (17.16)
\end{aligned}
$$

where we used the estimates (17.10) and (17.14), and the constant $\tilde{C} := CMNL^{1/q}$ is independent of E. Finally, using (17.16) once more,

$$
\begin{aligned}
\|A_E f\|^p_{L^p(W^\tau)} &= \int_{\mathbb{T}^d}\left(\sum_r |(UA_E f(x))_r|^2 \exp\left(\frac{2\tau}{p}\lambda_r\right)\right)^{\frac{p}{2}}\mathrm{d}x\\
&\le \int_{\mathbb{T}^d}\left(\sum_r (A_E(|(Uf)_r|)(x) + \tilde{C}\|f\|_{L^p(W)})^2 \exp\left(\frac{2\tau}{p}\lambda_r\right)\right)^{\frac{p}{2}}\mathrm{d}x\\
&\le \int_{\mathbb{T}^d}\left(\sum_r (2A_E(|(Uf)_r|)(x)^2\right.\\
&\qquad\left. + 2\tilde{C}^2\|f\|^2_{L^p(W)})\exp\left(\frac{2\tau}{p}\lambda_r\right)\right)^{\frac{p}{2}}\mathrm{d}x.
\end{aligned}
$$

Now we use Hölder's inequality with parameter $p/2 > 1$ in the finite sum over r,

$$
\begin{aligned}
&\le C'\int_{\mathbb{T}^d}\left(\sum_r (A_E(|(Uf)_r|)(x)^p + \tilde{C}^p\|f\|^p_{L^p(W)})\exp(\tau\lambda_r)\right)\mathrm{d}x\\
&\le C'\int_{\mathbb{T}^d} A_E(|(Uf)_r|)(x)^p\exp(\tau\lambda_r)\,\mathrm{d}x + C'\tilde{C}^p\|f\|^p_{L^p(W)}\sum_r \|e^{\tau\lambda_r}\|_{L^1(\mathbb{T}^d)}\\
&\le C'K\int_{\mathbb{T}^d} |(Uf)_r|^p\exp(\tau\lambda_r)\,\mathrm{d}x + C'\tilde{C}^p\|f\|^p_{L^p(W)}\sum_r \|e^{\tau\lambda_r}\|_{L^1(\mathbb{T}^d)},
\end{aligned}
$$

where we used (17.11) in the last step, and where $C' := C'(N, p)$ is a constant independent of E. We use the estimate (17.10) to conclude that

$$\|A_E f\|_{L^p(W^\tau)}^p \le C''\|f\|_{L^p(W^\tau)}^p, \tag{17.17}$$

with

$$C'' := C'\Big(K + \tilde{C}^p \sum_r \|e^{\tau\lambda_r}\|_{L^1(\mathbb{T}^d)}\Big)$$

independent of E. Hence, A_E extends to a uniformly bounded family of bounded operators on $L^p(W^\tau)$. However, we have chosen p such that $p > p_0$, so we may use Stein's interpolation theorem, see [23], to cover the case $L^{p_0}(W^\eta)$ for a suitable value of $\eta > 0$. Define the analytic class, for a definition see [23],

$$T_z := W^{\tau z/p} A_E W^{-\tau z/p}, \qquad \{z : 0 \le \mathrm{Re}(z) \le 1\},$$

and notice that (17.17) implies that T_1 is bounded on the (unweighted) vector-valued $L^p(\mathbb{C}^N)$. Also notice that $T_0 = A_E$ is bounded (uniformly in E) on $L^q(\mathbb{C}^N)$ according to Theorem 17.1 in the scalar case, where q is the dual exponent to p, i.e., $1/p + 1/q = 1$. Using the observation that for $z = x + iy$,

$$W^{\tau(x+iy)/p} A_E W^{-\tau(x+iy)/p} = W^{i\tau y/p} T_x W^{-i\tau y/p},$$

with $W^{\pm i\tau y/p}$ unitary matrices, one can check that the conditions in Stein's interpolation theorem are met. Hence, by Stein's interpolation theorem, T_θ is bounded on $L^{p_0}(\mathbb{C}^N)$, with an operator norm that does not depend on $E \in \mathcal{B}$, where we define $0 < \theta < 1$ by the equation $1/p_0 = (1-\theta)/p' + \theta/p$, which is possible since $\max\{2, p_0\} < p$. Therefore, $W^{\theta\tau/p} \in \mathbf{A}_p(\mathcal{B})$, and we may put $\eta = \theta\tau/p$. The operator norm of T_θ depends only on p_0 and the A_{p_0} constants of $\{e^{\lambda_i}\}_i$.

Finally, for D without L^1-normalization, we introduce $V = U^* B U$, where

$$B = \mathrm{diag}(\|e^{\lambda_1}\|_{L^1(\mathbb{T}^d)}^{-1}, \ldots, \|e^{\lambda_N}\|_{L^1(\mathbb{T}^d)}^{-1}).$$

Notice that $W' = VW = U^*(BD)U$, where BD has normalized elements. Using Lemma 17.1, $(W')^{\theta\tau/p} \in \mathbf{A}_{p_0}(\mathcal{B})$ if and only if $W^{\theta\tau/p} \in \mathbf{A}_{p_0}(\mathcal{B})$. This completes the proof. □

The following example shows how one can control the geometric condition (17.9) in a setup where we have a basis consisting of structured rectangles.

Example
Given a vector $\mathbf{a} \in (0, \infty)^d$, we consider the one parameter basis $\mathcal{B}$ generated by the associated anisotropic balls

$$B_r^{\mathbf{a}}(x) := \left\{ y \in \mathbb{T}^d : \sum_{j=1}^{d} |x_j - y_j|^{1/a_j} < r \right\}, \qquad x \in \mathbb{T}^d, r > 0.$$

One can verify that $|B_r^{\mathbf{a}}(x)| \asymp \min\{1, r^{a_1+\cdots+a_d}\}$. For small r, we have $\mathrm{diam}(B_r^{\mathbf{a}}(x)) \asymp r^{a_{\min}}$, with $a_{\min} := \min\{a_1, \ldots, a_d\}$. Hence, for r small,

$$|B_r^{\mathbf{a}}(x)| \asymp r^{a_1+\cdots+a_d} = r^{a_{\min}(a_1/a_{\min}+\cdots+a_d/a_{\min})} \asymp \mathrm{diam}(B_r^{\mathbf{a}}(x))^{d+\beta},$$

where $\beta := \frac{a_1}{a_{\min}} + \cdots + \frac{a_d}{a_{\min}} - d \geq 0$.

We conclude this section with the following corollary to Theorem 17.3 that provides an abundance of examples of log-preserving mactices.

Corollary 17.1 *Suppose $\mathcal{B}$ is a structured basis for $\mathbb{T}^d$. Let $\alpha \in (0, 1]$, Suppose $U = (u_{ij})$ is a unitary matrix function with entries $u_{ij} \in Lip_\alpha(\mathbb{T}^d)$, $i, j \in \{1, 2, \ldots, N\}$, for which there exist $C > 0$, and $0 \leq \beta < \alpha$, such that*

$$diam(E)^{d+\beta} \leq C|E|, \qquad E \in \mathcal{B}. \tag{17.18}$$

Then U is log-preserving.

Remark 17.2 The smoothness conditions imposed on the elements of U in Theorem 17.3 and Corollary 17.1 may not appear to be very natural at first, but it is in fact proven by Bloom in [2, Theorem 5.12], in the case $d = 1$, that any log-preserving $U = (u_{ij})$ matrix must satisfy $|u_{ij}| \in \mathrm{Lip}_\beta(\mathbb{T})$ for some $\beta := \beta(U) > 0$.

17.5 The Distance to a Matrix in BMO

One important application of the results of Sects. 17.3 and 17.4 is to the study of the BMO distance to L^∞ for matrices related to weights in $\mathbf{A}_2(\mathcal{B})$. This problem was studied in detail by the authors in [19] for the base of cubes Q.

For a structured basis $\mathcal{B}$ for $\mathbb{T}^d$, and a matrix weight $M : \mathbb{T}^d \to \mathbb{C}^{N\times N}$, satisfying $\exp(M) \in \mathbf{A}_2(\mathcal{B})$, we define

$$\epsilon(M) := \inf_{\lambda \in \mathbb{R}_+} \{\lambda : \exp(M/\lambda) \in \mathbf{A}_2(\mathcal{B})\},$$

where we notice that $\epsilon(M) \leq 1$. In the scalar case, using a reverse Hölder estimate, one can deduce that ϵ is always strictly less than one, but this self-improving type result is known to fail for matrix weights [2].

We consider the following notion of BMO distance to L^∞ for matrices.

Definition 17.6 Suppose $U\colon \mathbb{T}^d \to \mathbb{C}^{N\times N}$ is a log-preserving matrix. For $M = U^*DU$, with D a diagonal matrix in BMO, we define

$$\mathrm{dist}(M, L^\infty) := \inf\{\|D - D'\|_{BMO} : B = U^*D'U \in L^\infty_{N\times N}\},$$

where D' varies over the diagonal matrices in $L^\infty_{N\times N}$.

We now generalise Proposition 4.5 in [19] to general bases $\mathcal{B}$ using Theorems 17.2 and 17.3, where we mention that the geometric condition (17.9) stated in Theorem 17.3 ensures that the basis $\mathcal{B}$ satisfies the continuity condition required to apply Theorem 17.2. We have the following result, where we for the sake of brevity leave the proof for the reader.

Proposition 17.1 *Let $\mathcal{B}$ be a structured basis for $\mathbb{T}^d$ satisfying the geometric condition stated in Eq.* (17.9). *Suppose $U\colon \mathbb{T}^d \to \mathbb{C}^{N\times N}$ is a log-preserving matrix. Then there exist constants $c_1, c_2 > 0$, depending only on U, such that for $W = \exp(M) = U^*DU \in \mathbf{A}_2(\mathcal{B})$,*

$$c_1 \mathit{dist}(M, L^\infty) \leq \epsilon(M) \leq c_2 \mathit{dist}(M, L^\infty).$$

17.6 Almost Diagonal Matrix Weights

In this final section we consider another special class of matrix weights, the so-called almost diagonal weights. As the name suggest, almost diagonal weights are matrix weights having a certain diagonal dominance that allows one to infer information about the weight from the diagonal elements of the weight. We have the following definition.

Definition 17.7 Let $W : \mathbb{T}^d \to \mathbb{C}^{N\times N}$ be a matrix weight. Suppose $W = (w_{ij})_{i,j=1}^N$ and let $W^{-1} = (u_{ij})_{i,j=1}^N$. We say that W is almost diagonal provided

$$w_{ii}u_{ii} \in L^\infty(\mathbb{T}^d), \qquad i = 1, 2, \ldots, N.$$

The following proposition is the fundamental result about almost diagonal $\mathbf{A}_2(\mathcal{B})$-weights. The result was proven by Bloom [2] in the univariate case $d = 1$, and here we give a much simplified proof that also provides a generalization to any structured basis for $\mathbb{T}^d$.

Proposition 17.2 *Let $\mathcal{B}$ be a structure basis for $\mathbb{T}^d$, and let $W = (w_{ij}) : \mathbb{T}^d \to \mathbb{C}^{N\times N}$ be an almost diagonal matrix weight. Suppose $w_{ii} \in A_2(\mathcal{B})$ for $i = 1, 2, \ldots, N$. Then $W \in \mathbf{A}_2(\mathcal{B})$.*

Proof It is known, see [16, Theorem 3.1], that the almost diagonal condition implies the boundedness of the projection operators $P_j : L^2(W) \to L^2(w_{jj}\,\mathrm{d}x)$ given by $P_j f = \langle f, e_j\rangle$, with e_j the jth standard basis vector in $\mathbb{R}^d$. Now consider the averages A_E, $E \in \mathcal{B}$. By Theorem 17.1, the family A_E is uniformly bounded on $L^2(w_{jj}\,\mathrm{d}x)$, since $w_{jj} \in A_2(\mathcal{B})$. We now notice that for $f = (f_j)_{j=1}^N \in L^2(W)$, we have $A_E f = \sum_{j=1}^N (A_E f_j)e_j$. Hence, using the boundedness of the projections P_j, $j = 1, \ldots, N$,

$$\begin{aligned}
\|A_E f\|_{L^2(W)} &\le \sum_{j=1}^N \|(A_E f_j)e_j\|_{L^2(W)} \\
&= \sum_{j=1}^N \|A_E f_j\|_{L^2(w_{jj}\,\mathrm{d}x)} \\
&\le C\sum_{j=1}^N \|f_j\|_{L^2(w_{jj}\,\mathrm{d}x)} \\
&\le C'N\|f\|_{L^2(W)}.
\end{aligned}$$

We conclude that the family $\{A_E\}_E$ is uniformly bounded on $L^2(W)$ and it follows from Theorem 17.1 that $W \in \mathbf{A}_2(\mathcal{B})$. □

17.6.1 A₂ Matrix Multipliers

Let us now consider the problem of obtaining information about $\mathbf{A}_2(\mathcal{B})$-multipliers, i.e., we would like to characterise the matrices D satisfying that $DW \in \mathbf{A}_2(\mathcal{B})$ provided $W \in \mathbf{A}_2(\mathcal{B})$. The full matrix multiplier problem is currently out of reach, so here we restrict our attention to an almost diagonal setup. Let us first verify that the class of almost diagonal matrices is invariant under multiplication by diagonal matrices.

Lemma 17.3 *Let $W : \mathbb{T}^d \to \mathbb{C}^{N\times N}$ be an almost diagonal matrix weight, and let $D : \mathbb{T}^d \to \mathbb{C}^{N\times N}$ be a diagonal matrix weight, say $D = diag(\lambda_1, \lambda_2, \ldots \lambda_N)$. Then DW is an almost diagonal matrix weight.*

Proof Let us denote $W = (w_{ij})_{i,j=1}^N$ and $W^{-1} = (u_{ij})_{i,j=1}^N$. Then by the cofactor formula for the matrix inverse, we have

$$u_{jj} = \frac{1}{\det(W)} \cdot C_{jj},$$

where C_{jj} denotes the minor

$$C_{jj} := \det\big((w_{p\neq j,q\neq j})\big).$$

By the definition of the determinant, we have

$$C_{jj} = \sum_{\pi} \operatorname{sgn}(\pi) w_{1,\pi(1)} \cdots w_{j-1,\pi(j-1)} w_{j+1,\pi(j)} \cdots w_{N,\pi(N-1)},$$

where the summation is over all permutations on $\{1, \ldots, j-1, j+1, \ldots, N\}$. Hence, the assumption that W is almost diagonal is then expressed through the N conditions

$$F_j := \frac{w_{jj}}{\det(W)} \sum_{\pi} \operatorname{sgn}(\pi) w_{1,\pi(1)} \cdots w_{j-1,\pi(j-1)} w_{j+1,\pi(j)} \cdots w_{N,\pi(N-1)} \in L^\infty(\mathbb{T}^d),$$

for $j = 1, 2, \ldots, N$. Now, notice that $DW = (\lambda_i w_i)_{i,j=1}^N$. Put $(DW)^{-1} = (\eta_{ij})_{i,j=1}^N$. Then, by similar calculations as above, using $\det(DW) = \det(D)\det(W)$,

$$\begin{aligned}\eta_{jj} &= \frac{1}{\prod_{i=1}^N \lambda_i} \frac{1}{\det(W)} \sum_{\pi} \operatorname{sgn}(\pi) \lambda_1 w_{1,\pi(1)} \cdots \lambda_{j-1} w_{j-1,\pi(j-1)} \\ &\qquad \times \lambda_{j+1} w_{j+1,\pi(j)} \cdots \lambda_N w_{N,\pi(N-1)} \\ &= \frac{1}{\lambda_j} \frac{1}{\det(W)} \sum_{\pi} \operatorname{sgn}(\pi) w_{1,\pi(1)} \cdots w_{j-1,\pi(j-1)} w_{j+1,\pi(j)} \cdots w_{N,\pi(N-1)}.\end{aligned}$$

We conclude that

$$\lambda_j w_{jj} \eta_{jj} = F_j, \qquad j = 1, 2, \ldots, N,$$

so DW is almost diagonal since $F_j \in L^\infty(\mathbb{T}^d)$ for $j = 1, 2, \ldots, N$. □

We will need the following definition of the Calderón-Zygmund decomposition property in order to derive Proposition 17.3 below.

Definition 17.8 Let $\mathcal{B}$ be a basis for $\mathbb{T}^d$. We say that $\mathcal{B}$ has the Calderón-Zygmund (C-Z) decomposition property provided that there exists a constant C such that for any $B_0 \in \mathcal{B}$, any $f \in L^1(B_0)$, and any $\alpha > \frac{1}{|B_0|}\int_{B_0} |f(x)|\,\mathrm{d}x$,

- $B_0 = F \cup G$ with $F \cap G = \emptyset$
- $|f(x)| \leq \alpha$ almost everywhere for $x \in F$
- G is a union $G = \cup_k B_k$ of sets $B_k \in \mathcal{B}$, whose interiors are mutually disjoint so that for each B_k,

$$\alpha \leq \frac{1}{|B_k|}\int_{B_k} |f(x)|\,\mathrm{d}x \leq C\alpha.$$

It can be shown that, e.g., the family of cubes $\mathcal{B}$ and the family $\mathcal{R}$ of all rectangles aligned with the coordinate axes in $\mathbb{T}^d$ both satisfy the C-Z decomposition property. This can be derived directly from results in [8] and [13]. We refer to [18] for further discussion of the Calderón-Zygmund decomposition property. We can now give the following characterisation.

Proposition 17.3 *Let $\mathcal{B}$ be a structured basis for $\mathbb{T}^d$ that satisfies the CZ-decomposition property. Suppose $D : \mathbb{T}^d \to \mathbb{C}^{N\times N}$ be a diagonal matrix weight, with*

$$D = diag(\lambda_1, \lambda_2, \ldots \lambda_N).$$

Then, for any almost diagonal matrix weight $W = (w_{ij}) : \mathbb{T}^d \to \mathbb{C}^{N\times N}$ with $w_{jj} \in A_2(\mathcal{B})$, we have

$$DW \in \mathbf{A}_2(\mathcal{B})$$

if and only if

$$\lambda_j \in \mathcal{M} := \left\{ w : w, w^{-1} \in \bigcap_{p>1} A_p(\mathcal{B}) \right\}, \quad j = 1, 2, \ldots, N.$$

Proof Suppose that $\lambda_i \in \mathcal{M}$ for $j = 1, 2, \ldots, N$, and let $W = (w_{ij}) : \mathbb{T}^d \to \mathbb{C}^{N\times N}$ be any almost diagonal matrix with $w_{jj} \in A_2(\mathcal{B})$. By Lemma 17.3, DW is almost diagonal, with diagonal elements $(\lambda_j w_{jj})_j$. However, by Proposition 17.4, we have $\lambda_j w_{jj} \in A_2(\mathcal{B})$ since $\lambda_j \in \mathcal{M}$. It now follows by Proposition 17.2 that $DW \in \mathbf{A}_2(\mathcal{B})$.

Conversely, suppose $DW \in \mathbf{A}_2(\mathcal{B})$ for any almost diagonal $W = (w_{ij}) : \mathbb{T}^d \to \mathbb{C}^{N\times N}$ with $w_{jj} \in A_2(\mathcal{B})$. In particular, $DW \in \mathbf{A}_2(\mathcal{B})$ for any $W = \mathrm{diag}(w_1, \ldots, w_N)$ with w_j an arbitrary weight in $A_2(\mathcal{B})$. Consequently, $\lambda_j w_j \in A_2(\mathcal{B})$ for w_j an arbitrary weight in $A_2(\mathcal{B})$. It follows from Proposition 17.4 below that $\lambda_j \in \mathcal{M}$. This concludes the proof. □

Proposition 17.4 below gives a characterisation of scalar $A_2(\mathcal{B})$-multipliers. The problem of characterising A_p-multiplies in the scalar setting for the cube basis has been studied by Johnson and Neugebauer [12] and Cruz-Uribe [6].

Proposition 17.4 *Let $\mathcal{B}$ be a structured basis for $\mathbb{T}^d$ that satisfies the CZ-decomposition property. Consider the families,*

$$\mathcal{M} := \Big\{ w : w, w^{-1} \in \bigcap_{p>1} A_p(\mathcal{B}) \Big\}, \quad \mathcal{N} := \Big\{ u : u^N \in A_2(\mathcal{B}) \text{ for any } N \in \mathbb{N} \Big\},$$

and

$$\mathcal{R} := \{ \rho : \rho w \in A_2(\mathcal{B}) \text{ whenever } w \in A_2(\mathcal{B}) \}.$$

We have,

i. $\mathcal{M} = \mathcal{N}$.
ii. $\mathcal{R} = \mathcal{N} = \mathcal{M}$.

Proof Part (i.). Take any $w \in \mathcal{M}$, and fix $N \in \mathbb{N}$. Then $w \in A_{1+1/N}(\mathcal{B})$ and the fact that $w^{-1} \in A_{1+1/N}(\mathcal{B})$ implies that $w^N \in A_{N+1}(\mathcal{B})$, which can be verified using the Hölder inequality. We notice that $w \in A_{1+1/N}(\mathcal{B})$ implies, since $p = 1 + 1/N \Rightarrow 1/(p-1) = N$, for any $E \in \mathcal{B}$,

$$⨍_E w \, \mathrm{d}x \left(⨍_E w^{-N} \, \mathrm{d}x \right)^{1/N} \leq [w]_{A_{1+1/N}(\mathcal{B})},$$

so

$$\left(⨍_E w \, \mathrm{d}x \right)^N ⨍_E w^{-N} \, \mathrm{d}x \leq [w]^N_{A_{1+1/N}(\mathcal{B})},$$

and we obtain

$$⨍_E w^{-N} \, \mathrm{d}x \leq [w]^N_{A_{1+1/N}(\mathcal{B})} \left(⨍_E w \, \mathrm{d}x \right)^{-N}.$$

The fact that $w^N \in A_{N+1}(\mathcal{B})$ yields

$$⨍_E w^N \, \mathrm{d}x \left(⨍_E w^{-N \cdot \frac{1}{N}} \, \mathrm{d}x \right)^N \leq [w^N]_{A_{N+1}(\mathcal{B})},$$

so

$$\fint_E w^N \, dx \le [w^N]_{A_{N+1}(\mathcal{B})} \left(\fint_E w^{-1} \, dx \right)^{-N}$$
$$\le [w^N]_{A_{N+1}(\mathcal{B})} \left(\fint_E w \, dx \right)^{N},$$

where we used Hölder's inequality in the last step to obtain the estimate

$$1 \le \fint_E w \, dx \fint_E w^{-1} \, dx \Longrightarrow \left(\fint_E w^{-1} \, dx \right)^{-N} \le \left(\fint_E w \, dx \right)^{N}.$$

Hence,

$$\fint_E w^N \, dx \cdot \fint_E w^{-N} \, dx \le [w^N]_{A_{N+1}(\mathcal{B})} \left(\fint_E w \, dx \right)^{N}$$
$$\cdot A^N_{1+1/N}(w) \left(\fint_E w \, dx \right)^{-N}$$
$$\le [w^N]_{A_{N+1}(\mathcal{B})} [w]^N_{A_{1+1/N}(\mathcal{B})},$$

and since the right hand side is independent of E, we may conclude that $w^N \in A_2(\mathcal{B})$, and consequently, $w \in \mathcal{N}$.

For the converse statement, pick $w \in \mathcal{N}$. For any $1 < q \le 2$, we notice that $w^{1/(q-1)} \in A_2(\mathcal{B}) \Rightarrow w^{-1/(q-1)} \in A_2(\mathcal{B})$. Hence, it can be shown by direct computation that

$$w = w^{\frac{1}{q-1} \cdot (q-1)} \in A_q(\mathcal{B}), \quad \text{and} \quad w^{-1} = w^{-\frac{1}{q-1} \cdot (q-1)} \in A_q(\mathcal{B}).$$

We may therefore conclude that $w \in \mathcal{M}$, since $A_2(\mathcal{B}) \subseteq A_p(\mathcal{B})$ for $p > 2$.

Part (ii.). We claim that $\mathcal{R} \subseteq \mathcal{N}$. Pick any $\rho \in \mathcal{R}$, and notice that $1 \in A_2(\mathcal{B})$. Hence, $\rho \cdot 1 = \rho \in A_2(\mathcal{B})$. Then, $\rho \cdot \rho = \rho^2 \in A_2(\mathcal{B})$, and by repeating the argument, we obtain $\rho^N \in A_2(\mathcal{B})$ for any $N \in \mathbb{N}$. By (i.), $\rho \in \mathcal{N}$.

To complete the proof, we will show that $\mathcal{N} \subseteq \mathcal{R}$. Pick any $\rho \in \mathcal{N}$, and suppose $w \in A_2(\mathcal{B})$. By the self-improving property, based on the Reverse Hölder Inequality, there exists $\epsilon > 0$ such that $w^{1+\epsilon} \in A_2(\mathcal{B})$. We put $\theta = 1/(1+\epsilon) \in (0,1)$, and notice that $1 - \theta = \epsilon/(1+\epsilon)$. Hence, we have

$$w\rho = w^{(1+\epsilon)\theta} \rho^{\frac{1+\epsilon}{\epsilon} \cdot (1-\theta)} \in A_2(\mathcal{B}),$$

which follows by a direct computation and the fact that $\rho^{\frac{1+\epsilon}{\epsilon}} \in A_2(\mathcal{B})$ since $\rho \in \mathcal{N}$. Hence, $\rho \in \mathcal{R}$. This completes the proof. □

17.6.2 Matrix A_1-Weights

Following Frazier and Roudenko [7], we define the matrix $\mathbf{A}_1(\mathbb{T}^d)$-class as follows.

Definition 17.9 Let $\mathcal{B}$ be a structured basis for $\mathbb{T}^d$, and let $W : \mathbb{T}^d \to \mathbb{C}^{N\times N}$ be a matrix weight. We say that $W \in \mathbf{A}_1(\mathcal{B})$ provided that

$$\|W\|_{\mathbf{A}_1(\mathcal{B})} := \sup_{E\in\mathcal{B}} \operatorname*{ess\,sup}_{y\in E} \fint_E \|W(t)W^{-1}(y)\|\, dt < +\infty. \tag{17.19}$$

Remark 17.3 It is proven in [7, Lemma 2.1] in the case $\mathcal{B} = \mathcal{Q}$ that condition (17.19) is equivalent to having the scalar weights $t \to \|W(t)\mathbf{x}\|$ in A_1 with a uniform bound on the A_1-constants for constant vectors $\mathbf{x} \in \mathbb{C}^N\backslash\{0\}$. One can easily verify that the proof generalizes verbatim to arbitrary structured bases $\mathcal{B}$ for $\mathbb{T}^d$.

We have the following result that provides a constructive method to construct non-trivial examples of A_1-matrix weights.

Proposition 17.5 *Let $W = (w_{ij}) : \mathbb{T}^d \to \mathbb{C}^{N\times N}$ be an almost diagonal matrix weight, and let $\mathcal{B}$ be a structured basis for $\mathbb{T}^d$. Suppose $w_{ii}^{1/2} \in A_1(\mathcal{B})$ for $i = 1, 2, \ldots, N$. Then $W^{1/2} \in \mathbf{A}_1(\mathcal{B})$.*

Proof Let $\{\mathbf{e}_j\}_{j=1}^N$ denote the standard basis for $\mathbb{C}^N$, and let $\mathbf{x} \in \mathbb{C}^N\backslash\{0\}$. Fix $E \in \mathcal{B}$ and suppose $t \in E$. We first notice that

$$\begin{aligned}\fint_E \|W(u)^{1/2}\mathbf{x}\|\, du &\le \sum_{j=1}^N \fint_E \|W(u)^{1/2}\langle \mathbf{x}, \mathbf{e}_j\rangle \mathbf{e}_j\|\, du\\ &\le \sum_{j=1}^N |\langle \mathbf{x}, \mathbf{e}_j\rangle| \fint_E \|W(u)^{1/2}\mathbf{e}_j\|\, du.\end{aligned}$$

Now, $\|W(u)^{1/2}\mathbf{e}_j\|^2 = \langle W(u)^{1/2}\mathbf{e}_j, W(u)^{1/2}\mathbf{e}_j\rangle = \langle W(u)\mathbf{e}_j, \mathbf{e}_j\rangle = w_{jj}(u)$. Hence,

$$\begin{aligned}\fint_E \|W(u)^{1/2}\mathbf{x}\|\, du &\le \sum_{j=1}^N |\langle \mathbf{x}, \mathbf{e}_j\rangle| \fint_E \sqrt{w_{jj}(u)}\, du\\ &\le C\sum_{j=1}^N |\langle \mathbf{x}, \mathbf{e}_j\rangle| \sqrt{w_{jj}(t)},\end{aligned}$$

where we used that $\sqrt{w_{jj}} \in A_1(\mathcal{B})$. Next we notice that

$$\begin{aligned}|\langle \mathbf{x}, \mathbf{e}_j\rangle|\sqrt{w_{jj}(t)} &= |\langle W^{-1/2}(t)W^{1/2}(t)\mathbf{x}, \mathbf{e}_j\rangle|\sqrt{w_{jj}(t)} \\ &= |\langle W^{1/2}(t)\mathbf{x}, W^{-1/2}(t)\mathbf{e}_j\rangle|\sqrt{w_{jj}(t)} \\ &\le \|W^{1/2}(t)\mathbf{x}\| \cdot \|W^{-1/2}(t)\mathbf{e}_j\|\sqrt{w_{jj}(t)}.\end{aligned}$$

We have, using the same argument as above, $\|W(t)^{-1/2}(t)\mathbf{e}_j\| = \sqrt{u_{jj}(t)}$, where $W^{-1} = (u_{ij})$. Thus,

$$|\langle \mathbf{x}, \mathbf{e}_j\rangle|\sqrt{w_{jj}(t)} \le \|W^{1/2}(t)\mathbf{x}\|\sqrt{u_{jj}(t)}\sqrt{w_{jj}(t)} \le C_j\|W^{1/2}(t)\mathbf{x}\|,$$

since $w_{jj}u_{jj} \in L^\infty(\mathbb{T}^d)$, $j = 1, 2, \ldots, N$, so

$$⨍_E \|W(u)^{1/2}\mathbf{x}\|\, du \le C\Big(\sum_{j=1}^{N} C_j\Big)\|W^{1/2}(t)\mathbf{x}\|, \qquad t \in E.$$

We conclude by Remark 17.3 that $W^{1/2} \in \mathbf{A}_1(\mathcal{B})$. □

Remark 17.4 For $W = (w_{ij}) : \mathbb{T}^d \to \mathbb{C}^{N\times N}$ a matrix weight, we may apply Proposition 17.5 to the Gram weight, $G := WW^* = W^2$. The corresponding sufficient condition for W to be in $\mathbf{A}_1(\mathcal{B})$ then reads

$$\|W\mathbf{e}_j\|\|W^{-1}\mathbf{e}_j\| \in L^\infty(\mathbb{T}^d), \qquad j = 1, 2, \ldots, N,$$

and $\|W\mathbf{e}_j\| \in A_1(\mathcal{B})$, $j = 1, 2, \ldots, N$. In particular, a diagonal matrix weight $W = \mathrm{diag}(\lambda_1, \ldots, \lambda_N)$ is in $\mathbf{A}_1(\mathcal{B})$ provided $\lambda_j \in A_1(\mathcal{B})$, $j = 1, 2, \ldots, N$. A more involved example is given below.

Example
Consider the matrix function

$$W : t \to \begin{bmatrix} \frac{2}{\sqrt{|t|}} & |t|^{-1/4} \\ |t|^{-1/4} & \frac{2}{\sqrt{|t|}} \end{bmatrix}, \qquad t \in [-\pi, \pi).$$

The eigenvalues are $2|t|^{-1/2} \pm |t|^{-1/4}$, so it follows that W is positive definite a.e. and it is therefore a matrix weight. Notice that, $\|W(t)\mathbf{e}_j\| \asymp 2|t|^{-1/2} + |t|^{-1/4} \in A_1(Q)$, $j = 1, 2$, since (individually) $2|t|^{-1/2}, |t|^{-1/4} \in A_1(Q)$. By

(continued)

a direct calculation, we obtain

$$W^{-1}(t) = \begin{bmatrix} \frac{2|t|}{4\sqrt{|t|}-|t|} & -\frac{|t|^{\frac{5}{4}}}{4\sqrt{|t|}-|t|} \\ -\frac{|t|^{\frac{5}{4}}}{4\sqrt{|t|}-|t|} & \frac{2|t|}{4\sqrt{|t|}-|t|} \end{bmatrix},$$

which implies that

$$\|W(t)\mathbf{e}_j\|\|W^{-1}(t)\mathbf{e}_j\| \asymp \frac{(4\sqrt{|t|}+|t|)^2}{\sqrt{|t|}\left(|t|^{\frac{3}{2}} - 8|t| + 16\sqrt{|t|}\right)} = O(1), \qquad \text{as } t \to 0,$$

for $j = 1, 2$. Hence, $\|W(t)\mathbf{e}_j\|\|W^{-1}(t)\mathbf{e}_j\| \in L^\infty(\mathbb{T})$, and we conclude using Remark 17.4 that $W \in \mathbf{A}_1(Q)$.

We mention that matrix weights with a structure similar to the one considered in Example 17.6.2 have been studied for matrix A_2-classes in [1].

References

1. K. Bickel, K. Lunceford, N. Mukhtar, Characterizations of A_2 matrix power weights. J. Math. Anal. Appl. **453**(2), 985–999 (2017)
2. S. Bloom, Weighted norm inequalities for vector-valued functions. PhD Thesis, Washington University in St. Louis (1981)
3. S. Bloom, A commutator theorem and weighted BMO. Trans. Am. Math. Soc. **292**(1), 103–122 (1985)
4. R.R. Coifman, C. Fefferman, Weighted norm inequalities for maximal functions and singular integrals. Stud. Math. **51**, 241–250 (1974)
5. R.R. Coifman, R. Rochberg, G. Weiss, Factorization theorems for Hardy spaces in several variables. Ann. Math. **103**(3), 611–635 (1976)
6. D. Cruz-Uribe, Piecewise monotonic doubling measures. Rocky Mountain J. Math. **26**(2), 545–583 (1996)
7. M. Frazier, S. Roudenko, Matrix-weighted Besov spaces and conditions of A_p type for $0 < p \leq 1$. Ind. Univ. Math. J. **53**(5), 1225–1254 (2004)
8. J. García-Cuerva, J.L. Rubio de Francia, *Weighted Norm Inequalities and Related Topics*. North-Holland Mathematics Studies, vol. 116 (North-Holland, Amsterdam, 1985). Notas de Matemática [Mathematical Notes], 104
9. M. Goldberg, Matrix A_p weights via maximal functions. Pac. J. Math. **211**(2), 201–220 (2003)
10. R. Hunt, B. Muckenhoupt, R. Wheeden, Weighted norm inequalities for the conjugate function and Hilbert transform. Trans. Am. Math. Soc. **176**, 227–251 (1973)
11. B. Jawerth, Weighted inequalities for maximal operators: linearization, localization and factorization. Am. J. Math. **108**(2), 361–414 (1986)
12. R. Johnson, C.J. Neugebauer, Homeomorphisms preserving A_p. Rev. Mat. Iberoamer. **3**(2), 249–273 (1987)

13. A.A. Korenovskyy, A.K. Lerner, A.M. Stokolos, On a multidimensional form of F. Riesz's "rising sun" lemma. Proc. Am. Math. Soc. **133**(5), 1437–1440 (2005)
14. B. Muckenhoupt, Weighted norm inequalities for the Hardy maximal function. Trans. Am. Math. Soc. **165**, 207–226 (1972)
15. F.L. Nazarov, S.R. Treĭl', The hunt for a Bellman function: applications to estimates for singular integral operators and to other classical problems of harmonic analysis. Algebra i Anal. **8**(5), 32–162 (1996)
16. M. Nielsen, M.G. Rasmussen, Projection operators on matrix weighted L^p and a simple sufficient Muckenhoupt condition. Math. Scand. **123**(1), 72–84 (2018)
17. M. Nielsen, H. Šikić, On stability of Schauder bases of integer translates. J. Funct. Anal. **266**(4), 2281–2293 (2014)
18. M. Nielsen, H. Šikić, Muckenhoupt class weight decomposition and BMO distance to bounded functions. Proc. Edinb. Math. Soc. **62**(4), 1017–1031 (2019)
19. M. Nielsen, H. Šikić, Muckenhoupt matrix weights. J. Geom. Anal. **31**(9), 8850–8865 (2021)
20. F. Ricci, E.M. Stein, Harmonic analysis on nilpotent groups and singular integrals. I. Oscillatory integrals. J. Funct. Anal. **73**(1), 179–194 (1987)
21. A. Ron, Z. Shen, Frames and stable bases for shift-invariant subspaces of $L_2(\mathbf{R}^d)$. Can. J. Math. **47**(5), 1051–1094 (1995)
22. S. Roudenko, Matrix-weighted Besov spaces. Trans. Am. Math. Soc. **355**(1), 273–314 (2003)
23. E.M. Stein, G. Weiss, Interpolation of operators with change of measures. Trans. Am. Math. Soc. **87**, 159–172 (1958)
24. S. Treil, A. Volberg, Wavelets and the angle between past and future. J. Funct. Anal. **143**(2), 269–308 (1997)
25. A. Volberg, Matrix A_p weights via S-functions. J. Am. Math. Soc. **10**(2), 445–466 (1997)

Applied and Numerical Harmonic Analysis (111 volumes)

1. A. I. Saichev and W. A. Woyczyński: *Distributions in the Physical and Engineering Sciences* (ISBN: 978-0-8176-3924-2)
2. C. E. D'Attellis and E. M. Fernandez-Berdaguer: *Wavelet Theory and Harmonic Analysis in Applied Sciences* (ISBN: 978-0-8176-3953-2)
3. H. G. Feichtinger and T. Strohmer: *Gabor Analysis and Algorithms* (ISBN: 978-0-8176-3959-4)
4. R. Tolimieri and M. An: *Time-Frequency Representations* (ISBN: 978-0-8176-3918-1)
5. T. M. Peters and J. C. Williams: *The Fourier Transform in Biomedical Engineering* (ISBN: 978-0-8176-3941-9)
6. G. T. Herman: *Geometry of Digital Spaces* (ISBN: 978-0-8176-3897-9)
7. A. Teolis: *Computational Signal Processing with Wavelets* (ISBN: 978-0-8176-3909-9)
8. J. Ramanathan: *Methods of Applied Fourier Analysis* (ISBN: 978-0-8176-3963-1)
9. J. M. Cooper: *Introduction to Partial Differential Equations with MATLAB* (ISBN: 978-0-8176-3967-9)
10. Procházka, N. G. Kingsbury, P. J. Payner, and J. Uhlir: *Signal Analysis and Prediction* (ISBN: 978-0-8176-4042-2)
11. W. Bray and C. Stanojevic: *Analysis of Divergence* (ISBN: 978-1-4612-7467-4)
12. G. T. Herman and A. Kuba: *Discrete Tomography* (ISBN: 978-0-8176-4101-6)
13. K. Gröchenig: *Foundations of Time-Frequency Analysis* (ISBN: 978-0-8176-4022-4)
14. L. Debnath: *Wavelet Transforms and Time-Frequency Signal Analysis* (ISBN: 978-0-8176-4104-7)
15. J. J. Benedetto and P. J. S. G. Ferreira: *Modern Sampling Theory* (ISBN: 978-0-8176-4023-1)

E. Hernández et al. (eds.), *The Mathematical Heritage of Guido Weiss*,
Applied and Numerical Harmonic Analysis,
https://doi.org/10.1007/978-3-031-76793-7

16. D. F. Walnut: *An Introduction to Wavelet Analysis* (ISBN: 978-0-8176-3962-4)
17. A. Abbate, C. DeCusatis, and P. K. Das: *Wavelets and Subbands* (ISBN: 978-0-8176-4136-8)
18. O. Bratteli, P. Jorgensen, and B. Treadway: *Wavelets Through a Looking Glass* (ISBN: 978-0-8176-4280-80)
19. H. G. Feichtinger and T. Strohmer: *Advances in Gabor Analysis* (ISBN: 978-0-8176-4239-6)
20. O. Christensen: *An Introduction to Frames and Riesz Bases* (ISBN: 978-0-8176-4295-2)
21. L. Debnath: *Wavelets and Signal Processing* (ISBN: 978-0-8176-4235-8)
22. G. Bi and Y. Zeng: *Transforms and Fast Algorithms for Signal Analysis and Representations* (ISBN: 978-0-8176-4279-2)
23. J. H. Davis: *Methods of Applied Mathematics with a MATLAB Overview* (ISBN: 978-0-8176-4331-7)
24. J. J. Benedetto and A. I. Zayed: *Sampling, Wavelets, and Tomography* (ISBN: 978-0-8176-4304-1)
25. E. Prestini: *The Evolution of Applied Harmonic Analysis* (ISBN: 978-0-8176-4125-2)
26. L. Brandolini, L. Colzani, A. Iosevich, and G. Travaglini: *Fourier Analysis and Convexity* (ISBN: 978-0-8176-3263-2)
27. W. Freeden and V. Michel: *Multiscale Potential Theory* (ISBN: 978-0-8176-4105-4)
28. O. Christensen and K. L. Christensen: *Approximation Theory* (ISBN: 978-0-8176-3600-5)
29. O. Calin and D.-C. Chang: *Geometric Mechanics on Riemannian Manifolds* (ISBN: 978-0-8176-4354-6)
30. J. A. Hogan: *Time-Frequency and Time-Scale Methods* (ISBN: 978-0-8176-4276-1)
31. C. Heil: *Harmonic Analysis and Applications* (ISBN: 978-0-8176-3778-1)
32. K. Borre, D. M. Akos, N. Bertelsen, P. Rinder, and S. H. Jensen: *A Software-Defined GPS and Galileo Receiver* (ISBN: 978-0-8176-4390-4)
33. T. Qian, M. I. Vai, and Y. Xu: *Wavelet Analysis and Applications* (ISBN: 978-3-7643-7777-9)
34. G. T. Herman and A. Kuba: *Advances in Discrete Tomography and Its Applications* (ISBN: 978-0-8176-3614-2)
35. M. C. Fu, R. A. Jarrow, J.-Y. Yen, and R. J. Elliott: *Advances in Mathematical Finance* (ISBN: 978-0-8176-4544-1)
36. O. Christensen: *Frames and Bases* (ISBN: 978-0-8176-4677-6)
37. P. E. T. Jorgensen, J. D. Merrill, and J. A. Packer: *Representations, Wavelets, and Frames* (ISBN: 978-0-8176-4682-0)
38. M. An, A. K. Brodzik, and R. Tolimieri: *Ideal Sequence Design in Time-Frequency Space* (ISBN: 978-0-8176-4737-7)
39. S. G. Krantz: *Explorations in Harmonic Analysis* (ISBN: 978-0-8176-4668-4)

40. B. Luong: *Fourier Analysis on Finite Abelian Groups* (ISBN: 978-0-8176-4915-9)
41. G. S. Chirikjian: *Stochastic Models, Information Theory, and Lie Groups, Volume 1* (ISBN: 978-0-8176-4802-2)
42. C. Cabrelli and J. L. Torrea: *Recent Developments in Real and Harmonic Analysis* (ISBN: 978-0-8176-4531-1)
43. M. V. Wickerhauser: *Mathematics for Multimedia* (ISBN: 978-0-8176-4879-4)
44. B. Forster, P. Massopust, O. Christensen, K. Gröchenig, D. Labate, P. Vandergheynst, G. Weiss, and Y. Wiaux: *Four Short Courses on Harmonic Analysis* (ISBN: 978-0-8176-4890-9)
45. O. Christensen: *Functions, Spaces, and Expansions* (ISBN: 978-0-8176-4979-1)
46. J. Barral and S. Seuret: *Recent Developments in Fractals and Related Fields* (ISBN: 978-0-8176-4887-9)
47. O. Calin, D.-C. Chang, and K. Furutani, and C. Iwasaki: *Heat Kernels for Elliptic and Sub-elliptic Operators* (ISBN: 978-0-8176-4994-4)
48. C. Heil: *A Basis Theory Primer* (ISBN: 978-0-8176-4686-8)
49. J. R. Klauder: *A Modern Approach to Functional Integration* (ISBN: 978-0-8176-4790-2)
50. J. Cohen and A. I. Zayed: *Wavelets and Multiscale Analysis* (ISBN: 978-0-8176-8094-7)
51. D. Joyner and J.-L. Kim: *Selected Unsolved Problems in Coding Theory* (ISBN: 978-0-8176-8255-2)
52. G. S. Chirikjian: *Stochastic Models, Information Theory, and Lie Groups, Volume 2* (ISBN: 978-0-8176-4943-2)
53. J. A. Hogan and J. D. Lakey: *Duration and Bandwidth Limiting* (ISBN: 978-0-8176-8306-1)
54. G. Kutyniok and D. Labate: *Shearlets* (ISBN: 978-0-8176-8315-3)
55. P. G. Casazza and P. Kutyniok: *Finite Frames* (ISBN: 978-0-8176-8372-6)
56. V. Michel: *Lectures on Constructive Approximation* (ISBN : 978-0-8176-8402-0)
57. D. Mitrea, I. Mitrea, M. Mitrea, and S. Monniaux: *Groupoid Metrization Theory* (ISBN: 978-0-8176-8396-2)
58. T. D. Andrews, R. Balan, J. J. Benedetto, W. Czaja, and K. A. Okoudjou: *Excursions in Harmonic Analysis, Volume 1* (ISBN: 978-0-8176-8375-7)
59. T. D. Andrews, R. Balan, J. J. Benedetto, W. Czaja, and K. A. Okoudjou: *Excursions in Harmonic Analysis, Volume 2* (ISBN: 978-0-8176-8378-8)
60. D. V. Cruz-Uribe and A. Fiorenza: *Variable Lebesgue Spaces* (ISBN: 978-3-0348-0547-6)
61. W. Freeden and M. Gutting: *Special Functions of Mathematical (Geo-)Physics* (ISBN: 978-3-0348-0562-9)
62. A. I. Saichev and W. A. Woyczyński: *Distributions in the Physical and Engineering Sciences, Volume 2: Linear and Nonlinear Dynamics of Continuous Media* (ISBN: 978-0-8176-3942-6)

63. S. Foucart and H. Rauhut: *A Mathematical Introduction to Compressive Sensing* (ISBN: 978-0-8176-4947-0)
64. G. T. Herman and J. Frank: *Computational Methods for Three-Dimensional Microscopy Reconstruction* (ISBN: 978-1-4614-9520-8)
65. A. Paprotny and M. Thess: *Realtime Data Mining: Self-Learning Techniques for Recommendation Engines* (ISBN: 978-3-319-01320-6)
66. A. I. Zayed and G. Schmeisser: *New Perspectives on Approximation and Sampling Theory: Festschrift in Honor of Paul Butzer's* 85th *Birthday* (ISBN: 978-3-319-08800-6)
67. R. Balan, M. Begue, J. Benedetto, W. Czaja, and K. A. Okoudjou: *Excursions in Harmonic Analysis, Volume 3* (ISBN: 978-3-319-13229-7)
68. H. Boche, R. Calderbank, G. Kutyniok, and J. Vybiral: *Compressed Sensing and its Applications* (ISBN: 978-3-319-16041-2)
69. S. Dahlke, F. De Mari, P. Grohs, and D. Labate: *Harmonic and Applied Analysis: From Groups to Signals* (ISBN: 978-3-319-18862-1)
70. A. Aldroubi: *New Trends in Applied Harmonic Analysis* (ISBN: 978-3-319-27871-1)
71. M. Ruzhansky: *Methods of Fourier Analysis and Approximation Theory* (ISBN: 978-3-319-27465-2)
72. G. Pfander: *Sampling Theory, a Renaissance* (ISBN: 978-3-319-19748-7)
73. R. Balan, M. Begue, J. Benedetto, W. Czaja, and K. A. Okoudjou: *Excursions in Harmonic Analysis, Volume 4* (ISBN: 978-3-319-20187-0)
74. O. Christensen: *An Introduction to Frames and Riesz Bases, Second Edition* (ISBN: 978-3-319-25611-5)
75. E. Prestini: *The Evolution of Applied Harmonic Analysis: Models of the Real World, Second Edition* (ISBN: 978-1-4899-7987-2)
76. J. H. Davis: *Methods of Applied Mathematics with a Software Overview, Second Edition* (ISBN: 978-3-319-43369-1)
77. M. Gilman, E. M. Smith, and S. M. Tsynkov: *Transionospheric Synthetic Aperture Imaging* (ISBN: 978-3-319-52125-1)
78. S. Chanillo, B. Franchi, G. Lu, C. Perez, and E. T. Sawyer: *Harmonic Analysis, Partial Differential Equations and Applications* (ISBN: 978-3-319-52741-3)
79. R. Balan, J. Benedetto, W. Czaja, M. Dellatorre, and K. A. Okoudjou: *Excursions in Harmonic Analysis, Volume 5* (ISBN: 978-3-319-54710-7)
80. I. Pesenson, Q. T. Le Gia, A. Mayeli, H. Mhaskar, and D. X. Zhou: *Frames and Other Bases in Abstract and Function Spaces: Novel Methods in Harmonic Analysis, Volume 1* (ISBN: 978-3-319-55549-2)
81. I. Pesenson, Q. T. Le Gia, A. Mayeli, H. Mhaskar, and D. X. Zhou: *Recent Applications of Harmonic Analysis to Function Spaces, Differential Equations, and Data Science: Novel Methods in Harmonic Analysis, Volume 2* (ISBN: 978-3-319-55555-3)
82. F. Weisz: *Convergence and Summability of Fourier Transforms and Hardy Spaces* (ISBN: 978-3-319-56813-3)
83. C. Heil: *Metrics, Norms, Inner Products, and Operator Theory* (ISBN: 978-3-319-65321-1)

84. S. Waldron: *An Introduction to Finite Tight Frames: Theory and Applications*. (ISBN: 978-0-8176-4814-5)
85. D. Joyner and C. G. Melles: *Adventures in Graph Theory: A Bridge to Advanced Mathematics*. (ISBN: 978-3-319-68381-2)
86. B. Han: Framelets and Wavelets: *Algorithms, Analysis, and Applications* (ISBN: 978-3-319-68529-8)
87. H. Boche, G. Caire, R. Calderbank, M. März, G. Kutyniok, and R. Mathar: *Compressed Sensing and Its Applications* (ISBN: 978-3-319-69801-4)
88. A. I. Saichev and W. A. Woyczyński: *Distributions in the Physical and Engineering Sciences, Volume 3: Random and Fractal Signals and Fields* (ISBN: 978-3-319-92584-4)
89. G. Plonka, D. Potts, G. Steidl, and M. Tasche: *Numerical Fourier Analysis* (978-3-030-04305-6)
90. K. Bredies and D. Lorenz: *Mathematical Image Processing* (ISBN: 978-3-030-01457-5)
91. H. G. Feichtinger, P. Boggiatto, E. Cordero, M. de Gosson, F. Nicola, A. Oliaro, and A. Tabacco: *Landscapes of Time-Frequency Analysis* (ISBN: 978-3-030-05209-6)
92. E. Liflyand: *Functions of Bounded Variation and Their Fourier Transforms* (ISBN: 978-3-030-04428-2)
93. R. Campos: *The XFT Quadrature in Discrete Fourier Analysis* (ISBN: 978-3-030-13422-8)
94. M. Abell, E. Iacob, A. Stokolos, S. Taylor, S. Tikhonov, J. Zhu: *Topics in Classical and Modern Analysis: In Memory of Yingkang Hu* (ISBN: 978-3-030-12276-8)
95. H. Boche, G. Caire, R. Calderbank, G. Kutyniok, R. Mathar, P. Petersen: *Compressed Sensing and its Applications: Third International MATHEON Conference 2017* (ISBN: 978-3-319-73073-8)
96. A. Aldroubi, C. Cabrelli, S. Jaffard, U. Molter: *New Trends in Applied Harmonic Analysis, Volume II: Harmonic Analysis, Geometric Measure Theory, and Applications* (ISBN: 978-3-030-32352-3)
97. S. Dos Santos, M. Maslouhi, K. Okoudjou: *Recent Advances in Mathematics and Technology: Proceedings of the First International Conference on Technology, Engineering, and Mathematics, Kenitra, Morocco, March 26-27, 2018* (ISBN: 978-3-030-35201-1)
98. Á. Bényi, K. Okoudjou: Modulation Spaces: *With Applications to Pseudodifferential Operators and Nonlinear Schrödinger Equations* (ISBN: 978-1-0716-0330-7)
99. P. Boggiato, M. Cappiello, E. Cordero, S. Coriasco, G. Garello, A. Oliaro, J. Seiler: *Advances in Microlocal and Time-Frequency Analysis* (ISBN: 978-3-030-36137-2)
100. S. Casey, K. Okoudjou, M. Robinson, B. Sadler: Sampling: *Theory and Applications* (ISBN: 978-3-030-36290-4)

101. P. Boggiatto, T. Bruno, E. Cordero, H. G. Feichtinger, F. Nicola, A. Oliaro, A. Tabacco, M. Vallarino: *Landscapes of Time-Frequency Analysis: ATFA 2019* (ISBN: 978-3-030-56004-1)
102. M. Hirn, S. Li, K. Okoudjou, S. Saliana, Ö. Yilmaz: *Excursions in Harmonic Analysis, Volume 6: In Honor of John Benedetto's 80th Birthday* (ISBN: 978-3-030-69636-8)
103. F. De Mari, E. De Vito: *Harmonic and Applied Analysis: From Radon Transforms to Machine Learning* (ISBN: 978-3-030-86663-1)
104. G. Kutyniok, H. Rauhut, R. J. Kunsch, *Compressed Sensing in Information Processing* (ISBN: 978-3-031-09744-7)
105. P. Flandrin, S. Jaffard, T. Paul, B. Torresani, *Theoretic Physics, Wavelets, Analysis, Genomics: An Indisciplinary Tribute to Alex Grossmann* (ISBN: 978-3-030-45846-1)
106. G. Plonka-Hoch, D. Potts, G. Steidl, M. Tasche, *Numerical Fourier Analysis, Second Edition* (ISBN: 978-3-031-35004-7)
107. P. Alonso Ruiz, M. Hinz, K. Okoudjou, L. Rogers, A. Teplyaev, *From Classical Analysis to Analysis on Fractals: A Tribute to Robert Strichartz, Volume 1* (ISBN: 978-3-031-37799-0)
108. S. Casey, M. Dodson, P. Ferreira, A. Zayed, *Sampling, Approximation, and Signal Analysis: Harmonic Analysis in the Spirit of J. Rowland Higgins* (ISBN: 978-3-031-41129-8)
109. J. Feuto, B. A. Kpata, *Harmonic Analysis and Partial Differential Equations: Proceedings of the Workshop in Abidjan, Côte d'Ivoire, May 22-26, 2023* (ISBN: 978-3-031-66374-1)
110. S. Foucart, S. Wojtowytsch, *Explorations in the Mathematics of Data Science: The Inaugural Volume of the Center for Approximation and Mathematical Data Analytics* (ISBN: 978-3-031-66496-0)
111. E. Hernández, M. Peloso, F. Ricci, F. Soria, A. Tabaco, *The Mathematical Heritage of Guido Weiss* (ISBN: 978-3-031-76792-0)

For an up-to-date list of ANHA titles, please visit http://www.springer.com/series/4968

The manufacturer's authorised representative in the EU is Springer Nature Customer Service Centre GmbH, Europaplatz 3, 69115 Heidelberg, Germany. If you have any concerns regarding our products, please contact ProductSafety@springernature.com

Printed and bound by CPI Group (UK) Ltd, Croydon, CR0 4YY
01/10/2026
02225914-0012